CCTV

This page intentionally left blank

CCTV
From Light to Pixels

Vlado Damjanovski

Third Edition

ELSEVIER

AMSTERDAM • BOSTON • HEIDELBERG • LONDON
NEW YORK • OXFORD • PARIS • SAN DIEGO
SAN FRANCISCO • SINGAPORE • SYDNEY • TOKYO

Butterworth-Heinemann is an imprint of Elsevier

Acquiring Editor: *Brian Romer*
Senior Editorial Project Manager: *Amber Hodge*
Project Manager: *Punithavathy Govindaradjane*
Designer: *Alan Studholme*

Butterworth-Heinemann is an imprint of Elsevier
225 Wyman Street, Waltham, MA 02451, USA
The Boulevard, Langford Lane, Kidlington, Oxford, OX5 1 GB, UK

Library of Congress Cataloging-in-Publication Data
Damjanovski, Vlado, 1956-
    CCTV : from light to pixels / Vlado Damjanovski. -- Third edition.
       pages cm
    Includes bibliographical references and index.
    ISBN 978-0-12-404557-6
    1. Closed-circuit television. I. Title. II. Title: C.C.T.V.

    TK6680.D36 2014
    621.388'5--dc23
                              2013023447

British Library Cataloguing-in-Publication Data
A catalogue record for this book is available from the British Library.

ISBN: 978-0-12-404557-6

For information on all Butterworth–Heinemann publications
visit our website at store.elsevier.com

Printed and bound in United States of America

14 15 16 17 18   10 9 8 7 6 5 4 3 2 1

Contents

Preface to this edition

The first edition of this book, called simply "CCTV," was published back in 1995, by my previous company "CCTV Labs." Later on, my American publisher Butterworth-Heinemann (now an imprint of Elsevier) published two editions, one in 1999, and the other in 2005. So, although this latest edition you are reading is a third edition, it is actually a fourth edition if we count the very first one from 1995. At the time, there were hardly any books on this topic, which was the original reason for my "book idea." Today, it is very encouraging to see, the "written word" about CCTV is more common, both in printed and electronic formats. This book is an attempt to not only recap some of the known and important topics covered originally, but to expand and update on new technologies and standards. As with the original idea of the first edition, I have again attempted to make this book as complete as possible, and as contemporary as possible. For this reason, some very old chapters that are now completely obsolete (like VCRs or CRTs) have been removed and, of course, new content has been added.

The **Closed Circuit Television (CCTV)** technology has not only evolved, but also diversified. Most importantly - CCTV is now almost completely digital. Today, almost all new systems are using digital and IP technology, for encoding, transmitting, viewing, and recording.

The **CCTV** has changed so much that many are questioning if the term "CCTV" should still be used. There is a trend where "IP Surveillance" or "IP CCTV" is perhaps offered as a more appropriate, and this could well be true, at least for some. But, I have decided to keep the term CCTV first and foremost because it is a continuation of the original idea for a complete book on CCTV technology, and, also, because of the intention for this book not to be seen as exclusive for surveillance systems, but also for schools, universities, hospitals, and manufacturing plants who may not necessarily use CCTV just for security purposes. So even with digital and IP technology the Closed Circuit TV is still used by a confined (closed) circuit of people, rather than offering it as a stream to an open audience, for anybody on the network. For these reasons, I considered that the CCTV term still stands its ground, hence the reason to continue using it. So, even though the majority of readers would be looking at this book as a great help in understanding and designing surveillance systems, my intention was not to limit the topics to this area only.

If the previous edition, published in 2005, was badly needed because of the technology updates on the 1999 edition (digital was only starting then), then the 2005 edition was, after 8 years, even in a more desperate need for yet another update. The digital technology has further matured and diversified and new standard initiatives are under way, so this update was long overdue.

I am pleased to see that my books are still highly regarded by the readers, and rated with five stars on many web sites, including the popular Amazon.com. The previous books have also been translated into Russian, where they also enjoyed immense success. Such an acknowledgement can only motivate any author to be even more committed to making each and new edition even better and more informative.

I certainly could not change the contents of the basics of CCTV from the previous editions, but I did "fine tune" certain sections, and added some new text where it was needed. Most importantly, I enhanced the contents with new chapters on Digital and Networking in CCTV.

I have tried to cover the theory and practice of all components and fundamentals of CCTV. This is a very wide area and involves various disciplines and technologies from electronics, telecommunications, optics, fiber optics, programming, networking, and digital image processing. So, my intention was to have a new complete book which still encompasses the basic concepts of CCTV but also includes, explains, and de-mystifies the new trends in digital CCTV, video compressions, and networking concepts and principles.

Television is a complex technology, both analog and even more so digital, especially for people who have never had the opportunity to study it. Understanding digital is harder, if not impossible, without understanding analog television principles. So, if you are not familiar with the analog CCTV, do not think even for a moment that you can bypass it and go straight to digital and networking. Everything makes so much more sense in the digital once you know the analog CCTV.

As with the previous editions I had to read and learn new things myself, and then I tried to put everything into the same style and perspective as the previous chapters. Understandably, I did not want to reinvent the wheel, but I made efforts to simplify and explain the most important aspects of these new technologies. I would not have felt comfortable writing about these new subjects if I did not have some practical experience (though modest, at least so far), so that I tried to see it from a CCTV practical perspective. Should you be interested in more in-depth knowledge of networking and digital there are numerous books and courses I would recommend and some are listed in the Bibliography. I hope, however, that this book will give you a good overview of the relevant CCTV aspects.

I have deliberately simplified explanations of concepts and principles, made many illustrations, tables, and graphs for better understanding, and tried to explain them in a reasonably plain language. Still, a technical-minded approach is required.

This book is intended for, and will be very helpful to, installers, sales people, security managers, consultants, manufacturers, and everyone else interested in CCTV, providing they have some basic technical knowledge.

The specially designed CCTV test chart that was traditionally printed on the back covers of the previous editions, has now evolved to an SD/HD (Standard Definition / High Definition) version. Like with the previous CCTV dedicated test charts, this one too is a world-first, specifically designed for the CCTV industry. This, time though, I have decided not to print a copy of this chart at the back cover, but I have included a chapter explaining its usage. The reason for not printing it this time is simple — the new test chart has some very fine details and tonal grading which are difficult to achieve with the standard CMYK printing off-set technology. In addition, the size of the book cover and the glossy finish do not help in achieving what the chart is designed to do. So, anybody interested in obtaining a proper A3+ size of the new SD/HD test chart, can do so by visiting my company's web site (***www.vidilabs.com***).

These test charts have been accepted by many companies around the world, and they have been recommended in many standards and testing procedures by various governments, institutions, consultants, and companies. The test chart is a great tool to check the quality of a system and its components and compare them with others.

Near the end, I would like to also inform readers that for over eight years I have conducted CCTV seminars based on my books. They have proven to be invaluable in summarizing the CCTV technology and knowledge in a small and well-digested presentations. Typically these are one-day, two-day, or three-day intensive seminars, covering all the topics in the book, or part of it. If anybody is interested in this, please visit the CCTV Seminars web site (*www.cctvseminars.com*) and arrange a suitable training for yourself or your colleagues.

I would like to thank many readers who have already made numerous suggestions and corrections. Readers who themselves write technical articles would know that no matter how many times one goes through one's own text will still find things that could be corrected, or be said somehow differently, and unavoidably there will be some errors, although I did my best to eliminate them. Feel free to write to me if you find something needs to be changed or corrected for future editions.

This book has been made possible by my publisher, Elsevier, as well as the CCTV manufacturers who have believed in me and co-sponsored this edition. These are (in alphabetical order): Axis Communications, Axxon, C.R.Kennedy, and Dallmeier Electronic.

Thank you for purchasing the book, and I hope you will enjoy reading it and using it.

Vlado Damjanovski, B.E. Electronics

May, 2013, Sydney

E-mails: **vlado@vidilabs.com**

              vlado@cctvseminars.com

              vlado@damjanovski.com

Web Pages: **http://www.vidilabs.com**

              http://www.cctvseminars.com

              http://www.damjanovski.com

Introduction

This book has 14 chapters, and they are written in a logical order.

Chapter 1, **SI units of measurement,** introduces the basics of the units of measurement which I thought are important to mention, even though they are not only a CCTV subject, but rather a technical issue. Many products, terms, and concepts exist in the world of CCTV which sooner or later need to be referred to with a correct unit. SI units are suggested by the ISO (the International Standardization Organization), and if we accept these units as universal it will make our understanding of the products and their specifications clearer and more accurate. I have also listed the common metric prefixes because I have found a lot of technicians or consultants do not know them. If you are an engineer or have a good technical background you may find this chapter of no interest, so you can go directly to Chapter 2.

Chapter 2, **Light and television,** starts with a little bit of history so we can gain a wider perspective of the television revolution. Then we go to the very basics of human vision: light and the human eye. It is necessary to explain the human eye and how it works because television relies greatly on the human eye's physiology. It is interesting to compare the similarities between the eye's and the camera's operation.

Optics in CCTV is Chapter 3, which focuses on the first and important product used in CCTV, the lens. Apart from the discussion on how lenses work and what their most important features are, there is explanation on F-stop and depth of field. I also offer an explanation of how and what to adjust (ALC and Level) on a lens and how to determine a focal length for a particular angle of view. This requires explanation on various sensor sizes, which indirectly affects the angles of view. Back-focus adjustment is also covered, although this process may not be required in many modern CCTV cameras with factory prefitted lenses. C and CS-mount standards are also discussed and explained, as well as various chip sizes.

Chapter 4, **General about TV systems,** is very important, especially for readers without prior knowledge of how television works and how colors are produced with the red, green, and blue primaries. Both major analog standards PAL and NTSC are discussed with their timing and waveform. Certainly, there is description of the digital HD standards. General discussion on resolution is also included, and more importantly, the difference between a broadcast signal and CCTV video signal. Near the end of the chapter the most common instruments used in TV and what they measure are discussed.

Chapter 5, **CCTV cameras,** is probably the most interesting chapter in the book. It discusses at length the concepts of CCD and CMOS cameras, various designs, and camera specifications. Many projects and tenders depend on correct understanding of camera specifications and, for this reason, more space is devoted to these topics. Some special cameras which require mentioning have also been included in this chapter. A discussion on measurement and calculation of light coming onto a camera are discussed, as well as power supplies, and voltage drops. Although they seem trivial, a lot of problems have been caused by improper camera setting or powering (unregulated or overrated power supply, thin wires, high-voltage drop). I have also included, at the end of this chapter, a very practical checklist which you or your installers can use in order to make the CCTV installation trouble-free.

Displays are discussed in Chapter 6. The main topic is on generic digital displays, predominantly LCD, and the meaning of pixels in creating images. A little bit on the CRT monitors is mentioned also, but they have become old technology and slowly obsolete, so this is a stripped down section compared to the previous edition. A very important part in this chapter forms the theory and formulas given for the required displays to be able to show face recognition and identification details, as recommended by various CCTV standards. Although not strictly displays, printed pixels are also covered, so that the readers can see the difference between display and printed pixels. Obviously, you will find explanations on various important issues associated with monitors, like gamma, the impedance switch, and viewing conditions. At the end of this chapter, a description of some major new developments in the display technology is included. At the time of the release of the previous edition of this book, many of these technologies were only technical news, but today some of them have been or are being widely adopted.

Chapter 7, **Analog video processing,** is a chapter that covers, for the purposes of continuity and in reduced form, the now old analog matrix switchers. Also, it encompasses the "good old" sequential switchers, as representatives of the "analog" processing range, and, of course, quads, multiplexers, video motion detectors, and frame stores as representatives of the "early digital" range. Although slowly being forgotten, we have covered the VCR multiplexers in case somebody comes across this technology for maintenance purposes. Analog video format has also been used for video analytics. This is mentioned here too.

Chapter 8, **Digital CCTV,** is the core chapter of the new digital technology used in TV and CCTV. It includes information about the theory of why digital is better than analog, why we need to compress and how we compress, and image compression as well as video compression. It explains the basis of JPG and MPEG compression, using the discrete cosine transformation. This chapter covers the most common compression types used in CCTV, such as JPG, JPEG-2000, MPEG-2, MPEG-4, H.264, and even introduces the latest H.265 video compression which was only released beginning of this year, during the working on this book.

In Chapter 9, **Video management systems (VMS),** this is the chapter covering the modern CCTV equivalent of analog matrix switchers. In most cases the VMS are in actual fact computers with dedicated software, switching signals over the networks, recording them and decoding them. There are many VMS platforms on the market today, and individual models are not covered, as this can be found from each and separate manufacturer, but we have covered the generic hardware and software concepts one needs to know, various hard disk standards and redundant storage techniques, as well as operating systems and file formats. We have completed this chapter with a few words about the new initiatives for inter-operability.

Transmission media, Chapter 10, covers the transmission media used in CCTV, both for analog and digital signals. Coaxial cable was the most common and widely accepted for analog video, but category twisted pair cabling becomes now the preferred method. We have devoted some space on the actual termination techniques. Also included here are microwave, RF wireless, infrared, telephone lines, and, certainly, fiber optics as the most capable in terms of distances and bandwidth.

Chapter 11, **Networking in CCTV**, is probably now the biggest chapter in this book. Although networking as a concept could have followed the Digital CCTV chapter, I decided to place it after the Transmission media chapter because the same transmission media could be used for analog, as well as for digital. In the whole book concept we follow the logical flow of the video signal, from the light, through the optics, cameras, encoding, then transmitting over a media and displaying and recording. Media is used for transmission, but without Ethernet standards it can't be made digital. So, the Networking in CCTV chapter explains the theory and standards in digital communications, predominantly Ethernet, when using any of the transmission media in the previous chapter, but now dealing with binary signals. The Networking in CCTV chapter does not intend to substitute the more in-depth literature you can find on networking and IT technology (since there is plenty of it around) but it gives the "non-IT" reader some basic concepts and understanding of the increasingly more important information technology.

Chapter 12, **Auxiliary equipment in CCTV,** includes some auxiliary items and topics that could not be made a logical part of the previous chapters. These include the pan and tilt heads, housings, lighting, infrared lights, ground loop correctors, ingress protection standards, lightning protection, video amplifiers, and distribution amplifiers.

The previous twelve chapters focus on the equipment side of a CCTV system, so in Chapter 13, **CCTV System design**, we discuss how to design a CCTV system. This chapter is based purely on practical experience and on feedback from installers and users. It starts from understanding customers requirements, how to design a system, make some basic sketches which will lead to a more elaborate drawings, with quotation to follow. During this stage it also expected to consider the equipment, not just the camera end, but also the equipment end. For this reason, information on equipment racks and the standards used in CCTV are covered. To finalize a system design, we also have to consider commissioning, training, and handing over. Preventative maintenance is often forgotten, but it is an important part of a complete CCTV system offering. Even if preventative maintenance is done after the system is finished I think it is important for this to be listed here as part of the complete picture of CCTV.

The last, Chapter 14, **Video testing**, advises readers on a practical testing one can make with the ViDi Labs test chart. Many people found the test charts we have introduced since the first book in 1995 are extremely useful, and, not surprisingly, it has become a de facto industry standard. So, even if you don't have one, it might be interesting to know what it can measure and how to use it. If you are interested in obtaining a copy, please visit the web site mentioned at the end of the chapter.

Appendix A, **Common terms used in CCTV,** explains exactly what the heading says. I have tried to include all the terms, acronyms, and names one might come across in CCTV and accompanying fields.

In Appendix B, **Bibliography and acknowledgments**, you can find some interesting reference material and web sites, some of which I have used in the preparation of this book.

In Appendix C, **CCTV training based on this book**, we offer information on the training provided by the author based on this book.

Appendix D, acknowledges the **book co-sponsors** without which this book would have been impossible to prepare, and gives a short biography of the author.

1. SI units of measurement

The basic units

The laws of physics are expressions of fundamental relationships between certain physical quantities.

There are many different quantities in physics. In order to simplify measurement and to comply with the theory of physics, some of them are taken as basic quantities, while all others are derived from those basic ones.

Measurements are made by comparing the magnitude of a quantity with that of a given unit of that quantity.

In physics, of which electronics and television are a part, the *International System of Units,* known as **SI** (from the French *Système Internationale*), is used.

The following are **the seven basic units**:

Unit	Symbol	Measures
Meter	[m]	length
Kilogram	[kg]	mass
Second	[s]	time
Ampere	[A]	electric current
Kelvin	[K]	temperature
Candela	[cd]	luminous intensity
Mole	[mol]	amount of substance

These basic units are defined by internationally recognized standards.

The standard for meter, for example, was defined as a certain number of wavelengths of a specific radiation in the spectrum of krypton until 1983. In October 1983 it was redefined as the distance that light travels in vacuum during a time of 1/299,792,458 second.

The standard of kilogram, for example, is the mass of a particular piece of platinum-iridium alloy cylinder kept at the International Bureau of Weights and Measurements in Sèvres, France.

The basic unit of time, the second, was defined in 1967, as a "time required for a Cesium-133 atom to undergo 9,192,631,770 vibrations."

Kelvin degrees have the same scale division as Celsius degrees, only that the starting point of 0° K is equivalent to –273° C; this is called the ***absolute zero***.

All other units in physics are defined with some combination of the above-mentioned basic units. For example, an area of a block of land is defined by the equation:

$$P = a \times b$$

where a is the width of the block of land, and b is the length. If both a and b are expressed in meters [m], the product P will be expressed in [m²]. We should mention that in mathematics the multiplication is not always represented with the × sign as above, but very often a dot · is used in between the factors being multiplied, or sometimes even without a symbol at all.

We all know that speed, for example, is defined as [m/s], although we quite often use [km/h]. We can easily convert [km/h] into [m/s] by knowing how many meters there are in a kilometer and how many seconds there are in an hour.

SI units are almost universally accepted in science and industry throughout the world, and we should all be aware that measurements like "inches" for length, "miles per hour" for speed, and "pounds or stones" for weight should be used as little as possible. They often cause confusion in people from various professions and various parts of the world. If you use SI units, more people will understand you and your product. Also, it is easier to compare products from various parts of the world if they use the same units.

Another very important thing to clarify is that every symbol in the SI system has a precise meaning relative to the letter used (capital or small). So, a kilometer is written as [km], not [Km] or [klm]. A megabyte is written as [MB], not [mB]. A nanometer is written as [nm], not [Nm] and so on. As technical people involved in closed circuit television, we should stick to these principles.

Derived units

All other physical processes can be explained and measured using the basic units. We will not go into the details of how they are obtained, nor is it the purpose of this book to do so, but it is important to understand that there is always a fundamental relation between the basic and derived unit.

The following are some of the derived SI units, several of which will be used in this book:

Quantity	Unit	Symbol / Definition
Area	Square meter	m^2
Volume	Cubic meter	m^3
Velocity	Meters per second	m/s
Acceleration	Meters per second per second	m/s^2
Frequency	Hertz	$Hz = 1/s$
Density	Kilograms per cubic meter	kg/m^3
Force	Newton	$N = kg{\cdot}m/s^2$
Pressure	Pascal	$Pa = kg/m{\cdot}s^2$
Torque	Newton meter	$T = N{\cdot}m$
Energy, work	Joule	$J = N{\cdot}m$
Power	Watt	$W = J/s$
Electric charge	Coulomb	$C = A{\cdot}s$
Electric potential	Volt	$V = \Omega/A$
Electrical resistance	Ohm	$\Omega = V/A$
Electrical capacitance	Farad	$F = C/V$
Conductance	Siemens	$S = A/V$
Magnetic flux	Weber	$Wb = V{\cdot}s$
Magnetic field intensity	Tesla	$T = Wb/m^2$
Inductance	Henry	$H = Wb/A$
Illumination	Lux	$lx = lm/m^2$
Luminous flux	Lumen	$lm = cd{\cdot}steradian$
Luminance	Nit	$nt = cd/m^2$

Metric prefixes

When the number of units (i.e., the value) for a particular measurement is very high or very small, there is a convention for using certain symbols before the basic unit, and each has a specific meaning. The following are metric prefixes accepted by the international scientific and industrial community that you may find not only in CCTV but also in other technical areas:

Prefix	Multiple	Symbol
exa-	10^{18}	E
peta-	10^{15}	P
tera-	10^{12}	T
giga-	10^{9}	G
mega-	10^{6}	M
kilo-	10^{3}	k
hecto-	10^{2}	h
deca-	10	D
unity	$10^{0} = 1$	
deci-	10^{-1}	d
centi-	10^{-2}	c
milli-	10^{-3}	m
micro-	10^{-6}	μ
nano-	10^{-9}	n
pico-	10^{-12}	p
femto-	10^{-15}	f
atto-	10^{-18}	a

By using these prefixes, we can say 2 km, referring to 2000 meters. If we say 1.44 MB, we are thinking of 1,440,000 bytes.

A very common measurement of data transmission speed over networks is expressed in megabits per second (Mb/s), which is different from megabytes per second (MB/s).

One byte is equal to 8 bits, and they are denoted with lower case "b" for bits and capital "B" for bytes. A nanometer will be 0.000000001 meters. The frequency of 12 GHz would be $12 \cdot 10^9 = 12{,}000{,}000{,}000$ Hz and so on.

A very common unit used these days in CCTV when handling hard disk drives is gigabytes (GB). One gigabyte is equal to thousand megabytes, or a million kilobytes. The correct value for binary 1 GB megabytes is 1024 MB (which is 2^{10}), and the correct binary value for 1 MB is 1024 kB. When hard disk manufacturers write 300 GB on their disks, typically this represents a decadic 300,000,000,000 bytes. So when such a hard disk is installed in the computer, the operating system reports 279 GB. This is the real binary value, and it is obtained by dividing 300,000,000,000 with 1024 to get kB, then with 1024 again to get MB, and finally with 1024 again to get GB.

Now that we have established the basics of a technically correct discussion that is, introduced the basic units of measurement, we can start with the fundamentals of all visions, including photography, cinematography, television, and certainly CCTV – *light*.

2. Light and television

Let there be light.

A little bit of history

Light is one of the basic and greatest natural phenomena, vital not only for life on this planet, but also very important for the technical advancement and ingenuity of the human mind in the visual communication areas: photography, cinematography, television, and multimedia. The main source of light for our planet is our closest star - the sun.

Even though it is so "basic" and we *see* it all the time around us, light is the single biggest stumbling block of science. Physics, from a very simple and straightforward science at the end of the nineteenth century, became very complex and mystical. It forced the scientists in the beginning of the twentieth century to introduce the postulates of quantum physics, the "principles of uncertainty of the atoms," and much more – all in order to get a theoretical apparatus that would satisfy a lot of practical experiments but, equally, make sense to the human mind.

This book is not written with the intent of going deeper into each of these theories, but rather I will discuss the aspects that affect video signals and CCTV.

The major "problem" scientists face when researching light is that it acts as dual nature: it behaves as though it is a wave – through the effects of refraction and reflection – but it also appears as though it has particle nature – through the well-known photo-effect discovered by Heinrich Hertz in the nineteenth century and explained by Albert Einstein in 1905. As a result, the latest trends in physics are to accept light as a phenomenon of a "dual nature."

It would be fair at this stage, however, to give credit to at least a few major scientists in the development of physics, and light theorists in particular, without whose work it would have been impossible to attain today's level of technology.

Isaac Newton was one of the first physicists to explain many natural phenomena including light. In the seventeenth century he explained that light has a particle nature. This was until Christian Huygens, later in that century, proposed an explanation of light behavior through the wave theory. Many scientists had deep respect for Newton and did not change their views until the very beginning of the nineteenth century when Thomas Young demonstrated the interference behavior of light. August Fresnel also performed some very convincing experiments that clearly showed that light has a wave nature.

A very important milestone was the appearance of James Clerk Maxwell on the scientific scene, who in 1873 asserted that light was a form of high-frequency electromagnetic wave. His theory predicted the speed of light as we know it today: 300,000 km/s. With the experiments of Heinrich Hertz, Maxwell's theory was confirmed. Hertz, however, discovered an effect that is known as the *photo-effect*, where light can eject electrons from a metal whose surface is exposed to light. However, it was difficult to explain the fact that the energy with which the electrons were ejected was independent of the light intensity, which was in turn contradictory to the wave theory. With the wave theory, the explanation would be that more light should add more energy to the ejected electrons.

This stumbling block was satisfactorily explained by Einstein who used the concept of Max Planck's theory of quantum energy of photons, which represent minimum packets of energy carried by the light itself. With this theory, light was given its dual nature, that is, some of the features of waves combined with some of the features of particles.

This theory so far is the best explanation for the majority of light behavior, and that is why in CCTV we apply this "dual approach" theory to light. So, on one hand, in explaining the concepts of lenses we will be using, most of the time, the wave theory of light. On the other, the principles imaging chips operation (CCD or CMOS), for example, based on the light's particle (material) behavior.

Clearly, in practice, light is a mixture of both approaches, and we should always have in mind that they do not exclude each other.

Light basics and the human eye

We now know that, as James Clerk Maxwell asserted, light is an electromagnetic radiation. The human eye is sensitive to this radiation and it picks up the various frequencies as colors. Electromagnetic radiation obviously comes in all frequencies, i.e., wavelengths, as can be seen in the drawing below. The visible light occupies only a very little "window" in this range. Based on various testing on many viewers, this window seems to be between 380 nm and 750 nm. We take this, however, to be roughly from 400 nm to 700 nm, for easy remembering.

The 400 nm corresponds to violet and 700 nm to red color. There is a continuous color change from the violet to blue, green, yellow, orange, and red as the wavelength increases. In the many experiments and tests that have been conducted to check the sensitivity of an average human eye and, it has been found out that **not all colors produce the same effect on the eye's retina**. Green color, for example, excites the eye the most. In other words,

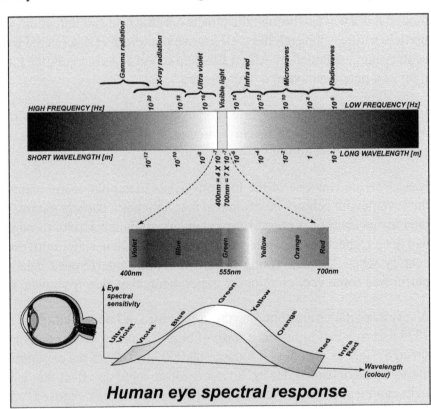

Human eye spectral response

if we have all the wavelengths of the light with an equal amount of energy, the green will produce the highest "output" on the retina. Frequencies higher than the violet (wavelengths shorter than 400 nm) and lower than the red (wavelengths longer than 700 nm) cannot be detected by an "average" human eye. I emphasize this "average" because human eye sensitivity is a statistical curve. Some people see all colors well, but there are people who are "color blind," which means their eye spectral sensitivity is different (usually narrower) from the one shown. Some "color blind" people cannot see red color, some cannot see blue. A trained, professional eye of a painter or a photographer may develop very high sensitivity for detecting various frequencies (colors) which might look the same to others. Some may even extend their minimum and maximum detectable frequency limit, that is, see deeper violet or red colors that are invisible to other individuals.

A very interesting question to ask ourselves is why is the eye's spectral sensitivity maximum in the green color area (at around 555 nm)? This can be associated with the fact that of all the sun's energy that penetrates the Earth's atmosphere, the biggest amount is contained in the wavelengths at around 555 nm.

After millions of years of evolution of life on this planet, we (like most animals) have developed vision using wavelengths that are most readily available (at least during the daytime). Some species have evolved in the other direction. An obvious alternative is the night vision eye characteristics of animals whose food targets are warm-blooded mammals. Body heat is nothing more than infrared radiation. Typical examples are snakes, cats, and owls. Some snakes, for example, apart from using the eyes for general vision, also have infrared sensitive pit organs with which they can detect temperature change of less than 0.5° C (1° F). Cats, including wild cats such as leopards, pumas, and other members of the cat family, are known for their good nighttime vision, which would mean that their near infrared response is far better than that of the human eye.

We will concentrate on the human eye, and it is very important to understand the "construction" of it. This will perhaps be of general interest, but we will also see a lot of conceptual similarities between the eye and the TV camera construction.

This cross section shows that the eye has a lens that focuses the image onto the retina. The retina is actually the "photosensitive area," which is composed of millions of cells, called **cones** and **rods**. These cells can be considered a part of our nervous system. The **cones are sensitive to the medium and bright intensity of light and colors.** The **rod cells are sensitive to lower light levels (night vision), and they do not distinguish colors.** We use rod cells to see at night, which means **when it is dark we cannot distinguish colors.**

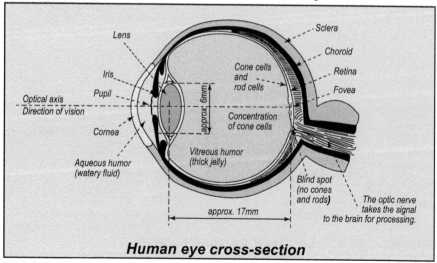

Human eye cross-section

There are approximately 10 million cones in each eye, and the number of rods is over 100 million. The cones are concentrated around the area where the optical axis passes. This area is colored with a yellowish pigment and is called the **fovea**. The fovea is the central area that our brain processes and although it is a small area, the concentration of cones there is approximately 50,000. The average focal length of an eye (i.e., the distance between the lens and the retina when an infinitely distant object is being viewed) is approximately 17 mm. This focal length gives an undistorted image in a solid angle of approximately 30°. This is also the size of the area most populated with the cone cells. **This is why an angle of about 30° is considered a standard angle of vision.**

The concentration of cones increases toward the center of the optical axis with the peak being at only 10°. Each of these cone cells is connected to the brain via **separate optic nerves, through which electrical pulses are sent to the brain.** The eye, of course, sees a much wider angle, since the retina covers nearly a 90° solid angle and there are cones outside the yellow area as well, but these other cones are connected to each nerve in groups. With this area we do not see as clearly as when we use the single nerve cones, and that is why this area is known as the **peripheral vision area**.

The brain's "image processing section" concentrates on 30°, although we see best at around 10°. This processing is further supported with the constant eye movement in all directions, which is equivalent to a pan/tilt head assembly in CCTV.

For a single lens reflex (SLR) camera the standard angle of view of 30° is achieved with a 50 mm lens,

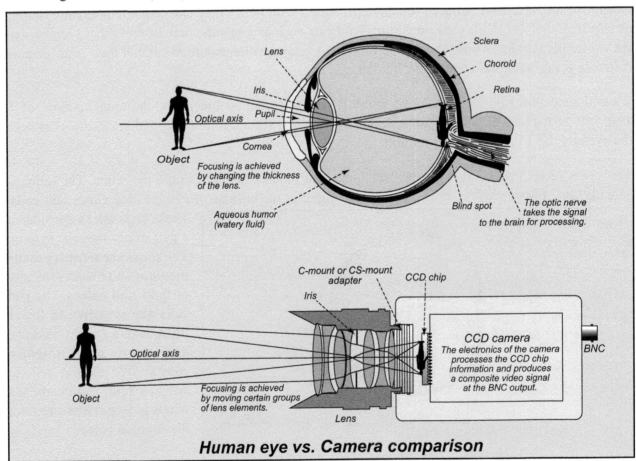

Human eye vs. Camera comparison

for a 2/3" camera this is a 16 mm lens, for a 1/2" camera a 12 mm lens, and for a 1/3" camera an 8 mm lens. In other words, images of any type of camera, taken with their corresponding standard lenses, will be of a very similar size and perspective as when seen through our eyes.

Lenses shorter in focal length give a wider angle of view and are called **wide-angle** lenses. Lenses with longer focal length narrow the view, and therefore they look as if they are bringing distant objects closer, hence the name **telephoto** ("tele" meaning distant). Another matter of interest associated with CCTV is that by knowing the focal length of the eye and the maximum iris opening of approximately 6 mm, we can find the equivalent "F-number" (discussed later in the book) of the eye:

$$F_{number\text{-}eye} = 17/6 = 2.8$$

With such a fully opened iris we can still see quite well in full moonlight (this is approximately 0.1 lux at the object). Have this number in mind when comparing the minimum illumination characteristics of different cameras.

The **focusing that the human eye** does in order to see objects at various distances **is achieved by changing the thickness of the eye lens**. This is done by the ciliary muscles. If the eye is normal, it should be able to focus from infinity down to a minimum distance of about 20 cm in early childhood, to 25 cm at age 20, to 50 cm at age 40, and to 5 m at age 60. When we look at something very far away, that is, eye focused on infinity, the ciliary muscles are relaxed and the eye lens is **thin**.

If the eye cannot focus at infinity that vision defect is called **nearsightedness,** or **myopia**. Such eyes require glasses to help the "defective" human eye lens focus the image on the retina. These glasses are sometimes called reducing glasses because they have a negative focus (or diopter).

A diopter is the inverse value of the focus of a lens, where the focus is expressed in meters. Reducing glasses have a negative diopter. So, "reducing" glasses with a diopter of – 0.5, for example, have a negative focus of $1/(-0.5) = -2$ m.

Another defect an eye may have is when it cannot focus on an image that is very close, that is, the eye's lens cannot be thickened enough for some reason.

This defect is called **farsightedness,** or **hypermetropia**.

People with hypermetropia need glasses to be able to see close objects sharply. These glasses would need to have the opposite characteristics from those in the previous case, that is, they would have to be magnifying glasses with positive focus and diopter.

Two eyes produce images that when mixed in our brain, give a stereoscopic (3D) impression of the volume of space. If we cover one eye, it is very hard to judge the "three-dimensionality" of the space in front of us.

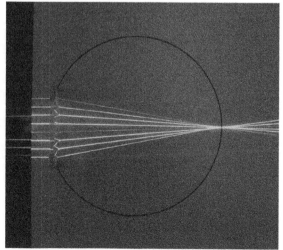

Simulation of how the eye works

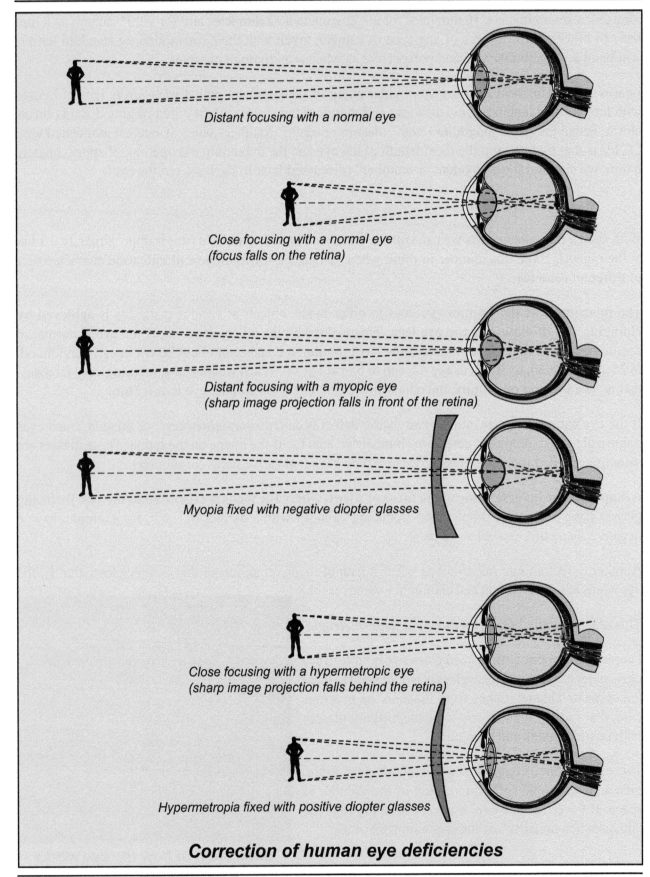

Distant focusing with a normal eye

*Close focusing with a normal eye
(focus falls on the retina)*

*Distant focusing with a myopic eye
(sharp image projection falls in front of the retina)*

Myopia fixed with negative diopter glasses

*Close focusing with a hypermetropic eye
(sharp image projection falls behind the retina)*

Hypermetropia fixed with positive diopter glasses

Correction of human eye deficiencies

The distance between the eyes (60–70 mm) ensures **our perception of three dimensions up to 10–15 m away**. After this distance it is very hard to judge which of two objects is closer. This can be experimented with by trying to see two objects in the air at different long distances, like balloons for example. If we are looking at, let us say, two distant trees, the brain brings a conclusion on the basis of the soil and perspective in front of us, but the perspective "decision" would not be concluded on the basis of the eye's "stereoscopic mechanism."

It is amazing when you think about the complexity of the eye and the brain's power for "image processing." We perform these operations hundreds of times a day without even thinking about it, not to mention the fact that the images that fall on the retina are upside down, owing to the nature of the optical refraction, and we also do not consider the eye movement in all directions when we follow something. All of these things are being deciphered and controlled by the brain.

The "eye–brain" configuration is far superior to any camera that the human mind has, or will ever invent. But, as technical people, we can say that by understanding how the eye "works" and using the ever improving visual technology, both in hardware and software, we are getting better images and more sophisticated information about the world around us and we can view things the human eye cannot see or monitor things in places where the human cannot be present.

With experiments and testing it has been found that **the most a human eye can resolve is no more than 5 ~ 6 lp/mm** (line pairs per millimeter) at around 0.3 m distance, like when reading a fine text. This equates to a minimum viewing angle of about one-sixtieth of a degree (1/60°), which is due to the size of the cone cells in the retina. So, **1/60° is considered the limit of angular discrimination for normal vision**. We can use this minimum angular vision for better understanding and optimizing the psycho-physiology of the viewing. This fact is also used by the Apple's retina display technology on iPhones and iPads, introduced in 2011 and 2012.

A known viewing distance parameter, from the Displays chapter later in the book (Chapter 6), recommends for CCTV viewing a distance of around seven times the monitor height for SD video and four times the monitor height for HD video. So, we should understand that the viewing distance is an important factor for the experience of seeing fine details in an image. It is of no use if a viewer gets closer to the monitor than recommended, but it is also not going to get any better if he is positioned further away from the monitor screen. This is discussed later in the book as "optimum viewing distances."

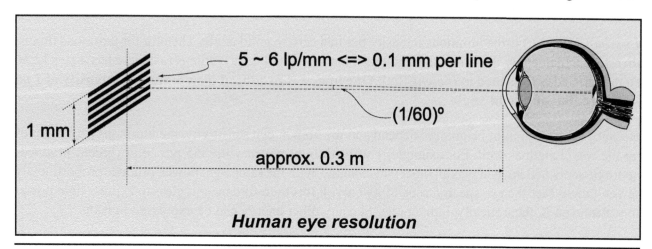

Human eye resolution

Light units

Light is a physical phenomenon but is interpreted by psychological processes in our brain. It is, therefore, a bit more complex to measure than other physical processes. Some prerequisites have to be established in order to make objective measurements. One of these is the bandwidth of the light frequencies considered, and this is usually from 400 nm to 700 nm. All of the frequencies contribute to the light energy radiated by the source.

Let us, first of all, make clear the kind of light sources we have. The basic division is into two major groups:

- Primary sources (the sun, street lights, tungsten lights, monitor CRTs, LEDs)

- Secondary sources (all objects that do not generate light but reflect it)

We do not apply the same type of measurement when measuring the amount of light radiated by a tungsten globe, for example, and the light reflected by an object. It is not the same if we are analyzing light radiated from a source in all directions, or just in a narrow solid angle. These are some of the reasons we have so many different units of light measurement.

The science that examines all these different aspects is called *photometry*, and the units defined are called *photometric units.*

Many different units have been defined by various scientists, depending upon the point of view taken. Because of this, CCTV camera specifications are even harder to understand and describe precisely. But let us try to shed some light on these units and explain what they mean. We will start in a logical order, that is, from the source of the light, traveling through space, falling onto an object, and finally as it is reflected from it.

Luminous intensity (I) is the illuminating power of a primary light source, radiated in all directions. The unit that measures this kind of light is the *candela* [cd]. **One candela is approximately the amount of light energy generated by an ordinary candle.** Since 1948 there has been a more precise definition of a candela as **the luminous intensity of a black body heated up to a temperature at which platinum converges from a liquid to a solid state.**

Luminous flux (F) is the luminous intensity but in a certain solid angle. The unit for luminous flux is, therefore, obtained by dividing the luminous intensity with 4π (pi) radians (a sphere has $4\pi = 12.56$ steradian) and is measured in *lumens* [lm]. **One lumen is produced by a luminous intensity of 1 cd in one radian of a solid angle.**

Because the sensation of brightness depends on the human eye sensitivity, the luminous flux depends on the wavelength as well. For example, 1 watt of light power with 555 nm color (green) produces approximately 680 lm, whereas all other wavelengths, with the same light power, produce proportionally fewer lumens (see the eye spectral sensitivity curve). It is therefore meaningless to express light power in watts, even if, theoretically, light energy like any other energy can be expressed in watts.

Illumination (E) is the most commonly used term in CCTV, especially when referring to the camera's minimum illumination characteristics. Illumination is very similar to the luminance except that we are now referring to objects that are secondary sources of light.

Therefore, **the illumination of a surface is the amount of luminous flux on a unit area**.

When luminous flux of 1 lumen falls on an area of 1 m² (square meter), it is measured in **lumens per square meter** or **meter-candelas**, but it is better known as *lux* [lx].

This means that if we have a sphere of 1 meter radius, a light source with luminous intensity of 1 candela inside this sphere will produce illumination on the internal surface of 1 lx.

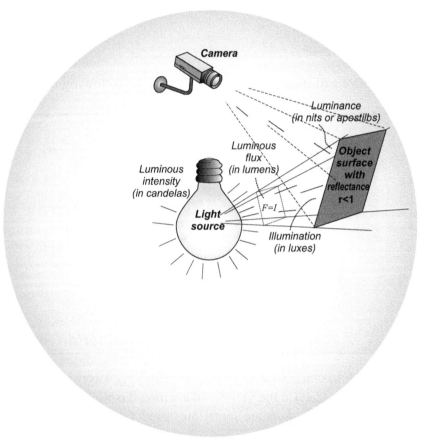

Light units and their meaning

Mathematically, this relation can be described as:

$$E = Flux\ /\ Area = F/A \qquad [lx] \qquad\qquad (1)$$

The flux F is, by definition, equal to luminous intensity times the solid angle, i.e.,

$$F = I \cdot \omega \qquad [lm] \qquad\qquad (2)$$

From the basics of volumetric trigonometry, and assuming a punctual source of light, we can express ω through the area A being lit and its distance from the source d:

$$\omega = A\ /\ d^2 \qquad [rad] \qquad\qquad (3)$$

When (2) and (3) are replaced in (1), we get

$$E = I\ /\ d^2 \qquad [lx] \qquad\qquad (4)$$

which means that **the illumination falls off with the square of the distance when the perpendicular area is being lit**. If, however, this area is at a certain angle to the incoming light, we can approximate the real area with the projection at an angle θ, as per the diagram shown here. In that case the formula (4) becomes:

$$E = I \cdot \cos \theta / d^{2} \qquad [\text{lx}] \qquad (5)$$

Typical levels of illumination are shown on the following drawing:

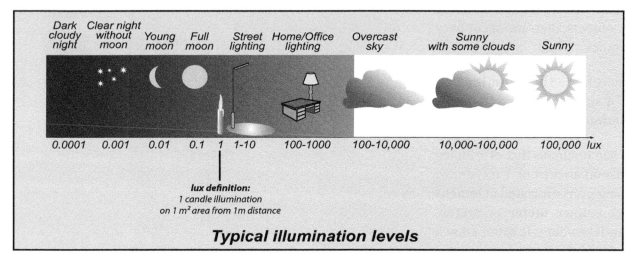

Typical illumination levels

Very rarely, in certain small areas and from very strong light sources, levels higher than 100,000 lx can be experienced (in the vicinity of a strong flashlight, for example). To describe such illuminations, higher units called *phots* are sometimes used. **One phot is equal to 10,000 lx.**

In American terminology, where square feet are still widely used instead of the SI units, illumination is expressed in **square-foot candelas**, or better known as *foot-candelas*. Because of the "square meter – square foot" ratio, equal to nearly 10 (or more precisely 9.29), it is reasonably easy to convert luxes into foot-candelas and vice versa. Basically, if an illumination is given in foot-candelas, just multiply it by 10 and the approximate value in luxes is obtained, and if a value is given in luxes, in order to convert it to foot-candelas, divide it by 10.

Luminance (L) describes the brightness of the surface of either a primary or a secondary source of light. Since brightness embeds subjective connotation, luminance is used as an objective, scientific term. Luminance depends both on the luminous intensity of the surface itself and on the angle at which it is being observed. It is therefore measured per unit of projected surface area perpendicular to that direction. There are quite a few units for luminance. The internationally preferred metric unit is **nit. One nit is equal to one candela per square meter of projected surface area** (*I/A*). If, instead of candelas, lumens are used to describe the luminous flux of a source, the luminance will then be expressed in **apostilbs** [asb]. Things get a bit more complicated when we have a surface where the luminous flux radiated, or reflected, in a direction θ to the normal is directly proportional to cos θ. Such a surface will appear equally bright when seen from all directions because both the reflected light and projected surface area follow the same cosine law. This type of surface is called a **lambert** radiator or reflector (depending on whether the surface is a primary or secondary source of light) and is usually described as a **perfectly diffusing surface**. For the purpose of measuring such light luminance in the metric system, a unit called **lambert** was introduced. The equivalent American unit would be the **foot-lambert**.

How much of the illumination is seen by the camera depends not only on the intensity of the source itself, but also on the reflectivity of the object being illuminated. Obviously, it is not the same if the object is white as opposed to black. With the same amount of light we can, naturally, see more if the objects are white. This is why we have to introduce another factor when talking about illumination, and this is the percentage of object *reflectivity*. The definition of reflectivity could be described with the following simple relation:

$$\rho = \text{light reflected from surface / light incident on surface} = E/L \quad [\%] \quad\quad (6)$$

Realistically, this percentage ranges from a very low 1% for black velvet to 32% for a typical soil surface and up to 93% for bright snow in the field of view. Caucasian human flesh has a reflectivity factor between 19% and 35%. The ViDi Labs test chart discussed later in the book has an approximate reflectivity factor of 60~70%.

The reflectivity is an important factor when stating a camera's minimum illumination because with the same level of illumination and various reflectivity factors, an object may appear more or less bright, indirectly affecting the camera performance.

Illuminometer (lux-meter)

As the name suggests, a *lux-meter* is an instrument designed to measure the illumination of objects of interest. Typically, these objects are what the camera should be viewing. We discussed the lux units in Chapter 3.

Lux is a unit for measuring reflected visible light. The word *visible* is very important to highlight, as **a lux-meter is not designed and cannot measure infrared light**. This is a very common mistake made by some manufacturers when stating their camera's minimum illumination capability, where low luxes are stated thinking of infrared light.

Another common error some installers are making is by using a lux-meter pointing it towards the sky, or towards the source of light in a room.

The lux-meter should be used pointing towards the objects of interest. Basically, **it should see the same angle of view and from the similar distance as the camera**. The illumination falls off with the inverse square of the distance, which means, it is not the same if reading the lux-meter very close to the object of surveillance, and the same measured from the camera position further away. Clearly, the further the camera is, the lower amount of measured lux-es will be detected. Furthermore, the reflectivity of the object plays a big role, so that the same light source will produce higher lux

A good quality lux-meter

reading if the surrounding is brighter. A lux-meter measures whatever visible light falls on its sensor, which should be similar to what the camera "sees" with it's own imaging sensor.

As we discussed under the human eye sensitivity in the *Light basics and the human eye*, illumination below 0.1 lux is practically invisible to human eye, and this falls in the so-called *"Scotopic"* vision area. There are no commercially available lux-meters that can measure at this, or below this level. Typical lux-meters go down to 1 lux, although some may measure even to 0.1 lux, but not below this. Manufacturers claiming that their CCTV cameras can see in as low light level as 0.00001 lux for example are not doing any favors to the industry. **Not only such light levels are impos-**

Minimum illumination tested with a candle light

sible to measure, but it is impossible for such low energy photons to generate electrons in the imaging sensor. There are many more thermally generated electrons at normal temperature then what illumination below 0.001 lux can produce, hence nothing can be seen.

One of the easiest and best method to see how a camera behaves in very low light illumination levels is by use of a candle, in a room with no additional lights. This would require even the light from the display devices (monitors) to be reduced, or simply removed, from the test room. Remember, the definition of 1 lux was that it is an illumination of a typical candle on one square meter surface area at one meter distance.

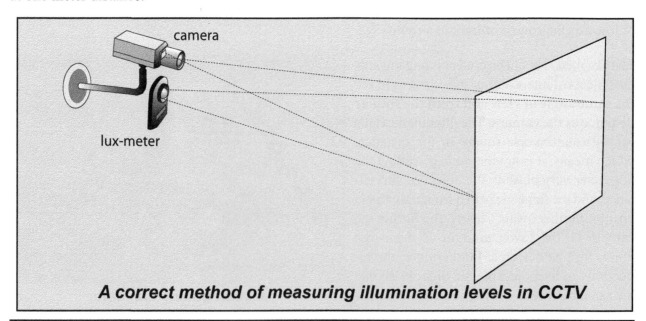

A correct method of measuring illumination levels in CCTV

An indirect method of measuring low light with 100x ND filter

An alternative method of measuring low light camera performance is the method where ND filters are used. This is an indirect method, perhaps not as accurate, but it offers a way of testing low light if the minimum measuring capability of a lux-meter don't allow it. First, a known attenuation ND filter needs to be used, and then illumination is measured behind the filter. For example, if the lux-meter measures 500 lux without the ND filter, than putting a 100 x ND filter, we know the light behind it would be approximately 5 lux (100 times less). Using the same ND filter in front of a camera under test, at around 1 lux (which we could measure) would show camera's response at 0.01 lux. Adding another ND filter of say 10 x we can measure the same response at 0.01 lux (1000 times less than 10 lux).

Smartphones lux-meters

Modern day smartphones have become powerful pocket-computers with built-in cameras and various sensors. Among the many applications (apps) there are some clever ones that simulate lux-meters by measuring illumination levels through the built-in camera(s). Such applications might be useful, but need to be used with caution since they perform as good as the program algorithm of converting camera data (exposure, f-stop, and ISO) is inside.

There are some applications which allow for calibration with a real lux-meter, in which case they may become more useful.

It is important to remember that the angle of measurement can not be wider than the lens of the smartphone camera itself. Also, strong spotlights within the field of view may influence the illumination reading by the smartphone. The example on the right shows an application which makes a reading with an error of less than 10% (49 lux vs. 46) after being calibrated, and it also shows the spot area of measurement.

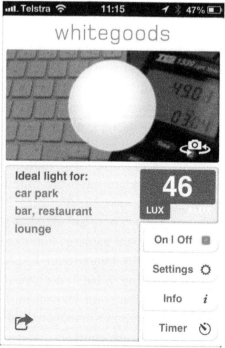

One of the many apps

Measuring illumination with an SLR or DSLR

Very often you may have to measure and quantify the object illumination in your CCTV system without having a lux-meter. In photographic world there are light meters, which are not necessarily lux-meters, as they show exposure, f-stop and ISO setting needed to achieve good exposure. These numbers are also known in the photographic world as *Exposure Values (EV)*. There are a number of known brands on the market, and although most of them will show you light measurement in photographic EV values, there are some that will show you readings directly in luxes.

We will show here that it is possible to use a light-meter, either directly on a typical Single Lens Reflex camera (SLR) or Digital SLR (DSLR), or from a stand-alone light meter that shows EV values. This could be an extremely useful tool, and here I will explain the principles and formula to calculate such lux measurement.

Please note that most DSLR cameras (since the introduction of digital photography, the SLRs are now called DSLRs) would have a light meter, while compact cameras may not necessarily have it. So if you cannot find any indicators for exposure and aperture on your camera, you may not be able to make use of this. Please also note that a logical prerequisite for a more accurate measurement is to have the DSLR camera see the same field of view as the potential (or existing) CCTV camera. For this reason the best lens to have on the DSLR camera is a zoom lens so that you can adjust the angle of view as close to the CCTV camera's view as possible.

A good quality light-meter

First, let us just refresh our memory about some basic rules of photographic exposure.

The exposure indicators on all photographic cameras are in seconds, or to be more precise, in fractions of a second. This means that when a camera light indicator shows 125, it actually indicates 1/125 of a second. If the exposure is longer than 1 second, it is usually denoted with a " symbol or "s" after the number; for example "2 s" indicates 2 seconds. Standard exposure numbers are 1; 2; 4; 8; 15; 30; 60; 125; 250; 500; and 1000. These are all parts of a second. There are cameras that can set the exposure for longer than 1 second and shorter than 1/1000. As you may notice, the values are chosen so that they represent approximately half of the previous exposure number.

A typical light measurement display in a modern SLR camera (showing 1/250 s at F-8)

The indicators for the lens iris opening, or aperture, are shown as "F-stop" values. So the number "5.6" indicates F-5.6. The higher this number is, the smaller the lens iris opening is. Typical F-stop numbers (as explained under the "Optics in CCTV" chapter) are 1.0; 1.4; 2; 2.8; 4; 5.6; 8; 11; 16; 22; 32; and 44. At each F-stop the opening is half the area size it was at the previous F-stop, that is, half the amount of light transmitted than the previous F-stop.

For a correct exposure of the imaging chip in your camera an internal light meter is used, which sets the correct time duration and aperture when exposing to light. In "Program" mode, both values are chosen automatically by the camera. In "Aperture-priority" mode, you set the F-stop and the camera computer selects the exposure. In "Exposure-priority" mode, it is the other way around; you select the exposure and the camera selects the aperture (i.e., F-stop).

Combinations of the exposure and F-stop can be such that they allow for equal amounts of light to get onto the chip. For example, if you or the camera selects 1/30 s and F-5.6, you will produce the same effect on the chip with 1/60 s and F-4. Of course, with the latter F-stop you would have a slightly narrower depth of field, but other than that the chip will be correctly exposed too. Because of this "equality" of

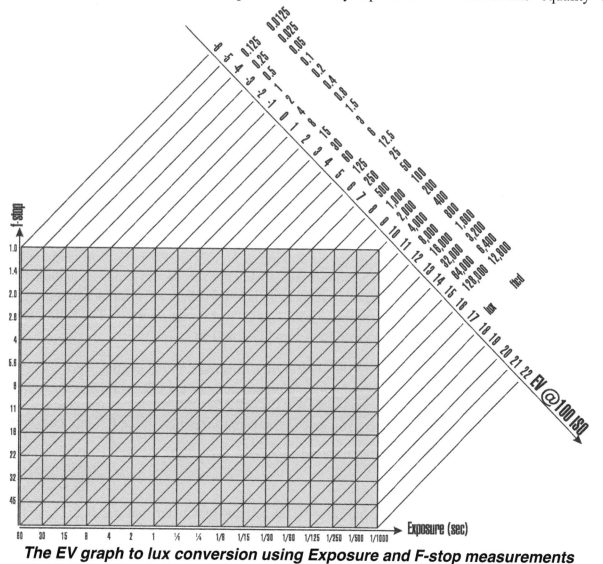

The EV graph to lux conversion using Exposure and F-stop measurements

the amount of light with different exposure/F-stop combinations, photographic experts have advised an Exposure Value (EV) rating for the amount of light that can be measured by the camera light meters. We will not be going into details of exactly how the light is measured inside the camera, as this would require a full book to cover all models, but in general there are "Averaging" light meters, "Spot" light meters, and "Multi-pattern" light meters. I will not discuss these in depth because they are beyond the scope of this book, but a majority of cameras would have at least the "Averaging" light metering. This is close enough for our CCTV applications where illumination levels can only be determined approximately.

Reference No. (RN_t)	Speed (s)	Aperture (F-stop)	Reference No. (RN_f)
0	1	1	0
1	1/2	1.4	1
2	1/4	2	2
3	1/8	2.8	3
4	1/15	4	4
5	1/30	5.6	5
6	1/60	8	6
7	1/125	11	7
8	1/250	16	8
9	1/500	22	9
10	1/1000	32	10
11	1/2000	45	11
etc...			

The relationship between reference numbers, the exposure, and the F-stop

When you buy your photo camera, you will usually find an EV graph somewhere inside the camera's manual, indicating its light measurement capability. This graph should look similar to the one shown in the picture on the previous page. Most of the time the EV graphs refer to a film (or CCD/CMOS chip in digital cameras) sensitivity of 100 ISO, which is a pretty standard sensitivity. For this reason, in our calculations in this text, we have assumed such setting on your camera of 100 ISO. Of course, any other sensitivity can be used; you just need to adjust the numbers accordingly.

The EV graph is very simple to read. For example, a combination of 1/30 s and F-5.6 makes an EV value of 10. The same exposure can be achieved with 1/60 s and F-4 since they also have a combined exposure value of 10.

The EV scale is put together by summing up the Reference Numbers of the exposure and the F-stop (RN_t and RN_f). This table is shown above and indicates that both of these, the exposure and the F-stop, have value "0" for exposure of 1 second and the aperture is F-1.0. Then, the reference number goes to 1 for the next smaller value, being ½ s for the exposure and F-1.4 for the aperture. The table continues like that, that is, reference number 2 is given to ¼ s and F-2, and so on.

EV values are obtained by summing up these reference numbers. For example, exposure of 1/30 s and F-2.8 have an equivalent EV of 8 because the reference number for 1/30 s is 5 and for F-2.8 is 3.

Here are simple formulas which will give you a very good approximation of the RN numbers with the simple use of a scientific calculator:

$$RN_f = 6.7\log(\text{F-stop}) \tag{7}$$

where F-stop is the number of the F-stop indicated by the camera light meter, that is, 5.6; 8; 11; and so on.

$$RN_t = -3.32\log t \tag{8}$$

where t is the absolute exposure time; that is, if the camera shows 1/125, this is what you put under t. If preferred, you could use just the number 125 (we will call it T) instead of the absolute time t but the minus sign in front of the logarithm disappears, that is, the second formula becomes:

$$RN_T = 3.32 \log T \tag{8a}$$

Please note: the logarithms are with base 10.

The EV is calculated by adding these two values:

$$EV = RN_f + RN_t = 6.7 \log(\text{F-stop}) - 3.32 \log t \tag{9}$$

or, if T is used instead of t:

$$EV = RN_f + RN_T \tag{9a}$$

Let us work out one example.

If my camera, being set to have 100 ISO sensitivity, shows 1/250 exposure setting and F-8, the reference numbers for F-stop and exposure can be calculated as:

$$EV = RN_f + RN_t = 6.7 \log 8 - 3.32 \log(1/250) = 6.7 \times 0.9 + (-3.32) \times (-2.398) = 6 + 8 = 14$$

where the result is rounded.

There is a simple connection between EV values and the camera measurements described by the following equation:

$$I_{lux} = 2.5 \times 2^{(RN_f + RN_t)} = 2.5 \times 2^{EV} \tag{10}$$

The right-hand side of the above equation is 2 to the power of the EV number and I_{lux} is obtained in luxes. For example, if the EV value of what the camera has measured is 15, this means the approximate illumination of the scene is:

$$I_{lux} = 2.5 \times 2^{15} = 81,192 \text{ lux}$$

Of course, such precision is impossible when measuring light, since many factors influence light measurement, including the reflectivity of the surrounding objects, the primary sources of light in the field of view (a light from a light-pole in the field of view will affect the average illumination dramatically), and so on. We would usually approximate the above result with 82,000 luxes.

Please note that the "dynamic range" of the light meter EV measurement may vary from camera to camera. Better cameras will have wider range. Also, do not forget to set 100 ISO when using these measurement instructions. Of course, if 200 ISO film is used, everything will be shifted for 1 EV value as the 200 ISO is twice as sensitive as the 100; 400 ISO is four times as sensitive, and the EV values will be shifted by two numbers. For example, if a measurement with 200 ISO gives 16 EV, this is equivalent to 15 EV light reading with 100 ISO.

To conclude this discussion, let us work out a practical example.

If a light measurement shows exposure of 1/15 s and F-2.8 (at the 100 ISO setting), this would give us:

$$EV_{(F-2.8+1/15)} = 6.7 \log 2.8 - 3.32 \log(1/15) = 3 + 4 = 7$$

$$I_{lux} = 2.5 \times 2^7 = 320 \text{ lux}$$

To convert this value in foot-candles, you need to divide the value with 10, which gives around 32 foot-candles.

It should be common knowledge that a bright sunny day will give illumination of around 100,000 lux, a typical office environment would have anything between 100 and 1000 lux, a full moon night should produce around 0.1 lux, and so on.

A bright sunny day will give an EV reading of around 15 or 16, while for comfortable CCTV monitoring at night street-lights should produce an EV value of around 3, which converts to around 20 luxes.

Be aware of your light meter EV range. Many cameras have EV range between 1 and 20 EV. This indicates that the lowest light illumination you could measure with such a camera is around 5 lux. This should be sufficient for a majority of CCTV projects, but if you want to measure even lower illumination I suggest you take a look at some of the professional light meters.

If you don't have any photo camera or lux-meter, and yet want to find out the minimum illumination capability of your CCTV camera, there is a simple and reasonably accurate way of testing this using the very very basic definition of the 1 lux illumination: a real candle, as defined previously under the heading "Light units."

Light up a candle, turn off the lights, and position a 1 m white surface at 1 m distance from the candle, that should give you around 1 lux when seen from the candle position. Certainly, if the surface is not white, the reflectivity percentage will reduce the appearance of 1 lux.

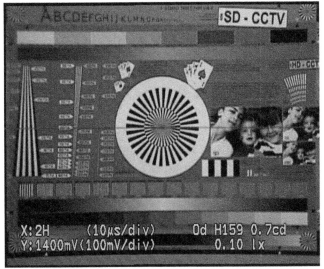

Low light measurement of the ViDi Labs chart SD portion, at 0.1 lux

Using this same reasoning, later on when explaining real testing with the ViDi Labs SD/HD test chart, we will understand why the illumination of the test chart, when in full field of view and illuminated by one single candle will produce approximately 0.07 luxes. This is a product of the chart surface which is around 12% of a square meter, and it's reflectivity, which is around 60%.

Light onto an imaging sensor

In order to fully understand the "light issue," as seen by the camera, we need to know how much light actually falls on the imaging area.

The illumination amount at the CCD (or CMOS) chip, E_{CCD}, depends mostly on the luminance L of the object, but also on the F-stop of the lens, that is, the light-gathering ability of the lens. **The lower the F-number (bigger iris opening) the more light will get through the lens**, as will be explained later in the book. It is also proportional to the **transmittance factor τ of a lens**. Namely, depending on the quality of the glass and its manufacture, as well as the inner walls of the lens mechanics, a certain percentage of the light will be lost in the lens itself.

All of the above factors can be combined into the following relation:

$$E_{CCD} = \pi \cdot \tau \cdot L / (4 \cdot F^2) \quad [\text{lx}] \tag{11}$$

where:

L = average luminance of the object (lux)

τ = transmittance of the lens (in percentage)

F = the actual F-stop of the lens used

$\pi = 3.14$

In the next few lines we will show how this relation is obtained and approximated, so that the technical people using these formulas can have a clear understanding of what is being assumed in order to get to formula (11). However, because these calculations involve slightly more complex mathematics, I suggest that readers with no interest, or without the background, should just directly use relation (11) as it is, knowing the values of L, τ, F, and π.

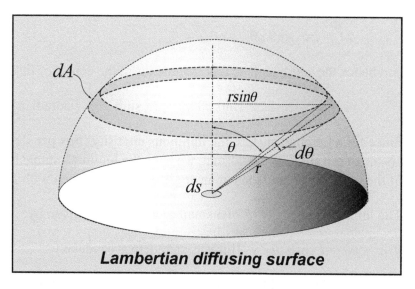

Lambertian diffusing surface

An object viewed by a camera, when lit by a light source, radiates light, more or less, in all directions, depending upon the reflectivity function. In practice, the majority of smooth surface objects can be approximated with a ***Lambertian perfectly diffusing surface***.

The flux, then, can be regarded as passing through a hemisphere of radius r and center ds. If we now consider the incremental angle $d\theta$ at an angle θ to the normal, the flux occupying the volume of a

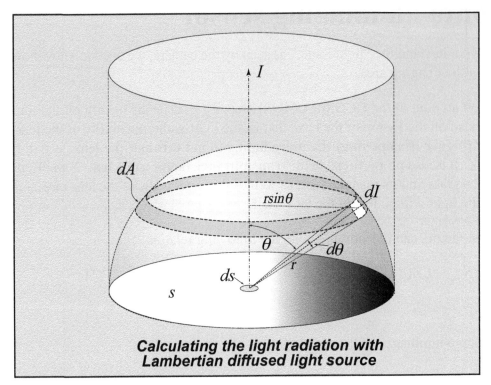

**Calculating the light radiation with
Lambertian diffused light source**

revolution swept out by the angle $d\theta$ passes through an annular ring on the surface of the sphere, with width $rd\theta$ and circumference $2\pi r\sin\theta$.

This elementary surface area is given by:

$$dA = 2\pi r^2\sin\theta \, d\theta \tag{12}$$

and hence the solid angle ω that it subtends at the center of the sphere is given by:

$$\omega = dA / r^2 = 2\pi \, r^2 \sin\theta \, d\theta / r^2 = 2\pi \sin\theta \, d\theta \quad \text{[steradian]} \tag{13}$$

Since for a Lambert surface the luminous intensity (flux per steradian) in a given direction falls as the cosine of the angle to the normal, we have the luminous intensity of the whole surface in the direction of the normal as I, and then at an angle θ it will be given with $I\cos\theta$.

The luminous intensity dI of a small area ds will be given by:

$$dI = I \cos\theta \, ds / s \quad \text{[lumens/steradian = candelas]} \tag{14}$$

Since I/s is the actual luminance L in the perpendicular direction, the above relation becomes:

$$dI = L \cos\theta \, ds \qquad \text{[cd]} \tag{15}$$

The elementary flux dF is equal to the elementary intensity dI times the solid angle:

$$dF = L \cos\theta \, ds \, 2\pi \sin\theta \, d\theta \quad \text{[lm]} \tag{16}$$

The total light emitted into a cone of an angle θ can be found by integration from 0 to θ:

$$F = \int 2\,\pi\,Lds\,\sin\theta\,\cos\theta\,d\theta = \pi\,L\,ds\,\sin^2\theta \qquad [\text{lm}] \qquad (17)$$

If we want to find the total flux radiated in all directions, we have to put 90° for the angle θ so that the total flux emitted in all directions will then be:

$$F_t = \pi\,L\,ds \qquad [\text{lm}] \qquad (18)$$

Now, if we have to calculate the flux emitted into a solid angle smaller than 90°, as may be the case when a camera is viewing an object, the total flux F_o is given by the formula:

$$F_o = \pi\,L\,ds_o\,\sin^2\theta_o \qquad [\text{lm}] \qquad (19)$$

If the lens transmission factor is τ, then the flux falling on the CCD or CMOS chip plane is:

$$F_{CCD} = F_o\,\tau = \pi\,\tau\,L\,ds_o\,\sin^2\theta_o \qquad [\text{lm}] \qquad (20)$$

The illumination of the imaging chip would be flux divided by the imaging chip area ds_{CCD}, that is,

$$E_{CCD} = \pi\,\tau\,L\,\sin^2\theta_o\,ds_o\,/ds_{CCD} \qquad [\text{lx}] \qquad (21)$$

The ratio (ds_{CCD}/ds_o) ,which is inverse in the preceding formula, is also known as the **magnification ratio of a lens m**. The magnification ratio can also be approximated as a ratio between the focal length of the lens and the distance to the object.

$$m = (f/D)^2 = ds_{CCD}/ds_o \qquad (22)$$

When we replace (18) in (17), it becomes:

$$E_{CCD} = \pi\,\tau\,L\,\sin^2\theta_o\,(D/f)^2 \qquad [\text{lx}] \qquad (23)$$

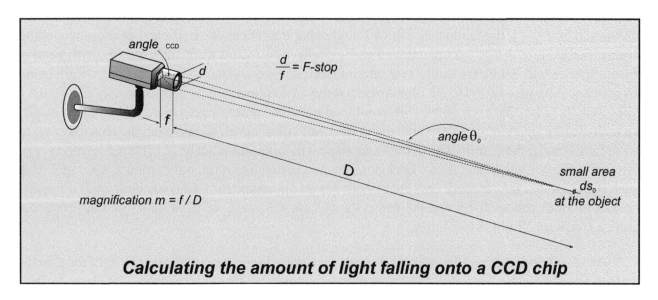

Calculating the amount of light falling onto a CCD chip

We need to introduce here another ratio in a lens (*d/f*), which is also known as the lens F-stop (this will be explained in more detail in Chapter 3). For objects at a reasonably long distance from the camera (again, this is typical in CCTV) we get the following to be true:

$$\tan g\ \theta_o = d/2D = \sin\theta_o / \cos\theta_o = \sin\theta_o \qquad (24)$$

Such an approximation can be made because for very long distance to objects the angle θ_o is very small and the cosine of such angles is very close to 1.

So, we can substitute $\sin^2\theta_o$ with $(d/2D)^2$, and thus equation (19) becomes:

$$E_{CCD} = \pi\ \tau\ L\ (d/2D)^2\ (D/f)^2 \quad [lx] \qquad (25)$$

If we sort this out we will have:

$$E_{CCD} = \pi\ \tau\ L\ (d^2/4D^2)\ (D^2/f^2) = \tau\ \pi\ L\ (d^2/4f^2) \qquad (26)$$

And finally this becomes the simplified formula for calculating the light amount falling onto an imaging device:

$$E_{CCD} = \pi\ \tau\ L/(4\ F^2) \quad [lx] \qquad (27)$$

This is a very useful formula because it uses only two variables (the luminance of an object and the lens F-stop) to calculate the approximate illumination that falls onto an imaging chip. But the approximation we made should not be forgotten, so it should be used only for rough calculations and in cases that correspond to the conditions of the approximation, that is, the camera looking at an object with diffused light, similar to Lambertian source (most of the real-life objects are like that, except mirrors and surfaces alike), at a reasonably long distance relative to its focal length lens. Usually, the lens transmittance factor τ ranges between 0.75 and 0.95. If you do not have the correct number from your lens manufacturer, for calculation purposes a realistic transmittance factor can be taken to be 0.8.

Let us work out an example. If the light at the object plane is around 300 lx, as in an average office area (this would be E_{object}), the luminance can be found using the reflection coefficient of the surrounding objects, that is, $L = \rho\ E_{object}$. As mentioned earlier, reflection factors vary substantially with various objects, but we will not be far from a real office situation if we assume 50%. If the lens we are using has an iris setting of, say, F-16, the illumination at the CCD plane will be approximately $E_{CCD} = 0.8 \cdot 3.14 \cdot 300 \cdot 0.5 / (4 \cdot 256) = 0.36$ lx. This, combined with the camera's automatic gain control (AGC), is a realistic illumination for a CCD chip plane for a full video signal. If, however, the lens iris is set to F-1.4, for example, the illumination of the CCD plane will be approximately 48 lx (using relation (17)). This is a far higher value than the CCD chip needs, and in practice it can only produce a recognizable video if an auto iris lens is used, or if the camera has an electronic (or CCD) iris built-in. If a manual iris lens is used with an F-1.4 and the camera's AGC is set to off, 48 lx at the chip will produce a saturated, or washed-out, white image.

A very basic rule of thumb is that even a lens with the lowest F-number attenuates the light for a factor of 10 +. The higher the F-number, the lower the amount of light that reaches the CCD plane. In fact, it is inverse proportional to the square of the F-number.

With the above conclusions we are actually tapping into a very interesting question raised with CCD cameras (especially B/W, i.e., cameras without infrared cut filters): If the object illumination is as at a full sunny day (approximately 100,000 lx), the F-number has to be very high in order to "stop-down" the light as required by the CCD chip. This is in the vicinity of $0.1 \sim 0.3$ lx (or close) for a full video. Such an F-number is actually so high that it requires the attenuation of the lens to be in the order of over 1,000,000 times. Using the approximated formula (27), assuming the same values for $\tau = 0.8$, $\rho = 0.5$, and assuming the camera CCD chip requires 0.2 lx for a 1 V_{pp} signal, we will get an F-number of 886.

This is an extraordinarily high F-number to be achieved by mechanical means (leaves shutter). The precision of the leaves' movement is limited, and, more importantly, an unwanted optical effect called a Fresnel Edge Refraction becomes noticeable with small iris openings. This means that, in practice, **very high F-stops cannot be achieved by using just mechanical methods**. So, special optical neutral density (ND) filters are used to "help" the leaves shutter achieve high F-stops as required by the sensitive CCD chips. The inferior optical precision of such filters could make an image appear less sharp in very bright light and yet quite good in lower or normal light conditions.

Similarly, sometimes it is required to allow much less light to get into the CCTV camera in order to minimize depth of field, for the purposes of back-focusing. Doing back-focusing at day-time is very tricky, since most of the auto iris lenses will close the iris automatically which will make lens focusing appear sharp, even if it is not at night time. In order to prevent this from happening, ND filters may be used to force the iris open, thus achieving narrow depth of field. Precise focusing, or back-focusing, can be achieved much easier then.

Variable ND filter can be very useful in adjusting camera back-focus or focus

As mentioned earlier, under the optical accessories, there are now special variable ND filters, where by rotating one of the rings on the filter, variable light attenuation density can be achieved, typically up to 100 times. Using such filters can also be helpful in determining minimum illumination capability of a camera, especially if the illumination levels are below the measuring ability of a lux meter (which is typically 0.1 lux). In such a case, illumination is first measured through the ND filter, calculate the attenuation, and then the same filter is used on the camera lens at lower light and apply the attenuation factor. If the ND filter attenuates linearly and visible light only, it is possible to measure the camera low light performance in levels lower than what a lux meter can measure. Combining two such filters it is possible to attenuate the illumination by a factor of up to 10,000.

Colors in television

Colors are a very important and complex issue in television, and certainly the same applies in CCTV. Color offers valuable additional information of the objects being viewed. Some times color alone can be the decisive clue about determining an intruder or stolen car type for example. More importantly, the human eye captures color information quicker than the fine details of an object. Color cameras are not so good in low-light levels when compared to the monochrome (B/W) cameras. The reason for this is the usage of the optical infrared cut filter on the color CCD chips, which attenuate the light and eliminate the invisible infrared portion of the projected image (more on this in Chapter 5). With the ever-improving CCD and CMOS technology, however, the color camera minimum illumination performance has improved dramatically. From 10 lx @ F1.4 at the object, of a few years ago, we now have cameras that can see down to less than 1 lx @ F1.4 at the object or even lower.

As already explained under the **Light Basics and the Human Eye** section earlier in this chapter, the colors we see are actually various wavelengths of light. When we see red, for example, it is a wavelength reflected from a red object when white light is shone on it. Black absorbs almost all wavelengths, whereas white reflects most of them.

The science of colors is very complex and becomes even more complicated when the natural colors around us are reproduced by the phosphor coating of the cathode ray tubes (CRTs), plasma, or LCD screens.

The concept of producing colors in television is by **additive mixing** of three primary color phosphor dots next to each other. These are tiny dots, representing parts of a mask that is on the inside of a monitor's CRT. A similar concept of mixing color is used on plasma and LCD monitors. The actual color mixing happens when we view monitors from a distance and a **resultant** color of the three colored dots appears in our eyes. The resultant color appears different to the individual red, green, and blue elements.

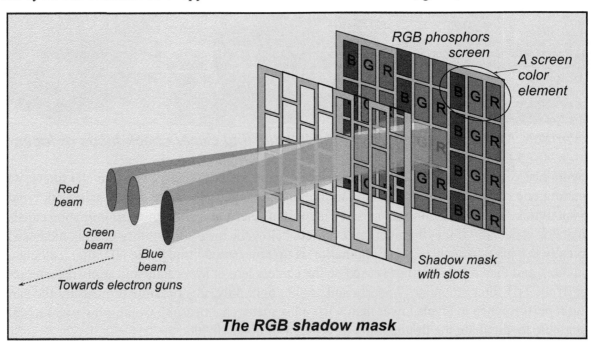

The RGB shadow mask

The additive color mixing in television is opposite to the one used in painting and printing technology, where colors are obtained by *subtractive mixing*.

In additive mixing, by adding the light produced by the color pixel elements makes the resultant color brighter. Therefore, to get white, all three primary color elements need to be present with their corresponding amounts. Resultant colors are obtained by adding and therefore the name additive.

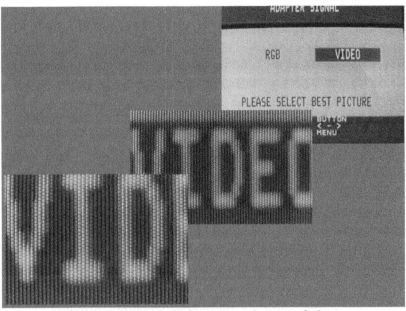

Color images on TV are made up of three phosphor mosaics (RGB).

With subtractive mixing of colors, when we use paper or canvas as a secondary source of light (reflected), colors are mixed in our eye after they are reflected from the surface. If we mix (add) all the primary colors, we produce darker colors instead of brighter. The colors are mixed by reflected light, whose color is defined by the pigment, which absorbs (subtracts) the wavelength its surface has.

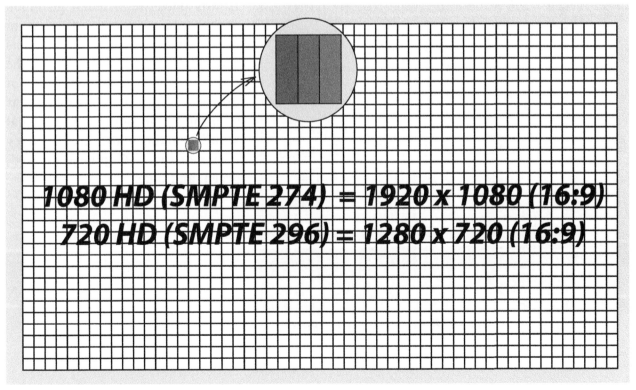

In an LCD screen each pixel is made of the same primary colors (RGB)

Getting back to television, three colors, as mentioned, are used as primary colors: red, green, and blue, usually referred to as RGB.

Television theory and experiments have shown that with these three primary colors **most** of the natural colors can be represented (but not all).

Inside a color CRT, there are **three different phosphor coatings**, each of which radiates its own color when bombarded by the electron beam. Very similar, in LCD screens, there are three different optical filters: red, green, and blue, which tint the white light of the light panel behind the screen. The polarity of the individual liquid crystal layers matrix in front of it blocks or transmits a percentage of each primary colored pixel light. When viewed from a distance this is perceived as a resultant color in our eyes.

In the (now) old CRT technology the three primary phosphor coatings have different luminosity properties, which means **equal beam intensity produces unequal brightness**. In order to compensate for these discrepancies of the primary phosphors, every CRT monitor has a special matrix circuit that multiplies each of the color channels with a different compensating number. This can be shown by the very well-known color TV luminance equation, which is electronically applied to the three primary signals in the CRT:

$$L_{screen} = 0.3R + 0.59G + 0.11B \qquad\qquad (28)$$

The blue phosphor produces more light than the other two; it therefore has to be multiplied by 0.11 in order to reduce the luminance to be equal to the other two components.

A close up of the RGB pixels on a screen

In this book we will not go much deeper into the theory of colors in television, for it requires a book on its own. It is important, however, for the reader to appreciate the complexity of the issue and accept that **all colors as seen on TV are obtained by visual additive mixing of the three primary colors of the CRT phosphor: red, green, and blue.**

Color temperatures and light sources

Very often in television, CCTV, and photography, the term *color temperature* is used when talking about light sources. **Color temperature refers to the temperature to which an imaginary perfectly black body is heated and consequently produces light.**

The theory of physics states that **the spectrum of light generated by heating is mostly dependent on the temperature of the body and *not on the material*.** This very important statement has been proven by the physicist Max Planck whose formula explains the relationship between the peak wavelengths radiated and the temperature to which the body is heated:

$$\lambda_m = 2896/T \hspace{6cm} (29)$$

In the above relation λ_m is the wavelength and T is the temperature in Kelvin degrees.

From the diagram below it can be noted that the peaks for different temperatures are outside of the visible spectrum, that is above 700 nm, in the infrared region. For tungsten (wolfram) filament light, the working color temperature is around 3000° K, and more than three-quarters of the energy is radiated in the infrared region in the form of heat. Heat is nothing more than infrared light. Higher temperatures for tungsten lights cannot be used because the melting point of wolfram is around 3500° K. Increasing the temperature to more than 2800° K will dramatically shorten the lifetime of the tungsten light. In today's tungsten globes, the air is extracted from inside the bulb in order to minimize the burning of the filament. Tungsten light is good for B/W cameras, since they are more sensitive to the infrared portion of the spectrum. Color cameras have to be compensated for the yellow/reddish color produced by a 2800° K light globe typically found in domestic lighting.

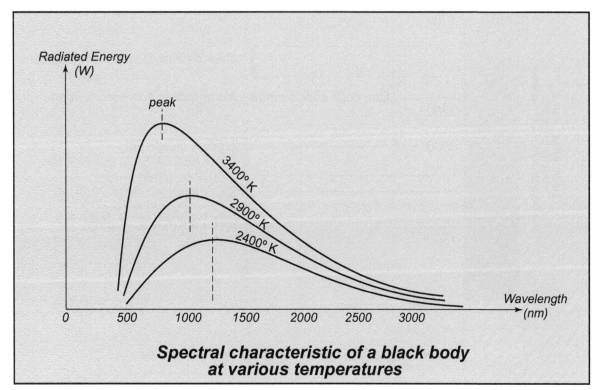

***Spectral characteristic of a black body
at various temperatures***

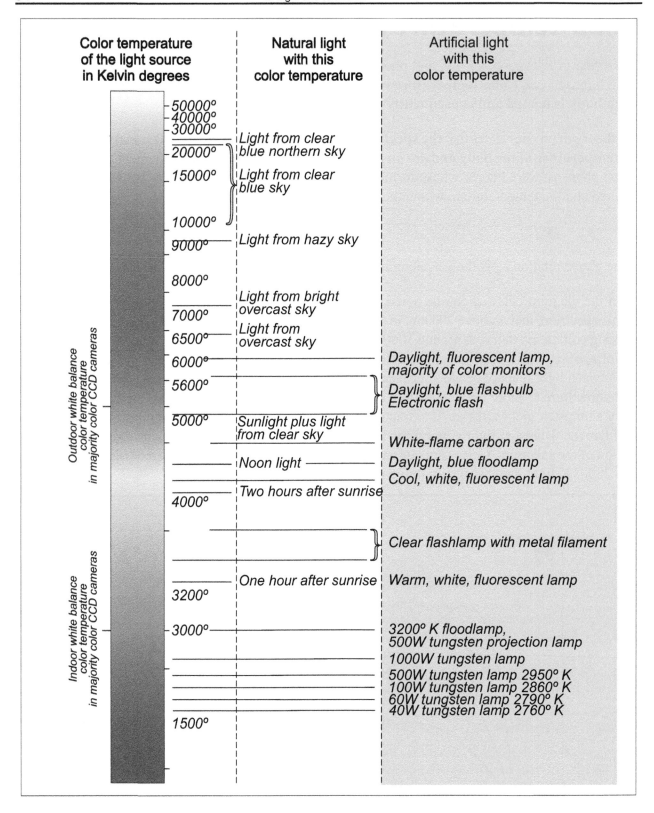

Color temperatures of various light sources

For accurate testing of cameras, very often a light source of around 3200° K is specified. Such lights can be purchased from professional photographic shops, but there is a general rule of thumb which can be used to calculate the color temperature and the lumens produced by such a light source:

500W tungsten => 3200° K (approximately 27 lumens/watt)

200W tungsten => 2980° K (approximately 17.5 lumens/watt)

75W tungsten => 2820° K (approximately 15.4 lumens/watt)

It is known that a tungsten light source produces a yellowish image on a photographic camera. In order to compensate for this blue optical filters (complementary color) can be mounted on the lens itself. Electronic cameras (CCTV and TV) compensate the yellowish color shift electronically by changing the primary colors' information by a certain percentage. Most of the CCTV cameras have the so-called *automatic white balance* (AWB) circuitry that adjusts its color temperature automatically upon powering the camera up and seeing a larger white area. A more advanced camera can readjust such a white balance "on the fly," that is, without powering the camera down and up again. This white balance is usually referred to as *automatic tracking white* (ATW) and is very practical especially when using pan/ tilt/zoom (PTZ) cameras covering a larger area, part of which might be an area with tungsten light, for example, and another with neon light.

A typical photographic tungsten light source with 3200° K

The sun, as a natural source of light, has a very high physical body temperature, but the equivalent light color temperature that we get on the Earth's surface varies with the time of the day and weather conditions. This is due to the light reflection and refraction through the atmosphere. As shown in the table of *Color temperatures of various light sources* on the next page, on a clear day, at noon, the color temperature reaches over 20,000° K, while on a cloudy day it drops down to nearly 6000° K. This is why photographs taken at sunset hours appear reddish. The lower the color temperature, the redder the pictures will appear, and the higher it is, the bluer they will appear.

Artificial sources of light have various color temperatures, depending on the source. The above-mentioned formula (29) applies to heat sources only, that is, sources of light where a metal is heated up to a high temperature. There are, however, gas sources of light, where light generation is of a different nature. Neon lights, or mercury vapor lights, for example, generate light when an electromagnetic field is applied to them. The atoms are excited by an energy sufficient to cause certain atom reactions, and energy is released in the form of light. **This light is of a discrete character due to the quantum behavior of the atoms**. The position(s) of the wavelength(s) will depend on the gas used. Some of the glass tubes used with such gases are coated on the inside with a fluorescent powder that might absorb certain primary wavelengths and then regenerate a continuous secondary spectrum of visible light.

Gas sources can also be described by their color temperature; only in this case we use a so-called *correlational color temperature.*

For the purposes of having a reference point and correct color reproduction, ***standard sources of white light*** have been defined. There are a few definitions (standards) used in practice. These standard sources of white light are marked as **A, B, C, D**$_{6500}$, and **W.**

Source **A** is the most natural standard as it represents a tungsten (wolfram) light globe, filled with some gas to reduce burning of the filament. That is why most of the other later developed standards are based on source A. As mentioned earlier, at a certain temperature, the characteristics of a wolfram light coincide a great deal with the radiation of a black body. This means **the spectrum of source A, at a certain temperature, can be represented by only *one* detail – the temperature, which is equal to the temperature of the black**

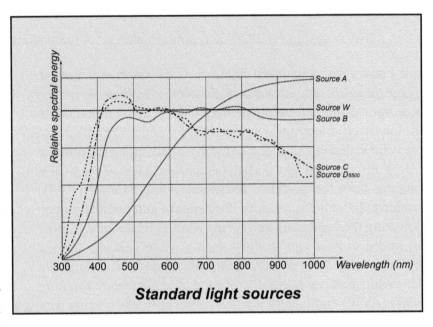

Standard light sources

body. To be precise, the real temperature of the wolfram and the black body at which their spectrums are supposed to be identical is not exactly the same. The black body is hotter by approximately 50° K. The spectrum characteristic of the standard source A is defined as a color temperature of 2854° K, while the real filament temperature is approximately 2800° K. This is an insignificant difference, however, and the theoretical approximation is valid and accepted as a descriptive factor for the color temperature of such sources.

Standard source **B** radiates white light, similar to direct sunlight at noon. Source B can be obtained by filtering the light from source A through a special light filter.

Similarly, by using another type of light filter, standard light source **C** can be obtained. The characteristics of sources B and C cannot be represented with the color temperature of a black body, as can be seen on the diagram above. However, if the color of a black body looks similar to either of the sources B or C, we use the term *correlational color temperature.* So, the correlational temperature of source B is 4880° K, and for source C it is 6740° K.

The ***International Committee for Light*** (**CIE**) in 1965 suggested a new standard source of light, which is supposed to represent an average daylight color temperature and is represented as the **D** standard. The recommended correlational color temperature for the standard D is 6500° K, so the standard is marked as **D**$_{6500}$. This source of light cannot be obtained by modifying source A, but its spectral characteristic can be approximated with some other physical sources, as is the case with a correct mixture of the three phosphor coatings of the CRT of a color monitor. An important fact to remember is that D_{6500} is often used as a reference for color monitors.

Last, there is another, fictitious, light source with a uniform distribution of radiated energy, which looks like a flat horizontal line. This is only for calculating purposes, and the code of this light source is **W**. The human eye adapts to the color temperature differences quite easily, and our brain automatically compensates

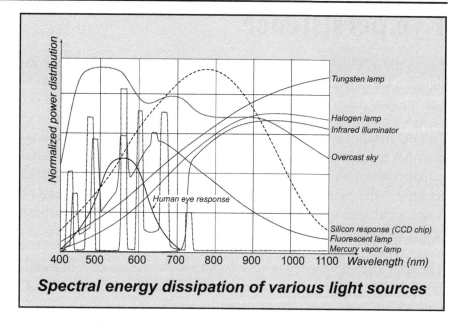

Spectral energy dissipation of various light sources

the color variation due to different light sources. Film emulsions, tubes, and camera CCD or CMOS chips are a bit different. When using a film camera, special films or optical filters have to be used if color temperature needs to be corrected. With TV cameras this is achieved by electronic compensation, which can be either manual or automatic.

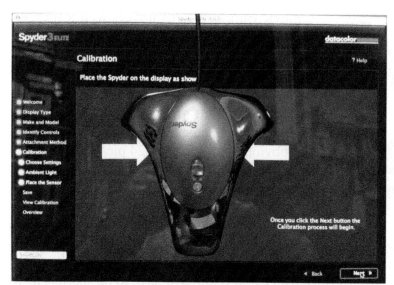

Electronic calibration of display is of paramount importance for graphic design and publishing

Finally, and as already mentioned, color temperature of the monitor screen has to be taken into account when trying to reproduce perfect colors. The majority of monitors these days can be set up for equivalent color temperature between 5000° K (referred to as D50) and 6500° K (referred to as D65), but some of them might have higher (9300° K). These color temperatures are close to the color temperature of the average overcast daylight, as shown in the previous table.

Professional photographers or cinematographers working extensively with accurate colors use monitor calibrating devices, such as the Spyder shown on this photograph.

Eye persistence

For us in CCTV, it is very important to know how the human eye works, and as we will see further in the text, we actually use an anomaly of the human eye in order to "cheat" the brain into thinking we see "motion pictures" despite pictures being static. This anomaly is the ***persistence*** of the human eye.

Eye persistence is the most important "eye defect" used in cinematography and television. The eye does not react instantly to the changes of light intensity. There is a delay of more than a few hundred milliseconds during which the brain gets the information about the object we are watching. This delay increases with an increase of the object's illumination. Not all parts of the retina have equal persistency. The central area around the fovea has longer persistency. Eye persistence also depends on the spectral characteristics of the light source, that is, its color and brightness.

The above eye deficiency is very important for the concept of motion pictures. As can be seen on the graph below, the persistence depends very much on the intensity of the light, or the brightness of the area we are looking at. The brighter the area is, the faster we have to change the pictures if we do not want to notice the flicker. The first movies from the beginning of the twentieth century, cartoons, and even the cartoon "flipping books" we used to play with as kids are based on the concept of persistency.

When individual pictures with a logical consecutiveness are played in front of our eyes at a speed equal to or faster than the persistency of the eye, we will see continuous moving pictures even though the pictures are **still,** individually.

An old film projector with a Maltese cross

A movie camera records images with a speed of 24 pictures/second. This is usually enough for the film to be projected with a very low light intensity projector, as in the beginning of the cinema revolution. For bigger audiences, bigger and stronger light projectors were needed, as well as brighter screens (as we have today). So it was obvious that the initial 24 pic/s speed needed increasing.

From a photographic point of view, which is very similar to the cinematographic one, it is impractical to increase the frame rate of the movie camera from 24 pic/s to a higher rate because the exposure time of every film frame will have to be shortened. To achieve that, the film either has to be of a higher sensitivity, which is reflected in the bigger grain structure of the film, or the iris of the lens needs to be opened more, which results in not-so-good pictures at lower light levels as well as a reduced depth of field.

Neither of these two suggestions was acceptable for cinematographers, so the solution was found in increasing the projection frequency (not the recording) from 24 to 48 with a simple but clever design. This was achieved with the so-called Maltese Cross shutter, which is a circular blade that is cut in the shape of the Maltese Cross. This rotates in front of the projection light bulb and not only blocks the light when the film moves from one frame to another (so the viewers do not see black lines between each film frame), but it also interrupts the projection while the frame is stationary (for the duration of 1/24 s) and produces two flashes of the same frame. As a result, we have a projection of 48 frames/s, which is flicker-free to the eye. Clearly, there are only 24 **different** pictures recorded each second, but the cross produces 48 of the same, and our brain perceives flicker-less continuous moving pictures.

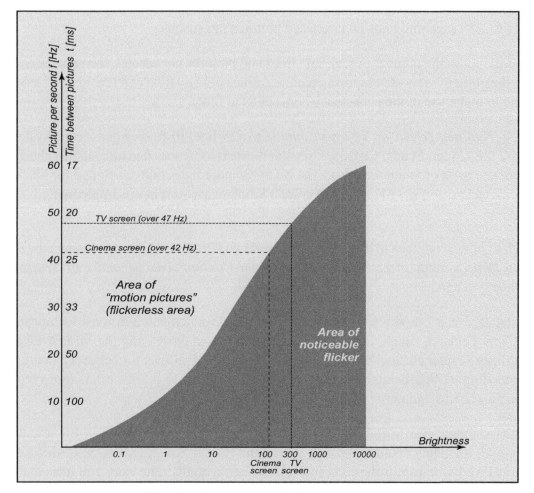

The human eye persistence curve

Television uses the same principles of eye persistence to achieve the illusion of motion by using so-called interlaced scanning. The conceptual difference is in composing the images not by using a light projector through a celluloid film, but with **electronic scanning** of a CRT screen. In television, still images are created by scanning, where a picture is formed line by line, in the same manner as when reading a book, from left to right and from top to bottom. These principles shall be explained in more detail later in the book.

It is important for the reader to understand that **television also projects static images which, when displayed fast enough, are seen as "motion pictures."** Whether this is done by interlaced or progressive scanning is irrelevant at this stage, but it should be noted that the television technology today is at such a stage that it can use improved "tricks" for the eye's motion illusion to be even better.

In the world today there are three basic television systems that differ in the number of pictures per second, the number of lines that compose each picture, and the method of color encoding. But in all of them the concept of producing motion is the same.

> PAL: 625 scanning lines/50 interlaced pictures per second

> NTSC: 525 scanning lines/60 interlaced pictures per second

Although different in the number of scanning lines and pictures per second, the general concept is the same from the point of view of composing picture frames field by field and line by line, scanning them at a fast rate to make use of the persistency concept as in film.

The NTSC's (*National Television Systems Commitee*) 525-line, 30-frames-per-second system is shared primarily by the United States, Canada, Greenland, Mexico, Cuba, Panama, Japan, the Philippines, Puerto Rico, and most of South America. The NTSC standard was first developed for black and white (monochrome) television in 1941. The first color TV broadcast system was implemented in the United States in 1953.

More than half of the countries in the world use one of two 625-line, 25-frame systems: the PAL (*Phase Alternating Line*) system or the SECAM (*Sequential Couleur Avec Memoire* or *Sequential Color With Memory*) system.

The PAL standard was introduced in the early 1960s and implemented in most European countries, Australia, New Zealand, China, India, and many countries in Africa and the Middle East. The PAL standard utilizes a wider channel bandwidth than NTSC, which allows for better picture quality. Also, the color encoding in PAL, being designed after the introduction of NTSC, offers more accurate color reproduction and better immunity to noise.

The SECAM standard was introduced in the early 1960s and implemented in France; it is used in parts of Europe, including countries in and around the former Soviet Union. SECAM uses the same bandwidth as PAL but transmits the color information sequentially. The extra 100 lines in the SECAM and PAL systems add significant detail and clarity to the video picture, but the 50 fields per second (compared to 60 fields in the NTSC system) means that a slight flicker can sometimes be noticed. It is interesting to note that although Russia, for example, uses SECAM for broadcast TV, the CCTV industry there uses PAL.

With the introduction of the new digital high definition TV standards (HD) it is possible to have both interlaced and progressive scanning. These are usually denominated with a lower case "i" or "p" next to the standard. For example, "1080i" refers to HDTV with 1920 × 1080 pixels and interlaced scanning. With the HD TV however, the differences between various countries are eliminated and the same number of pixels are used world-wide (1920x1080).

This page intentionally left blank

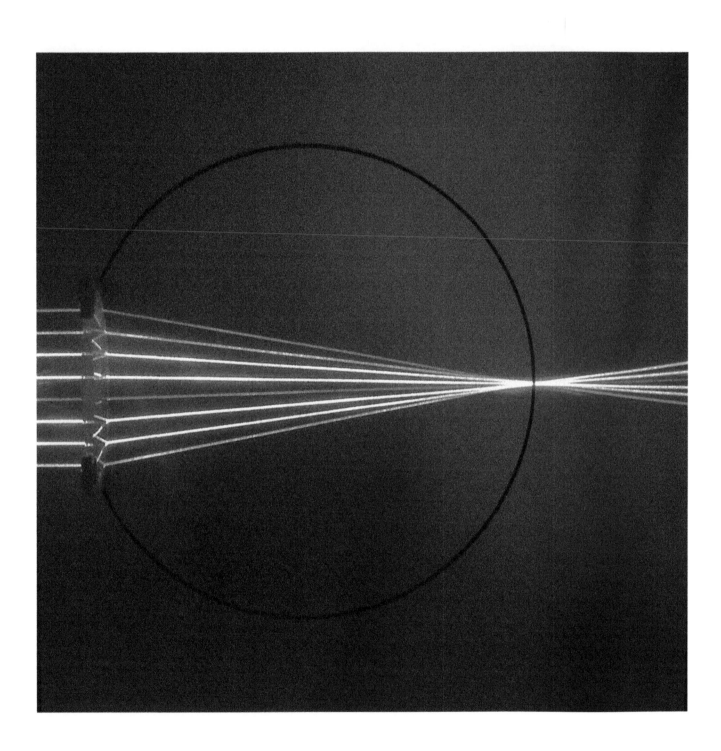

3. Optics in CCTV

Some people take optics quality in CCTV for granted. With the camera resolution development, as well as the miniaturization of CCD and CMOS chips and increase in pixels, we are coming closer to the limits of optical resolution and we need to know a bit more than an average technician. This chapter discusses, again, in a simplified way, the most common optical terms, concepts, and products used in CCTV.

Refraction

The very first and basic concept we have to understand is the concept of **refraction** and **reflection**.

When a light ray traveling through air or a vacuum enters a denser medium, like glass or water, it reduces its speed by a factor **n** (always bigger than 1) known as the **index of refraction**. Different media (which are transparent to light) have different indices of refraction. For example, the speed of light in air is 300,000 km/s and almost the same as in vacuum. If a light ray enters glass, for example, which has an index of 1.5, the speed is reduced to 200,000 km/s.

According to the wave theory of light, the reduction of the light speed is reflected in its shortened wavelength. This phenomenon represents the base of the concept of refraction. If a light ray enters the glass perpendicularly, the wavelength of the light ray shortens, but when the ray exits the glass it resumes to normal speed, that is, returns to the original "air wavelength" and continues its travel in the same direction. If, however, the light ray enters the glass at any angle other than the perpendicular, interesting things happen: the light ray (considered to be of a wave nature in this case) has a front that does not enter the glass media at the same time because it comes under an angle. The parts of the front that enter the glass first are "slowed down" first.

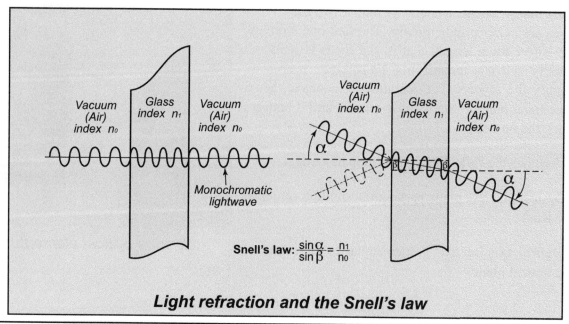

Light refraction and the Snell's law

The end result is the refraction of the light ray; the ray does not continue in the same direction but deflects slightly. This deviation depends on the density of the media.

The denser the media – that is, the higher the index of refraction – the greater the inclination of the original direction.

There is a very simple relation between the angles of incidence and refraction and indices of refraction between the two different media. This relation was discovered by the Dutch physicist Willebrord Snell in the early seventeenth century. By using a very

Laser light refraction through a prism

simple calculation, we can determine the angles of refraction in various media. As we shall see later on, the same concepts are used when calculating the angles of total reflection and numerical aperture in fiber optics.

The basics of refraction are graphically explained in the diagram on the previous page, where it is assumed a monochromatic (single frequency) light ray enters the glass. The bottom drawing also shows that a percentage of the incident light is always reflected back into air (or vacuum); in the case of glass this percentage is very small.

The refraction and reflection theories will be used in the next headings when explaining lens and fiber optics concepts.

Lenses as optical elements

There are many optical components, but the two basic types of lenses are *convex* and *concave*. The first one, convex, has a positive focal length; that is, the focus is real, and we usually call it a magnifying glass, since it appears to magnify the objects. The second one, concave, has a negative focal length; the focus is virtual, and it appears to reduce the objects.

Various optical elements

Every lens has the following important parameters:

- Optical plane (a plane passing through the center of the lens)

- Optical axis (an axis perpendicular to the center of the optical plane)

- Focus (a point where rays falling parallel to the optical axis converge)

• Focal length (the distance between the optical plane and the focus, in meters)

• Diopter (an inverse value of the focal length, where the focal length is stated in meters)

In respect to the physical size and the type of surface of the lens, there are many different types, such as plano-convex, convex-concave, and plano-concave. The name describes the physical appearance of the lens, where *plano* means one of the two surfaces is a plane.

Different types of lenses have been put together in order to correct various distortions (aberrations) caused by different factors.

As an example of why this is necessary, let us examine a sun ray falling onto a prism.

We all know the rainbow effect produced on the other side of the prism. This happens because the "white" rays coming from the sun

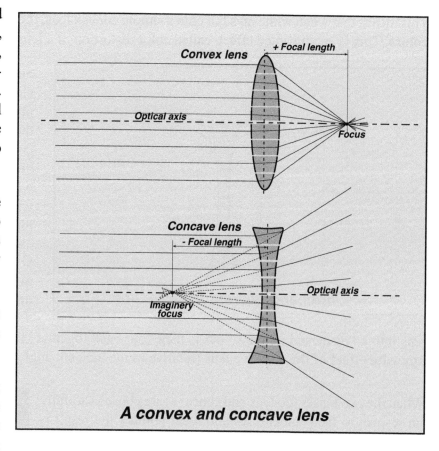

A convex and concave lens

are composed of all the wavelengths (that is, colors) the human eye can see. Because they all enter the glass prism with the index of refraction $n_1 > n_0$, different wavelengths are changed at slightly different "rates" (proportional to their frequency), thus producing the rainbow at the other end of the prism. This is actually a **decomposition of the white light**. The color red has the longest wavelength (lowest frequency); therefore, it is refracted least. The color violet has the shortest wavelength (highest frequency); therefore, it is refracted the most.

A very similar effect is the fabulous rainbow after the rain, which is actually the refraction and reflection of sun rays through the raindrops.

No matter how impressive this effect looks, it is an unwanted effect in a lens design.

A convex lens can be approximated with

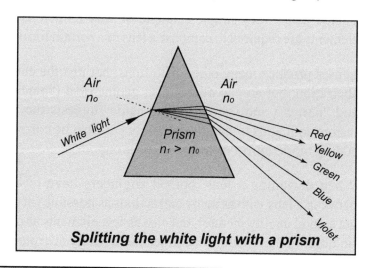

Splitting the white light with a prism

many little prisms next to each other, forming a mosaic. It is, then, obvious that the image created by such a lens using daylight (which is actually most common) will be decomposed into the basic colors as is the case with the prism light decomposition.

This means that when white rays fall onto a simple convex lens, **the focal point will vary for different colors.** This is an unwanted effect, called color distortion of a lens, or a *chromatic aberration.*

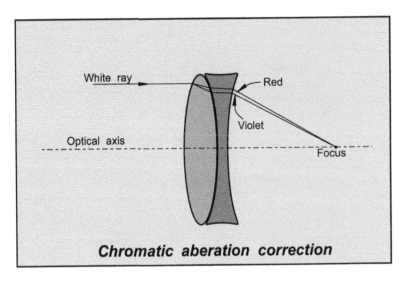

Chromatic aberation correction

So, it should be clearly understood that chromatic aberration happens not so much because of the imperfection of the lens manufacture (although this is not excluded), but rather because of the physical process of decomposing white light into the basic wavelengths when the light passes through a single piece of lens.

Chromatic aberration **can** be minimized by **combining convex and concave** lenses together, where a white ray is first split by the convex lens into a "dispersed rainbow" but is then "put back together" by the concave lens **because of the opposite effect of the concave lens** (relative to the incident angle).

When the two lenses (convex and concave) are chosen carefully (in respect to their thicknesses and focal points), the result is that all the colors come together in the focus and form a single focusing point. This is achieved with a proper selection of the convex-concave pairs, preserving the wanted combined focal length as in the single-piece lens. A special transparent glue is used to join the two lenses.

This is just a very simple example of why numerous optical elements are required to compose a lens of a certain focal length.

Lenses produce many other distortions, not just the chromatic aberration, but among others, also geometrical ("pincushion" and "barrel") and spherical. The name suggests the type of distortion it adds to the image. These can also be corrected by adding more optical elements to the group.

When designing a lens, optical engineers have to balance between a lens with as many corrections as possible (in order to get a good quality picture), but also as few elements as possible (in order to be economical and technologically acceptable).

A cross-section of a lens

One can imagine how many combinations are possible when designing a lens with a particular focal length with half a dozen (or more) different optical elements. Earlier, optical engineers used to work together with mathematicians when designing a lens with a certain focal length and size, and they used to do hundreds and hundreds of calculations and iterations manually. The physical size, the focal length, and the absolute and relative positions of every element are all variables. The only way to find such a combination of a known focal length was by painfully long iterations.

Typical marking on a CCTV lens

Obviously, the desired result was to get a good quality lens without going overboard with the number of optical elements. Since this was quite a challenging task, manufacturers used to register the particular lens design with their "recipe" of how many lenses, what focal length and at what positions they were placed. That is why in cinematography and photography we may still see the lenses of a certain manufacturer with registered names like Planar and Xenar. These names are actually patented designs of lenses for a particular lens size and focal length.

Today, in the computer era, there are many professional programs for computerized optical simulations. Within a few minutes optimum results are obtained, suggesting only as many optical elements as necessary, yet correcting as much visible distortions as possible, for the given target price.

This is why lenses of a certain focal length are available with different costs and sizes, all giving the same viewing angle but different picture quality.

Lens quality depends on many factors, and one should not take it for granted. This is especially important with zoom lenses, as there are so many variables in their design. Zoom lenses are widely used in most of the bigger CCTV systems, so we should be very careful when choosing them.

There is no simple rule, so the best suggestion, again, is to do some testing and comparisons.

The factors that determine the lens quality can be summarized by the following points:

1. Lens design

> • Number of elements
>
> • Relative position
>
> • Aberration correction in the design stage

2. Lens elements manufacture

> • Glass type
>
> • Technology and type of glass manufacturing (heating, cooling, cleanness)
>
> • Precision of grinding and polishing (very important)
>
> • Anti-reflection coatings of the glass (micrometer layers for minimizing losses)

3. Lens mechanical composition

> • The lens's positional fixing and stability (shock, temperature)
>
> • The lens's moving mechanics (especially zooming, focusing, iris leaves)
>
> • Internal light reflections (matte black absorption)
>
> • Gears used for motorized lenses (plastic, metal, precision)

4. Electronics (refers to auto iris and motorized lenses)

> • Auto iris electronics quality (gain, stability, precision)
>
> • Electric consumption (auto iris – usually low, but some older models may require more than a

Zoom lens mechanics

Disassembled zoom lens

camera can give since the camera powers the auto iris)

• Zoom and focus control circuitry (voltages: 6, 9, or 12 volts, three- or four-wire control)

Geometrical construction of images

Images can be constructed by using simple optical and geometrical rules. As can be seen on the following drawing, at least two rays are used to create the image of an object.

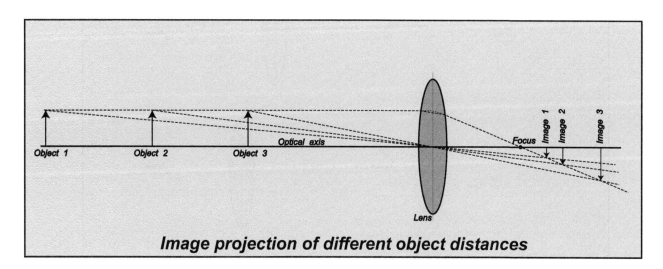

Image projection of different object distances

There are three basic rules to follow:

• Objects taken at various distances touch the optical axis with one end.

• By definition, rays that pass through the center of the lens do not change direction, that is, in the center, a lens behaves like parallel glass and no refraction occurs.

• By definition, all rays parallel to the optical axis pass through the focus.

There is a very basic lens formula, worth mentioning, which we use when calculating the light falling onto a CCD chip:

$$1/D + 1/d = 1/f \tag{30}$$

where D is the distance from the object to the lens, d is from the lens to the image, and f is the focal length of the lens. Note that d refers to a "non-infinite" distance object image and that is why it is bigger than f, whereas if the object is at an infinite distance, d would be equal to f.

Please note the position of images for various distance objects. **Lens focusing is achieved by changing the distance between the lens and the image plane** (which is where the CCD/CMOS chip is located). So, *only* **when a lens is focused at an infinitely far object does the image projection coincide with**

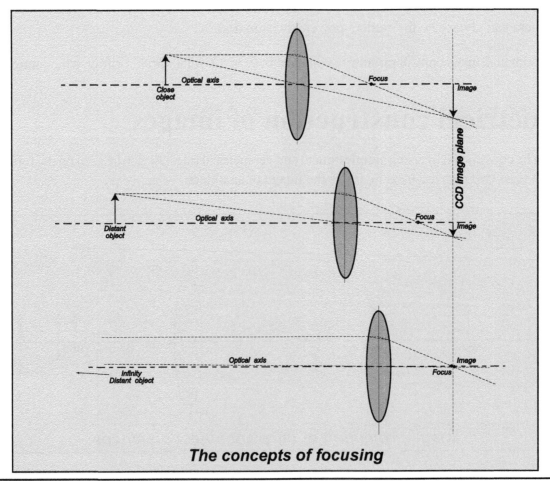

The concepts of focusing

the focus plane. In all other cases the distance between the lens and the image is bigger than the focal length of the lens (the lens is pushed away from the imaging chip).

It should also be noted that in practice, a lens is composed (as discussed earlier) of many optical elements. Therefore, they are represented by an equivalent single-element lens located at the principal point. The following drawing explains this.

A lens composed of many optical elements (single thin lens) has **two principal points** called **primary** and **secondary principal points**. **For a thin lens, these points coincide and they are located at the center of the lens.**

The planes that pass through these principal points and are perpendicular to the optical axis are called *principal planes*.

The principal planes have the following properties:

> • A ray incident to the primary principal plane (and parallel to the optical axis) will leave the secondary principal plane at the same height, traveling toward the focal point (focus).

> • An incident ray directed toward the primary principal point will leave the secondary principal point at the same angle.

> • The focal length of such a lens is measured from the secondary principal plane to the focus.

Using the above properties, we can construct a geometrical image in the same manner as was shown with the single optical element.

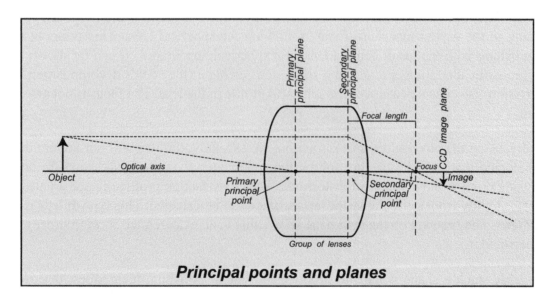

Principal points and planes

The secondary principal point may fall outside the group of lenses. This is the case with very short focal length lenses. The shorter the focal length is, the more optical elements have to be added for correcting various distortions, making the lens more expensive. With the CCD chip reduction (2/3" down to 1/2", then to 1/3", and now to 1/4"), shorter focal length lenses have to be manufactured in

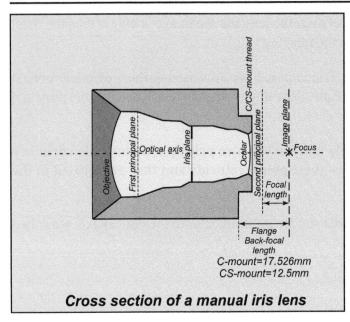

Cross section of a manual iris lens

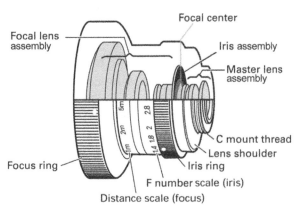

Cross section of manual iris lens

order to preserve the same wide angle as the preceding chip sizes. This, in turn, has forced the industry to reduce the C-mount 17.5-mm back-flange distance in order for the optics to get simpler, smaller, and cheaper. The new format of back-flange distance is 12.5 mm, and since it is smaller, it is referred to as the CS-mount standard.

Aspherical lenses

As mentioned earlier, spherical aberration is a common distortion that appears in the majority of lenses of a spherical type. Spherical-type lenses are the most common since they are produced by grinding and polishing in the easiest mechanical way, following the spherical laws. This refers to a circular machine polishing with the result being a lens of a spherical appearance. It can be shown that apart from the chromatic aberrations present in a single-lens element (the "color decomposition" of white light), aberration also occurs because of the spherical profile of the lens. The focus is not a very precise single point.

Theoretically, using the physical laws of refraction, we can show (but we will not go into the details) that a bell-shaped lens, which does not follow the spherical law, is the ideal shape for obtaining a single focusing point without spherical distortions. The cross-section profile of such a lens is a curve that deviates slightly from a circular shape, appearing more bell shaped. This type of lens is called an *aspherical lens*. The drawing on the next page shows this in an exaggerated form in order to help the reader understand it.

Understandably, such a shape is hard to produce by regular polishing techniques, but, if properly manufactured, it offers quite a few advantages over the conventional spherical lenses, including **higher iris openings** (which is reflected in a lower F-stop), **wider angles of view, shorter minimum object distances,** and **fewer optical elements** because there are fewer aberrations to correct (thus resulting in lighter and smaller lens designs).

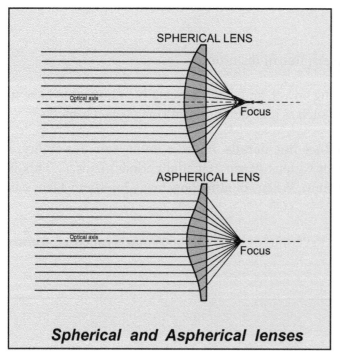

Spherical and Aspherical lenses

Aspherical lens elements

This technology is more expensive due to the aforementioned complex polishing techniques.

Many optical companies are now manufacturing **molded** aspherical lenses, avoiding the critical process of grinding. This process does not offer the same glass quality as the regular one, but it does offer a solution for more economical production of aspherical lenses. The quality of such lenses is yet to be proven, but they do exist and are available in the CCTV market as well.

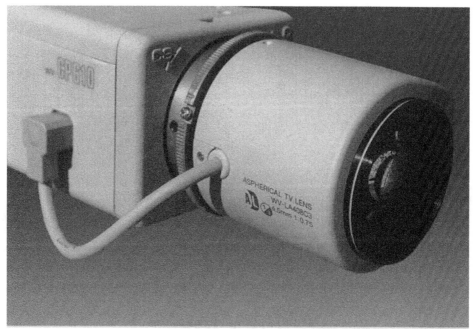

Aspherical auto iris lens

CTF and MTF

What we want from a lens is sharp and clear images, free of distortions.

As already mentioned, lenses have limited resolving power, and this is especially important to have in mind when using them in high-resolution systems, such as high definition and mega pixel cameras.

Resolution refers to the lens's ability to reproduce fine details. In order to measure this ability, a chart that consists of black and white stripes with various density (spatial periods) is used. **This is usually expressed in lines per millimeter** (lines/mm). When counting how many lines/mm a lens can resolve, we count both black and white lines.

A characteristic that shows the "response" of a lens to various densities of lines/mm is called a *Contrast Transfer Function* (CTF).

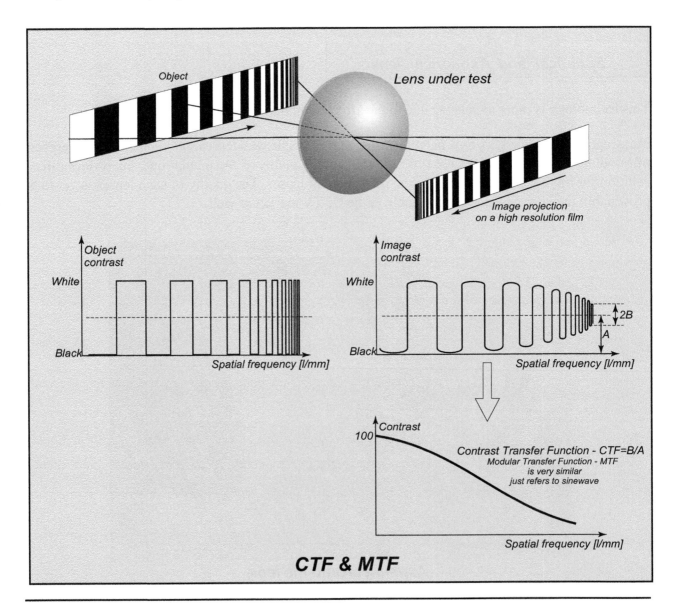

Theoretically, it is better to know the lens characteristics for a continuous variation of black to white (in the form of a sine wave), and not just for stripes that abruptly change from black to white. This would be especially suitable for TV lenses since the optical signal is converted into an electrical signal with which sine waves are easier to represent and evaluate. This characteristic is known as a ***Modulation Transfer Function*** (MTF).

In practice, however, it is much easier to produce a test chart with just black/white stripes rather than the sine wave variation between black and white. CTF is not exactly the same as MTF, but it is much easier to measure and is **good enough to describe the lens's global characteristics.**

The easiest analogy of MTF to understand would be the spectral response of an audio system. In an audio system we usually describe the output level (voltage or sound pressure) versus the audio frequency. In optics it is similar, where MTF is expressed in contrast values (from 0 to 100%) versus spatial frequency (expressed in lines/mm), as can be seen on the previous page.

Different lenses have different MTF characteristics, depending on the quality of the glass, optical design, and application. For example, a photographic lens will have a better MTF than a CCTV lens. The reason for this is simple: the photographic film structure can register over 120 lines/mm and manufacturers need to produce better lenses in order to minimize picture deterioration when film is blown up to a poster size.

CCD and CMOS chips have a lower resolution than the film crystal structure. Technically, **there is no need** to go to the "expense" of producing a lens with much higher resolution than a CCD chip. With the miniaturization of chips, however, and especially their increase in pixels for the same chip size, we are actually coming closer to the film resolution limits, so lenses need to feature better characteristics.

An average 1/3" imaging chip of an analog camera, for example, has approximately 768 pixels (picture elements) in the horizontal direction. When we take into account the physical width of the 1/3" imaging chip (4.8 mm), we can conclude that the maximum number of vertical lines (black and white pairs) we can have is (768:4.8):2 = 80 lines/mm. This resolution is easily achieved with most TV lenses, since the optical technology can produce over 80 lines/mm. But for the same size chip of 1/3" size, made for the high definition CCTV (1920 x 1080), we actually need a lens of about (1920:4.8):2 = 200 lines/mm. This means that a 1/3" HD camera **demands much more** from the lens resolution than a standard definition video from the same size chip. For these reasons, the CCTV industry has started recognizing the need for a higher quality lenses when mega-pixel cameras (IP cameras with more than million pixels) are used, and typically such lenses can be found with "Mega Pixel" lens written on them.

This statement should not be taken on face value, as there are "mega-pixel" lenses and there are "mega-pixel" lenses. Real life testing would be the only way to tell which one is

Photo courtesy of Tamron

A mega-pixel lens

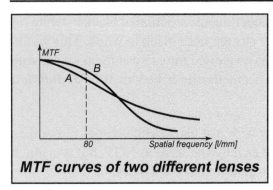

MTF curves of two different lenses

better than other.

Different lenses have different MTF characteristics, and sometimes it may be necessary to decide which one to use on the basis of these characteristics.

The diagram presented here shows such an example. We can evaluate this in the following way. Lens *A* has its MTF extending into the high spatial frequency range, which means it can resolve finer details than lens *B*. Lens *B*, however, has better response in the lower frequencies. If we need a lens for a high-resolution output, like HD chip, for example, lens *A* will be a better choice, but for an SD CCTV purposes, where the chip cannot see more than 80 lines/mm, we are better off with lens *B* since we will have better contrast with it.

F and T numbers

In addition to the MTF and CTF characteristics of a lens, the ***F-number*** (more commonly, the ***F-stop***) is also a very important parameter.

The F-number indicates the brightness of an image formed by a lens. This is usually written (engraved) on the lens itself as F-1.4, for example, or sometimes in another form, such as 1:1.4. The F-number depends on the **focal length** of the lens and the **effective diameter** of the area through which the light rays pass. This area can be controlled by a mechanical leaves assembly, which we usually refer to as the *iris*.

It is important to note that the **effective diameter** of a lens is not the actual lens diameter, but rather the **diameter of the image of the iris as seen from in front of the lens.** The first lens diameter is usually called the ***entrance pupil.*** There is also an ***exit pupil***, as shown on the diagram below. The actual iris diaphragm is positioned between these two pupils, which also happens to be between the two principal points.

The lower the F-number, the bigger the iris opening is, and that means more light is transmitted through the lens. The lowest number for a particular lens is the

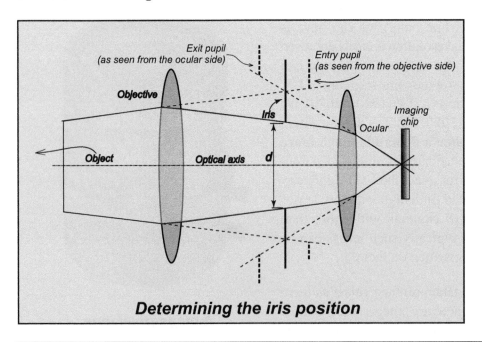

Determining the iris position

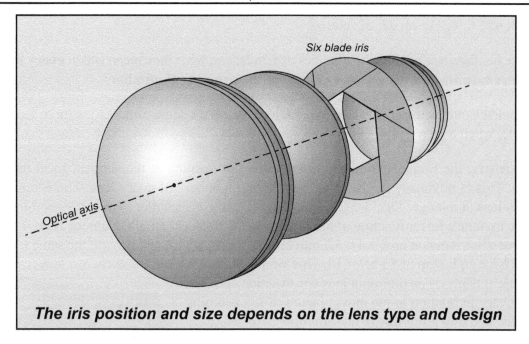

Six blade iris

Optical axis

The iris position and size depends on the lens type and design

number engraved (or written) on the lens itself, representing the **light-gathering ability** of that lens.

Often, the lower F-stop lenses are called *faster lenses*. The reason for this is that, in the early days of photography, by increasing the amount of light (lower F-stop), the film exposure time needed to be shortened, thus allowing pictures with fast action to be taken without losing any sharpness because of camera movement.

If a 16 mm lens has a minimum F-stop of 1.4, for example, it is usually written as 16 mm/1.4, or sometimes as 16 mm 1:1.4. The maximum effective iris opening is equivalent to a circle with a diameter of 16/1.4 = 11.43 mm – equivalent because the iris leaves would usually make a triangle, a square, a pentagon, or a hexagon opening.

In order to understand the consecutiveness of the F-numbers, we will have to do some simple calculations.

Starting with the example of a 16 mm/1.4 lens, let us find the area when the iris is fully open (that is, at F-1.4):

$$A_{1.4} = (d/2)^2 \cdot \pi = (11.43/2)^2 \cdot \pi = 32.66 \cdot 3.14 = 102.5 \text{ mm}^2 \qquad (31)$$

Let us halve this area – that is, take 51.25 mm² as a new area, and let us calculate what the iris opening is:

$$A_x = (x/2)^2 \cdot \pi \Rightarrow x = 2 \cdot \sqrt{(A_x / \pi)} = 8 \text{ mm} \qquad (32)$$

where √ is a square root of the elements in the brackets. Now, the F-stop with an 8 mm iris opening would be 16/8 = 2, that is, F-2.

Here we have F-2 representing an area that is exactly half of the F-1.4. If we proceed with the same logic, we will get the following familiar numbers:

    2.8; 4; 5.6; 8; 11; 16; 22; 32; etc.

All of these numbers are common to all types of lenses, and what they mean is that **every next higher F-number transmits half the amount of light of the previous F-number**.

Now it should be much clearer why a 16 mm/1.0 lens makes the same camera look more sensitive than, for example, when a 16 mm/1.4 lens is used.

For zoom lenses, the F-numbers quoted refer to the iris opening at the shortest focal length of the zoom lens. This is obviously the best "light-gathering number" of every lens. The F-number of the same zoom lens at a longer focal length setting (tele) is always smaller than at the shorter end. But it is wrong to assume a linear function of the F-stop versus the focal length. Namely, if an 8–80 mm/1.4 lens is in question, it makes an 8/1.4 = 5.7 mm effective iris opening, while with the same iris at 80 mm we should have an F-stop of 80/5.7 = 14. This simply is not the case because it depends on the zoom lens construction. The iris plane may vary in relation to the moving parts of the zooming components, obeying a **nonlinear law**. In most cases we have much better values for the F-stop at the higher focal length than indicated, but they are still worse than at the lower focal length.

It is fair to say that every piece of glass, no matter how good it is, introduces some light loss. These losses might be a very small percentage of the total light energy, but they should be considered if accurate lens characteristics need to be taken into account. An indication of lens's level of light transmission is shown by the *transmittance factor*, which is always less than

Vari-focal lenses have become very popular.

100%. This is why many professionals prefer to use T-numbers instead of F-numbers.

The definition of a T-number takes the F-stop and the lens transmittance into account:

$$T\text{-number} = 10 \cdot F\text{-number}/\sqrt{(\text{Transmittance})} \tag{33}$$

where the symbol $\sqrt{}$ means square root. Since the transmittance of a lens is, as mentioned, always less than 100% (usually 95 to 99%), it is obvious that the T-number will be a bit higher than the F-number. For example, if a 16 mm/1.4 lens has a transmittance of 96%, the T-number will be equal to 1.43.

Depth of field

When a lens is focused on an object, theoretically, the whole plane passing through the object and perpendicular to the optical axis should be in focus.

Practically, objects slightly in front of and behind the object in focus will also appear sharp. This "extra" depth of sharpness is called *depth of field*.

A wide depth of field might be an undesired feature, as it is, for example, when we want an object we are photographing to be isolated from the foreground and the background. This is very characteristic when taking portrait shots with a telephoto lens, where the depth of field is very narrow.

In CCTV, however, we often want the opposite effect. We want to have as many objects in focus as possible, no matter where the real focusing plane is.

The depth of field depends on the focal length of the lens, the F-stop, and the format size of the lens (2/3", 1/2", etc.). A general rule is **the shorter the focal length, the wider the depth of field; the higher the F-stop, the wider the depth of field, and the smaller the lens format, the wider the depth of field.**

The depth of field effect is explained by the so-called ***permissible circles of confusion***. The permissible circle of confusion is a projected circle of the depth of field area. If the smallest picture element (pixel) of the imaging chip is equal to or bigger than the permissible circle of confusion, then it is obvious that we cannot see details smaller than that circle. In other words, all objects and their details that appear within the circle will look equally sharp, since that is the actual size of the pixels. From this it is clear that the size of the permissible circles of confusion for a CCTV camera is determined by the pixel size of the imaging chip – in other words, the chip resolution.

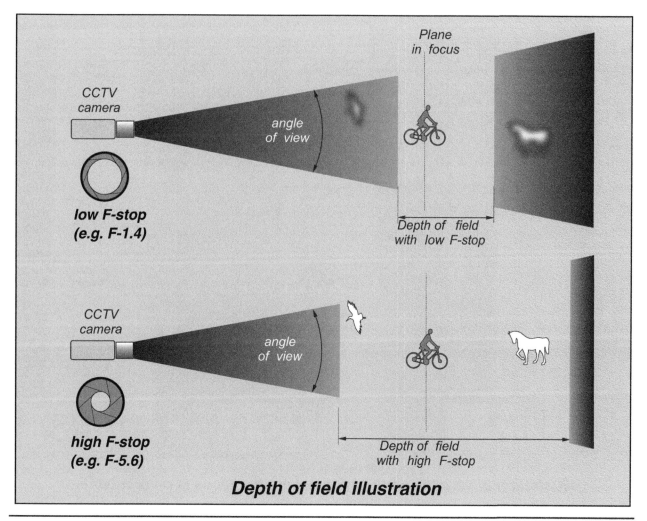

Depth of field illustration

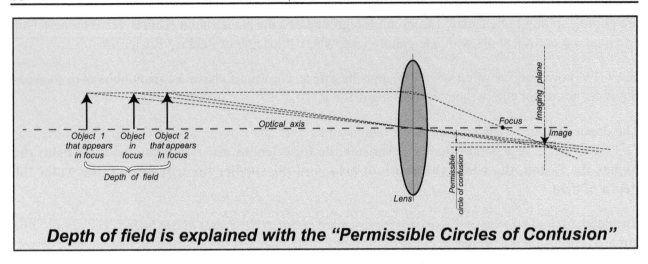

Depth of field is explained with the "Permissible Circles of Confusion"

It may now be understood why some short focal length lenses in CCTV, such as 2.6 or 3.5 mm, **do not have a focusing ring** at all but only an iris adjustment. This is because even with the lowest F-stop for that lens (be it 1.4 or 1.8) the depth of field is so wide that it actually shows sharp images from a couple of centimeters in front of the lens up to infinity. There is literally no need for focusing.

Illustration of various depth of field where the focus is on number 8.

As shall be explained later in the book, the depth of field is an effect of which we should be very aware, especially when adjusting the so-called back-focus. If the back-focus is not adjusted properly, and a camera is installed at daylight (that is when the auto iris of the lens closes the iris as much as possible, due to excessive light), the depth of field will produce sharpness even in areas that are not really in focus.

Practical experience shows that depth of field applied in this way (when the back-focus is not done correctly) is the biggest source of frustration for a 24-hour operating system. The reason is obvious: at night, when the iris opens due to a low light level (providing the AI functions properly), the depth of field narrows down and shows the images out of focus even if they were in focus during the day. When an operator complains to the installer or service people, not knowing the cause of such a problem, he or she usually gets the service to visit during the daytime. Obviously, the problem will not be there then, thanks to a wide depth of field that reappears "inexplicably" at nighttime.

The moral of the above is that the back-focus adjustment (discussed later in the book) should be done when the iris is fully opened. The easiest way to have the iris opened is when low light levels reach it, either at the end of the day (or at night) or by artificially reducing the daylight with external neutral density filters (usually placed in front of the lens objective). All this is in order to reduce the depth of field and consequently make back-focus adjustment easier and more accurate.

Quite often, when cameras with infrared lights are used another effect is present. Because of the extremely long wavelength of the infrared light (compared to normal light) and the lesser angle of refraction, we get the focused image plane slightly behind the imaging chip. Refer to the heading Lenses as Optical Elements for further explanation of this phenomenon. If an image is sharp at day, then at nighttime objects of the same distance will be out of focus. This might be a quite noticeable and unwanted effect. In order to minimize it, a lens should be designed with a special compensation for infrared viewing (some manufacturers have special glass lenses for this purpose). Another practical and common solution would be to have the camera back-focused at night with an infrared light on, in which case the depth of field is minimal but the objects are in focus. At day, the depth of field will increase the sharpness to a wider area, compensating for the difference between the infrared and the normal light focus.

The last note I would like to make is about the vari-focal lenses and the back-focus adjustment with them. Namely, vari-focal lenses started to prevail with fixed cameras lately, despite the vari-focal concept being one of the lowest optical quality when compared to manual focal length lenses, or even zoom lenses. The fact is that they offer the easiest method to decide the angle of view on the spot, during the installation process. But it is also a fact that such vari-focal lenses have typically two optical elements groups which are positioned until the desired angle of view becomes sharp, then this is screw-locked and such an image stays fixed with that choice of view. Back-focus for such lenses is not of great importance. There is one exception to this rule, and this is the Pentax's design of "Vari-focal+" series of lenses, where in fact the "vari-focus+" is a manual zoom lens design, so that by correct back-focusing, a sharpened image, stays sharp for all the angles of view such a lens offers.

A very special vari-focal

Neutral density (ND) filters

Earlier, when we discussed F-stops, we also mentioned some F-numbers – 1.4, 2, 2.8, 4, 5.6, 8, 11, 16, 22, 32, and so on. This list continues – 44, 64, 88, 128, and so on. The higher the F-number, the smaller the iris opening is.

For photographic or movie film, F-32 is considered quite a high number. The film emulsion is so sensitive that even on the sunniest days, this F-stop, combined with the available shutter speed, is enough to compensate for the excessive light.

Film sensitivity is measured in ISO units, and the most common film we use for everyday purposes has a sensitivity of 100 ISO units.

CCD and CMOS chips are much more sensitive than a 100 ISO film, especially in chips with larger pixels (the larger the pixel the more electrons it may contain resulting in better light performance). Translated into everyday language, this means that modern imaging chips are so sensitive that the low light level situation is not really a problem (although you would have a lot of customers asking you, "How sensitive is your camera?"), but rather the strong light. Typically, for "live" viewing television cameras use one exposure

ND film filters in an auto iris lens

speed only, 1/50 s in PAL, and 1/60 in NTSC, which means we can only manipulate the F-stop to reduce the amount of light.

An average imaging chip requires 0.1 lx at the chip to produce a full video signal. A bright sunny day at the beach, or on the snow, produces more than 100,000 lx at the object. To reduce this to 0.1 lx, very high F-stops, in the order of up to F-1200, need to be used. Using the basic definition for F-stop, for an average 16 mm/1.4 lens, we will get F-1200 to be an effective iris opening of 16/1200 = 0.013 mm.

Mechanically, this is impossible to produce due to not only the very small size and precision required, but also because with such a small iris we would introduce new problems, such as edge diffraction of light (known as the Fresnel effect), which will affect the picture quality.

The solution was found in the use of ***internal neutral density*** (ND) ***filters***. These are very thin films of circular, neutral color coatings, positioned in the middle of the lens, close to the iris plane. The filters get less transparent toward the middle of the concentric circles. The F-stop is thus achieved by a combination of the mechanical iris (leaves) and the optical ND filter (optical attenuation). This is a very simple and efficient way of battling strong light. The filters are called neutral because they attenuate all wavelengths (colors) evenly, therefore not changing the color composition of the image.

The optical precision of such thin films is very important in order to preserve the lens's MTF characteristics as the F-stop increases. Theoretically, the resolving power of any lens is best in the middle of the mechanical iris setting, and it reduces as the F-stop goes lower or higher (this is different from the depth of field effect), but the ND filters may reduce it

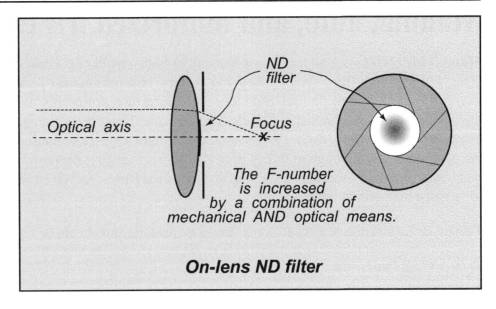

ND filter

Optical axis

Focus

The F-number is increased by a combination of mechanical AND optical means.

On-lens ND filter

even further. Whether or not this will be obvious depends on the quality of the lens in general.

Apart from the internal ND filters, there are also *external ND filters,* which are not so sophisticated. These are just precise semitransparent pieces of glass, or optical filters if you like, that attenuate the light × number of times. This may be 10, 100, or 1000 times. Two or three of these can be combined, so, for example, 10 with 1000 times will result in an ND filter with 10,000 times attenuation.

Sometimes, and probably more correctly, the attenuation of the external ND filters is expressed in F-stops. Knowing that every next F-stop will divide the light gathering ability by 2 (50% of the previous number), we can establish the following logic: 100 times ND filter is divided by 100, which is halfway between 2^6 and 2^7 ($2^6 = 64$, $2^7 = 128$). This means 100 times attenuation is approximately 6.5 F-stops. One thousand times attenuation is close to 2^{10}, which means approximately 10 F-stops.

These types of ND filters are very handy, as already explained, for minimizing the depth of field for the purposes of back-focus adjustments or AI level adjustments during the daytime.

In the last few years (between the time of the previous edition of this book and the current one) there are some new *external variable ND filters* available, where two pieces are put together and by their rotation a different density of ND filtering is achieved. This could be a useful tool for when adjusting back-focusing or testing low light performance of cameras.

Photo courtesy of Tiffen

Variable ND filter suitable for back-focusing

Manual, auto, and motorized iris lenses

Manual iris (MI) lenses adjust the iris manually (that is, by hand). These lenses are very common in areas with constant light, such as shopping centers, underground car parks, and libraries. Basically, these are areas where natural light does not interfere noticeably with ambient light, so we have almost constant illumination levels. Eventual small variations are compensated by the camera's automatic gain control (AGC). Larger variations of light levels, like outdoor usage, can be handled by the electronic-iris, although in theory optical auto iris is always preferred for such cases. The fact is that many IP cameras these days come with manual iris lenses, and electronic-iris - not optical, is used to produce the optimal image.

Two major factors decide at what F-stop (iris) a manual iris lens should be set for optimum performance:

- Light intensity

- Depth of field

These factors contradict each other, and that is why MI settings are always a compromise. When using it in very low light level situations, or when using not so sensitive cameras, the general tendency is to open the iris (low F-number) as much as possible. Obviously, in such cases the depth of field, as well as the MTF, as explained previously, will be minimal. We should not forget that apart from the depth, the lens resolution at the lowest F-stop is usually the poorest. A compromise is often the best solution (if the camera's minimum illumination characteristic allows for it), and the lens is set to one or two F-stops higher than the lowest (e.g., F-2, F-2.8).

Photo courtesy of Cosmicar (Asahi Precision Co.Ltd.)

Manual iris fixed focal length lenses

At this point, it should also be noted there are claims of some new lens designs with practically no iris, just fully opened aperture, which are capable of high depth of field due to their patented design. Despite requesting information from such manufacturers for the purposes of illustrating them in this book, I had no success in obtaining such details, and hence I cannot comment on their claimed capability.

A manual iris and an auto iris lens

Auto iris (AI) lenses have electronic circuitry that processes the video signal coming out of the camera and (the circuitry) decides, on the basis of the video signal level, whether the iris should open or close.

Auto iris works as automatic electronic-optical feedback. If the video signal is low, the electronics tells the iris to open, and if it is too high, it tells it to close.

In order to do this, the AI lens **takes power from the camera** (usually 9 V DC), **as well as the video signal** and references the electronics of the lens and the camera with a third common wire (called zero, negative, or common). Quite often, you will find lenses with shielding as well. This is to protect the video signal wire from strong external electromagnetic interference. Usually, this wire does not have to be connected to the camera body because the connection is already made with the lens's metal ring when fitted on the camera. By keeping the AI cable as short as possible, the amount of unwanted

Photo courtesy of Cosmicar (Asahi Precision Co.Ltd.)

Auto iris fixed focal length lenses

interference induced in the video signal is minimal. This goes hand in hand with the ever decreasing camera size. Be aware, however, of plastic C/CS-mount adaptors that will not common the lens case with the camera's body.

Following are some color codes for the AI wires that are widely accepted in the industry:

- Black is usually used for common,

- Red for power (derived from the camera), and

- White for video.

Some manufacturers, in order to lower manufacturing costs, have started using two-wire AI cables (red-power and white-video) with a shielding used as the common wire.

Often, lenses with four-wire cables can be found, where the fourth wire is usually green. In most cases this is an unused wire, but in some lenses it offers remote control of the iris, usually known as ***motorized iris*** (MRI) control. When such control is wanted, the iris opens and closes as instructed by the voltage from a site driver (controlled by an operator), much in the same way as zoom and focus are controlled.

The latter type of lens is the preferred one in systems with electronic-iris cameras. The reason for this is that electronic iris and auto iris do not work well together. If the two of them are enabled, the electronic iris usually works faster, and by the time the mechanical auto iris responds to the light fluctuations, the electronic iris has already reduced the shutter exposure, forcing the auto iris to open more. The end result is a widely opened iris and a very short electronic exposure. This gives a full video output signal as expected, but the **depth of field is minimal and vertical smearing is more noticeable** because of the very short exposure of the imaging chip.

Because of this, when auto iris lenses are used it is suggested that the electronic iris be switched off. The electronic iris is, however, quicker and more reliable since there are no moving parts (only electronics), although it **does not control the depth of field.**

So, to gain the benefits of both, motorized iris lenses are now recommended with electronic-iris cameras. With such lenses, operators can adjust the iris according to the light-level situation and required depth of field.

The current consumption of the AI circuitry is very small, typically below 30 mA, and it does not represent any noticeable load on the camera power supply. Be aware, however, as mentioned earlier, that older lenses (especially bigger zoom lenses) may demand more current drive, in which case (if a camera output current is not sufficient), a separate 9 V DC power supply has to be used for the auto iris electronics inside the lens.

Video- and DC-driven auto iris lenses

The division of lenses gets a bit more confusing in respect to the processing circuitry when auto iris lenses are in question. Namely, apart from the "normal" AI lenses we have in the majority of cases, where the electronics are built inside the lens itself and which we call *video-driven AI* lenses (since they require a video signal from the camera), we can also find so-called *DC-driven AI* lenses. These lenses are similar to the video-driven ones, with the exception that **the processing electronics are not inside the lens but rather inside the camera.** The lens, in that case, has only the motor and the iris mechanism. Clearly, when DC-driven lenses are used, the camera has to be designed to have such an output. Instead of having power, video, and common wires, we will have power, DC level, and common connection. Often, these types of lenses are called Galvanometric auto iris lenses.

A DC-driven lens cannot be used on a camera that does not have that type of connector, and vice versa. If a camera has a DC auto iris connector, you will usually find level and ALC adjustments (explained in the following paragraphs) on the camera itself, instead of on the lens.

DC-driven lenses seem to have prevailed since the previous edition of the book, so these days it is very rare to come across a video-driven lenses.

Video-driven AI lenses, both fixed and zoom, have two potentiometers for adjusting the response and type of operation: **level** and **ALC** (automatic light compensation). This also applies to DC-driven lenses, only in that case, as mentioned above, the settings are on the camera itself.

Level adjusts the iris opening on the basis of the average level of the signal. The level is also known as sensitivity adjustment because of its appearance on the monitor screen as brightness variation of the object. When the level

A fixed auto iris lens

potentiometer is adjusted, iris operation should be checked both daily and nightly. If the working point is shifted too high, the picture may look okay at day but very dark at night. The opposite is also true:

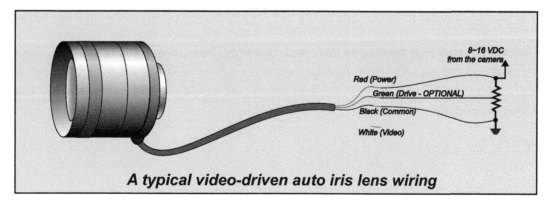

A typical video-driven auto iris lens wiring

if the working point is shifted too low, it may be acceptable at night but too bright at daylight. To make sure that this does not happen, the best adjustment is achieved in the late afternoon with a little help from a torch. First, make sure the picture is as good at low light as it can be (that is, iris fully opened). Then, shine the torch at the lens and see if the iris closes sufficiently to see the torch filament only.

ALC and level pots

If tests cannot be conducted in the late afternoon, the alternative is to use some external ND filters. These filters can be selected to attenuate the daylight to the level equivalent to a low light level situation, which is usually a couple of luxes. Then, instead of using a torch, all it requires is to remove the ND filters and see whether and how the iris reacts.

ALC, as we have noted, stands for automatic light compensation. **The ALC is a photometric adjustment of the iris, and it should be thought of as "automatic backlight compensation."** The ALC part of the auto iris circuit decides on which portion of the video signal level the auto iris should react. ALC adjusts the video reference point for the iris operation depending on the picture contrast. In most cases, when the signal is "rich" with details from the darkest to the brightest (0 to 0.7 V), the reference level is in the middle. If very bright spots appear in the picture, they will participate in the calculation of the reference point and will force the auto iris to close to produce a video signal with "full dynamic" range. The visual appearance then will be a high-contrast picture. So, very bright objects (e.g., sun reflections, bright lights, windows, etc.) will force the iris to close, making the dark objects even darker, sometimes too dark to distinguish any details. In such situations we may change the ALC setting from the factory default to the extreme position to make the iris disregard the bright areas and open more than usual. This allows for the objects in shadow to be more distinguishable.

This adjustment is equivalent to the backlight compensation found in many camcorders. The backlight compensation is used, as the name suggests, to fight against the backlight. The idea is to tell the lens electronics to disregard the very bright areas of the image and open the iris more in order to see details of the darker objects in the foreground.

This is very useful when positioning the camera in hallways, for example, looking through glass doors and against a bright background. If a person walks in the hallway, he or she will be a silhouette. When the ALC is adjusted, the iris can be forced to open by one or two F-stops more, thus brightening the face of the person. Similarly, the ALC can be adjusted to do the opposite job, that is, close the iris more than it should in order to see details of the very bright background, as through the hallway door.

The ALC setting has two ends marked as Peak and Average. The first example above would correspond to Peak setting, and the second to the Average setting. Factory defaults are usually in the middle of these two positions. Please note that, in order to see the effects of the ALC adjustments, a very high-contrast scene is needed.

Auto iris lens electronics

As the optics quality of a lens cannot be taken for granted, neither should the electronics of an auto iris lens. Different circuit designs offer different quality and precision of operation. This, combined with the mechanical construction of the iris shutter, determines whether a lens is good, average, or bad.

The responsiveness of the iris to abrupt light changes is not instant and ranges anywhere from half a second to two seconds. This needs to be taken into account when adjusting level and/or ALC settings on a lens. The delay depends on the feedback, that is, the electronic and mechanical combination. The electronics has its automatic gain control (AGC), but how effectively this combination works depends on the camera's electronics, including the AGC.

The combination of the two can be such that they may produce oscillation in the auto iris operation, which is usually called ***ringing*** or ***hunting***. The ringing appears as a pulsating picture, depending on the camera viewing direction and light conditions. It is especially common when looking against strong light. To minimize it, usually level adjustment is sufficient, and sometimes ALC or both. There are unfortunate camera/lens combinations, however, where ringing cannot be eliminated. The solution is usually found in replacing the lens with the similar of another brand. Some newer auto iris lenses come with an additional potentiometer for adjusting the level of the lens's AGC.

As mentioned earlier, the auto iris lens cable is usually protected with a shielding that is often not connected to the auto iris. The shielding's purpose is to protect the video signal wire from picking up noise. In order for it to be effective, it is sufficient for one end of the shielding to be connected to the common of the signal electronics, which happens to be done through the lens body (the C- or CS-mount ring) and the camera C-mount thread. With camera miniaturization, the cables are getting shorter, further minimizing the risk of unwanted external noise interfering with the operation.

A disassembled fixed autp iris lens

Finally, let's remember that the AI current consumption is very low, typically below 30 mA.

Image and lens formats in CCTV

A lens sees objects with the same angle of vision in all directions, that is, the angle of vision has a conical shape. Therefore, the image area projected by a lens has a circular shape, but the camera's sensitive area (CCD or CMOS chip in our case) is a rectangle **within** the imaging circle.

In today's standard definition (SD) television, this rectangle is with the **aspect ratio of 4:3**, that is, the standard is 4 units in width by 3 units in height. As mentioned at the beginning of the book, this aspect was adopted for the film format in the early days of television.

The all-new high-definition television (HDTV) system, which is the current HD standard, has an aspect ratio of 16:9. The idea is to have better movie presentations.

The "imaging rectangles" are within the image circles, which have all (or at least the majority) of the aberrations corrected.

There is no point in making a lens that produces a much bigger image circle than is required. Therefore, the lenses are made to suit the image format, no less and no more. There are exceptions, such as when lenses made for other purposes, photography, for example, are used on a CCTV camera with a special C-mount adaptor.

Today in CCTV, we have quite a few different chip sizes: 2/3", 1/2", 1/3", and 1/4". High-definition cameras and some special-application cameras may have 1/2.5" or even larger chip sizes. In order to understand this variety, we should know a little bit of the history of TV.

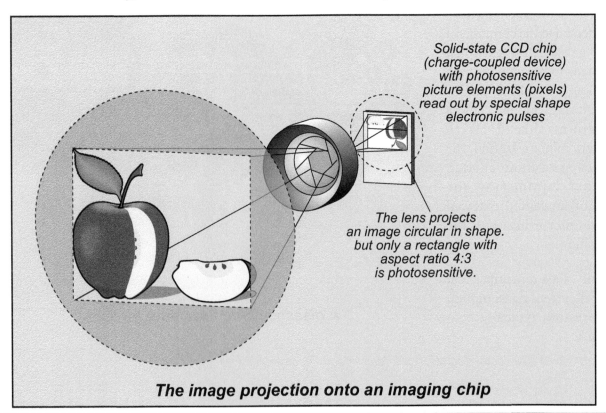

Solid-state CCD chip
(charge-coupled device)
with photosensitive
picture elements (pixels)
read out by special shape
electronic pulses

The lens projects
an image circular in shape.
but only a rectangle with
aspect ratio 4:3
is photosensitive.

The image projection onto an imaging chip

The very first TV cameras used imaging tubes of a certain diameter and were referred to as 1" Vidicon or 2/3" Newvicon cameras. **These dimensions referred to the actual diameter of the imaging tube and the projected image circle diameter.** The imaging area is a rectangle with a 4:3 aspect ratio, and this rectangle has a diagonal that is **smaller** than the actual tube diameter. Therefore, a 2/3" tube camera has an imaging area, scanned by the electron beam, of approximately 8.8 × 6.6 mm. This area gives a diagonal length of approximately 11 mm. **This is not equal to 2/3", which, converted into millimeters, is 17 mm.** So, do not think that the CCD/CMOS chip reference has the same meaning of the diagonal size, as is with TV screens. **Imaging chip size reference is not the diagonal of such a chip.**

When we say a 2/3" CCD chip, we are really referring to a device that has an imaging area equal to what a 2/3" tube would have, or it's lens projected image circle.

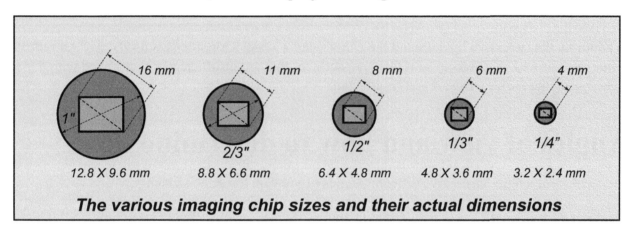

The various imaging chip sizes and their actual dimensions

When the first CCTV solid state cameras were made, the common tube size was 2/3". The image area of such tubes, as mentioned previously, was 8.8 × 6.6 mm, so the CCD chips designed in those days were of the same imaging area size and they were called 2/3" chips. The idea was to use the same lenses as tube cameras did.

With the evolution of technology, CCDs, and later on CMOS chips, were getting smaller, and the new chip size called 1/2" measured an imaging area of only 6.4 × 4.8 mm. The compatibility with the 2/3" lenses was preserved (using the same C-mount), but of course, the angle of view changed: it got smaller compared to when the same type of lens was used on a 2/3" camera.

So, new lenses were designed for the 1/2" chips, which did not project as big an image as with 2/3" chips. In other words, owing to this reduction of the imaging area, lenses were designed to have the desired focal length but with a smaller imaging circle projected, that is, a circle with a diameter sufficient to cover a 1/2" chip but not necessarily 2/3". These new lenses are called 1/2" lenses. They still have the C-mount ring, but they are smaller, and consequently cheaper, than their 2/3" counterparts.

The same development later on happened with 1/3" chips, where 1/3" lenses were made to produce an image circle sufficient in diameter to cover only the 1/3" chips.

An obvious problem that will occur if a 1/3" lens is used on a 1/2" chip is that the image corners will be cut off, as shown in the following illustration.

The same applies when a 1/2" lens is used on a 2/3" chip. There is no problem, however, if a bigger lens is used on a smaller chip. Since a lens of a bigger format will project an image circle much larger than the actual chip size, there will be no corners cut off or any other deformation.

It should be taken into consideration, however, that the reduction in the imaging pickup area may result in a relative resolution reduction, since a smaller area is used (see the discussion on MTF and CTF).

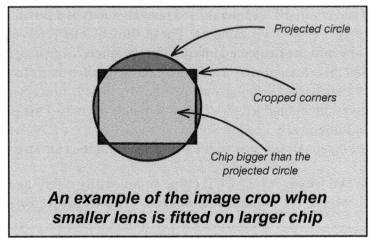

An example of the image crop when smaller lens is fitted on larger chip

In addition, the excessive light around the chip (when a larger format lens is used) may get reflected inside the lens and CCD block, so if there are surfaces that are insufficiently neutralized with a black matte finish, the usable image might be affected.

Angles of view and how to determine them

Different focal length lenses give different angles of view on the same size chip.

Also, same focal length lenses give different angles of view on the different size chips.

The above statements were complicated enough for the analogue cameras, where the chip is of the known size for a given camera. This, however, has become even more complicated with the introduction of multi-format mega-pixel chips, where the one camera with one 5 mega-pixel (MP) chip for example, can be switched to SD mode, HD mode, 3 MP or 4 MP mode. One has to be very careful if correct angle of view needs to be calculated before installation happens.

We quite often use the horizontal angle of view as a reference since the vertical can be found from it, knowing the video signal aspect ratio (4:3 for SD or 16:9 for HD).

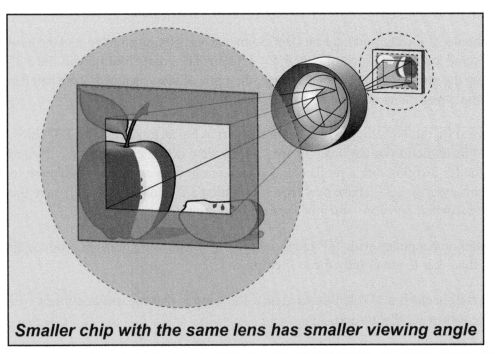

Smaller chip with the same lens has smaller viewing angle

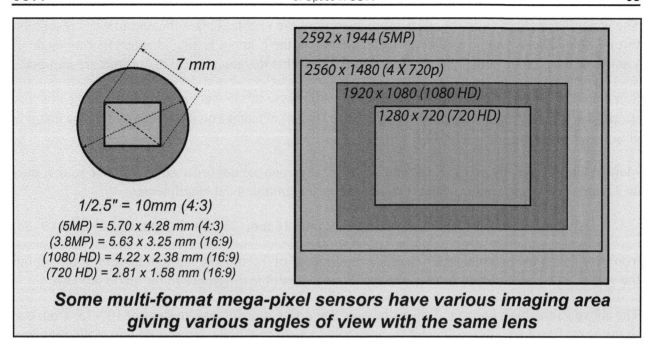

2592 x 1944 (5MP)
2560 x 1480 (4 X 720p)
1920 x 1080 (1080 HD)
1280 x 720 (720 HD)

7 mm

1/2.5" = 10mm (4:3)
(5MP) = 5.70 x 4.28 mm (4:3)
(3.8MP) = 5.63 x 3.25 mm (16:9)
(1080 HD) = 4.22 x 2.38 mm (16:9)
(720 HD) = 2.81 x 1.58 mm (16:9)

***Some multi-format mega-pixel sensors have various imaging area
giving various angles of view with the same lens***

There are some very basic rules to follow when analyzing the angles of view:

- The shorter the focal length, the wider the angle of view is.

- The longer the focal length, the narrower the angle of view is.

- The smaller the chip, the narrower the angle of view (with the same lens) is.

- The vertical angle of view can be easily determined if the horizontal is known.

As mentioned earlier, **approximately 30° is considered a standard angle of view for whatever size the image format is**. Just to refresh our memory, 30° is taken as standard because it corresponds to human perspective impression and what the human eye sees as "normal."

The following are image formats with their corresponding standard lenses for a 30° horizontal angle of view:

1" = 25 mm

2/3" = 16 mm

1/2" = 12 mm

1/3" = 8 mm

1/4" = 6 mm

In CCTV, the widest angle of view that manufacturers offer is approximately 94°, which is achieved with 4.8 mm for a 2/3" CCD camera, 3.5 mm for a 1/2", and 2.6 mm for a 1/3".

The wider the angle of view, the more barrel distortions are visible. This is the reason when resolution testing of a camera is to be conducted (as explained in the Camera testing chapter) narrow angle of view lenses should be used (typically less than 30°), so that the geometrical distortions are minimal.

Some unique "*fish-eye*" lenses offering almost a 180° angle of view are available, but these are very specialized and show only a circular (thus the name "fish-eye") image on the screen (within the imaging chip image area).

Manual lenses typically come in discrete values; that is, one cannot order any value one wants, such as 5.8 mm or 14 mm. So it is useful to know the most common focal length lenses:

   2.6 mm, 3.5 mm, 4.8 mm, 6 mm, 8 mm, 12 mm, 16 mm, 25 mm, 50 mm, and 75 mm

You may find some manufacturers have 3.7 mm instead of 3.5 mm, or 5.6 mm instead of 6 mm, but the values are very close and there is practically no perceptible difference in the angle of view.

The above values have horizontal angles of view that differ, more or less, in steps of 10°–15° from one to the next. These are quite sufficient to cover all practical situations, but should you really require a special focal length that is not listed above, inquire at your supplier as some manufacturers do have manually variable-focus lenses (both MI and AI) where the focal length can be varied from 2.6–8 mm, 6–12 mm or perhaps from 8–40 mm. The optical quality of such lenses, however, is not as good as that of fixed lenses, due to the limited precision and simplicity of the moving mechanics. But, again, the quality in most cases goes with the price.

What focal length lens should be used for a particular application?

This is probably the most commonly asked question when designing a CCTV system. Many techniques can be used to determine the angles of coverage, and which one you are going to use is entirely up to you, as long as the result is what your customer will be happy with.

Here is a listing of all practical methods. These are:

• *Viewfinder calculator*. This is usually a circular-shaped calculator, or a slide ruler type, supplied by the lens manufacturers (ask your supplier for one), where, in order to find the lens, three things need to be known: the CCD chip size, the distance between the camera and the object, and the width of the

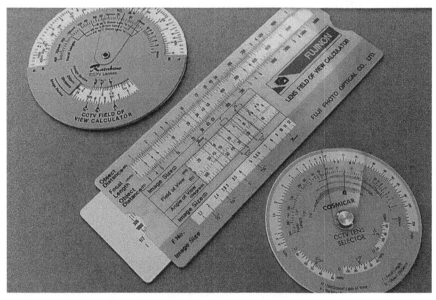

Various lens calculators

object. By adjusting these few things, the calculator should give you the focal length in mm. There are also ruler-shaped calculators with the same concept.

• ***Optical viewfinder***. This device looks like a zoom lens, but it is used not on a camera but by human eye. When you are on site, you can manually zoom in and out and set the view to what your customer requires. A scale indicator on the viewfinder shows the focal length of the lens that will give you the same view on the particular type of camera (2/3", 1/2", or 1/3"). In order to see the same view that the camera would see, you have to position yourself close to where the camera would be installed. Although tool like this doesn't have all the available chip sizes that are available today, it would be relatively easy to work out the focal length by scaling

Optical lens-viewfinder

the focal length up or down, proportional to the chip size (this will be explained further in the book)

An iPhone lens-finder app

• ***Smart phone App tools***. Since the last issue of this book (2005) so much has changed in the modern world, including the appearance of a whole new smart phone industry. These phones, such as Apple's iPhone and Google's Android-based phones, are pocket computers and have thousands of applications available to install. Some of these applications are calculators for finding angles of view.

One such free application for the iPhone, at least at the time of writing this book, is the Fujinon OpticalCalculator shown on the left.

• ***A simple lens formula***. The following formula may seem to be a complicated way of determining angles of view, but it is actually the simplest. This formula uses the geometrical similarity of triangles, as shown in the figure on the next page. It is easy to understand, and therefore it can be easily recalled whenever necessary.

The only thing you need to memorize are the CCD/CMOS chip widths of the most commonly used cameras: 6.4 mm for 1/2", 4.8 mm for 1/3", and 3.4 mm for a 1/4" chip.

This lens formula gives you the focal length of the lens directly into millimeters.

$$f = s_{CCD} \cdot d / w_{object} \qquad (34)$$

where f is the lens focal length we are looking for (in mm), s_{CCD} is the imaging sensor width (in mm), d is the distance from the camera to the object (in meters), and w_{object} is the width of the object we wish to view (in meters).

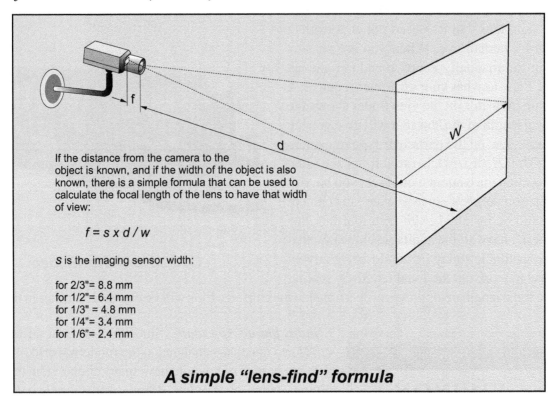

If the distance from the camera to the object is known, and if the width of the object is also known, there is a simple formula that can be used to calculate the focal length of the lens to have that width of view:

$$f = s \times d / w$$

s is the imaging sensor width:

for 2/3"= 8.8 mm
for 1/2"= 6.4 mm
for 1/3" = 4.8 mm
for 1/4"= 3.4 mm
for 1/6"= 2.4 mm

A simple "lens-find" formula

The same formula can be used if we want to find what focal length lens we need, to see a certain object's height, in which case instead of w_{CCD} and w_{object} we will be working with h_{CCD} and h_{object}, where h stands for height.

We can add to this list the chip portion of the HD mode of the 5MP chip shown on page 89, which is found from the proportions between the total horizontal pixels (2592) which is 5.70 mm, versus the unknown portion of the HD mode of 1920 pixels.

$$2592 : 5.70 = 1920 : x \implies x = 13,039 / 2592 = 5 \text{ mm}$$

The above simple calculation is the basis of finding out other portions of a mega-pixel chip.

• **A more complicated formula**. This formula gives the resulting angle of view in **degrees**. It is based on elementary trigonometry and requires a scientific calculator or trigonometric tables.

$$\alpha = 2 \cdot \arctan (w_{object}/2d) \tag{35}$$

where α is the angle of view (in degrees), *arctan* is an inverse *tangent* trigonometric function (you need a scientific calculator for this), which is sometimes written as *tan*[-1], w_{object} is the object width (in meters), and d is the distance to the object the camera is looking at.

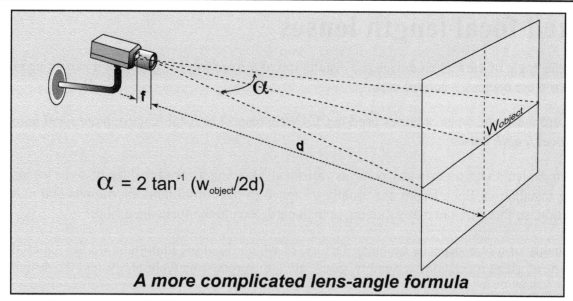

$$\alpha = 2 \tan^{-1} (w_{object}/2d)$$

A more complicated lens-angle formula

• *A table and/or graph.* This is easy to use as it does not require any calculations, however, it requires a table or graph to be handy. The table below gives only the horizontal angle of view for a given lens, because this is most commonly required. Vertical angles are easily found by applying the aspect ratio rule, that is, divide the horizontal angle by 4 and then multiply it by 3 for an SD camera, or divide by 16 and multiply by 9 for an HD camera.

Approximate horizontal angles of view with various CCD/CMOS chips (in degrees)				
Focal length	**1/4 "**	**1/3"**	**1/2"**	**2/3"**
2.0 mm	82°	-	-	-
2.6 mm	57°	86°	-	-
4.0 mm	47°	67°	77°	-
4.8 mm	40°	57°	67°	83°
6.0 mm	32°	48°	56°	70°
8.0 mm	25°	36°	44°	56°
12 mm	17°	25°	30°	39°
16 mm	13°	17°	23°	30°
25 mm	8°	12°	15°	18°
50 mm	4°	6°	7°	10°

In all of the above methods, we have to take into account monitor over-scanning as well (typical of CRT). In other words, most CRT monitors do not show 100% of what the camera sees. Usually, 10% of the picture is hidden by the over-scanning by the monitors.

Some professional monitors offer the under-scanning feature. If you get hold of such a monitor you can use it to determine the amount of over-scanning by the normal monitor. This is very important to know when performing camera resolution tests, as will be described later. Of course, the above doesn't apply to the LCD and plasma monitors where it is expected that the image fills the whole screen with minimal over-scanning.

Fixed focal length lenses

Once the angle of view is set during the installation of a fixed focal length lens it cannot be changed later from the operator's control desk.

There are two basic types of lenses used in CCTV (in respect to focal length): *fixed focal length* and *vari-focal length* lenses.

Fixed focal length lenses may also come as vari-focal, allowing for the focal length to be set manually during installation. Such lenses are usually designed to have minimum aberrations and maximum resolution, so there are not many moving optical parts, except the focusing group.

The quality of a lens depends on many factors, of which the most important are the materials used (the type of glass, mechanical assembly, gears, etc.), the processing technology, and the design itself.

When manufacturers produce a certain type of lens, they have in mind its application and use. The lens quality aimed for is dictated by practical and market requirements. As mentioned previously, when MTF and CTF were discussed, there is no need to go to the technical limits of precision and quality (and consequently increase the cost) if that cannot be seen by the imaging device. This, however, does not mean that there is no difference among different makes and models of the same focal length. Usually, the price goes hand in hand with the quality.

More than two decades ago, when 1" tube cameras were used, 25 mm lenses offered a normal angle of view (approximately 30° horizontal angle).

With the evolution of the formats (that is, with their reduction), the focal length for the normal angle of view was reduced, too. The mounting thread, however, for compatibility purposes, remained the same.

C-mount thread

With the C-mount format this thread was defined as **1"-32UN-2A**, which means it is **1" in diameter with 32 threads/inch**. When the new and smaller CS format was introduced, the same thread was again kept for compatibility, although the back-flange distance was changed. This will be explained later in this chapter.

In respect to the iris, there are two major groups of fixed focal length lenses: manual iris (MI) and automatic iris lenses (AI), and these were described under the previous heading.

Finally, let us mention the vari-focal group of lenses. These lenses should be classified as fixed focal length lenses, because once they are manually set to a certain angle of view (focal length) they have to be re-focused, unlike zoom lenses, which once focused, stay in focus if the angle of view is changed.

Vari-focal lenses can be cla-sified as manually adjustable fixed focal lenses.

Mention should be made here about a new vari-focal lenses made by Pentax (and we hope others will follow) called "vari-focal +" which in fact are manual zoom lenses offered for the price of a typical vari-focal lens. Once a correct back-focus is made and a good sharpness is set the lens does not change by changing the focal length (angle of view), which is not the case with a typical vari-focal construction. This makes the setup of camera angle coverage during the installation easier. Once focused, widening or narrowing the angle of view keeps the image sharp.

Zoom lenses

A very special vari-focal

In the very early days of television, when a cameraman needed a different focal length lens he would use a specially designed barrel, fitted with a number of fixed lenses that rotated in front of the camera. Different focal lengths were selected from this group of fixed lenses.

This concept, though practical compared to manually changing the lenses, lacked continuity of length selection, and, more importantly, optical blanking was unavoidable when a selection was being made.

That is why optical engineers had to come up with a design for a continuous focal length variation mechanism, which got the popular name *zoom*. The zoom lens concept lies in the simultaneous movement of a few groups of lenses. The movement path is obviously along the optical axis but with an optically precise and **nonlinear correlation**. This makes not only the optical but also the mechanical design very complicated and sensitive. It has, however, been accomplished, and as we all know today, zoom lenses are very popular and practical in both CCTV and broadcast television.

With a special ***barrel cam mechanism***, usually two groups of lenses (one called ***variator*** and the other ***compensator***) are moved in relation to each other so that the zooming effect is achieved **while preserving the focus** at an object. As you can imagine, the mechanical precision and durability of the moving parts are especially important for a successful zooming function.

For many perfectionists in photography, zoom lenses will never be as good as fixed ones. In the absolute sense this is very true because the moving parts of a zoom lens must always have some tolerance in its mechanical manufacture that introduces more

Zoom lenses

aberrations than a fixed lens design. Hence, the absolute optical quality of a certain focal length setting in a zoom lens can never be as good as a well-designed fixed lens of the same focal length.

For CCTV applications, however, compromises are possible with good results. Continuous variation of angles of views, without the need to physically swap lenses, is extremely useful and practical. This is especially the case where cameras are mounted in fixed locations (as on a pole or on top of a building) and resolution requirements are not as critical as in the film industry. It should not be assumed, however, that in their evolution zoom lenses will not come very close to the optical quality of the fixed ones.

Zoom lenses are usually represented by their ***zoom ratio***. This is the ratio of the focal length at the telephoto end of the zoom and the focal length at the wide angle end. Usually, the telephoto angle is narrower than the standard angle of vision, and the wide angle is wider than the standard angle of vision. Since the telephoto end always has a longer focal length than the wide angle, the ratio is a number larger than one.

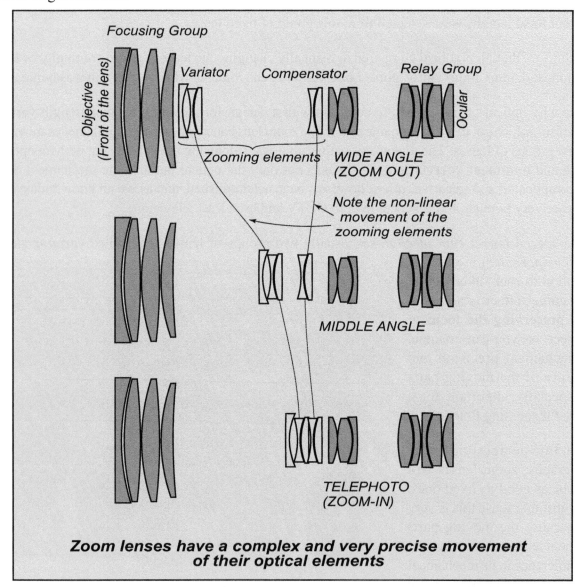

Zoom lenses have a complex and very precise movement of their optical elements

The most popular zoom lenses used in CCTV are:

- 6×: Six times, with 6–36 mm, 8–48 mm, 8.5–51 mm, and 12.5–75 mm being the most common.

- 10×: Ten times, with 6–60 mm, 8–80 mm, 10–100 mm, 11–110 mm, and 16–160 mm as the most common examples.

- 15×: Fifteen times, with 6–90 mm, 8–120 mm.

Other ratios are available, such as 20×, or even 44× and 55×, but they are much more expensive and, therefore, not very common.

A manual zoom lens

In the last five to ten years, the miniature PTZ domes have become very popular. Most of them have integral zoom lenses with optical zoom range of 12×, 16×, or even 20× zoom range. Usually, digital zooming of at least half a dozen times is added onto this, which makes these little domes extremely powerful. The digital zooming is not the same as optical, but in some cases it may help see distant objects a bit better. These PTZ dome cameras can have such powerful optical zooming, and yet look so small (the typical diameter of a PTZ dome module is around 12 cm), because they are based on 1/4" CCD chips. The smaller the chip, the smaller the optics is. This is one of the main reasons for chip reduction, besides the manufacturing cost. It should be made quite clear that the precision of manufacturing zoom lenses for 1/4" CCD chips is more optically demanding because pixels are denser.

Zoom lenses are also characterized by their F-stop (or T-number). The F-stop, in zoom lenses (as already mentioned when F-numbers were discussed) refers to the shortest focal length. For example, for a 8–80 mm/1.8 lens the F-1.8 refers to the 8 mm. **The F-stop is not constant throughout the zoom range.** It usually stays the same with the increase of focal length only until a certain focal length is reached, after which a so-called *F-drop* occurs. The focal length at which this F-drop occurs depends on the lens construction. The general rule, however, is the smaller the entrance lens, the higher the likelihood of an F-drop. This is one of the main reasons lenses with bigger zoom have to have bigger front lens elements (called *objective*), where the intention is to have a minimal F-drop.

Zoom lenses, as is the case with fixed lenses, come with a manual iris, automatic iris, or motorized iris. Even though AI was explained in the previous section with fixed lenses, because there is an additional and common subgroup with motorized iris, we will go through this again.

A *manual iris zoom lens* would have an iris ring, which is set manually by the installer or by the user. This is a very

PTZ mini-dome camera

rare type of lens in CCTV, and it is used in special situations, such as when doing demonstrations or camera testing.

An ***automatic iris zoom lens***, often called auto iris (AI), is the most common type of zoom lens. This lens has an electronic circuit inside, which acts as an electronic-optical feedback. It is usually connected to the back of the camera where it gets its power supply (9 V DC) and its video signal. The lens's electronics then analyze the video signal level and act accordingly: if the signal exceeds the video level of 0.7 V, the lens closes the iris until a 0.7 V signal is obtained from the camera AI terminal. If, however, the signal is very low, the iris opens in order to let more light in and consequently increases the video level.

Two adjustments are available for this type of lens (as with fixed lenses): **level** and **ALC**.

Level, as the name indicates, adjusts the reference level of the video signal that is used by the electronics of the lens in order to open or close the iris. This affects the brightness of the video signal. If it is not adjusted properly (that is, adequately sensitive for daylight and low light situations), a big discrepancy between the day and night video signals will occur. Obviously, the camera sensitivity has to be taken into account when adjusting the iris level for low light level situations.

ALC adjustment refers to the automatic light compensation of the iris. This is in fact very similar to the backlight compensation (BLC) found in many camcorders (as we have already explained in the fixed lenses section). This light compensation is usually applied when looking at scenes with very high contrast. The idea behind BLC operation is to open the iris more (even if there is a lot of light in the background) **so as to see details of the objects in the foreground.** A typical example would be when a camera is looking through a hallway (with a lot of light in the background) trying to see the face of a person coming toward the camera. With a normal lens setting, the face of the person will appear very dark because the background light will cause the iris to close. A proper ALC setting could compensate for such difficult lighting conditions. The bright background in the example above will become white, but the foreground will show details. The ALC setting actually adjusts the reference level relative to the video signal average and peak values. This is why the marks on the ALC of a lens show Peak and Average.

Remember that, when you strat adjusting the ALC, a very high-contrast scene needs to be viewed by the camera. If the opposite (low-contrast scene) is seen, no visible change of the video signal will occur. So, by tweaking the ALC pot in a scene with normal contrast, a misalignment may occur that will be visible only when the picture light changes.

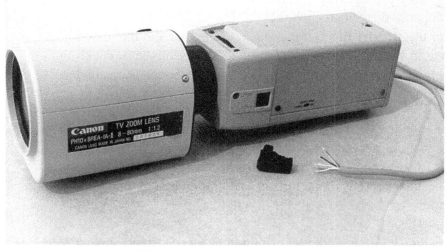

Auto iris connection of a zoom lens

All of the above refer to the majority of AI lenses, which are driven, as described, by the video signal picked up from the AI connector at the back of the camera. Because of this, and because there is another subgroup of AI zoom lenses that are not driven by the video signal taken from the camera, we also call this AI type *video-driven AI*.

The other subgroup of the AI group of lenses are the ***DC-driven AI*** zoom lenses.

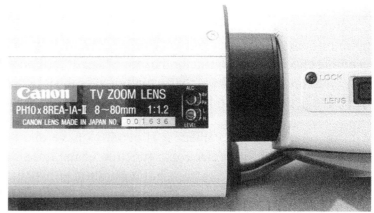

Auto iris zoom lens with ALC and level pots

The DC-driven AI lenses do not have all the electronics for video processing, only the motor that opens and closes the iris. **The whole processing, in DC-driven auto iris lenses, is done by the camera's AI electronic section.** The output from such a section is a DC voltage that opens and closes the iris leaves according to the video level taken from inside of the camera. Cameras that have DC AI output also have the level and ALC adjustments, but in this case on the camera body and not on the lens.

It should be clearly noted that video-driven AI zoom lenses cannot be used with cameras that provide DC AI output, nor can DC AI be used with a video AI output camera. Some cameras can drive both these types of AI designs, in which case a switch or separate terminals are available for the two different outputs. Pay attention to this fact, for it can create problems that initially seem impossible to solve. In other words, make sure that both the camera and the lens are of the same type of AI operation.

The advantage of video-driven AI zoom lenses is that they will work with the majority of cameras. The advantage of DC-driven AI zoom lenses is that they are cheaper and are unlikely to have the "hunting" effect as the camera processes the gain. The disadvantage is that not all cameras have a DC-driven AI output. Video-driven AI lenses were more common until half a dozen years ago, but now DC-driven prevail.

Motorized iris lenses belong to the third lens subgroup, if selection on the basis of the iris function is made. This is an iris mechanism that can be controlled remotely and set by the operator according to the light conditions. This type of zoom lens has become increasingly popular in the last few years, especially with the introduction of the electronic-iris.

In order to open or close the iris, instead of an AI circuit driving the iris leaves, a DC voltage, produced by the PTZ site driver, controls the amount of opening

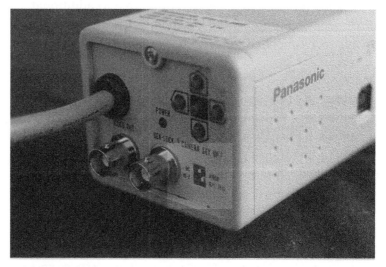

Some cameras can "drive" both types of AI lenses.

or closing. Although there are still PTZ site drivers for camera/lens/pan/tilt head combinations, due to increase of integrated PTZ mini-domes this is becoming a rarity, since all such electronics are inside the mini-domes. But, to put it very simply, the site drivers are electronic circuits that are capable of receiving encoded digital data for the movement of the Pan/Tilt head, as well as the zoom lens functions, and converting it into voltage that actually drives the PTZ assembly. In the case of motorized iris lenses, the PTZ site driver has to have an output to drive the iris as well.

With the electronic-iris camera it is better to have this type of lens iris control than an automatic one. The electronic-iris (electronic function of the imaging chip) is a faster and more reliable light-controlling section of the camera, but it **does not substitute the depth of field** effect produced by the high F-stops of an optical iris. **Optical and electronic irises cannot function properly if they are working simultaneously**. The video camera usually balances with a low F-stop (high iris opening), which results in a very narrow depth of field, and a high electronic shutter speed, which produces a less efficient charge transfer (that is, high smear). This is especially obvious when such a camera/lens combination comes across a high-contrast scene. To avoid a low-quality picture, and yet use the benefits of a fast and reliable electronic-iris function, and even more, have depth of field, motorized iris lenses are the solution. It will obviously require an operator's intervention, but that does not have to happen until the picture demands it, since the CCD-iris will be functioning constantly to compensate for the abrupt light variations.

Finally, let us mention the *vari-focal* lenses again. Vari-focals do not have the same functionality as the zoom lenses. This is why we made their classification in the fixed focal lenses group. They have become very popular since the decision on what focal length lens is required can be made on the spot by the varying the rings and achieving sharp image during commissioning. They are practical from the installers point of view, only one type of lens is carried around, and they are practical when customer doesn't know what angle of coverage is required. The downside is that they have to always be manually re-focused once the angle of view (that is, the focal length) is changed, and again, the optical quality of vari-focal lenses rarely achieves or exceeds the quality of a fixed focal length lens. This is because it is more difficult to produce the same optical resolution due to additional movement when compared to fixed focal lenses. Trials are always recommended for better judgment.

One more thing that needs to be mentioned here is the proliferation of the so-called mega pixel lenses. These are typically vari-focal lenses made with better optical quality to suit the higher density of the high definition imaging chips. The fact is that a typical HD chip may have physical size of a 1/3" chip, but instead of 768 active horizontal pixels (typical for a PAL SD camera) they have 1920 pixels for the HD signal. This is nearly three times the density of the SD chip, hence the optical quality of the lens needs to be that much better. Again, not all lenses are equal, and one has to be reminded again and again, that in HD CCTV

Photo courtesy of Fujinon

A "mega pixel" lens

Four different "mega pixel" lenses may have very different qualities

the lens quality is even more critical.

Above on this page there are some samples of images taken with a HD camera looking at the ViDi Labs test chart, with four different vari-focal lenses. All lenses are claimed to be "mega-pixel" vari-focal lenses. The settings on the camera are the same (compression type, compression amount, illumination, etc.).

Admittedly, the quality of reproduction of the images in this book is somewhat reduced (it is not the same as viewing it live on a HD screen), but it still shows the difference in quality.

This is just one more example trying to reiterate the fact that not all optics are of the same quality. The best way to find out is to test it yourself.

C- and CS-mount and back-focus

"***Back-focusing***" is what we call the adjustment of the lens back-flange relative to the CCD/CMOS image plane. Back-focusing is very important in CCTV. Currently, there are two standards for the distance between the back-flange of a lens and the CCD/CMOS image plane:

> • ***C-mount***, represented with 17.5 mm (more precisely, 17.526 mm).

This is a standard mounting, dating from the very early days of tube cameras. It consists of a metal ring with a 1.00/32 mm thread and a front surface area at 17.5 mm away from the image plane.

> • ***CS-mount***, represented with 12.5 mm.

This is the latest standard intended for smaller camera and lens designs. It uses the **same thread** of 1.00/32 mm as the C-mount, but it is approximately 5 mm closer to the image plane. The intention is to preserve compatibility with the old C-mount format lenses (by adding a 5 mm ring) and yet allow for cheaper and smaller lenses, to suit smaller imaging chip sizes.

Since both of the above formats use the thread type of lens mounting, there might be small variations in the lens's position relative to the CCD chip when mounted (screwed in), hence the need for a little variation of this position (back-focus adjustment).

In photography, for example, we never talk about back-focusing simply because most of the brands come with a bayonet mount, which has **only one fixed position of the lens relative to the film plane**. Camcorders, for that matter, come with lenses as an integral part of the unit, so the back-focus is already adjusted and never changes.

In CCTV, because of the modular concept of the camera/lens combination and the thread mount, it is a different story.

Back-focus adjustment is especially important and critical when zoom lenses are used. This is because the optics-to-chip distance in zoom lenses has to be very precise **in order to achieve good focus throughout the zoom range**.

Obviously, the back-focusing adjustment applies to fixed lenses as well, only in that case we tend not to pay attention to the distance indicator on the lens ring when focusing. If we want to be more accurate, when the back-focus is adjusted

Some cameras don't require C/CS adaptor.

correctly on a fixed lens, the distance indicator should show the real distance between the camera and objects. Most installers, however, do not pay attention to the indicator on the lens since all they want to see is a sharp image on the monitor. And this is fine, but if we want to be precise, the back-focus adjustment should apply to **all** lenses used in CCTV. With zoom lenses this is more critical.

An important factor to be taken into account when doing back-focus adjustment is the effect of depth of field. The reason is very simple: if a CCTV camera is installed at daytime (which is most often the case), and if we are using an AI lens, it is natural to see the

If a lens is C-mount and the camera is CS, a C/CS adaptor ring is required.

iris set at a high F-stop to allow for a good picture (assuming the AI is connected and works properly). Since the iris is at a high F-stop, we have a very high depth of field. The image seems sharp no matter where we position the focusing. At nighttime, however, the iris opens fully owing to the low light level situation, and the operator sees the picture out of focus.

This is actually one of the most common problems with new installations. When a service call is placed, usually the installer comes during the daytime to see what the problem is, and if the operator cannot explain exactly what he or she sees, the problem may not be resolved, since the picture looks great

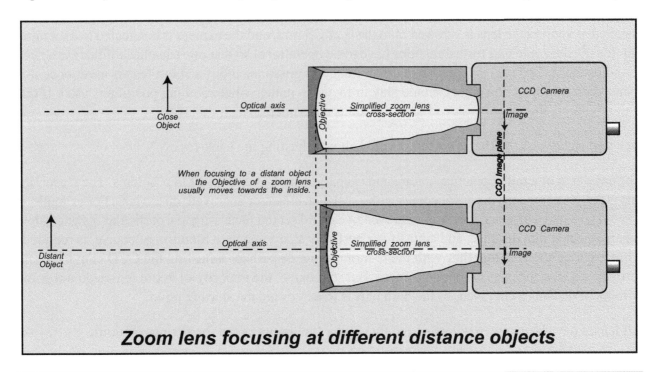

Zoom lens focusing at different distance objects

with a high F-stop.

The moral of the above is: always do the back-focus adjustment with a low F-stop (largest lens iris opening).

How do you make the iris open to maximum? The following different methods can be used:

• Adjust the back-focus at low light levels in the workshop (easiest).

• Adjust the back-focus in the late afternoon, on site.

• Adjust the back-focus at daytime, by using external ND filters.

There is one exception to the above: if a camera with electronic-iris is used, then the optical iris can be opened fully even at daylight because the electronic-iris will compensate for the excessive light.

On-camera C/CS ring

This means that with electronic-iris cameras, the back-focus can easily be adjusted even at daylight without being confused with the depth of field and without a need for ND filters. Obviously, you should not forget to switch the electronic-iris off after the adjustment if you decide to use auto iris.

Back-focus adjustment

In the following few paragraphs we will examine a procedure for proper back-focus adjustment. This discussion is based on practical experience and by no means the only procedure, but it will give you a good understanding of what is involved in this operation. We should also clarify that often, with a new camera and zoom lens setup, there might not be a need for back-focus adjustment. This can be easily checked as soon as the lens is screwed onto the C or CS ring, and the camera is connected to a monitor. Obviously, the zoom and focus functions have to be operational so that one can check if there is a need for adjustment. The idea is to get as sharp an image as possible, and if a zoom lens is used, once it is focused on an object, the object should stay in focus no matter what zooming position is used. If this is not the case, then there is a need for back-focus adjustment.

One will rightly ask: "What is so complicated about adjusting the back-focus?"

The answer is of a rather practical nature and is apart from the depth-of-field problems. The reason for this is that no zoom lens in CCTV comes with a distance indicator engraved on the lens. For example, if a zoom lens had a distance indicator engraved, we could set the focus ring to a particular distance, then set an object at that distance, and adjust for a perfectly sharp image (a monitor, of course, is required) while rotating the lens **together** with the C-mount ring, or perhaps adjusting the CCD chip back and forth by a screw mechanism on the camera. But, of course, the majority of zoom lenses do not come with these distances engraved, so the hard part is to determine the starting point.

All lenses have two known points on the focus ring (the limits of the focus ring rotation):

- Focus infinity "∞" (no lens focuses past this point)

- Focus at the minimum object distance (MOD)

The second point varies with different lenses; that is, we do not know what the minimum focusing distance of a particular lens will be unless we have the manufacturer's specification sheet, which is usually not supplied with the lens or, quite often, is lost in the course of installation.

This leaves us with only the focus infinity as a known point. Obviously, infinity is not literally an infinite distance, but it is big enough to give a sharp image when the lens is set to the "∞" mark.

The longer the focal length of the lens, the longer the infinity distance that has to be selected. For a typical CCTV zoom lens of 10× ratio, which is usually 8–80 mm, 10–100

Some CCD cameras use miniature hexagonal screws to secure the C-mount ring to the camera.

mm, or 16–160 mm, this distance may be anything from 200–300 m onward. From this, we can see it is impossible to simulate this distance in the workshop, so the technician working on the back-focus needs to point the camera out through a window, in which case *external ND filters* are required to minimize the effect of depth of field (unless a electronic-iris camera is used, of course).

The next step would be to set the focus to the infinity mark. To do this, a PTZ controller would be required. However, this could be impractical, and if it is, the following procedure is suggested.

In this situation, I recommend simulating the zoom and focus control voltages by using a regular 9 V DC battery and applying it to the focus and zoom wires. Do not forget, the lens focus and zoom control voltages are from ±6 V DC to ±9 V DC, and the lens has a very low current consumption, usually below 30 mA. A

Focus-adjusting tool

9 V battery has plenty of capacity to drive such motors for a considerable time, at least long enough for the adjustment procedure to be completed.

There is no standard among manufacturers for the lens wire color coding, but, quite often, if no information sheet is supplied with the lens, the black wire is common, the zoom wire is red, and the focus wire is blue. This is not a rule, so if in doubt it is not that difficult to work it out by using the same battery and monitor. Instead of a monitor, an even more

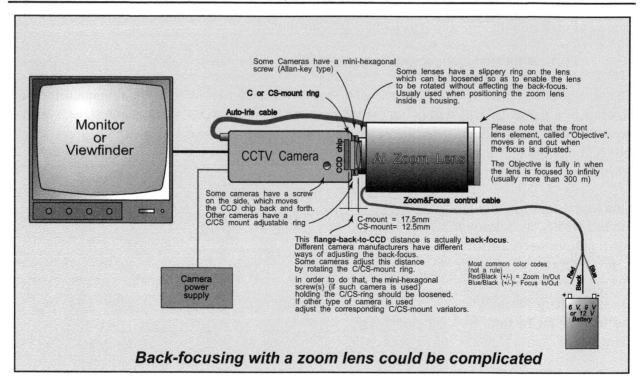

Back-focusing with a zoom lens could be complicated

practical tool would be a **viewfinder**; some call it a focus adjuster. This is a little monitor that is battery operated, with a rubber eyepiece to protect from excessive daylight. On a bright sunny day, if a normal monitor is used in the field it would be almost impossible to see the picture on the screen, so it is highly recommended that you use a viewfinder instead.

If no monitor is available at the point of adjustment, a distinction should be made between which optical parts move when focusing and which when zooming. This is not so naive, since zoom lenses are enclosed in black (or beige) boxes and no moving parts are visible. A rule of thumb, however, would be zooming elements are not visible from the outside, while focusing is performed by the first group of lenses, called the *objective*. When focusing is done, the objective rotates around its optical axis and at the same time moves along the optical axis toward either the inside or the outside of the lens. All lenses have this common concept of the objective moving toward the outside when focusing to closer distances and moving toward the inside of the lens when focusing to infinity.

So, even if the zoom lens does not have any visible markings for distances and zoom factors, using the above logic we can start doing the back-focus adjustment.

With the battery applied to the focus wire we need to focus the lens to infinity. Even if we do not have a monitor, this will be achieved when the **lens objective goes to the end position on the inside of the lens**.

The next step is to point the camera to an infinity object, at a distance we have already mentioned. The infinity objects can be trees, roof-tops or antennas on the horizon.

Now, without changing the focus, zoom in and out fully. If the picture on the monitor looks sharp throughout the zooming range, **back-focusing is not necessary.**

If, however, the camera's C- or CS-mount ring is out of adjustment, we will not see a sharp picture on the monitor for all positions of the zoom.

Then, we proceed with adjustment by either rotating the lens together with the C-ring (if the camera is of such a type) or by shifting the imaging chip with a special back-focus adjustment screw or in some cameras by rotating a large ring with C & CS written on it.

The first type of camera is the most common. In this case, the C- or CS-mount ring is usually secured with miniature hexagonal locking screws. These need to be loosened prior to the adjustment but after the zoom lens is screwed in tightly.

Then, when the focus is to be adjusted (after we did the battery focusing and pointing to infinity) we need to rotate the zoom lens but now **together** with the ring (that is why we have loosened the ring). Again, some cameras may have a special mechanism that shifts the imaging chip back or forth, in which case it is easier since we do not have to rotate the lens. Also note, from the previous edition of this book, there are some modern cameras that have been designed with an electronic moving imaging chip which adjust the back-focus position automatically.

By doing one of the above, the distance between the lens and the CCD chip changes until the picture becomes sharp. Do not forget, because we have made the depth of field minimal by opening the iris as much as possible (with low light level simulation), the sharpness of the objects in the distance should be quite easily adjusted. Once we find the optimum we should stop there.

Please note that the focus wires are not used yet; that is, we still need to have the zoom lens focused at infinity. We are only making sure that while zooming, the lens stays focused at infinity throughout the zoom range. Also, we need not be confused when the objects at infinity are getting smaller while zooming out; because of the image size reduction, they might give the impression that they are going out of focus.

The next step would be to zoom by using the 9 V battery. Watch the video picture carefully and make sure that the objects at infinity stay in focus while zooming in or out. If this is the case, our back-focus is nearly adjusted.

In order to confirm this, the next step would be to point the camera at an object that is only a couple of meters away from the camera. Then we zoom in on the object and use the focusing wires to focus on it. When focused precisely, use the zoom wires and zoom out. If the object stays in focus, that will be **confirmation** of a correct back-focus adjustment.

The last step would be to tighten the little hexagonal screws (if such a camera is used) and secure the C/CS mount ring on the camera.

If the above procedure does not succeed from the very first go a couple of iterations might be necessary, but the same logic applies.

As one can imagine, the mechanical design and robustness of the C-mount and chip combination is very important, especially the precision and "parallelness" of the C-ring and the imaging chip plane.

Little variations of only one-tenth of a millimeter at the image plane may make a focus variation of a couple of meters at the object. With bad designs, such as locking the C-ring with only one screw or poor mechanical construction, problems might be experienced even if the above procedure is correct. So it is not only the lens that defines the picture quality, but the camera's mechanical construction as well.

We have mentioned that a monitor is required when doing the back-focus adjustment, which is not a surprise. This is fine when the adjustments are done in the workshop, but when back-focusing needs to be performed on site it is almost impossible to use a normal monitor. The reason for this is not so much the impracticality of the need for a main supply (240 VAC or whatever the country you are in has), but more so because of the bright outdoor light compared to the brightness produced by a monitor. This is why I have recommended the use of a viewfinder monitor (like the ones used on camcorders) with a rubber eyepiece that protects from external light and allows for comfortable use. In addition, these little viewfinder monitors are battery operated and very compact. Some manufacturers have viewfinder focus adjusters specially made with a flicker indicator to show when objects are in focus.

Small and practical tools like this one make a difference between a good and a bad CCTV system installation and/or commissioning.

Optical accessories in CCTV

Apart from fixed and zoom lenses in CCTV, we also have some optical accessories.

One of the more popular is the *2× teleconverter* (also known as an *extender*). The teleconverter is a little device that is usually inserted between the lens and the camera. The 2× converter multiplies the focal length by a factor of 2. In fact, this means a 16 mm lens will become 32 mm, a zoom lens 8–80 mm will become 16–160 mm, and so on. It is important to note, however, that the F-number is also increased for one F-stop value. For example, if a 2× converter is used on a 16 mm/1.4, this becomes 32 mm/2. Back-focusing a lens with a 2× converter may be more complicated. It is recommended that you first do the back-focusing of the zoom lens alone, and then just insert the converter. Some zoom lenses come with a teleconverter built-in but removable with a special control voltage. For this purpose the auxiliary output from a site driver can be used. In general, the optical resolution of a lens with a converter is reduced, and if there is no real need for it, it should be avoided. It should be noted that 1.5× converters also exist.

Another accessory device is the *external ND filter*, which comes with various factors of light attenuation – 10×, 100×, or 1000×. They can also be combined to give higher factors of attenuation. As we have already described, external ND filters are very helpful in back-focusing and AI adjustments. Since they come as loose pieces of glass, you may have

Optical accessories

to find a way of fixing them in front of the lens objective. Some kind of a holder could be made for better and more practical use of the filters. As mentioned earlier under the internal ND filters, there are some new circular ND filters where by rotating of the one ring a continuous increase of the optical density is achieved. These filters might be more expensive than the single blocks with 10×, 100×, or 1000× attenuation, but a certainly more practical

A 100X neutral density filter

Polarizing filters are sometimes be required when using a CCTV camera to view through a window or water. In most cases, reflections make it difficult to see what is beyond the glass or water surface. Polarizing filters can minimize such an affect. However, there is little drawback in the practicality of this, since a polarizing filter requires rotation of the filter itself. If a fixed camera is looking at a fixed area that requires a polarizing filter, that might be fine, but it will be impossible to use it on a PTZ (pan/tilt/zoom) camera because of constant camera repositioning and objective rotation when focusing.

For special purposes, when the camera needs to have a close-up (macro) view of a very small object, it is possible to focus the lens on objects much closer than the actual MOD (Minimum Object Distance) as specified by the lens manufacturer. This can be achieved with special sets of *extension rings* that can be purchased through some lens suppliers. It is much easier and also more practical to use surplus CS-mount adaptor rings. By combining one or more of them, and depending upon the focal length in use, macro views can be obtained. This might be useful for inspecting surface

A circular variable neutral density filter

mount PCB components and stamps, detecting fake money, and monitoring insects or other miniature objects.

Extension rings

Lux-meter

As the name suggests, a *lux-meter* is an instrument designed to measure the illumination of objects of interest. Typically, these objects are what the camera should be viewing. We discussed the lux units under the chapter Light and television, so we will just repeat what was said there.

Lux is a unit for measuring reflected visible light. The word *visible* is very important to highlight, as **a lux-meter is not designed and cannot measure infrared light**. This is a very common mistake made by some manufacturers when stating their camera's minimum illumination capability, where low luxes are stated thinking of infrared light.

Another common error some installers are making is by using a lux-meter pointing it towards the sky or towards the source of light in a room.

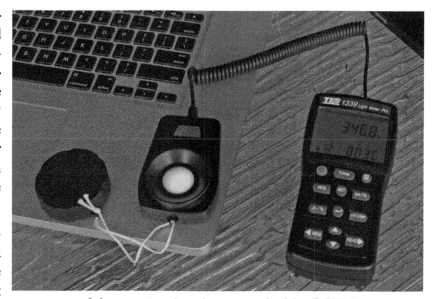

A lux-meter is often needed in CCTV

The lux-meter should be used pointing towards the objects of interest. Basically, **it should see the same angle of view and from the similar distance as the camera**. The illumination falls off with the inverse square of the distance, which means, it is not the same if reading the lux-meter very close to the object of surveillance, and the same measured from the camera position further away. Clearly, the further the camera is, the lower amount of measured lux-es will be detected. Furthermore, the reflectivity of the object plays a big role, so that the same light source will produce higher lux reading if the surrounding is brighter. A lux-meter measures whatever visible light falls on its sensor, which should be similar to what the camera "sees" with its own imaging sensor.

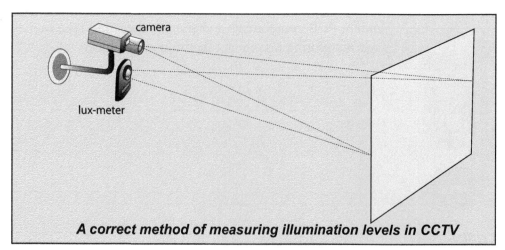

A correct method of measuring illumination levels in CCTV

As we discussed under human eye sensitivity, illumination below 0.1 lux is practically invisible to human eye, and this falls in the so-called *"Scotopic"* vision area. There are no commercially available lux-meters that can measure at or below this level. Typical lux-meters go down to 1 lux, although some may measure even to 0.1 lux, but not below this. Manufacturers claiming that their CCTV cameras can see in as low light level as 0.00001 lux for example are not doing any favors to the industry. **Not only such light levels are impossible to measure, but it is impossible for such low energy photons to generate electrons in the imaging sensor.** There are many more thermally generated electrons at normal temperature then what illumination below 0.001 lux can produce, hence nothing can be seen.

One of the easiest and best methods to see how a camera behaves in very low light illumination levels is by use of a candle in a room with no additional lights. This would require even the light from the display devices (monitors) to be reduced or simply removed from the test room. Remember, the definition of 1 lux was that it is an illumination of a typical candle on one square meter surface area at one meter distance.

An alternative method of measuring low light camera performance is the method where ND filters are used. This is an indirect method, perhaps not so accurate, but offers a way of testing low light if the minimum measuring capability of a lux-meter don't allow it. First, a known attenuation ND filter needs to be used, and then illumination is measured behind the filter. For example, if the lux-meter measures 500 lux without the ND filter, than putting a 100× ND filter, we know the light behind it would be approximately 5 lux (100 times less). Using the same ND filter in front of a camera under test at around 1 lux (which we could measure) would show camera's response at 0.01 lux. Adding another ND filter of say 10× we can measure the same response at 0.01 lux (1000 times less than 10 lux).

4. General about TV systems

This chapter discusses the theoretical fundamentals of video signals, their bandwidth, and resolution. It is intended for technical people who want to know the limits of the television system in general and CCTV in particular.

A little bit of history

In order to understand the basic principles of television, we have to refer to the effect of eye persistence (see Chapter 2).

Television, like cinema, uses this effect to cheat our brain, so that by showing us still images at a very fast rate, our brain is made to believe that we see "motion pictures."

In 1903, the first film shown to the public was *The Great Train Robbery* which was produced in the Edison Laboratories. This event marked the beginning of the motion picture revolution. Although considered younger than film, the concept of television has been under experimentation since the late nineteenth century. It all began with the discovery of the element selenium and its *photoelectricity* in 1817 by the Swedish chemist Jons Berzelius. He discovered that the electric current produced by selenium, when exposed to light, would depend on the amount of light falling onto it. In 1875, G.R. Carey, an American inventor, made the very first crude television system, in which banks of photoelectric cells were used to produce a signal that was displayed on a bank of light

Baird's television receiver, 1923

bulbs, every one of which emitted light proportional to the amount of light falling onto the photo cells.

Zworykin's iconoscope

A few minor modifications were made to this concept, such as the "scanning disk" presented by Paul Nipkow in 1884, where elements were scanned by a mechanical rotating disk with holes aligned in a spiral. In 1923, the first practical transmission of pictures over wires was accomplished by John Baird in England and later Francis Jenkins in the United States of America. The first broadcast was transmitted in 1932 by the BBC in London, while experimental broadcasts were conducted in Berlin by the Fernseh Company, led by the cathode ray tube (CRT) inventor Professor Manfred von Ardenne. In 1931, a Russian born engineer, Vladimir Zworykin, developed the first TV camera known as the iconoscope, which had the same concept as the later developed tube cameras and the CRT.

Both of these technologies, film and TV, produce many static images per second in order to achieve the illusion of motion. In TV, however, instead of projecting static images with a light projector through a celluloid film, this is achieved with electronic beam scanning. Pictures are formed line by line, in the same manner as when reading a book, for example, from left to right and from top to bottom (as seen from in front of the screen). The persistency of the phosphor coating of the monitor's CRT is playing an important role in the whole process.

As of a few years ago, the CCTV industry, following the broadcast television digitisation, is converting to fully digital. This might not be the case with existing systems, installed in the past, but with the new systems there is almost no exception — all new CCTV projects include digital cameras, digital recorders and digital (non-CRT) displays. Digital is the reason for yet another new book on CCTV. The digital technology however, doesn't change the concept of human eye persistence, despite that we may no longer use CRT displays, but rather LCD or plasma technology. **The principles are the same**. We perceive the multitude of subsequent images as moving pictures, as long as they are more than certain critical number, below which we see flicker in our eyes or motion appears jerky. Above this critical persistence frequency number (which depends on the size of the screen and it's brightness) we see continuous motion.

The very basics of analog television

In the analog television world, there were three different television standards used worldwide. CCIR/ PAL recommendations were used throughout most of Europe, Australia, New Zealand, most of Africa, and Asia. A similar concept was used in the EIA/NTSC recommendations for the television used in the United States, Japan, and Canada, as well as in the SECAM recommendations used in France, Russia, Egypt, some French colonies, and Eastern European countries. The major difference between these standards is in the number of scanning lines and frame frequency.

Before we begin the television basics, let us first explain the abbreviation terminology used in the technical literature discussing analog television:

CCIR stands for ***Committée Consultatif International des Radiotelecommuniqué***. This was the committee that recommended the standards for B/W television accepted by most of Europe, Australia, and others, which later on was renamed to ITU. Hence, equipment that complied with the B/W TV standards was referred to as **CCIR compatible**. The same type of standard, but later extended to color signals, was called **PAL**. The name comes from the concept used for the color reproduction by alternate phase changes of the color carrier at each new line – hence, ***Phase Alternating Line*** **(PAL).** Majority of CCIR/PAL systems are based on 625 scanning lines and 50 fields/s, although there are variations with 525 lines.

EIA stands for ***Electronics Industry Association***, an association that created the standard for B/W television in the United States, Canada, and Japan, where it was often referred to as **RS-170**, the recommendation code of the EIA proposal. When B/W TV was upgraded to color, it was named by the group that created the recommendation: ***the National Television Systems Committee*** **(NTSC)**. The EIA/NTSC systems are based on 525 scanning lines and 60 fields/s.

SECAM comes from the French "***Séquentiel Couleur avec Mémoire***" which actually describes how the color was transmitted by a sequence of chrominance color signals and the need for a memory device in the TV receiver when decoding the color information. Initially patented in 1956 by Henri de France, the SECAM was actually the first analog color television proposal, based on 819 lines and 50 fields/s. Later on, SECAM switched to 625 lines.

All of these analog TV standards' recommendations have accepted the picture ratio of the TV screen to be 4:3 (4 units in width by 3 units in height). This is due mostly to the similar film aspect ratio of the early days of television. The different number of lines used in different TV standards dictates the other characteristics of the system, such are the signal bandwidth and resolution.

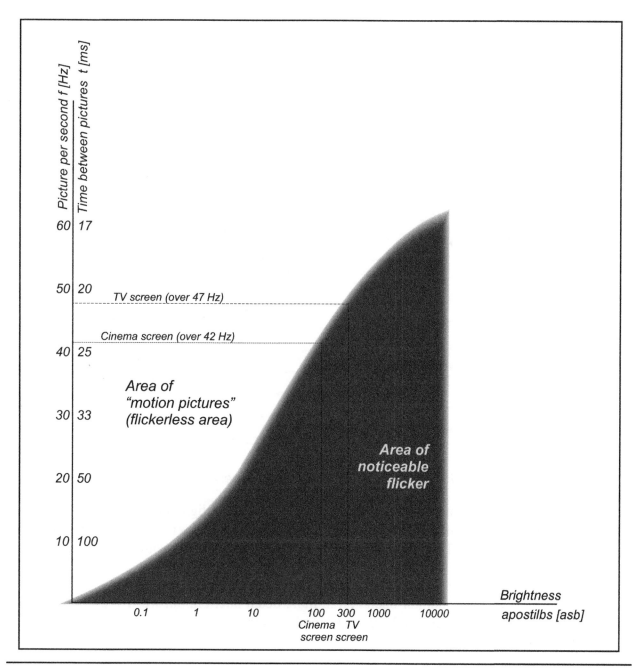

Regardless of these differences, all of the systems use the same concept of composing pictures with electron beam scanning lines, one after another. We refer here to electron beam, as this was the original signal created by a tube camera of the past and its CRT display monitor. These technologies can hardly be found these days, but the concept is the same using CCD or CMOS imaging chips when they are composing pictures pixel by pixel, merged into lines, and lines merged into TV frames. The following description refers to the original technology used in CCTV only until recently, and it is useful to understand these principles before we switch to digital. Everything makes more sense in the digital technology once we know its analog predecessor.

When a video signal, as produced by a camera, comes to the monitor input, the voltage fluctuations are converted into current fluctuations of electrons in the electron beam that bombards the phosphor coating of the ***cathode ray tube*** (CRT) as it is scanning line by line. The phosphor coating produces light proportional to the amount of electrons, which is proportional to the voltage fluctuation. This is, of course, proportional to the light information falling onto the camera (tube in the past, and CCD/CMOS chip at present); thus, the monitor screen shows an image of what the camera has seen.

Simplified representation of the interlaced scanning

The phosphor coating of the monitor has some persistency as well — light produced by the beam does not immediately disappear with the disappearance of the beam. It continues to emit light for another few milliseconds. This means the TV screen is lit by a bright stripe that moves downward at a certain speed.

This is obviously a very simplified description of what happens to the video signal when it comes to the monitor. We will discuss monitor operation in more detail later on, but we will use the previous information as an introduction to the television principles for the readers who do not have the technical background.

Many factors need to be taken into account when deciding the number of lines and the picture refresh rate to be used. As with many things in life, these decisions have to be a compromise, a compromise between as much information as possible, in order to see a faithful reproduction of the real objects, and as little information as possible, in order to be able to transmit it economically and receive it by a large number of users who can afford to buy such a TV receiver.

The more lines used, combined with the number of pictures per second, the wider the frequency bandwidth of the video signal will be, thus dictating the cost of the cameras, processing equipment, transmitters, and receivers. With today's digital terminology, the same could be said, where frequency bandwidth would be replaced with digital streaming bandwidth.

The refresh rate of still images, that is, the number of pictures composed in 1 second, was decided on the basis of the persistence characteristic of the human eye and the luminance of the CRT. Theoretically, 24

Timing intervals of TV signals	NTSC (µs)	PAL (µs)
Field period	16,683	20,000
Line period	63.5	64
Line blanking interval	10.7 ~ 11.1	11.8 ~ 12.3
Front porch interval	1.4 ~ 1.6	1.3 ~ 1.8
Line synchronization pulse interval	4.6 ~ 4.8	4.5 ~ 4.9
Field blanking interval	20 H + 1 H	25 H + 1 H
Duration of field synchronization pulse sequence	3 H	2.5 H
Duration of pre-equalizing pulse sequence	3 H	2.5 H
Duration of post-equalizing pulse sequence	3 H	2.5 H
Equalizing pulse interval	2.2 ~ 2.4	2.2 ~ 2.4
Interval between field synchronizing pulses	4.6 ~ 4.8	4.5 ~ 4.9
Start of color burst, from leading edge of line sync pulse	5.2 ~ 5.4	5.5 ~ 5.7
Color subcarrier burst duration (NTSC 9 cycles, PAL 10 cycles)	2.5	2.0 ~ 2.5
Duration of burst blanking pulse (per field)	9 H	9 H

pictures per second would have been ideal because it would have been very easy to convert movies from a film (used widely at the time of television's beginning) to electronic television. Practically, however, this was impossible because of the very high luminance produced by the phosphor of the CRT, which led to the flicker effect as it depends on the viewing distance and the screen luminance, as shown on the diagram on the previous page.

With many experiments it was found that more than 24 pictures per second are required, and at least 48 for the flicker to be eliminated. This would have been a good number to use because it was identical to the cinema projector frequency and would be very practical when converting movies into television format. Still, this was not the number that was accepted in the early days of television. The television engineers opted for 50 pictures per second in European and 60 in American recommendations. These numbers were sufficiently high for the flicker to be undetectable to the human eye, but more importantly they coincided with the mains frequency of 50 Hz used

Early projector design with Maltese cross

all over Europe and 60 used in the United States, Canada, and Japan. The reason for this lies in the electronic design of the TV receivers that were initially very dependent on the mains frequency. Should the design with 48 pictures have been accepted, the 2 Hz difference for CCIR and 12 Hz for EIA, would have caused a lot of interference and irregularities in the scanning process.

The big problem, though, was how to produce 50 (PAL) or 60 (NTSC) pictures per second, without really increasing the initial camera scan rate of 25 (that is 30) pictures per second. Not that the camera scan rate could not be doubled, but the bandwidth of the video signal would have to be increased, thus increasing the electronics cost, and bandwidth, as mentioned previously. Also, broadcasting channels were taken into account, which would have to be wider, and therefore fewer channels would be available for use, without interference, in a dedicated frequency area.

All of the above forced the engineers to use a trick, similar to the Maltese Cross used in film projection, where 50 (60) pictures would be reproduced without increasing the bandwidth. The name of this trick is *interlaced scanning*.

Instead of composing the pictures with 625 (525) horizontal lines by progressive scanning, the solution was found in the alternate scanning of odd and even lines. In other words, instead of a single TV picture being produced by 625 (525) lines in one progressive scan, the same **picture was broken into two halves**, where one-half was composed of only **odd lines** and the other of only **even lines**. These were scanned in such a way that they precisely fitted in between each other's lines. This is why it was called interlaced scanning. All of the lines in each half – in the case of the CCIR signal 312.5 and in NTSC 262.5 – form a so-called *TV field*. There are 25 odd fields and 25 even fields in the CCIR and SECAM systems, and 30 in the EIA system – a total of 50 fields per second, or 60 in EIA, flicking one after the other, every second.

An odd field together with the following even field composes a so-called *TV frame*. Every CCIR/PAL signal is thus composed of 25 frames per second, or 50 fields. Every EIA/NTSC signal is composed of 30 frames per second, which is equivalent to 60 fields.

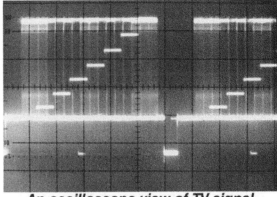

An oscilloscope view of TV signal in line mode (with horizontal sync)

The actual scanning on the monitor screen starts at the top left-hand corner with line 1 and then goes to line 3, leaving a space between 1 and 3 for line 2, which is due to come when even lines start scanning. Initially, with the very first experiments, it was hard to achieve precise interlaced scanning. The electronics needed to be very stable in order to get such oscillations that the even lines fit exactly in between the odd lines. But a simple and very efficient solution was soon found in the selection of an odd number of lines, where every field would finish scanning with half a line. By preserving a linear vertical deflection (which was much easier to ensure), the half line completes the cycle in the middle

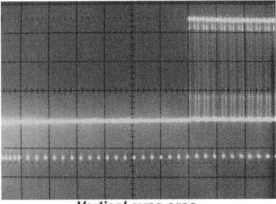

Vertical sync area as seen on an oscilloscope

of the top of the screen, thus finishing the 313th line for CCIR (263th for EIA), after which the **exact interlace was ensured for the even lines.**

When the electron beam completes the scanning of each line (on the right-hand side of the CRT, when seen from the front), it receives a *horizontal synchronization pulse* (commonly known as *horizontal*

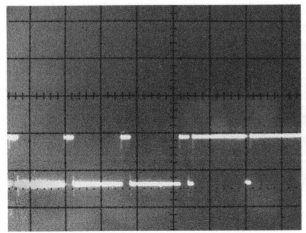

The vertical sync can be seen

The vertical sync pulses shown on an oscilloscope (left) and on a monitor with V-Hold adjustment (right)

sync). This sync is embedded in the video signal and comes after the active line video information. It tells the beam when to stop writing the video information and to **quickly** "fly" back to the left at the beginning of the new line. Similarly, when a field finishes a *vertical sync* pulse "tells" the beam when to stop "writing" the video information and to **quickly** fly back to the beginning of the new field. The fly-back period of the electron beam scanning is faster than the actual active line scanning, and it is only positional. That is, no electrons are ejected during these periods of the picture synthesis.

In reality, even though the scanning system is called 525 TV lines (or 625 for PAL), **not all of the lines are active, that is, visible on the screen**. As can be seen on the NTSC and PAL TV Signal Timing Chart (on the previous two pages), some of the lines are used for vertical sync equalization, others for some special test or messages signals (remember the teletext), so that not all of the lines as defined by the television standards are actually active lines.

In PAL, we cannot see more than 576 active TV lines and in NTSC not more than 480 active TV

A test pattern generator signal and its waveform on an oscilloscope

lines. These are the limits of analog PAL and NTSC television, by design. Remember these numbers for the active lines (576 and 480) as we will come across them when we introduce digital. They form the basis of the so-called "4CIF" (for PAL) and "4SIF" (for NTSC) digital pixel representation of an analog image converted to digital.

Some of the "invisible" lines are used, as mentioned earlier, for other purposes quite efficiently. In the PAL Teletext concept, for example, lines 17, 18, 330, and 331 are used, where 8-bit digital information is inserted. The Teletext decoder in your old TV or VCR can accumulate the fields' digital data, which contain information about the weather, exchange rates, Lotto, and so on.

In some NTSC systems, line 21 carries closed captioning (i.e. subtitling information). Some of the other invisible lines are used for specially shaped *Video Insertion Test Signals* **(VITS),** which when measured at the receiving end, give valuable information on the quality of the transmission and reception in a particular area. In CCTV, some manufacturers use the invisible lines to insert camera ID, time and date, or similar information. When recorded or converted to digital, the invisible lines may also be recorded although they would be invisible on the display screen but the information is there, embedded in the video signal. This type of information during the analog VHS recording was used as a security feature where a special TV line decoder was used when necessary, revealing the camera ID together with the time and date of the particular camera.

The video signal and its spectrum

This heading discusses the theoretical fundamentals of the video signal's limitations, bandwidth, and resolution. This is a complex subject with its fundamentals involving higher mathematics and electronics, but I will try to explain it in plain and simple language.

Most of the artificial electrical signals can be described by a mathematical expression. Mathematical description is very simple for signals that are periodical, like the main power, for example. A periodical function can always be represented with a sum of sine-waves, each of which may have different amplitude and phase. Similar to a spectrum of white light, this is called *spectrum of an electrical signal*. The more periodical the electrical signal is, the easier it can be represented and with fewer sine-wave components. Each sine-wave component can be represented with discrete value in the frequency spectrum of the signal. The less periodical the function is, the more components will be required to reproduce the signal. Theoretically, even a non-periodical function can be represented with a sum of various sine-waves, only that in such a case there will be a lot more sine-waves (with various frequencies) to sum up in order to get the non-periodical result. In other words, the spectral image of a non-periodical signal will have a bandwidth more densely populated with various components. **The finer the details the signal has, the higher the frequencies will be in the spectrum of the signal.** Very fine details in the video signal will be represented with high-frequency sine-waves. This is equivalent to high-resolution information. **A signal rich with high frequencies will have wider bandwidth.** Even a single, but very sharp, pulse will have a very wide bandwidth.

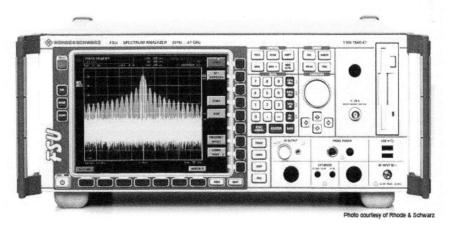

A typical Spectrum Analyzer

Photo courtesy of Rhode & Schwarz

The above describes, in a very simplified way, the very important *Fourier spectral theory*, which states that **every signal in the time domain has its image in the frequency domain**. The Fourier spectral theory can be used in practice – **wide bandwidth periodical electrical signals can be more efficiently explored by analyzing their frequency spectrum**. Without going deeper into the theory itself, CCTV users need to accept the concept of the *spectrum analysis* as very important for examining complex signals, such as the video itself. The video signal is perhaps one of the most complex electrical signals ever produced, and its precise mathematical description is almost impossible because of the constant change of the signal in the time domain. The video information (that is, luminance and chrominance components) changes all the time. Because, however, we are composing video images by *periodical* beam scanning, we can **approximate** the video signal with some form of a periodical signal. One of the major components in this periodicity will be the line frequency – for PAL, $25 \times 625 = 15,625$ Hz; for NTSC, $30 \times 525 = 15,750$ Hz.

It can be shown that the spectrum of a simplified video signal is composed of *harmonics* (multiples) of the line frequency around which there are companion components, both on the left- and right-hand sides (*sidebands*). The inter-component distances depend on the contents of the video picture and the dynamics of the motion activity. Also, it is very important to note that such a spectrum, composed of harmonics and its components, is **convergent,** which means the harmonics become smaller in amplitude as the frequency increases. One even more important conclusion from the Fourier spectral theory is that **positions of the harmonics and components in the video signal spectrum depend only on the picture analysis** (4:3 ratio, 625 interlaced scanning). The video signal energy distribution around the harmonics depends on the contents of the picture. The harmonics, however, are **at exact positions because they only depend on the line frequency**. In other words, the video signal dynamics and amplitude of certain components in the sidebands will vary, but the harmonics locations (as sub-carrier frequencies) will remain constant.

This is a very important conclusion. It helped find a way, in broadcast TV, to reduce the spectrum of a video signal to the minimum required bandwidth without losing too many details. There is always a compromise, of course, but since **the majority of the video signal energy is around the zero frequency and the first few harmonics, there is no need and no way to transmit the whole video spectrum**. Scientists and engineers have used all of these facts to find a compromise, to find how little of the video bandwidth need be used in a transmission, without losing too many details. As we already mentioned when discussing different TV standards, the more scanning lines that are used in a system the wider the bandwidth will be, and the higher the resolution of the signal is the wider the bandwidth will be.

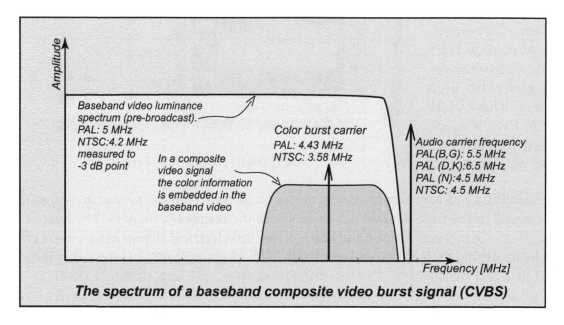

The spectrum of a baseband composite video burst signal (CVBS)

Taking into account the electron beam's limited size (which also dictates the smallest reproducible picture elements), the physical size of the analog TV screens, viewing distances, and the complexity and production costs of domestic TV sets, it has been concluded that for a good reproduction of a broadcast signal, 5 MHz of video bandwidth is sufficient. **Using a wider bandwidth is possible, but the quality gain factor versus the expense is very low**. As a matter of fact, in the broadcast studios, cameras and recording and monitoring equipment are of much higher standards, with spectrums of up to 10 MHz.

This is for internal use only, however, for quality recording and dubbing. Before such a signal is RF modulated and sent to the transmitting stage, it is cut down to 5 MHz video, to which about 0.5 MHz is added for the left and right audio channels. When such a signal comes to the TV transmitter stage it is modulated so as to have only its vestigial side band transmitted, with a total bandwidth, including the separation buffer zone, of 7 MHz (for PAL). But please note that the actual *usable video bandwidth* in broadcast reception is only 5 MHz. For the more curious readers we should mention that in most PAL countries, the video signal is modulated with amplitude modulation (AM) techniques, while the sound is frequency modulated (FM).

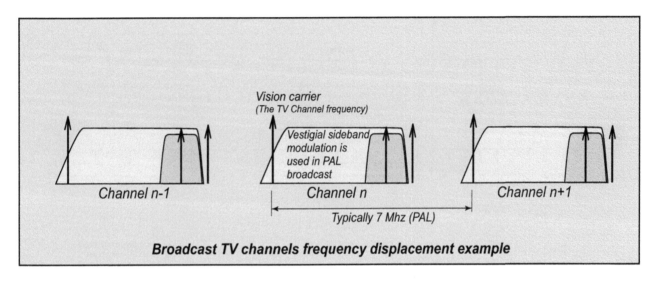

Broadcast TV channels frequency displacement example

Similar considerations apply when considering NTSC signals, where the broadcasted bandwidth is around 4.2 MHz.

In CCTV, with the majority of system designs, **we do not have such bandwidth limitations** because we do not transmit an RF-modulated video signal. We do not have to worry about interference between neighboring video channels. In CCTV, we use a raw video signal as it comes out of the camera, which is a **basic bandwidth video,** or usually called *baseband video*. This usually bears the abbreviation *CVBS*, which stands for *composite video burst signal*. The spectrum of such a signal, as already mentioned, ranges from 0 to 10 MHz, depending on the source quality.

The spectral capacity of the coaxial cable, as a transmission medium, is much wider than this. The most commonly used 75 Ω coaxial cable RG-59B/U, for example, can easily transmit signals of up to 100 MHz bandwidth. This is applicable to a limited distance of a couple of hundred meters of course, but that is sufficient for the majority of CCTV systems. Different transmission media imply different bandwidth limitations, some of which are wider and some narrower than the coaxial one, but most of them are considerably wider than 10 MHz.

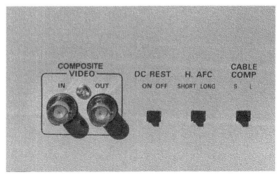

Bayonet Neill-Concelman (BNC) is the most common composite video input connector in CCTV.

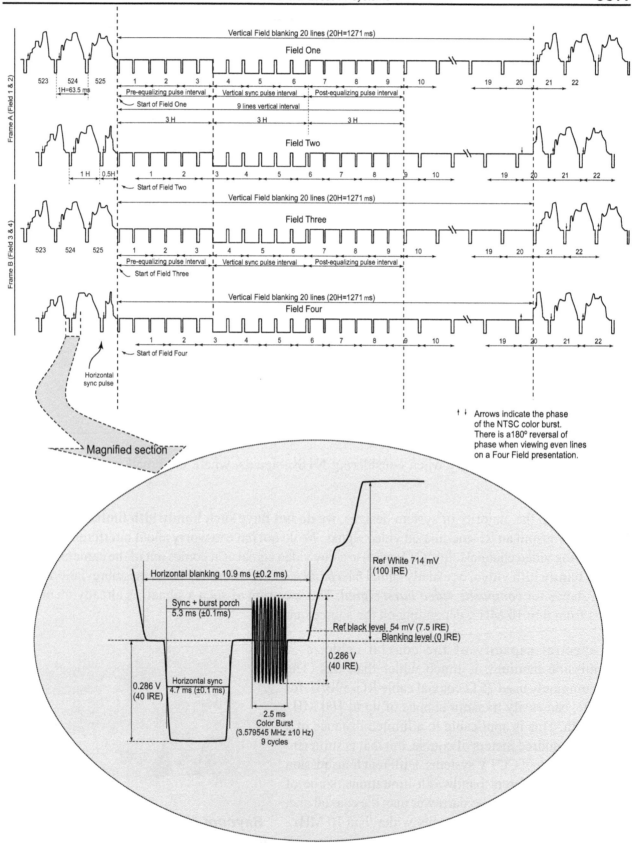

Field One

Vertical Field blanking 20 lines (20H=1271 ms)

523   524   525   1   2   3   4   5   6   7   8   9   10   19   20   21   22

1H=63.5 ms

Pre-equalizing pulse interval   Vertical sync pulse interval   Post-equalizing pulse interval

Start of Field One   9 lines vertical interval

3 H   3 H   3 H

Field Two

1 H   0.5H   1   2   3   4   5   6   7   8   9   10   19   20   21   22

Start of Field Two

Vertical Field blanking 20 lines (20H=1271 ms)

Field Three

523   524   525   1   2   3   4   5   6   7   8   9   10   19   20   21   22

Pre-equalizing pulse interval   Vertical sync pulse interval   Post-equalizing pulse interval

Start of Field Three

Vertical Field blanking 20 lines (20H=1271 ms)

Field Four

1   2   3   4   5   6   7   8   9   10   19   20   21   22

Start of Field Four

Horizontal
sync pulse

Magnified section

Frame A (Field 1 & 2)

Frame B (Field 3 & 4)

↑ ↓ Arrows indicate the phase
of the NTSC color burst.
There is a180° reversal of
phase when viewing even lines
on a Four Field presentation.

Horizontal blanking 10.9 ms (±0.2 ms)

Ref White 714 mV
(100 IRE)

Sync + burst porch
5.3 ms (±0.1ms)

Ref black level 54 mV (7.5 IRE)
Blanking level (0 IRE)

0.286 V
(40 IRE)

0.286 V
(40 IRE)

Horizontal sync
4.7 ms (±0.1 ms)

2.5 ms
Color Burst
(3.579545 MHz ±10 Hz)
9 cycles

NTSC TV signal timing chart

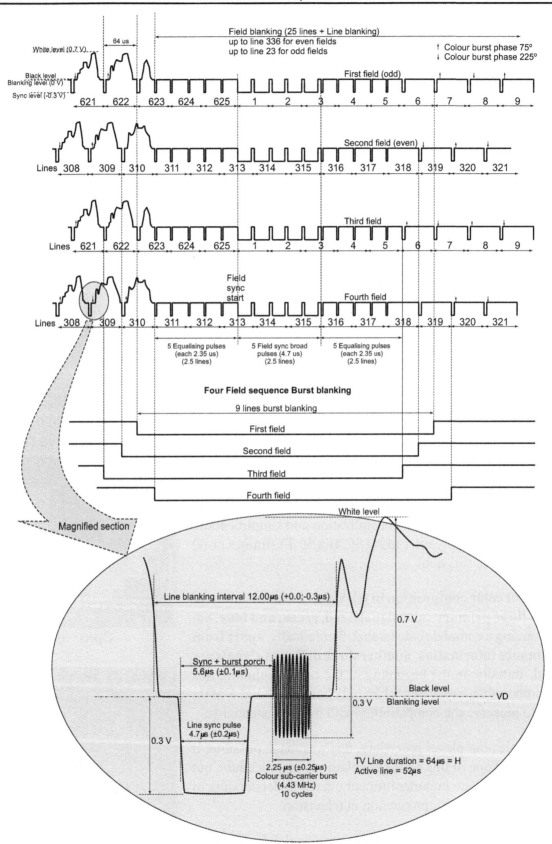

PAL TV signal timing chart

Color video signal

When color television was introduced it was based on monochrome signal definitions and limitations. Preserving the compatibility between B/W and color TV was of primary importance. The only way color information (chroma) could be sent together with the luminance without increasing the bandwidth was if the color information was modulated with a frequency that fell exactly in between the luminance spectrum components. This meant that the spectrum of the chrominance signal needed to be interleaved with the spectrum of the luminance signal in such a way that they do not interfere. This color frequency was called a *chroma sub-carrier* and the most suitable frequency, for **PAL**, was found to be **4.43361875 MHz**. In **NTSC**, using the same principle, the color sub-carrier was found to be **3.579545 MHz**.

At this point we need to be more exact and highlight that NTSC is defined with 29.97 TV frames per second exactly, not the rounded number 30 (!) typically stated. The reason for this is the definition of color signal in NTSC, as proposed by the RS170A video standard, which is based on the exact sub-carrier frequency of 3.579545 MHz. The horizontal scanning frequency is defined as 2/455 times the burst frequency, which makes 15,734 Hz. The vertical scanning frequency is derived from this one, and the NTSC recommends it as 2/525 times the horizontal frequency. **This produces 59.94 Hz for the vertical frequency (i.e., the field rate).** For the purpose of generalization and simplification, however, we will usually refer to NTSC as a 30 TV-frames or 60 TV-fields signal in this book.

The basics of color composition in television lie in the additive mixing of three primary color signals: red, green, and blue. So, for transmitting a complete color signal, theoretically, **apart from the luminance information, another three different signals are required**. Initially, in the beginning of the color evolution, this seemed impossible, especially when only between 4 and 5 MHz are used to preserve the compatibility with the B/W standards.

With a complex but clever procedure, this was made possible. It is beyond the scope of this book to explain such a procedure, but the following facts are important for our overall understanding of the complexity of color reproduction in television.

In a real situation, apart from the *luminance* signal, which is often marked as $Y = U_Y$, **two more signals** (not three) are combined.

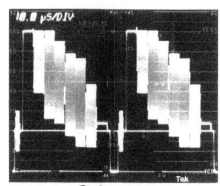

Color bars

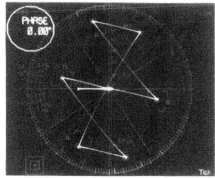

A vectorscope display of the color bars above (NTSC)

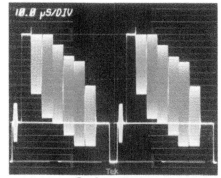

Color bars

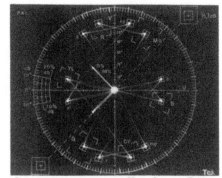

A vectorscope display of the color bars above (PAL)

These signals are the so-called ***color differences*** $V = U_R - U_Y$, and $U = U_B - U_Y$, which means the difference between the red and the luminance signal and between the blue and the luminance. **Color differences are used instead of just plain values for R, B (and G) because of the compatibility with the B/W system.** Namely, it was found that when a white or gray color is transmitted through the color system, only a luminance signal needs to be present in the CRT. In order to eliminate the color components in the system the color difference was introduced.

The next few formulas may not be important for a CCTV technical person to know, but it shows how color television theory managed to reduce the additional signals in a composite video from three to two. It all starts from the known basic relationship among the three color signals:

$$U_Y = 0.3U_R + 0.59U_G + 0.11U_B \tag{36}$$

where it can be show that **all three primary color signals can be retrieved using the luminance and color difference signals only**:

$$U_R = (U_R - U_Y) + U_Y \tag{37}$$

$$U_B = (U_B - U_Y) + U_Y \tag{38}$$

$$U_G = (U_G - U_Y) + U_Y \tag{39}$$

For white color $U_R = U_G = U_B$, thus $U_Y = (0.3 + 0.59 + 0.11)U_R = U_B = U_G$. The green color difference is not transmitted, but it is obtained from the following calculation (again using [36]):

$$U_G - U_Y = -0.51(U_R - U_Y) - 0.19(U_B - U_Y) \tag{40}$$

This relation shows that in color television, **apart from the luminance, only two additional signals would be sufficient for successful color retrieval. That is the red and the blue color difference (V and U)**, and they are embedded in the CVBS signal.

Because the R, G, and B components are derived from the color difference signals by way of simple and linear matrix equations, which in electronics can be realized by simple resistor networks, these arrangements are called ***color matrices***.

It should be noted here that the two discussed TV standards, NTSC and PAL, base their theory of color reproduction on two different exponents of the CRT phosphor (called ***gamma***, which will be explained in Chapter 6). The NTSC assumes a gamma of 2.2, and PAL 2.8. This assumption is embedded in the signal encoding prior to transmission.

In practice, gamma of 2.8 is a more realistic value, which is also reflected in a higher contrast picture. Of course, the reproduced color contrast will depend on the monitor's phosphor gamma itself, but most modern LCD have adjustments that compensate for various *gamma*, providing one knows such adjustments are needed. In larger systems, where there are multiple monitors of different sizes and makes, the first thing that needs to be done is adjusting all monitor parameters, using a reference signal from an electronic test pattern generator.

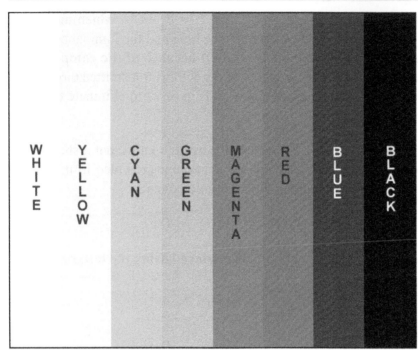

Standard order of color bars in television

In order to combine (modulate) these color difference signals with the luminance signal for broadcast transmission, a so-called *quadrature amplitude modulation* is used where the two different signals (V and U) modulate a single-carrier frequency (color sub-carrier). This is possible by introducing a phase difference of 90° between the two, which is the reason for the name *quadrature modulation*.

In the PAL color standard, we have another clever design to minimize the color signal distortions. Knowing that **the human eye is more sensitive to color distortions than to changes in brightness**, a special procedure was proposed for the color encoding so that distortions would be minimized, or at least made less visible. This is achieved by the color phase change, of 180°, in every second line.

So, if transmission distortions occur, which are usually in the form of phase shifting, they will result in a color change of the same amount. But because **the electronic vector representation of colors is chosen so that complementary colors are opposite each other, the errors are also complementary and, when "errored" lines next to each other are seen from a viewing distance, the errors will cancel each other out**. This is the reason for the name phase alternating line (PAL).

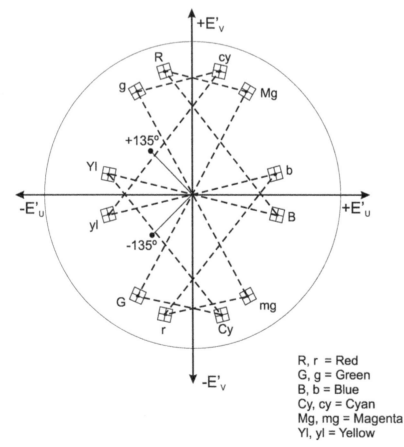

R, r = Red
G, g = Green
B, b = Blue
Cy, cy = Cyan
Mg, mg = Magenta
Yl, yl = Yellow

PAL color vectors

Resolution

Resolution is the property of a system to display fine details. The higher the resolution, the more details we can see. The resolution of a TV picture depends on the number of active scanning lines, the quality of the camera, the quality of the monitor, and the quality of the transmitting media.

Since we use two-dimensional display units (imaging sensors and display panels), we distinguish two kinds of resolutions: vertical and horizontal.

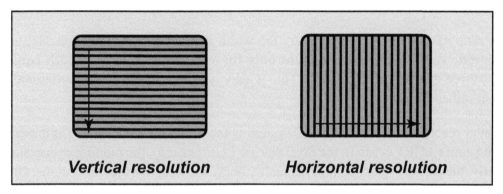

Vertical resolution   **Horizontal resolution**

The vertical resolution is defined by the number of vertical elements that can be captured on a camera and reproduced on a monitor screen. When many identical vertical elements are put together in the scanning direction, we get very dense horizontal lines. This is why we say **the vertical resolution tells us how many horizontal lines we can distinguish**. Both black and white lines are counted, and the counting is done vertically. Clearly, this is limited by the number of scanning lines used in the system – we cannot count more than 625 lines in a CCIR system or 525 in an EIA system. If we subtract from these the duration of the vertical sync and the equalization pulses, the invisible lines, and so on, **the number of active lines in CCIR comes down to 576 lines and about 480 in EIA.**

This is the absolute maximum as defined by the standards. In practice, the resolution is measured with a certain patterned image in front of the camera, so there are a lot of other factors to take into account. One is that the absolute position of the supposedly high-resolution horizontal pattern can never exactly match the interlaced lines pattern. Also, the monitor screen overs-canning cuts a little portion of the video picture, the thickness of the electronic beam is limited (if CRT is used as a display), and for color reproduction the "grill mask" is limited.

As early as 1933, Ray Kell and his colleagues found by experimenting that a **correction factor** of 0.7 should be applied when calculating the "real" vertical resolution. This is known as the *Kell Factor,* and it is accepted as a pretty good approximation of the real resolution. This

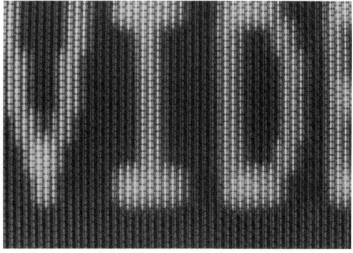

Close-up of a color screen

means that 576 has to be corrected (multiplied) by 0.7 to get the **practical limits** of the vertical resolution for PAL, which then becomes approximately 400 TV lines. The same calculation applies for the NTSC signal, which will give us approximately 330 TV lines of vertical resolution. This is all true in an ideal case, that is, with excellent video signal transmission.

Horizontal resolution is a little bit of a different story. **The horizontal resolution is defined by the number of horizontal elements that can be captured by a camera and reproduced on a monitor screen.** And, similar to what we said about the vertical resolution, **the horizontal tells us how many vertical lines can be counted.**

The aspect ratio of analog TV is 4:3, where the width is greater than the height. So, to preserve the natural proportions of the images, **we count only the vertical lines of the width equivalent to the height** (i.e., three-quarters of the width). This is why we do not refer to the horizontal resolution as just lines but rather *TV lines*.

The horizontal resolution of a display TV system is theoretically only limited to the cross section of the electron beam (if CRT is used), the pixel size (if LCD is used), the monitor processing electronics, and, naturally, the camera specifications. In reality, there are a lot of other limitations. One is the video bandwidth applicable to the type of transmission. Even though we may have high-resolution cameras in the TV studio, we transmit only 5 MHz of the video spectrum (as discussed earlier); therefore there is no need for television manufacturers to produce (analog) TV monitor with a wider bandwidth. In CCTV, however, the video signal bandwidth is dictated mostly by the camera itself. During the use of B/W monitors (no color grill mask) very high resolution (of up to 1000 TV lines) could be achieved. In fact this type of monitors were recommended when testing resolution of a CCTV camera. A monochrome (B/W) monitor is limited only by it's electronic quality, of which the most important are the electron beam precision and cross section.

A color system has an additional barrier, and that is the physical size of the color mask and its pitch. The color mask is in the form of a very fine grille. This grille is used for the color scanning with the three primary colors: red, green,

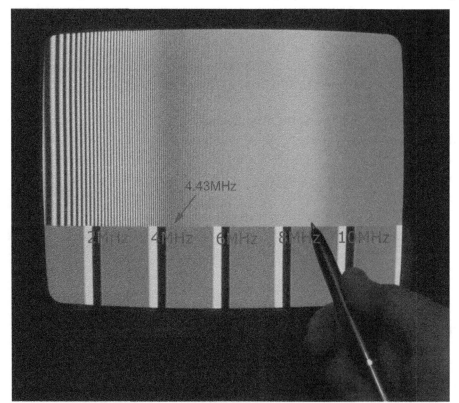

A 12 MHz sweep generator is used to check the bandwidth of a high-resolution monitor (shown 9 MHz = 700 TVL).

4. General about TV systems

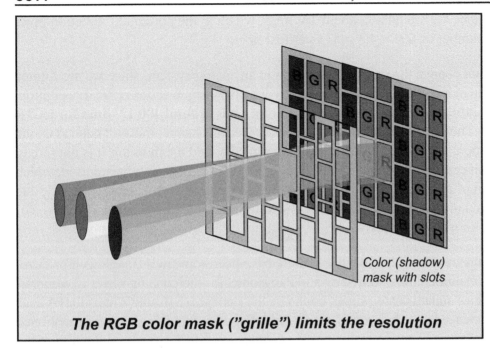

Color (shadow) mask with slots

The RGB color mask ("grille") limits the resolution

and blue. The number of the grille's color picture elements (RGB dots) is determined by the size of the monitor screen and the quality of the CRT. In CCTV, anything from 330 TV lines (horizontal resolution) up to 600 TV lines is possible (depending on the grille). During the era of CRTs (still used in many systems) the most common surveillance monitors were the standard 14" monitors with around 400 TV

SYSTEM	PAL B,G,H	PAL I	PAL D	PAL N	PAL M
Line/Field	625/50	625/50	625/50	625/50	525/60
Horizontal Frequency	15.625 kHz	15.625 kHz	15.625 kHz	15.625 kHz	15.750 kHz
Vertical Frequency	50 Hz	50 Hz	50 Hz	50 Hz	60 Hz
Color Sub Carrier Frequency	4.433618 MHz	4.433618 MHz	4.433618 MHz	3.582056 MHz	3.575611 MHz
Video Bandwidth	5.0 MHz	5.5 MHz	6.0 MHz	4.2 MHz	4.2 MHz
Sound Carrier	5.5 MHz	6.0 MHz	6.5 MHz	4.5 MHz	4.5 MHz

SYSTEM	NTSC M
Lines/Field	525/60
Horizontal Frequency	15.734 kHz
Vertical Frequency	60 Hz
Color Subcarrier Frequency	3.579545 MHz
Video Bandwidth	4.2 MHz
Sound Carrier	4.5 MHz

SYSTEM	SECAM B,G,H	SECAM D,K,K1,L
Line/Field	625/50	625/50
Horizontal Frequency	15.625 kHz	15.625 kHz
Vertical Frequency	50 Hz	50 Hz
Video Bandwidth	5.0 MHz	6.0 MHz
Sound Carrier	5.5 MHz	6.5 MHz

lines of resolution. Remember, we are talking about TV lines, which in the horizontal direction gives us an absolute maximum number of 400 × 4/3 = 533 vertical lines.

So, to summarize, we cannot change the vertical resolution in an analog system, since we are limited to the number defined by the scanning standard. That is why we rarely argue about vertical resolution. **The commonly accepted number for realistic vertical resolution is around 400 TV lines for CCIR and 330 TV lines for EIA.** The horizontal resolution, however, we can change; this will depend on the camera's horizontal resolution, the quality of the transmission media, and the monitor. It is not rare in CCTV to come across a camera with theoretical limit of 576 TV lines of horizontal resolution, which corresponds to a maximum of approximately 570 × 4/3 = 768 lines across the screen. In reality, due to the Kell factors, practical horizontal resolution is measured to be around 480 TV lines. This type of camera is considered a high-resolution camera.

At the time of writing this updated edition of the book, there have been some new imaging chips developed by the major camera manufacturers, like Sony and Panasonic, for their analog series of cameras. These cameras still produce a standard analog signal bound by the standard (576 active lines for PAL for example), but they have increased their horizontal pixel count. So, where imaging chips as mentioned above are designed to have 768 horizontal pixels (making "square" pixels, i.e. horizontal width of the pixels is equal to the vertical), the new chips are quoted to have 976 active horizontal pixels, making them "non-square" pixels. This yields, in theory, 732 TV lines (976 × 3/4), or around 600 TV lines in practice, although many advertised such cameras as 650 TVL. The reality is this resolution applies only

for a live video signal, and only if the monitor is of highest quality possible (found in some broadcast quality monitors). Recording such a signal, after it is being converted to digital, is still going to be limited to around 450 TVL, which is the limit of the A/D conversion when using ITU-601 standard, about which we'll be talking in the Digital CCTV section.

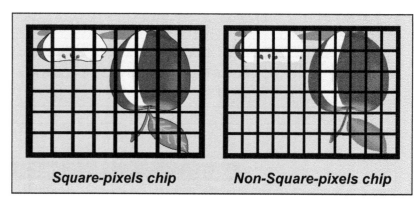

Square-pixels chip   *Non-Square-pixels chip*

There is a simple relation between the bandwidth of a video signal and the corresponding number of lines. If we take one line of a video signal, of which the active duration is 57 μs, and spread 80 TV lines across it, we will get a total of 80 × 4/3 = 107 lines. These lines, when represented as an electrical signal, will look like sine waves. So, a pair of black and white lines actually corresponds to one period of a sine wave. Therefore, 107 lines are approximately 54 sine waves. A sine wave period would be 57 μs/54 = 1.04 μs. If we apply the known relation for time and frequency (i.e., $T = 1/f$), we get $f = 1$ MHz. The following is a very simple rule of thumb, giving us the relation between the bandwidth of a signal and its resolution: **approximately 80 TV lines correspond to 1 MHz in bandwidth.**

440 TVL (5.5MHz)   500 TVL (6.3MHz)   565 TVL (7.1MHz)   628 TVL (7.8MHz)

MHz and TVL are interconnected values

The very basics of digital TV

Some of the greatest advancements in electronics at the end of last century and the beginning of this one is the switch from analog to digital. The reason is very simple — **digital signals are virtually immune to noise if conditions in which they are designed to operate are preserved.** This is explained in more details later on in the book. This switch to digital applies not only to telephony for voice communication, audio recording and distribution, but also to video. There is hardly any electronic technology today that does not employ digital signal processing or transmission.

As it will be shown later in the book, under the chapter Digital CCTV, the analog signal may be converted to digital (in fact this is what we do with most digital recorders that work with analog cameras nowadays) and we refer to this as *Standard Definition* digital signal, or short *SD*. But the more important advancement in the Broadcast industry, and hence in CCTV too, is the official acceptance of the *High Definition* television standard *(HD)*. In fact, one of the main reasons for this book update is exactly this — the introduction of digital and high definition technology. So, although analog TV and CCTV were the main part of this chapter, we can not bypass the HD technology now.

Near the end of the last century, the broadcast engineers from all over the world started discussing high definition television. One of the leaders in this area is certainly NHK from Japan, although their early adaptation was an analog form of HD TV. Since then, there has been long and hard debates between three global leaders in this area: Europe, Japan and the U.S., so that finally, at the beginning of the 20th century, the first agreed upon HD specifications started getting crystallized. **The important fact is that HD is now a world-wide standard, and not country based.** There are, admittedly, some minor differences between countries in the number of TV frames per second, being either 25, 30, 50 or 60, but the pixel count format is internationally compatible, meaning a TV from US will work in Europe or Australia, and other way around. This was not the case with analog TV.

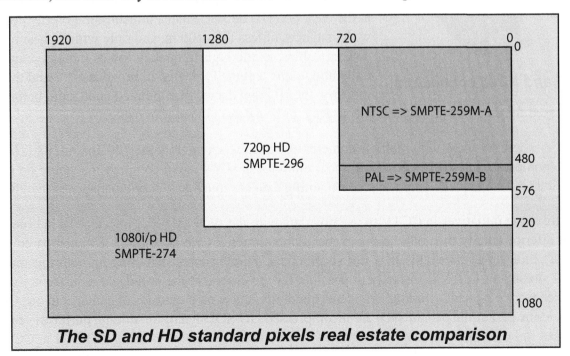

The SD and HD standard pixels real estate comparison

There are a few small variations of the HD standard, but the key points are: the signal is digital, it contains many more elements (resolution) than the analog and it's aspect ratio is 16:9, as opposed to 4:3.

The standard that defines the SD digital video is known as *SMPTE-259* (SMPTE stands for **Society of Motion Pictures and Television Engineers**). The standards that define HD digital video are known as *SMPTE-296,* refering to a high definition video with picture frame made of *1280 x 720* pixels, and the *SMPTE-274*, referring to a high definition video with picture frame made of *1920 x 1080* pixels. The 1080 HD is also known as *Full HD* and has around 5 times the details of the analog picture.

The SMPTE-296 comes as 720p, meaning it defines 1280 x 720 pixels only in "progressive" mode, while SMPTE-274 comes as 1080i and 1080p, meaning it can either be "interlaced" or "progressive" scanned image. The idea behind interlaced is similar as in analog, to reduce the bandwidth of data produced by such a signal.

HD produces a digital stream of live video, whose uncompressed spectrum is 1.485 Gb/s when 720p or 1080i mode is used, and 2.970 Gb/s when 1080p mode is used. The latter is often referred to as 3G signal, not to be confused with 3G for the 3rd generation of mobile telephony communication. The stated HD streams are referring to the raw data, before the compression.

The most common interface in HD television is the *DVI* (Digital Video Interface) and the *HDMI* (High Definition Multimedia). The HDMI is the more common as it deals with video and audio, as well as the copyright. For longer distances, typical of a broadcast studio, and where there are existing coaxial cables, the transmission is done over the one coaxial cable via the so called *Serial Digital Interface* (HD-SDI). **This is also known as SMPTE292M standard**, and it is in the form of a BNC connector at the back of a HD device. Such a transmission is only possible with distances of less than 100 m, and only with highest quality coaxial cables, as the high frequency losses increase exponentially above this length. Typically, in broadcast TV and the film industry, the HD raw data is manipulated and stored before it is

DVI and HDMI connectors

compressed for distribution via DVD, Blue-Ray, or broadcasted via air or cable.

In CCTV, where we work with multiple cameras and channels, typically the 720 and/or 1080 HD signals are compressed at the camera itself, most often using H264 video compression (more on this in the Digital CCTV chapter). This way, the streaming data has much lesser, but manageable bandwidth.

There are some initiatives in CCTV to use raw data with the same HD SDI interface as in broadcast, but the inferior quality of cables and poor installation practices we have in CCTV (compared to the broadcast industry), makes this idea not so practical. Furthermore, switching and storing multiple raw HD streams in CCTV is virtually impossible for the costs that the broadcast industry is ready to pay. Since we are typically working with tens, hundreds, and even thousands of cameras in one CCTV system, at the end of the path of such an uncompressed HD SDI signal, for storage purposes - the raw signal has to be compressed anyhow. The only really attractive point in using HD SDI in CCTV is the use of existing coaxial cabling infra-structure a customer may have, but only if the cable length is within the maximum prescribed distance HD SDI requires. This is rarely the case in practice, except for very small businesses needing HD CCTV system, in which case the cost is typically higher than using the main trend of non-SDI components.

So, for a typical HD CCTV system, the preferred method is compressing the HD signal at the camera end and making the signal available as IP stream (TCP, UDP or RTSP). This way, installations with hundreds and even thousands of cameras is possible over digital networks, and so is their long term storage. It is true to say that the compressed video streams have slightly lower video quality compared to the uncompressed HD signal, but with good encoding such a loss is imperceptible for the pruposes of CCTV recognition and

A HD camera with an uncompressed SDI output

identification of objects, far outweighing the gain in bandwidth and storage. In addition, compressed HD signal can also be transmitted over existing coaxial cable by using media converters (Cat-5/6 to Coax). In fact, because a compressed HD video stream could well occupy below 10 Mb/s (compressed from the raw 1.485 Gb/s) much longer distances can be reached over existing coaxial cables. Furthermore, multiple cameras can be transmitted over the same coax, since adding even a dozen of HD cameras might require no more than 100 Mb/s.

A typical CCTV HD camera with RJ-45 LAN output connector

Today, HD is an adopted standard and most of the broadcast stations already transmit HD over their existing ("analog") antennas and the same allocated spectrum. Most of the countries around the world will or would have their analog broadcast turned off by the time you are reading this, so there will be no more analog TV. Wanted us or not, the same destiny will apply to CCTV. As a little sister to broadcast, CCTV has already started following suit, so that many new projects have been proposed or installed with HD cameras and monitors.

Digital video formats	Standard	Signal mVpp	Active pixels	Live uncompressed bandwidth	Compressed with CCTV neglectable visual losses
PAL SD 576i	SMPTE-259M-B	800	720 x 576	177 Mb/s	1~4 Mb/s
NTSC SD 480i	SMPTE-259M-A	800	720 x 480	143 Mb/s	1~4 Mb/s
HD 720p	SMPTE-296	800	1280 x 720p	1.485 Gb/s	2~10 Mb/s
HD 1080i	SMPTE-274	800	1920 x 1080	1.485 Gb/s	2~10 Mb/s
HD 1080p	SMPTE-274	800	1920 x 1080p	2.970 Gb/s	4~20 Mb/s

HD signal representation

The switch of the television technology from analog to digital has changed the signal appearance too. The electronic waveform representation is no longer what we used to see on oscilloscopes, where the video levels were easily seen and measured. Digital signals are pretty monotonous, as they are composed of zeros and ones only. The explanation here refers to electronic representation of SD as well as HD signals. The only difference is in the frequency, i.e. bit rate. Typical voltage levels near the source are around 800 mV (as opposed to 1 Vpp of analog), whereby zeros and ones are swinging from -400 mV to 400 mV. Certainly, like with analog signals, there is a voltage drop at longer distances.

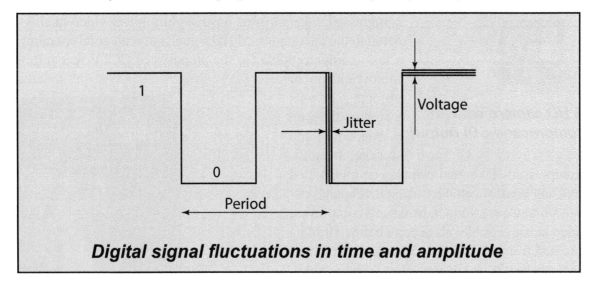

Digital signal fluctuations in time and amplitude

The digital pulses representing the bits of HD stream might be affected by *jitter*, which is the deviation of the signal from the absolute perfect timing. It occurs mostly due to cross-talk and reflected signals due to bad termination (yes termination plays a role in digital too). Another unwanted effect of raw digital transmission is the *drift*, which occurs due to drifting of the source device clock (camera for example) and the receiving device clock (HD TV for example). And, finally, voltage fluctuation of digital zeros and ones due to interference and cable quality also play role in creating the eye pattern.

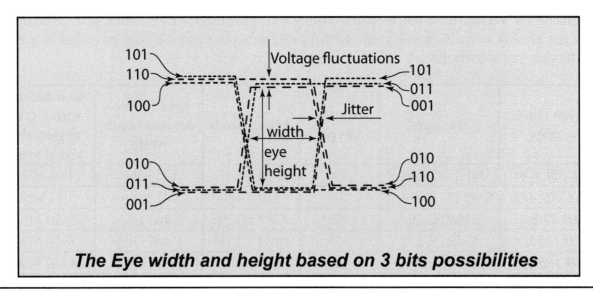

The Eye width and height based on 3 bits possibilities

The main parameters that can be measured are the fluctuations of ones and zeros bits through the so called *"eye-pattern."* The name comes from the it's which looks like an eye.

A digital video analyzer (often referred to as ***Bit Error Rate Analyzer***) accumulates three periods the main clock frequency and overlaps them on the screen, thus giving a good visual indication of how good a signal is. The eye window appearing in the centre of such a measurement indicates the

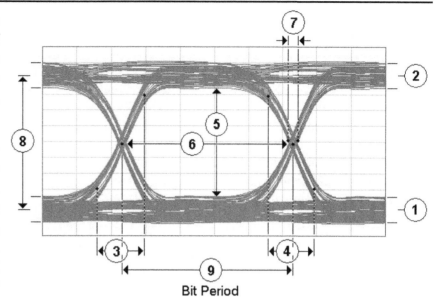

Bit Period

The eye measuring points

quality of the transmission. **The wider and the taller the eye shape is, the better the signal is**. In layman terms, the more open the eye is the better the signal is. A closed eye cannot see much. The picture above shows an actual eye measurement of an HD signal and it's corresponding measurement points:

1 - Zero level fluctuations
2 - One level fluctuations
3 - Rise time (10% to 90%)
4 - Fall time (90% to 10%)
5 - Eye height
6 - Eye width
7 - Jitter
8 - Eye amplitude
9 - Bit rate / period

All vertical units are measured in ***mV*** (for a HD signal this is typically a few hundred millivolts, where the full signal at the source is 800 mV) and the horizontal units in ***seconds*** (for a SD signal of 177 Mb/s this is typically 7.4 ns, and for HD of 1.485 Gb/s this yields 673 ps). This is obtained from inverting the bit rate. So for example, a 1.485 Gb/s will show a period of:

$$(1/1.485) \times 10^{-12} = 673.4 \text{ ps (pico seconds)}$$

The same applies to SD digital signals.

It is unlikely that in CCTV we would measure such signals, since such instruments are beyond a typical CCTV tool set, but it might be important to know what the key parameters are and what they represent.

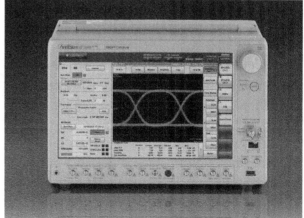

A Bit Error Rate (BER) analyzer

Megapixel CCTV

One of the many new terms used in CCTV would be the term *"megapixel"* (MP) referring to cameras that have anything above one million pixels. Technically, the HD SMPTE 274 format falls under this category, since it is a 2 megapixels (1080 HD is 1920 x 1080), but in CCTV, the term megapixel usually refers to all other camera formats that are above a million pixels and not necessarily HD.

At the time of writing this book, there were a number of different megapixel approaches, the simplest of which is a megapixel camera that may produce anything from a couple of megapixels up to 30 MP. Some MP cameras in CCTV produce streams of JPG images, one or a couple of images per second, offering detailed view of an entire scene whereby from one single shot it is possible to determine somebody's face identity, or vehicle's licence plate. The current technological limitation of large pixel count MP cameras is that the imaging sensor can not produce more than an image or two per second, due to the extreme amount of data collected from the pixels. MP cameras typically

A typical CCTV megapixel camera

require better lenses than a standard CCTV lens could offer. Some use photographic type of DSLR (Digital Single Lens Reflex) camera lenses due to their chip sizes being much larger format than 1/3". With the current compression technology, it is only possible to compress such images with Motion JPEG (M-JPG) compression, producing quite a large stream (of up to 100 Mb/s) when compared to the very efficient video compression such as H264 types (more on this in the chapter dealing with compression). Strictly speaking, such MP cameras with larger pixel count are closer to photography then television.

Another different, but very innovative, approach at the time of writing this book is the use of an array of HD cameras that are optically and mechanically positioned so as to make a larger and continuous image where each camera looks at a smaller section of it. Such an array of cameras produces live streams of each and every individual high definition camera which are then "merged together" into a seamless panorama that may offer as much as 60 MP in total. Objects at long distance can be digitally zoomed in without pixelation as they are within the range of the given design. A powerful computer runs smart software that does the merging of the views together. The total bandwidth of such an array of cameras typically stays well below 100 Mb/s, with the real time motion. This is possible due to the very high efficiency of the H264 video compression used by individual cameras.

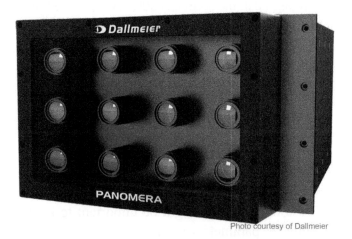

A multi-sensor panoramic camera

Beyond HD

Despite HD standard being only recently adopted, the broadcast and cinema industry have already progressed to some new and larger formats. One such new format is the so-called *4k HD*, or *Quad HD*. The 4k format works with imaging sensors four times the HD resolution, 3,840 x 2,160 pixels, making a total of over 8 megapixels, with real time streams. They have already been used for producing some of the newer movies, and chances are they will be adopted in broadcast TV as well. Some of the known camera manufacturers

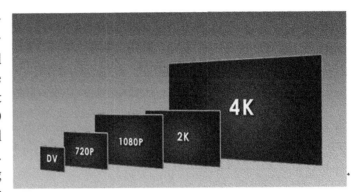

Various formats comparison

have already produced commercially available models. Clearly, in order to be able to view such video streams, some of the leading display manufacturers have already produced TV displays with 4k resolution. Cost of this equipment is still prohibitively high and there will be some time until this technology becomes available in CCTV, but it certainly out there. The type of compression this technology uses is said to be JPEG-2000 streaming with 250 Mb/s. A two hour, full featured, movie would require around 225 GB of storage even with such a compression.

One of the first commercially available 4k cameras

At the time of writing this book, an even larger format is being experimented with — the so-called *8k* or *Super High Vision (SHV)* format, displayed on *Ultra HD* TV. The Super Hi Vision camera is capable of streaming real-time video from unbelievable imaging sensor of 7,680 x 4,320 pixels (over 32 MP). The main proof of concept was the London Olympics 2012, where BBC and NHK engineers conducted a live broadcast from the London to Japan, via internet by bundling many broadband channels into one. Special display panels, combined of four 4k displays will be used for showing the signal in Japan and around London.

A 4k display

Instruments for CCTV

A *multi-meter* is certainly a very basic, but important, instrument for any electrician. Typically used to measure voltage of a power source, but also current that a camera or a system can draw. It is not possible, however, to determine any of the video signal electrical properties (be that analog or digital signal) with a typical electronic multi-meter. There are, however, specialized instruments that, when used correctly, can describe the tested video signal precisely.

In analog CCTV, these instruments include *oscilloscopes*, *spectrum analyzers*, and *vectorscopes*. In most cases an oscilloscope will be sufficient, and it is strongly recommended that the serious technician or engineer invest in it.

In digital video, as described earlier, it gets a bit more complicated as there are more measuring devices that might be needed. One instrument could be for testing the video quality of the most common IP cameras, where combined quality of the imaging sensor, optics and compression are evaluated. At the time of writing this book, the *ViDi Labs test chart* is the closest thing to be able to test all of these, and we will describe it in the chapter about Video testing. The second instrument is a *network analyzer*, used for testing the last that could be mentioned is for testing uncompressed video (either from an SDI output, or from a decoder) and testing its bit error rates, a so-called *BER instrument*.

Last, but not least, a *lux-meter* is an important instrument and can be used in both analog and digital CCTV.

Multi-meter

There are a few electrical units that an installing technician needs to understand clearly, since they are a part of every day CCTV job. These are *Volts, Amperes, Ohms, Watts, and Hertz*, all of which can be measured with a multi-meter.

The *Volt* [V] is one of the most common units of measuring and quantifying properties of any electrical device. **A *Volt* is defined as the electric potential difference across a conductor when a current of one Ampere dissipates one Watt of power.** If electricity was water then voltage would be pressure. The typical source of electrical power in the US, Canada and Japan is 120 VAC of alternating current (hence *VAC*), in Europe 220 VAC and in Australia 230 VAC.

A typical AA type of battery has 1.5 VDC of direct current (hence *VDC*), and a typical car battery has 12 VDC voltage. **Electric potential (often referred to as "voltage") is always measured in parallel to the terminals delivering the voltage.** When using a Voltmeter, and measuring DC voltage, the red probe is usually connected to the positive terminal and the black one on the negative. When measuring AC voltage, there is no constant polarity as the voltage alternates, hence it does not matter which color probe is connected to which terminal. All voltmeters have switch between AC or DC measurement. A voltmeter has very high input impedance, meaning it hardly draws any current from the source. Still, be very careful when measuring non-safe voltages (above 50 V) and make sure you have good probes with insulation. In all installation there should be a common rule of which of two wires is positive and which one negative. Typically, the wires will be colored red and black or one of the wires will be marked with a stripe. If common rules are used throughout your practice, mistakes with reverse polarity connections can be eliminated, or at least minimized.

Another important point in regards to the AC is that one of two wires delivering voltage is an active wire, while the other is the neutral. For most appliances it does not matter which wire is connected to which terminal of the device being powered, but when using line-locked 24 VAC cameras it might be important to know that the same "polarity" of wires is connected across all cameras. If all cameras are connected the same way, then once cameras are line-locked in the workshop, they should be line-locked when connected to real power source.

Cables carrying alternating power can easily be detected by cable tracers, also known as power probes, because alternating current produces electromagnetic field along the wires. If wires are not energized using a tone generator tool can be helpful when determining which wire is which.

A good quality multi-meter

The unit for measuring electric current, or amount of electric charge per second, is called *Ampere* [A] or sometimes abbreviated I for Intensity of electrical current. Current is always measured in series with the source of electricity.

The input impedance of the Ampere meter is close to zero. This means, when using an Ampere meter the probes are connected so that they interrupt the current between the source and the device they supply with current. If a DC current is measured, the red probe is connected first to the positive terminal, and if AC is measured the polarity is not important. The very important setting on the Ampere meter is the range of expected current measurement. The current should not exceed the maximum current range on the meter. Make sure you have some idea of how much current can the measured device pull, as if it is a short circuit, it may blow a fuse on your meter. Typical CCTV camera draws current typically well below 1 A.

Another parameter often measured in CCTV is the ***Ohm* [Ω]. An Ohm is the electrical resistance offered by a current-carrying element that produces a voltage drop of one Volt when a current of one Ampere is flowing through it.** Resistance increase with longer wires, thinner wires or warmer wires. Measuring resistance is reasonably easy and simple, but it should always be measured without power applied to it. When measuring resistance, an Ohm-meter delivers small current into it and measures the current drawn from the internal batteries, thus calculating and showing the resistance value in Ohms. The lower the Ohm value of the resistance is, the more current is drawn when connected to a power source. When checking cables for their continuity, two possibilities can be measured: short circuit or open circuit. When a cable has a short circuit at the other end the multi-meter measures very low resistance in Ohms. Depending on the cable length the number shown on a digital multi-meter can be anything from 0 to a couple of ohms. When a cable is open at the other end, then the resistance is infinite (very high) and this is usually displayed as 0.L MOhms. Many multi-meters would have the physical symbol for Ohms - Ω.

A **Watt** [W] is a unit of power. Watts are product of the voltage and the current (Watts = Volts x Amps). The low voltage transformers will normally list the voltage followed by the VA (Volt-Ampere) or Watts. There are special Watt-meters for measuring wattage, but in CCTV and security we don't use them readily; we usually calculate the wattage by measuring current consumption at a known voltage.

A Hertz [Hz] is the unit of frequency.

It describes the number of cycles an event does in one second, hence it can also be expressed as [cycles/s]. The power grid in the US and Canada produces AC power with 60 Hz, while Europe and many other countries use 50 Hz.

There are instruments to measure all of the above units in one test equipment, often called multi-meters, or Digital Volt Ohm Meters (DVOM). Measuring frequency is not that common for multi-meters, but there are some that can do that too.

Another important advantage of AC powered cameras is that they are usually designed to be able to lock their vertical TV field rate (60Hz) to the frequency of the power grid. When all cameras in a system are switched on to follow the grid frequency, we call this in CCTV - Line Locking (LL). The end result is that all cameras will be synchronized to the line power (providing they are on the same phase) and thus synchronized between themselves. This make video signal switching between various cameras smooth and without picture-roll. It also helps some recorders record faster as they do not need to do time base correction (TBC) internally. When using LL cameras, the technician installer should also be aware of the three phase power system.

Each of these phases deliver 120VAC relative to ground, but their phases are displaced by 120° between each other (this is a result from the way three phase generators produce electricity). This means, if one LL camera is connected to one phase, and another remote camera to another, they cannot be synchronized between themselves unless there is a way of adjusting the trigger point for locking the video signal to the line power to compensate for the 120° phase difference. Luckily, this is the case with most LL cameras.

Oscilloscope

The change of a signal (time-wise) can be slow or fast. What is slow and what is fast depends on many things, and they are relative terms. **One periodical change of something in one second is defined as Hertz.** Audio frequency of 10 kHz makes 10,000 oscillations in one second. The human ear can hear a range of frequencies from around 20 Hz up to 15,000–16,000 Hz. An analog video signal, as defined by the aforementioned standards, can have frequencies from nearly 0 Hz up to 5–10 MHz.

The higher the frequency, the finer the detail in the video signal.

How high the frequency can go depends, first of all, on the imaging sensor but also on the transmission (coaxial cable, microwave, fiber optics), and the processing/displaying media (codec, hard disk, monitor). A time analysis of any electrical signal (as opposed to a frequency analysis) can be conducted with an electronic instrument called an **oscilloscope**. The oscilloscope works on principles similar to those of a TV monitor; only in this case the scanning of the electron beam follows the video signal voltage in the vertical direction, while horizontally, time is the only variable. With the so-called time-base adjustment, video signals can be analyzed from a frame mode (20 ms) down to the horizontal sync width (5 µs).

Oscilloscope measurements have the most objective indication of the video signal quality, and it is strongly recommended to anyone seriously involved in CCTV. First, with an oscilloscope it is very easy to see the quality of the signal, bypassing any possible misalignment of the brightness/contrast on a monitor.

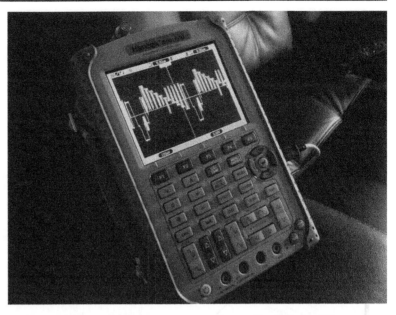

A practical and inexpensive handheld oscilloscope

Sync/video levels can easily be checked and can confirm whether a video signal has a proper 75-Ω termination, how far the signal is (reduction of the signal amplitude and loss of the high frequencies), and whether there is a hum induced in a particular cable. Correct termination is always required for proper measurements. That is, the input impedance of an oscilloscope is high and whichever way the signal is connected, it needs to see 75 Ω at the end of the line. A few examples of how an oscilloscope is to be connected for the purposes of correct video measurement are shown on the diagram on the next page.

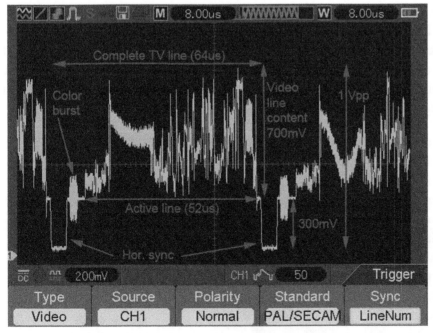

A typical single line of video signal with sync pulses

One of the most common problem with analog video signal over coax is the ground loop. A ground loop can be seen with an oscilloscope, when the time base is set to see the TV frame. The appearance is usually a waveform with a visible sinusoidal component instead of a straight line, which is an indication of mains frequency current induced in the video signal. The ground loop is most commonly a result from the induction of AC current from power cables being too close to the coaxial cables in a cable tray installation.

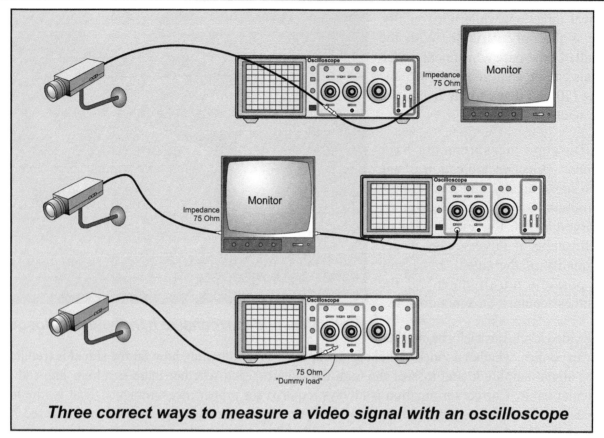

Three correct ways to measure a video signal with an oscilloscope

Spectrum analyzer

Every electrical signal that changes (time-wise) has an image in the frequency domain, as already discussed by the Fourier theory. The frequency domain describes the signal amplitude versus frequency instead of versus time. The representation in the frequency domain gives us a better understanding of the composition of an electrical signal. The majority of the contents of the video signal are in the low to medium frequencies, while fine details are contained in the higher frequencies. An instrument that shows such a spectral composition of signals is called a ***spectrum analyzer***.

A spectrum analyzer is an expensive device and is not really necessary in CCTV. However, if used properly, when combined with a test pattern generator with a known spectral radiation, a lot of valuable data can be gathered, especially about transmission media capability. Video signal attenuation, proper cable equalization, signal quality, and so on can be precisely determined. In broadcast TV, the spectrum analyzer is a must for making sure that the broadcast signal falls within certain predefined standard margins.

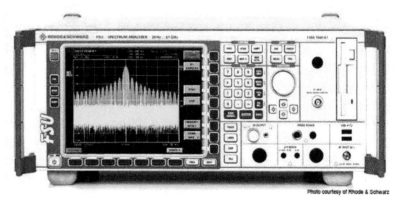

A typical Spectrum Analyzer

Vectorscope

For measuring the color characteristics of a video signal an instrument called a *vectorscope* is used. A vectorscope is a variation of an oscilloscope, where the signal's color phase is shown. The display of a vectorscope is in the polar form, where primary colors have exact known positions with angles and radii. The vectorscope is rarely used in CCTV but could be necessary when specific colors and lighting conditions need to be reproduced. It may especially be useful when determining the camera white balance.

In most cases, a color camera will have an automatic white balance that, as discussed earlier in the color temperature section, compensates for various color temperature light sources. Sometimes, however, with manual white balance cameras, a color test chart may need to be used, and with the help of a

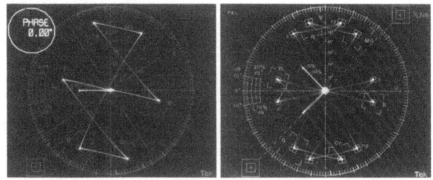

Vectorscope display of NTSC and PAL color bars

vectorscope, colors can be fine-tuned to fall within certain margins, marked on the screen as little square windows around primary (RGB) and complementary (CMY) colors (as it can be seen on the right photo).

It should be noted that a vectorscope display of the same image in NTSC is slightly different from the

vectorscope display in PAL, and this is because of the difference of color encoding in the two systems. PAL has vertically symmetrical color vectors, as it can be seen in the photos above.

The HD video signals have also color vectors which can be seen using HD vectorscope.

There are some instruments which might be more practical to some because they combine more instruments in one.

Some, like the device shown on the picture here, can do waveform analysis as an oscilloscope, vectorscope color analysis, spectrum analysis and show a display of the analyzed video signal. All of these

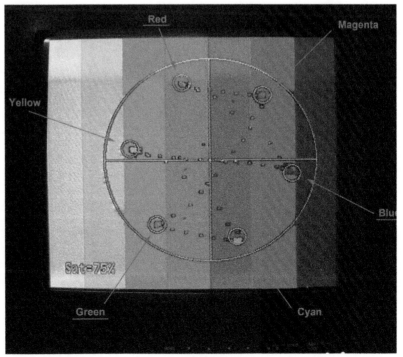

A Vectorscope representation of the TV color bars

measurements are made via a capture interface box, connected to the computer via a USB cable, and running its own software.

Furthermore, such products can check and measure not just analog SD video, but high definition uncompressed as well, connected via an SDI input. If you wish to check your HD output from an HDMI socket, you can use HDMI to SDI converter.

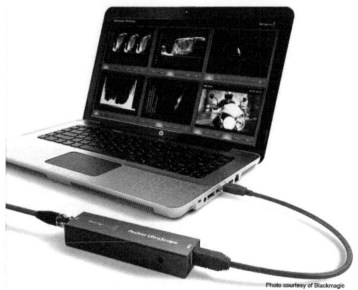

Photo courtesy of Blackmagic

A computer based multi-instrumental device

This page intentionally left blank

5. CCTV cameras

The very first and most important element in the CCTV chain is the element that captures the images – the camera.

General information about cameras

The term *camera* comes from the Latin *camera obscura*, which means "dark room."

This type of room was an artist's tool during the Middle Ages. Artists used a lightproof room, in the form of a box, with a convex lens at one end and a screen that reflected the image at the other, to trace images and later produce paintings.

In the nineteenth century, "camera" referred to a device for recording images on film or some other light-sensitive material. It consisted of a lightproof box, a lens through which light entered and was focused, a shutter that controlled the duration of the lens opening, and an iris that controlled the amount of light that passed through the glass.

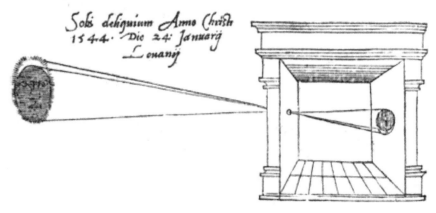

Joseph Nicéphore Niépce produced the first negative film image in 1826. This is considered the birth of photography. Initially, such photographic cameras did not differ much from the *camera obscura* concept. They were in the form of a black box, with a lens at the front and a film plate at the back. The initial image setup and focusing were done on an upside-down projection, which a photographer could see only when he or she was covered with a black sheet.

The first commercial photographic cameras had a mechanism for manual transport of the film between exposures and a viewfinder, or eyepiece, that showed the approximate view as seen by the lens.

Today, we use the term *camera* in film, photography, television, and multimedia. Cameras project images onto different targets, but they all use light and lenses.

To understand CCTV you do not need to be an expert in cameras and optics, but it helps if you understand the basics. Many things are very similar to what we have in photography, and since every one of us has been, or is, a family photographer, it will not be very hard to make a correlation between CCTV and photography or home video. In photographic and film cameras, which are now an obsolete technology, we converted the optical information (images) into a chemical emulsion imprint (film). In television cameras, we convert the optical information into electrical signals. They all use lenses with certain focal lengths and certain angles of view, which are different for different formats.

Lenses have a limited resolution and certain distortions (or aberrations) may be noticed on higher resolution images, be that on film or modern megapixel sensors. In the period from the last edition of this book (2005) and this one (2012), photography has completely changed and converted to digital, exceeding film in details and quality. This was considered impossible less than years ago.

One of the very early television cameras from 1931

To illustrate this, a high-resolution mega-pixel sensors in CCTV these days may have as high as 30 million pixels (picture elements) and over, while 100 ISO 35 mm color negative film has a resolution equivalent to around 16 million elements or film grains (source: *en.wikipedia.org*).

Tube cameras

The first experiments with television cameras, as mentioned earlier, were made in the 1930s by the Russian-born engineer Vladimir (Vlado :-) Zworykin (1889–1982). His first camera, made in 1931, focused the picture onto a mosaic of photoelectric cells. The voltage induced in each cell was a measure of the light intensity at that point and could be transmitted as an electrical signal. The concept, with small modifications, remained the same for decades.

Those first cameras were made with a glass tube and a light-sensitive phosphor coating on the inside of the glass. For this reason they were called *tube cameras*.

Tube cameras worked on the principles of *photosensitivity*, based on the photo-effect. This means the light projected onto the tube phosphor coating (called the *target*) had sufficient energy to cause the ejection of electrons from the phosphor crystal structure. The number of electrons was proportional to the light, thus forming an electrical representation of the light projection.

There were basically two main types of tubes used in the early days of CCTV: Vidicon and Newvicon.

Vidicon was cheaper and less sensitive. It had a so-called *automatic target voltage control*, which effectively controlled the sensitivity of the Vidicon and, indirectly, acted as an electronic iris control, as we know it today on CCD and CMOS cameras. Therefore, Vidicon cameras worked only with manual iris lenses. The minimum illumination required for a B/W Vidicon camera to produce a signal was about 5 ~ 10 lux reflected from the object when using an F-1.4 lens.

Newvicon tube cameras were more sensitive (down to 1 lux) and more expensive, and required auto iris lenses. Their physical appearance was the same as the Vidicon tube, and one could hardly determine which type was which by just looking at the two. Only an experienced CCTV

A studio tube camera from 1952

technician could notice the slight difference in the color of the target area: the Vidicon has a dark violet color, while the Newvicon has a dark bluish color. The electronics that control these two types of tubes are different, and on the outside of the camera the Newvicon type has an auto iris connection.

All tube cameras use the principles of electromagnetism, where the electron beam scans the target from the inside of the tube. The beam is deflected by the locally produced EMF which is generated by the camera electronics. The more light that reaches the photoconductive layer of the target, the

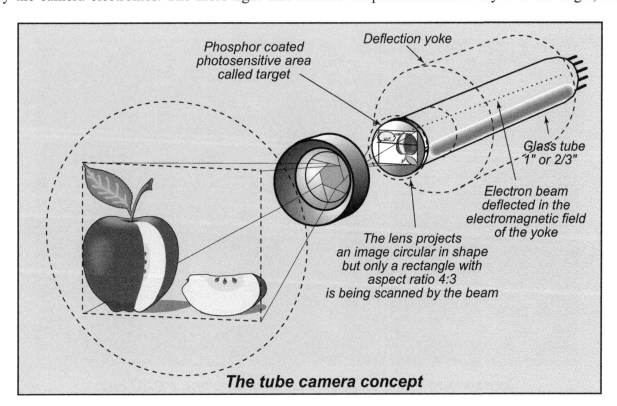

The tube camera concept

lower its local resistance will be. **When an image is projected, it creates a potential map of itself by the photosensitivity effect**. When the analyzing electron beam scans the photosensitive layer, it neutralizes the positive charges created, so that a current flow through the local resistor occurs. When the electron beam hits a particular area of the potential map, an electrical current proportional to the amount of light is discharged. This is a very low current, in the order of pico-Amperes (pA = 10^{-12} A), which is fed into a very high-input impedance video preamplifier, from

The inside of a tube camera

which a video voltage signal is produced. For a tube camera it is important to have a thin and uniform photo layer. This layer produces the so-called *dark current*, which exists even if there is no image projected by the lens (iris closed).

After a signal has been formed, the rest of the camera electronics add sync pulses, and at the output of the camera we get a complete video, known as a *composite video signal*.

There are a few important concepts used in the operation of tube cameras, which we need to briefly explain in order to appreciate the differences between this and the solid state (CCD and CMOS) technology.

The first concept is the **physical bulkiness of the camera**; due to the glass tube, electromagnetic deflection yoke around the tube, and the size of the rest of the electronic components in the era when surface mount components were unknown. This made tube cameras quite big.

Comparison of the physical size of a tube and a CCD chip

The second concept is the **need for a precise alternating electromagnetic field** (EMF) which will force the electron beam to scan the target area as per the television recommendations. To use an EMF to do the scanning means the external EMF of some other source may affect the beam scanning, causing picture distortions.

Third is the **requirement for a high voltage** (up to 1000 V), which accelerates the electron beam and gives it straight paths when scanning. Consequently, high-voltage components need to be used in the camera, which are always a potential problem for the electronic circuit's stability. Old and high-voltage capacitors may start leaking, moisture can create conductive air around the components, and electric sparks may be produced.

Fourth, there is the need for a phosphor coating of the target, which converts the light energy into electrical information. The phosphor as such is subject to constant electron bombardment that wears it out. Therefore, **the life expectancy of a tube phosphor coating is limited**. With constant camera usage, as is the case in CCTV, a couple of years will be the realistic life expectancy, after which the picture starts to fade out, or even an imprinted image will develop if the camera constantly looks at the same object. As a result, we can see pictures from a tube camera where, when people move, they appear as ghostlike figures, since they are semitransparent to the imprinted image.

And the fifth feature, conceptually different from the solid state cameras used today and inherently part of the tube camera design itself, consists of **geometrical distortions** due to the beam hitting the target at various angles. The path of the electron beam is shorter when it hits the center of the target as compared to when it scans the edges of the tube. Therefore, certain distortions of the projected image are present. This feature is certainly another drawback. In a lot of tube camera designs, we will find some magnetic and electronic corrections for such distortions, which means every time a tube needs to be replaced, all of these adjustments have to be re-made.

With the new CCD and now CMOS technology, none of the above problems exists in cameras. One tube's feature, however, was very hard to beat in the early days of the CCD technology. This was the resolution of a good tube camera. Vertical resolution is dependent on the scanning standard, and this would be, more or less, the same at both, CCD and tube cameras, but the horizontal resolution (i.e., the number of vertical lines that can be reproduced) depends on the thickness of the electron beam. Since this can be quite successfully controlled by the electronics itself, very fine details can be reproduced (i.e., analyzed while scanning).

Initially, with the CCD design, microelectronics technology was not able to offer picture elements (pixels) of the CCD chip smaller than the beam cross section itself. This means that in the very beginning of CCD technology, the resolution lagged well behind that of the tube cameras.

CCD camera

CCD cameras

It all started in 1969 with the discovery of the ***Charge Coupled Device*** (CCD), by George Smith and Willard Boyle from Bell Labs. Their discovery, initially thought of being used as a memory device, proved to be so valuable to the imaging industry today, that they were honoured with a Nobel Prize for Physics in 2009, 40 years after their invention. Interestingly enough, the CCD principles are based on the photoelectric effect, which were explained by Albert Einstein back in the early 1900-es, for which Einstein too received his Nobel prize in 1921, not for his theory of relativity, like many would have believed.

Albert Einstein

Photoelectric effect explains how light energy (photons) are converted into electrical signal in the form of free electrons. The base material of an imaging chip is silicon, a semiconductor that "decides" it's conductive state based on electric field around it. The name semiconductor in electronics indicates exactly that - a material that sometimes acts as a conductor (like copper) and sometimes as insulator (like rubber). When light photons "hit" the silicon area of the CCD (which we call ***picture element*** - or short "***pixel***") it frees up electrons proportional to the light intensity. **The more light that falls on the sensor the more electrons will be ejected from the silicon pixel.** The basic principle of CCD operation is the storing of the information of electrical charges in the elementary cells and then, when required, shifting these charges to the output stage. Then, these electron charges are shifted out vertically and/or horizontally by use of voltage pulses, in the same manner as shift registers in digital electronics shift binary values. For this reason they are sometimes referred to as ***mobile packets***. The quantity of the electrons in these mobile packets can vary widely, depending on the applied voltage and on the capacitance of the storage elements. **The amount of electric charge in each packet can represent information (except for the thermally generated noise electrons).** This is basically how light information (a scene projected by a lens onto the surface of the imaging sensor) is converted into electrical signal. This is a very useful process for converting light information into electrical information.

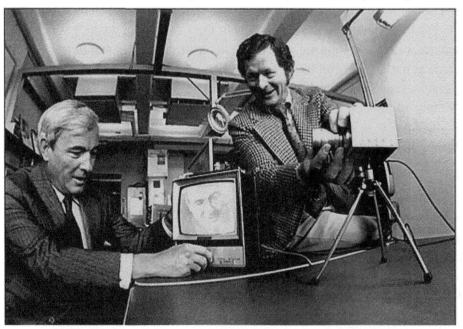

The fathers of CCD: George Smith and Willard Boyle

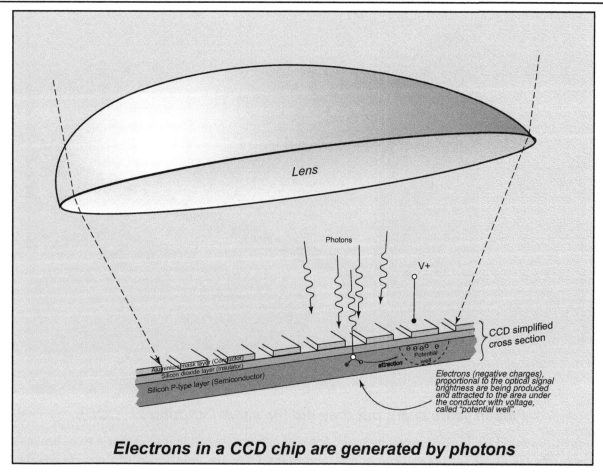

Lens

Photons

V+

CCD simplified cross section

Aluminium mask layer (Conductor)
Silicon dioxide layer (Insulator)
Silicon P-type layer (Semiconductor)

Potential well

attraction

Electrons (negative charges), proportional to the optical signal brightness are being produced and attracted to the area under the conductor with voltage, called "potential well".

Electrons in a CCD chip are generated by photons

One of the first commercially available solid state cameras, based on the CCD technology, was the Sony's "Mavica." The name Mavica was short for Magnetic Video Camera, because it stored its images on a magnetic floppy diskette. It appeared in 1981 and it was using a CCD chip with only 570 x 490 pixels.

So, for the first time in photography and television, the CCD sensor became the solid state camera's electronic eye. It revolutionized photography, television, and film, as light could now be captured electronically instead of on film. CCD technology is also used in many medical applications, e.g. imaging the inside of the human body, both for diagnostics and for microsurgery. Digital photography has become an irreplaceable tool in many fields of research. The CCD has provided new possibilities to visualize the previously unseen. It has given us crystal clear images of distant places in our universe as well as the depths of the oceans.

Mavica, the first solid state camera

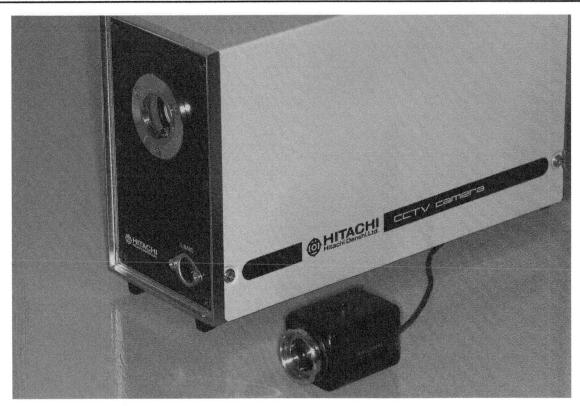

Fifteen years apart, but they did the same job (tube vs. CCD).

The construction of CCD sensors can be in the form of either a line (linear CCD) or a two-dimensional matrix (array CCD). It is important to understand that **CCDs are composed of discrete pixels, but they are not digital devices as such**. Each of these pixels can have any number of electrons it is capable of storing, proportional to the light that falls onto it, thus representing **analog information**.

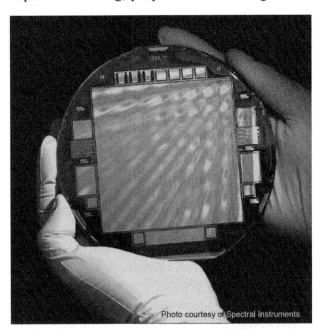

Silicon wafer with various CCD sizes

It is possible that further in the camera the signal may be converted to digital format (this is what IP cameras are designed to do), but the **CCD sensor output values are analog.**

So, despite their original design intention, today CCDs are not used as memory devices, but almost exclusively as imaging sensors. They can be found in many human industries and activities, not just CCTV cameras which is the prime interest of this book. They can be used as two-dimensional sensors but linear as well. The typical places you will find them are bar code scanners, facsimile machines, picture and document scanners, DSLR camera auto-focusing, geographic aerial and survey inspection (Google Earth and Maps), spacecraft planet scanning, industrial inspection of materials and much more.

Linear sensors are used in applications where there is only one direction of movement by the object, such as with facsimile machines, image scanners, and aerial scanning. So it is either the sensor that is moving or the object of interest is moving. Using one dimension only the complexity of transferring charges is dramatically reduced, and high resolution images with variable length of scans can be achieved. The precision of scanning in linear sensor systems depend on the mechanical movement precision, and the speed depends on the data processing speed of such an assembly.

It goes without saying now that CCD cameras have many advantages (in design) over the tube cameras, although, as mentioned earlier, in the early days of the technology it was hard to achieve high-resolution and pixel uniformity.

Today, as at the writing of this book, the largest single two-dimensional sensor chip may have over 120 megapixels; its only limited by the silicon wafer size.

Line scan CCD sensors are used in satellite imaging.

As we said earlier, CCDs come in all shapes and sizes, but the general division is into linear and two-dimensional matrices.

One could comment that the new prevailing technology today is the *Complementary Metal Oxide Semiconductor* (CMOS), and we will talk about this in the next heading, but in essence they are very similar solid state devices as the CCDs.

Photo courtesy of Teledyne Dalsa

A variety of sensor formats and sizes

In CCTV we are only interested in the two-dimensional sensor matrices with various sizes, most often referred to as 2/3", 1/2", 1/3", or 1/4" sizes. With the introduction of HD and megapixel technology, other sensor sizes are available as well. As mentioned earlier, these numbers are **not the diagonal sizes of the image sensor**, as many assume, but rather they are the sizes of the projection circle a suitable optics would have on the sensor. When tube cameras were invented this diameter was the same as the tube that would produce such an image.

Photo courtesy of Spectral Instruments

Currently largest single chip camera with 100MP

How charges are transferred in a CCD sensor

The quality of a signal as produced by the CCD chip depends also on the pulses used for transferring charges. These pulses are generated by an internal *master clock* crystal oscillator in the camera. The drawing below shows how video signal sync pulses are created using the master clock. This frequency depends on many factors, but mostly on the number of pixels the CCD chip has, the type of charge transfer (FT, IT, or FIT), as well as the number of phases used for each elementary shifting of charges; namely, the elementary shifting can be achieved with a two-phase, three-phase, or four-phase shift pulse. In CCTV, cameras with three-phase transfer pulses are the most common.

As you can imagine, the camera's crystal oscillator needs to have a frequency at least a few times

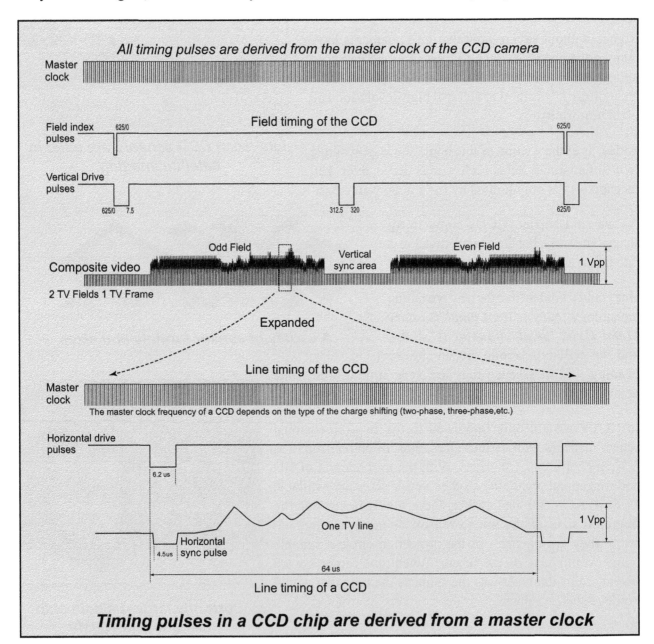

Timing pulses in a CCD chip are derived from a master clock

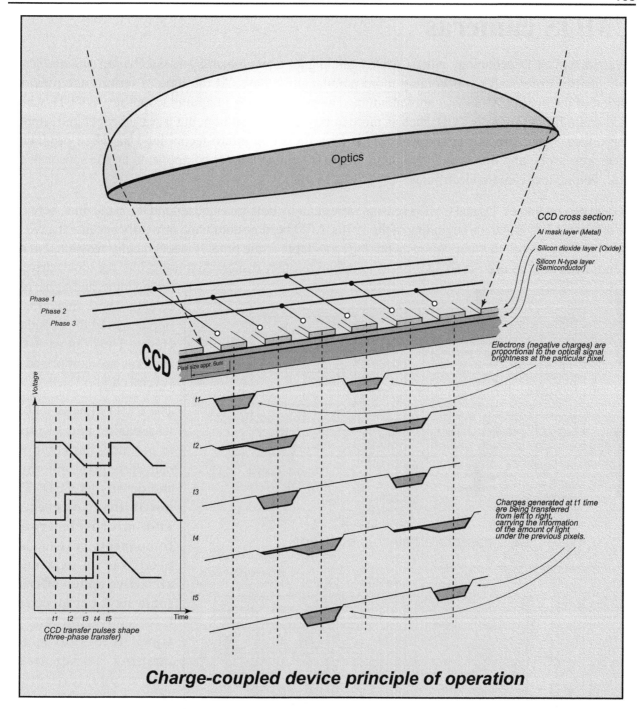

Charge-coupled device principle of operation

higher than the signal bandwidth that a camera produces. All other syncs, as well as transfer pulses, are derived from this master frequency. The drawing above shows conceptually how this charge transfer is performed with the three-phase shift on the CCD sensor. The pulses indicated with ϕ_1, ϕ_2, and ϕ_3 are low-voltage pulses (usually between 0 and 5 V DC), which explains why CCD cameras have no need for high voltage, as was the case with the tube cameras.

This is only one of many examples, but it clearly shows the complexity and number of pulses generated in a CCD camera.

CMOS cameras

A variation of CCD technology called CMOS (standing for ***Complimentary Metal Oxide Semiconductor***, as explained previously) is becoming more popular these days. At the time of writing the previous edition of the book, CMOS was only mentioned as an alternative imaging technology to CCD, which had certain limitation over CCD, such as more noise, lesser resolution, but it was cheaper and simpler to produce. The main advantage was that it uses the same micro-technology as when producing micorprocessors, and this made possible to integrate semiconductor components on the sensor itself, thus needing less camera electronics.

It could be said that CCDs and CMOS sensors were actually both invented around the same time, between the end of the 1960s and the beginning of the 1970s. CCD became dominant, primarily because it gave far superior images with the fabrication technology available at the time. It is technically feasible, but not economically so, to use the CCD process to integrate other camera functions, like the clock-drivers, timing logic, and signal processing. These are normally implemented in secondary chips. Thus, most CCD cameras are comprised of several chips.

CCD camera electronic concept

NOTE: Drawings based on Teledyne Dalsa whitepaper

CMOS camera electronic concept

The CMOS image sensors was easier and cheaper to produce, but could not deliver the pixel uniformity and density as CCD. **The possibility, however, to add micro-electronic components next to each pixel was an extremely attractive option offering some new perspectives.** Even converting analog signal to digital at the imaging sensor itself was a possibility. In addition, **CMOS power consumption was much lower than the CCD**. These reasons were sufficient to get imager manufacturers keep searching for improved CMOS technologies.

CMOS imagers sense light in the same way as CCD,

but from the point of sensing onward, everything is different. The charge packets are not transferred, but they are instead detected as early as possible by charge-sensing amplifiers, which are made from CMOS transistors. In some CMOS sensors, amplifiers are implemented at the top of each column of pixels – the pixels themselves contain just one transistor which is used as a charge gate, switching the contents of the pixel to the charge amplifiers. These passive pixel CMOS sensors operate like analog dynamic random access memory (DRAM). Conceptually, the weak point of CMOS sensors is the problem of matching the multiple different amplifiers within each sensor. Some manufacturers have overcome this problem by reducing the residual level of fixed-pattern noise to insignificant proportions. With the initial CMOS designs and prototype cameras, there were problems with low-quality, noisy images that made the

Photo courtesy of Sony

CMOS complete camera sensors

technology somewhat questionable for commercial applications. Chip process variations produce a slightly different response in each pixel, which shows up in the image as noise. In addition, the amount of chip area available to collect light is also smaller than that for CCDs, making these devices less sensitive to light.

Essentially, both types of sensors are solid state and convert light into electric charge and process it into electronic video signals. In a CCD sensor, **every pixel's charge, in form of electron current, is transferred through output nodes, then converted to voltage, and sent off-chip as an analog signal.** All of the pixel area in CCD can be devoted to light capture, and the output's uniformity is high. In a CMOS sensor however, **each pixel has its own charge-to-voltage conversion, and the sensor often includes electronic components** such as amplifiers, noise-correction, and digitization circuits, so that the chip may output digital bits rather than just analog values. **With each pixel doing its own conversion, uniformity is lower.** In addition, **the micro-components added around the pixels take up space too, thus reducing the CMOS light sensing area.** But the advantage is that

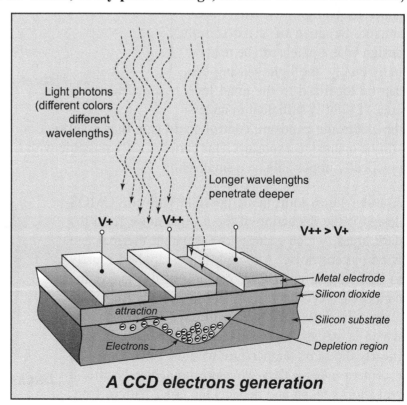

A CCD electrons generation

the imaging sensor can be built to require less off-chip circuitry for basic operation. This is especially critical with the clock requirement on CCDs, needing multiple nonstandard supply voltages and high power consumption. With CMOS technology this is much simpler and less power demanding. The current micro-electronic lithography is at the point where CMOS can be produced with much better uniformity than twenty years ago.

One of the very popular CMOS designs was (and still is) a sensor with pixel controlled exposure which improves the apparent dynamic range. In CCD sensors, because of how images are read, light sensing circuitry (electronic exposure) can only work on a full TV field area by averaging the illumination values (electrons generated by light). With the CMOS sensor, because of the different design where pixels can be read out individually, the light sensing area can be localized to the pixel level area. If light is sufficient in an area

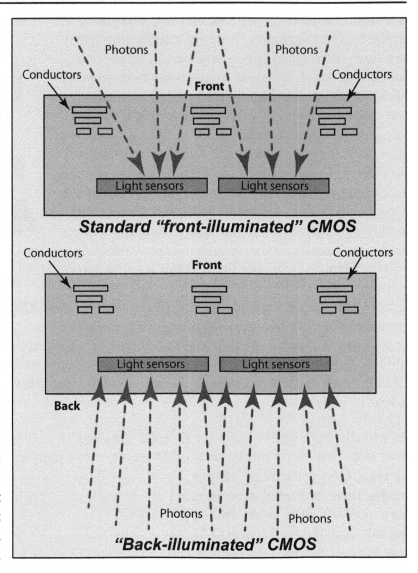

Standard "front-illuminated" CMOS

"Back-illuminated" CMOS

the electronic exposure control could shut that pixel section while allowing darker areas pixels to still be opened to exposure. This concept, only possible with CMOS, produces dynamic range that is practically impossible on a single chip.

Another innovation made possible with the CMOS design is the deduction of the inverse noise from the dark areas, which reduces the apparent thermal and fixed-pattern noise. Although this is applied mostly to DSLRs and not CCTV cameras at this stage, the concept is very interesting and ingenious. An image taken with low light areas will have a noise pattern which is statistically similar if it taken close in time (nearly the same temperature with the same chip). So a camera takes a dark exposure immediately after a real shot is taken and deducts the noise pattern from

Back-illuminated CMOS is used on the iPhone 4s

A single-chip color CMOS camera

the dark areas, producing a lower level of noise. This is how high some end DSLRs reduce their noise and, as a consequence, increase a camera's low-light capability.

The CMOS sensors have increased their presence, especially in today's smart phones. The first and most important reason is certainly their low power consumption compared to CCDs. Slowly but surely their quality, driven by the mass market production, is improving, not only by increasing the number of pixels, but also by improving their low light performance and dynamic range.

This is done by using the light sensing area from the back of the sensor (hence called ***back-illuminated***), as illustrated on the previous page. Since CMOS sensors have the semi-conductor components on their top side, which is the side that is typically exposed to light, this reduces the available light exposed area The back-illuminated CMOS sensors increases light-gathering ability without compromising on the pixel count.

Last, but not least, we should also mention the so-called "rolling shutter" of the CMOS sensor. This is an effect that shows moving or rotating object distorted in an unusual or unrealistic way. For example, if you record a movie with your smart phone while being driven in a car and record vechicles you are passing, they will appear slanted. Similarly, if you record a rotating fan blade or a plane propeller with a CMOS sensor, the resulting image is a bit unusual to say the least. This is a result from the different method a CMOS image is read. Because all of the sensors have direct access by the read-out electronics, the rolling shutter principle is basically reading out separate strips of the same image frame, rather than reading out the whole frame, which would be the case with CCD sensors. One should be aware of these artifacts, especially when recording high speed motion.

Rolling shutter effect

Spectral response of imaging sensors

The spectral sensitivity of imaging sensors varies slightly with various silicon substrates, but the general characteristic is a result of the photoelectric effect phenomenon: **longer wavelengths generate more electrons into the sensor's silicon structure**. How many electrons a packet of photons can generate in imaging electronics is called *Quantum efficiency* of the sensor. In layman's terms, this refers to higher sensitivity of the sensor to the red and infrared light. A typical and generalized imaging sensor spectral response curve, compared to human eye, is shown on the drawing below.

Even though this infrared "penetration" may seem beneficial (the imaging sensor may appear more sensitive), there are reasons for preventing these frequencies from getting inside the sensor silicon substrate. Namely, **a color camera needs to "see" visible light with similar spectral sensitivity as human eyes**. Only then can we make the camera produce colors as we see them. So, although the imaging sensors can

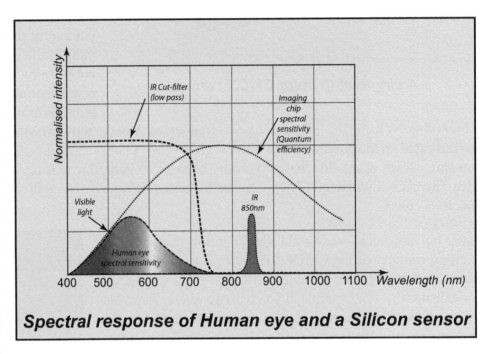

Spectral response of Human eye and a Silicon sensor

pick up infrared light and colors by their design, we need to remove such wavelengths in order to see colors. This is done by use of so-called *optical infrared cut (IRC) filter.* These filters are optically precise plan-parallel pieces of glass, mounted on top of the imaging sensor. As the name suggests, they behave as optical low-pass filters, where the cutting frequency is near 700 nm, that is, near the color red.

Color cameras *must* use an IRC filter if we are to see colors.

If the sensor needs to see infrared light, like the ones we often use in very low-light CCTV surveillance, we must remove the IRC filter. This is typically done by mechanically removing it from the top of the sensor. When IRC is removed, not only does the sensor "see" the infrared light, but it increases its low light performance response, making the camera appear more sensitive to low light. Of course, if IRC is removed color cannot be reproduced.

A rough rule of thumb is that a sensor without an IRC filter can produce usable video at much lower light levels, of up to 10 times of lower light levels (without the color information, of course). When such a sensor is exposed to infrared light, it can "see" not only the visible light, but also wavelengths that are invisible to human eye (beyond red wavelength). This is often used in surveillance for security purposes.

There are three reasons why a sensor with IRC filter removed "sees" better in low light: one is the removal of the IRC glass, which means more photons can pass through, second, it is the deeper penetration of the longer wavelengths (thus generating more electrons), and the last reason is the so-called Bayer color filter of each pixel, usually splitting into four segments for the RGB colors. However, when IRC filter is removed all of the sections (red, green, and blue pixel quadrants) are combined.

While we are at the infrared response of sensors, a few more things should be noted.

The first note is very simple: **color cameras without having the IRC filter removed cannot see infrared light**. Cameras stated to be Day/Night cameras would usually need to have a removable IRC filter. This function requires some sort of electromechanical mechanism, which inevitably undergoes some wear and tear. There have been some efforts to produce optical filters for both visible and infrared light so that they do not need to be removed to see both bands, i.e. the visible light and IR band-pass filter, around 850 nm (which is a typical LED infra red light used in CCTV). But, so far, there haven't been many cameras on the CCTV market with such a design, although attempts by some manufacturers have been made.

Single-chip color sensor with an IRC filter on top

Cameras with non-removable IRC filter that are claimed to be good low light performers, can only achieve this by going into so-called **integration** mode. This is basically a long exposure mode of the electronic shutter, where "long" in CCTV is considered anything longer than 1/25 sec. Clearly, when using such a mode, moving objects (like people) become blurrier when the exposure is long. But, some times, this is better than not being able to see anything in the dark.

The second note refers to the lens projection of the infrared wavelengths onto the chip. Not every lens can project infrared wavelengths on the sensor correctly. Because the infrared wavelengths go beyond the visible light wavelengths **their focusing point don't necessarily fall exactly on the same plane as the visible light**. This follows straight from the Fresnel optical law of refraction, which means longer wavelengths are bent less when they pass through the lens glass, and they will focus **behind** the sensor imaging plane. This means **the images will appear out of focus, unless the lens is infrared-corrected.** So, when designing a system with cameras intended to see infrared light, suitable **infrared corrected lenses** need to be used.

Photo courtesy of Tamron

Infrared corrected lens

The last, but not least, important note is that when cameras are used with infrared light the **illumination**

in luxes cannot be stated since infrared light is not part of the visible spectrum. The definition and **measuring of luxes can only be made with the visible spectrum sensors designed for that purpose.** Manufacturers quoting minimum illumination at 0 lux levels, thinking of infrared light, are stating contradicting facts.

Infrared light cannot be stated in luxes.

Courtesy of PDG Helicopters

High-voltage lines helicopter camera

We should also mention that there are special purpose built CCD or CMOS cameras with spectral sensitivity extended to ultra-violet portion of the spectrum, which is the opposite to the infra red portion. Such imaging sensors are specially developed to respond to wavelengths below the violet colors, referred to as ***ultra-violet***.

They are used in some special applications like high-voltage lines checking or machine vision production control.

Seeing colors

Color television is a very complex science in itself. The basic concept of producing colors in television is, as described earlier, generated by combining the three primary colors: red, green, and blue. The color mixing actually happens in our eyes when we view the monitor screen from a certain distance. The discrete color pixels (R, G, and B) are so small that we actually see a resultant color produced by the additive mixing of the three components. As mentioned earlier, this is called *additive mixing*, as opposed to *subtractive*, because by adding more colors we get more luminance, and with a correct mixture of the primary colors, white can be obtained.

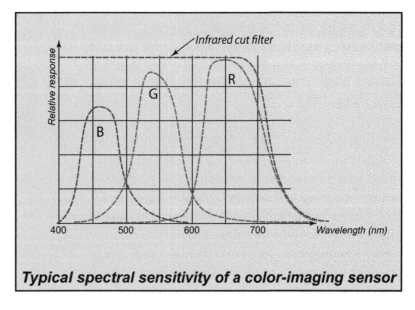

Typical spectral sensitivity of a color-imaging sensor

A color camera can produce RGB colors with two basic methods.

One is by using *optical split prism* and three separate sensors, one for each of the primary colors. The second method is using one sensor only and dividing each pixel to three primary *colors sub-pixels* (referred to as *Bayer mosaic filter*). Most broadcast color TV cameras are using the first method, which is of better quality, but more expensive, hence CCTV color cameras are typically using the single-chip color sensors.

The split-prism is a very expensive and precisely manufactured optical block with dichroic mirrors. Cameras using such technology are called *three-chip color cameras* and are not often used in CCTV. In fact, there are only few models in CCTV that split colors with this method. The main reason is certainly the cost, as it requires not only a split-prism, but three sensors instead of one, and more complex electronics. They do, however, offer a higher-resolution and superior color performance when compared to the *single-chip color cameras*.

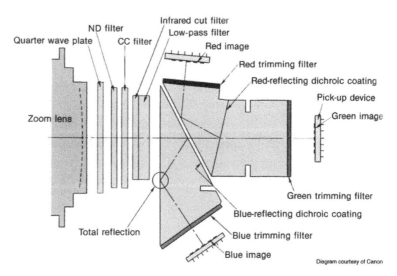

A split-prism block on a 3-CCD sensors camera

So, in CCTV, single-chip color cameras are the most common. There are two types of color filtering at the pixel level, either by using red, green, and blue colors (RGB), or by using cyan, magenta, yellow, and green (CyMgYeGr) components. Of the two, the first one is most common, so we will concentrate on explaining the color mosaic, also known as *Bayer mosaic filter*.

The illustration on the right demonstrates that each pixel is basically divided into four quadrants, where green color takes two diagonally opposite quadrants, the red and blue the remaining two. Now, clearly, the pixels and the subdivisions are not sensing different wavelengths of colors differently. In other words, is wrong to assume that there are red, green, and blue electrons, so to speak. These quadrants have

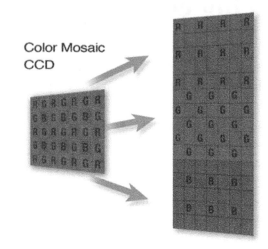

The Bayer mosaic filter

optical filters that pass only certain potion of wavelengths, where the primary colors are red, green, and blue. The filters spectral response is shown in the diagram "Typical spectral sensitivity of a color imaging sensor" on the previous page. The reason there are two green quadrants in each pixel is simply because the human eye is most sensitive to the green color, and the green color actually carries most of the luminance information in an image.

If the above concept is compared to the three-chip color cameras with split-prism it becomes obvious that the pixels of the latter one will have four times the pixel area, because these chips don't have divided pixels. This results in not only better low-light response and better dynamic range, but also better resolution, as there is no division of details as in the RGB mosaics.

It should be noted that the light-sensitive pixels are of the same silicon structure and are not different for different colors as some may think. To put it in layman's terms: there are no red or green or blue electrons. It is the CFA filter that splits the image into color components.

New developments are continually improving the CCD (and the new CMOS) imaging technology, and one of them is worth mentioning here. This is the multi-layered single-chip color developed by Foveon Inc. Instead of having pixels for each primary color separately they have invented a layering technique where colors are separated in depth, as they penetrate on the same pixels. The result is better color reproduction and higher resolution. This is perhaps similar to the three-chip split-prism color cameras, but without the split prism. Cameras with Foveon's X3 chip are already on the photographic market, and it will not be a surprise if similar cameras appear on CCTV cameras.

White balance

From color cameras we require, apart from the resolution and minimum illumination, a good and accurate color reproduction.

In the early days of color CCD cameras manufacturers used an external color sensor (usually installed on top of the camera) whose light measurement would influence the color processing of the camera. This was called *automatic white balance* **(AWB)**, but lacked precision owing to the discrepancy of the viewing angle between the white sensor and the camera lens. In modern cameras, we have a **through-the-lens** automatic white balance (TTL-AWB). Most of the AWB cameras today use this method.

Generally, the initial calibration of the camera is done by exposing the sensor to white on power up. This is used in cameras with AWB method. This is achieved by putting a white piece of paper in front of the camera and then turning the camera on. This stores correction factors in the camera's memory, which are then used to modify all other colors. In a way, this depends very much on the color temperature of the light source upon power up, in the area where the camera is mounted.

Many cameras have an AWB reset button that does not require camera powering down. How good, or sophisticated, these corrections are depends on the CCD chip itself and the white balance circuit design.

White balance setting usually can be automatic or manual.

Although the majority of cameras today have AWB, most models will have *manual white balance* (MWB) adjustments. In MWB cameras there are usually at least two settings (switch selectable): indoor and outdoor. Indoor is usually set for a light source with a color temperature of around 2800° K to 3200° K, while the outdoor is usually around 5600° K to 6500° K. These correspond to average indoor and outdoor light situations. Some simpler cameras, however, may have potentiometers accessible from the outside of the camera for continuous adjustment. Setting such a color balance might be tricky without a reference camera to look at the same scene. This gets especially complicated when a number of cameras are connected to a single switcher, quad, or multiplexer.

Newer design color cameras have *automatic tracking white balance* (ATW), which continually adjusts (tracks) the color balance as the camera's position or light changes. This is especially practical for PTZ cameras and/or areas where there is a mix of natural and artificial light. In a CCTV system where pan and tilt head assemblies are used, it is possible while panning for a camera to come across areas with different color temperature lights, like an indoor tungsten light at one extreme and an outdoor natural light at the other. ATW tracks the light source color temperature dynamically, that is, while the camera is panning. Thus, unless you are using ATW color cameras, you have to be very wary of the lighting conditions at the camera viewing area, not only the intensity but the color temperature as well.

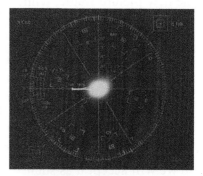

A Vectorscope showing good white balance

Last, as mentioned earlier, do not forget to take into account the monitor screen's color temperature. The majority of color CRTs are rated as 6500° K, but some of them might have higher (9300° K) or even lower (5600° K) color temperature.

Types of charge transfer in CCDs

CCD sensors, as used in CCTV, can be divided into three groups based on the techniques of charge transfer.

The very first design, dating from the early 1970s, is known as *frame transfer (FT)*. This type of CCD sensor is effectively divided into two areas with an equal size, one above the other, an imaging and a masked area.

The **imaging area is exposed to light for 1/50 s** for a PAL standard video (1/60 s for NTSC). Then, **during the vertical sync period, all photogenerated charges** (electronically representing the optical image that falls on the CCD sensor) **are shifted down to the masked area**. Basically, the whole "image frame" comes down.

Note the upside-down appearance of the projected image, since that is how it looks in a real situation; that is, the lens projects an inverted image and the bottom right-hand pixel is recreated in the top left-hand corner when displayed on a monitor.

For the duration of the next 1/50 s, the imaging area generates the electrons of the new picture frame, while the electron packets in the masked area are shifted out horizontally, line by line. The electron packets (current) from each pixel are put together in one signal and converted into voltage, creating a TV line information.

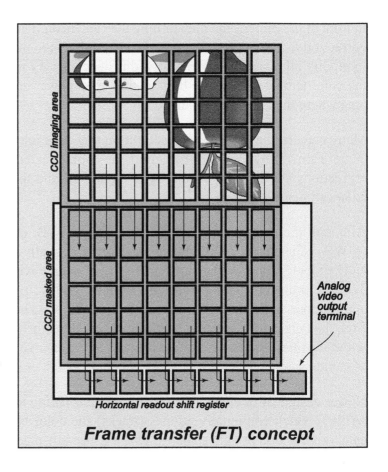

Frame transfer (FT) concept

Technically, perhaps, it would be more precise to call this mode of operation "field transfer" rather than "frame transfer," but the term *frame transfer* has been used since the early days of CCD development, so we will accept it as such.

The first design of the CCD sensor was good. It had surprisingly better sensitivity than Newvicon tubes and much better than Vidicon, but it came with a new problem that was unknown to tube cameras: *vertical smearing*. In the time between subsequent exposures when the charge transfer was active **nothing stopped the light from generating more electrons**. This is understandable since electronic cameras do not have a mechanical shutter mechanism as photographic or film cameras do. So where intense light areas were present in the image projection vertical bright stripes would appear.

To overcome this problem, design engineers have invented a new way of transference called *interline transfer (IT)*. The difference here is that the exposed picture is not transferred down during the vertical sync pulse period, but it is **shifted to the left masked area columns**. The imaging and masked columns are next to each other and interleave, hence the name, interline. Since the masked pixel columns are immediately to the right of the imaging pixel columns, the shifting is considerably faster; therefore, there is not much time for bright light to generate an unwanted signal, the smear.

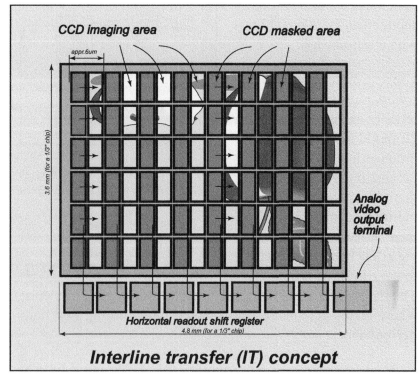

Interline transfer (IT) concept

To be more precise, the smear is still generated but in a considerably smaller amount. As a result, we also have a much better S/N ratio.

There is one drawback to the IT transfer chips, which is obvious from the concept itself: in order to add the masked columns next to the imaging columns on the same area as the previous FT design, the size of the light-sensitive pixels had to be reduced. This reduces the sensitivity of the chip. Compared to the benefits gained, however, this drawback is of little significance.

Another interesting benefit of IT design is the **possibility of implementing an electronic shutter in the CCD design**. This is an especially important feature, where the natural exposure time of 20 ms (1/50 s) for PAL, 17 ms (1/60 s) for NTSC, can be electronically controlled and reduced to whatever shutter speed is necessary, still producing a 1 V_{pp} video signal. This is perhaps one of the most important benefits from the interline transfer design, which is why it is widely used in CCTV.

Initially, with the IT chip, manual control of the electronic-shutter (exposure) was offered, but almost all cameras these days have *automatic electronic shutter*. This is known as automatic *CCD-iris*, or *electronic iris*. The electronic iris replaces the need for AI controlled lenses. So an MI lens can be used with an electronic iris camera even in an outdoor installation. It should be noted, however, that an **electronic iris cannot substitute the optical depth of field function** produced by the lens iris. Also, it should be remembered that when the electronic iris switches to higher shutter speeds noise and the smear increases due to lower charge transfer efficiency.

So when the electronic iris is enabled it will automatically switch from a normal exposure time of 20 ms (17 ms) to a shorter one, depending upon the light situation. This should not be mistaken with the increase of the frame rate. The number of images produced each second are still 25 (or 30) but the duration of each frame exposure becomes shorter than the 20 ms (17 ms) with electronic iris.

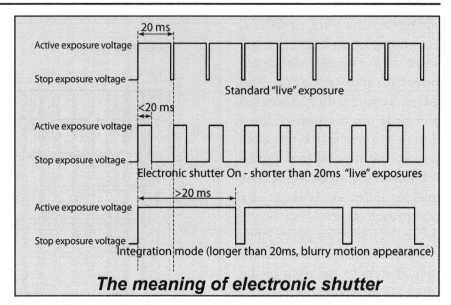

The meaning of electronic shutter

Longer exposure, called ***integration***, is also possible. Theoretically, exposures longer than 20 ms (17 ms for NTSC) don't produce live motion, simply because if they become longer than the speed of movement motion becomes blurry. But, then, at least it is possible to see in low light without having to have infrared illuminators. This is especially helpful these days with megapixel cameras, where pixels get smaller and smaller, thus limiting low light performance.

Reducing the pixel size in the interline transfer design indirectly reduces the chip's minimum illumination performance. This problem has been addressed with a very simple concept (technologically not as easy, however) of putting ***microlenses*** on top of every pixel. Microlenses concentrate all of the light that falls on them to a smaller area, that is, actually the pixel itself, and effectively increase the minimum illumination performance.

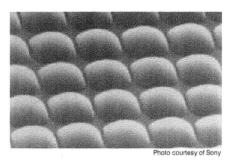

An electronic microscope photo of the on-chip micro lens structure

The most common types of CCD cameras in CCTV today have interline transfer sensors.

A typical cross section of an IT CCD chip with a micro lens on top of every pixel is shown on the drawing here. As you can see, the microstructure of the chip becomes quite complex when a high-quality signal needs to be produced.

The best CCD design so far is the so-called ***frame interline transfer (FIT)***, offering all the features of the interline transfer plus even less smear and a better S/N ratio.

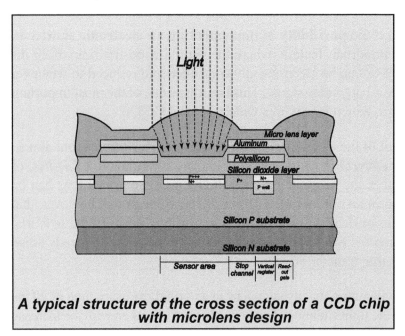

A typical structure of the cross section of a CCD chip with microlens design

As can be concluded from the simplified drawing, the FIT CCD works as an interline transfer in the top part of the chip, thus having the electronic iris control, but instead of holding the image in the masked columns for the duration of the next field exposure, it is shifted down to the better protected masked area.

With such a design, there is even less smearing, and in addition, there is also gain in the S/N ratio. Microlenses are also used here to increase the minimum illumination performance.

FIT CCD sensors have an even further advanced micro structure, with a lot of cells and areas designed to prevent spills of excessive charges to the area around, trap the thermally generated electrons, and so on.

With all these fine tune-ups, FIT CD sensors have a very high dynamic range, low smear, and high S/N ratio, which makes them

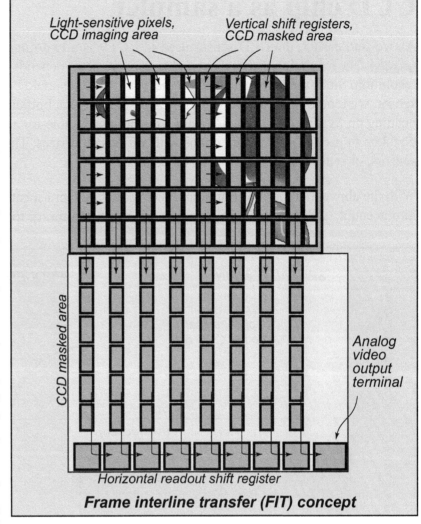

Frame interline transfer (FIT) concept

ideal for external camera shooting and news gathering in broadcast TV. These types of cameras, in broadcast TV, are usually referred to as ***electronic news gathering*** (ENG) cameras. These chips are relatively expensive for CCTV, so their main use is in broadcast TV.

In the end, we should point out that no matter how good the camera electronics is, if the source of information – the CCD sensor – is of an inferior quality, the camera will be inferior too. The opposite statement is also true. That is, even if the CCD sensor is of the best quality, if the camera electronics cannot process it in the best possible way the total package becomes second class.

It should also be noted that most of the handful of sensor manufacturers have CCD products of the same type divided into a few different categories, depending on the pixel quality and the amount of possible dead pixels. Different camera manufacturers may use different categories of the same sensor type. This is, in the end, reflected not only in the quality but also in the price of the camera.

CCD chip as a sampler

As we said earlier, the CCD sensor used in CCTV is a two-dimensional matrix of picture elements (*pixels*). The resolution that such a pixel-matrix produces depends on the number of pixels and the lens resolution. Since the latter is usually higher (or at least it should be) than the resolution of the imaging sensor, we tend to not consider the optical resolution as a bottleneck. However, as mentioned in the heading on MTF, lenses are made with a resolution suitable for a certain image size, and care should be taken to use the appropriate optics with various chip sizes. This is especially critical with the HD sensors, or sensors of higher megapixel count.

With the above in mind, it is important to consider another aspect of the sensor resolution to be taken into account, and this is the TV line non-continuity. Contrary to the "old" tube technology where a

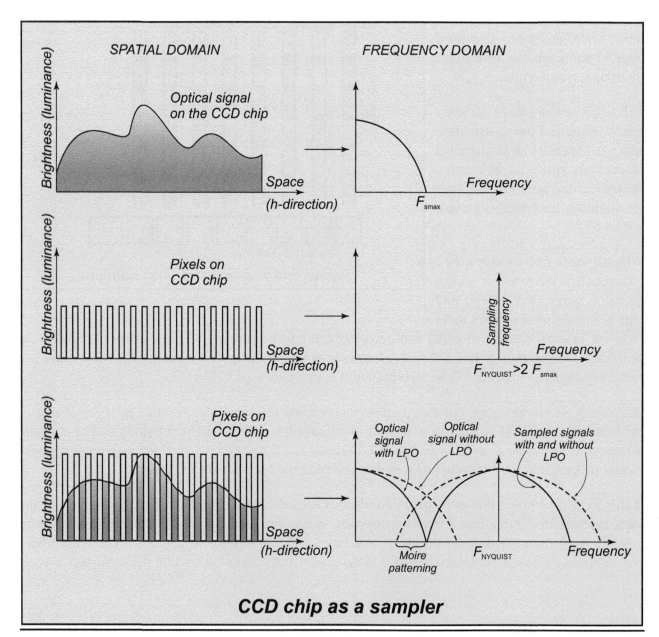

CCD chip as a sampler

line was produced by a **continuous** beam scanning along the line, CCD and CMOS sensors have **discrete** pixels and therefore the information contained in one TV line is composed of **discrete** values from each pixel. These discrete values do not represent digital information (as some may think) but rather discrete analog samples of the lens projected image. **In a way, the image sensors are optical samplers**. As with other samplers (like in music), we do not get the total information of each line, but only discrete values at positions equivalent to the pixel positions.

To some, it may seem impossible to reproduce a continuous signal from only portions of the same. In 1928, however, Nyquist showed that **a signal can be reconstructed perfectly, without any loss of information, if the sampling frequency is at least twice the bandwidth of the signal**.

Samples of the signal in between the sampled points are not necessary.

This is a great theory, proven correct and used in many electronic samplers such as in CD-audio and video. The sampling frequency, which is equivalent to two times the bandwidth, is called the *Nyquist frequency*.

There is, however, an unwanted by-product of the image sensor "periodically regular" sampling. This is the well-known *Moiré pattern* that occurs when taking shots of higher-resolution objects.

This is usually obvious with, for example, a news reader wearing a coat or shirt with a very fine pattern. This can mathematically be described as a fold-over frequency around the sampling one. Since the spatial sampling frequency should be twice the highest frequency in the optical image F_{smax}, we can represent it, in the frequency domain, with a single frequency located at the Nyquist frequency $F_{NYQUIST}$. The basic bandwidth spatial spectrum of the optical signal will be modulated around this frequency, very similar to an amplitude modulation side bands spectrum. If a high spatial frequency exists in the optical image projected on the CCD chip, and if this frequency is higher than half of the $F_{NYQUIST}$ frequency, the side bands (after the sampling is done) will fold over into the visible basic bandwidth and we will see the result as an unwanted pattern, known as the Moiré pattern.

The Moiré frequency is lower than the highest frequency of the camera ($F_{NYQUIST}/2 - F_{smax}$).

To minimize this unwanted effect, *low-pass optical (LPO) filtering* has to be done. These filters are usually part of the CCD chip glass mask and are formed by combining several birefringent quartz plates. The effect is similar to blurring the fine details of an optical image.

Some modern DLSRs remove the LPO filters in order to produce sharper images in appearance, but the side-effect of this is a more prominent Moire patterning.

Correlated double sampling (CDS)

The noise in an imaging sensor chip has several sources. The most significant is the thermally generated noise, but a considerable amount can be generated by the impurities of the semiconductors and the quality of manufacture. High noise reduces the image sensor's dynamic range, which in turn degrades image quality.

A careful CCD device design and fabrication can minimize the noise, although it cannot eliminate it completely. Also, low operating temperature can reduce thermally generated noise. Unfortunately, in practical CCTV systems used in surveillance, the user rarely has control over these parameters.

There is, however, a signal processing technique that can be implemented in the design of the CCD camera that reduces this noise considerably. This technique is called ***correlated double sampling (CDS).*** The term ***sampling*** here refers to the sensor signal output sampling.

The concept of CDS is based on the fact that the same noise component is present in the valid video as in the reference signal in between charge transfers. When the output stage of the CCD chip transfers packets of electrons, they are converted to an output voltage. To do this, CCD devices typically use a floating sensing diffusion to collect signal electrons as they are shifted out of the chip. As the electrons are transferred out of the CCD, the voltage on the sensing diffusion area drops. This voltage represents the valid data and is amplified on-chip by a thermally compensated amplifier. Before the next packet of signal electrons can be transferred into the diffusion area, it must be cleared from the previous packet. This represents a reference reset signal that has a thermal noise component of the same type. By extracting these two values, a less noisy signal is obtained.

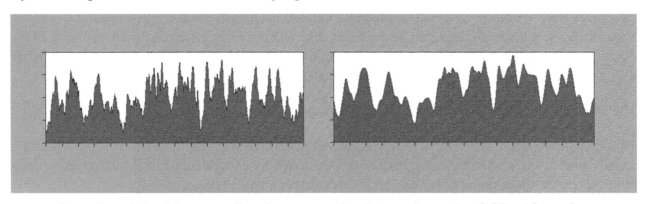

Correlated double sampling is one method to reduce the CCD chip noise.

CDS is best accomplished using two high-speed sample and hold circuits connected to the image sensor's output signal through a low-pass filter.

We will not go into more details on how these circuits are designed, as it is beyond the scope of this book, but it should be remembered that the CDS circuits are part of the camera electronics and not of the imaging sensor.

This is a good point in time to highlight the importance of good electronics surrounding the imaging sensor (CCD in this case). There are many low-cost cameras on the market using known manufacturers'

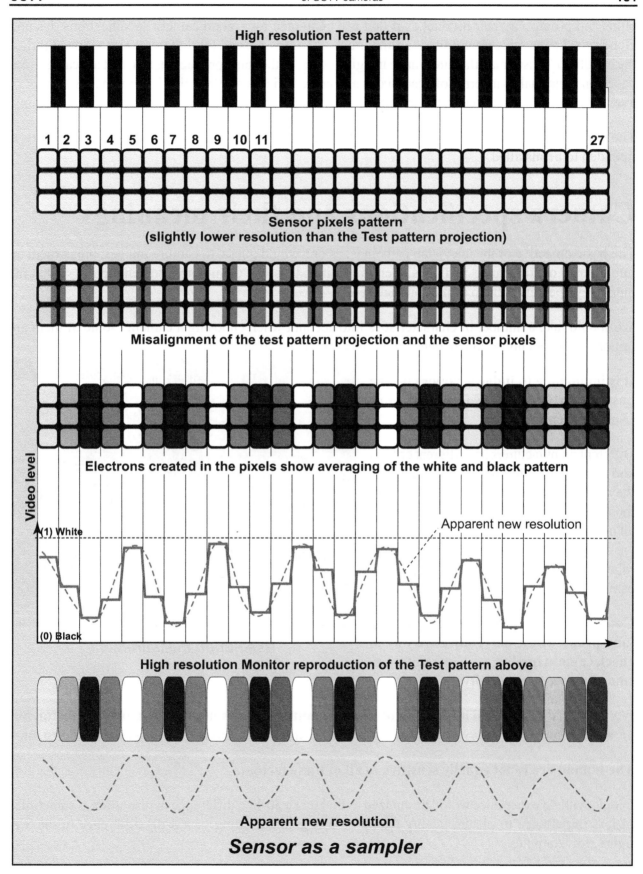

High resolution Test pattern

Sensor pixels pattern
(slightly lower resolution than the Test pattern projection)

Misalignment of the test pattern projection and the sensor pixels

Electrons created in the pixels show averaging of the white and black pattern

Apparent new resolution

High resolution Monitor reproduction of the Test pattern above

Apparent new resolution

Sensor as a sampler

chips and promoting their product as if that is a guarantee for a good camera specifications. It's not. Good signal processing electronics is as important as the sensor itself. Furthermore, as mentioned earlier, the handful of sensor manufacturers have imaging chips, of the same production run, that are catgeorized based on their quality in terms of dead pixels and noise variations. When this is combined with poor or average processing electronics, the end result could be much more inferior than what one might think.

The best way, in fact, the only way, to find out about a camera is to test it in the environment that it is intended to be installed.

Camera specifications and their meanings

The basic objective of the television camera is, as one would guess, to capture images, break them up into a series of still frames and lines, then transmit and display them onto a screen rapidly so that the human eye perceives them as motion pictures.

We should take a number of characteristics into account when choosing a camera. Some of them are important, others not so much, depending on the intended application.

It is impossible to judge a camera on the basis of only one or two specifications from a brochure.

Different manufacturers use different criteria and evaluation methods, and sometimes they measure specifications differently. In most cases, even if we know how to interpret all of the numbers from a specification sheet, we still have to evaluate the picture itself, relative to the picture taken with another camera.

Comparison tests are quite often the best (and probably the only) objective way to check camera quality, such as resolution, smear, noise, or sensitivity.

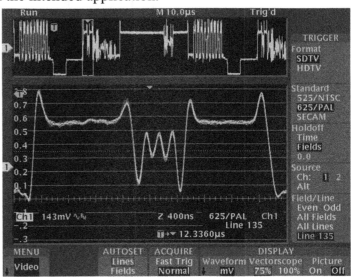

Resolution measurement

It is a fact that **the general impression of a good quality picture is not based on just one factor, but on a combination of many attributes – resolution, smear, sensitivity, noise, gamma, and so on.**

The human eye is not equally sensitive to all of these factors.

People with no experience would be amazed to find that a 50-line difference in resolution is sometimes of less importance to picture quality than a correct gamma setting or a 3-dB difference in the S/N figure, for example.

We will go through some of the most important features:

- Camera sensitivity

- Minimum illumination

- Camera resolution

- S/N ratio

- Dynamic range

Other, less important, but not wholly insignificant, features include gamma settings, dark current, spectral response, optical low-pass filtering, AGC range in dB, power consumption, and physical size.

Camera Sensitivity

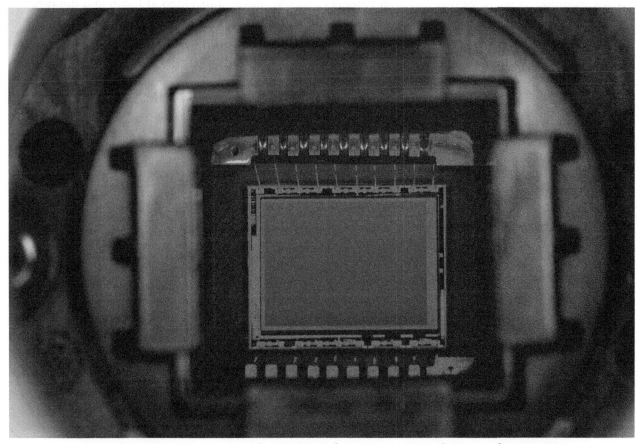

Imaging sensors are only a part of camera complete performance

The *sensitivity* of a camera, though clearly defined in broadcast TV, is quite often misunderstood in CCTV and is usually confused with minimum illumination.

Sensitivity is represented by the minimum iris opening (maximum F-stop) that produces a *full* 1V$_{pp}$ (1 V peak-to-peak) video signal of a gray scale test chart, when that same test chart is illuminated with 2000 lx from a 3200° K color temperature of the source.

The test chart used to check camera's sensitivity has to have a gray scale with tones from black to white, and an overall reflection coefficient of 90% for the white portion of the gray scale. One of the standard test charts used for such purposes is the EIA gray scale chart shown below in the middle. The white peak level needs to be 700 mV and the pedestal level around 20 mV. Gamma also plays a role in the proper reproduction of the grays and needs to be known that because of the non-linear human vision of luminance, typically a gamma of 0.45 is applied to the linear response (gamma=1) of the imaging sensor, which is then inverted by a typical display gamma of 2.2. In order to establish the sensitivity of the camera, a manual iris lens, usually of 16 to 50 mm (depending on the sensor size), is required. In order to get a realistic measurement, the camera's AGC should be switched off.

When all of the above is done, the manual iris lens is closed just until the white peak level starts droping from 700 mV, relative to the blanking level. The value obtained from the lens's iris setting, like F-4 or F-5.6, represents the camera sensitivity. The higher the number is, the more sensitive the camera is. So, for example, if we have two different cameras and one shows full video at F-5.6 and the other at F-8, we would say the latter camera (with F-8) has better sensitivity. This is because the F-8 transmits less light than F-5.6, so that the camera (sensor) sensitivity is higher to be able to produce full video with less illumination.

It is important to consider using the same light source and gray scale chart when comparing different cameras.

The example above illustrates a camera sensitivity of F-5.6 for a gray scale test chart signal reproduced as full 1 V$_{pp}$ video.

Minimum illumination

A camera's ***minimum illumination***, contrary to the sensitivity, is not clearly defined in CCTV. It usually refers to **the lowest possible light at the object at which a chosen camera gives a recognizable video signal**. It is therefore expressed with luxes at the object, at which such a signal is obtained. The term ***recognizable*** is very loosely used, and depending on the manufacturer, it may or may not be defined correctly. This represents one of the biggest loopholes in CCTV. Most manufacturers do not specify

what video level we should get at the camera output for the light amount specified as the minimum illumination. This level could be 30% (of the 700 mV), sometimes 50%, and, for some, even 10% might be acceptable.

The usual wording when describing minimum illumination would be, for example: "0.1 lx at the object with 80% reflectivity using an F-1.4 lens."

Have in mind, however, that with high AGC circuitry in the camera, even 10% of video (70 mV) could be pumped up to appear as a much higher value than what it really is. This could obviously be misleading as electronically amplified signal does not represent

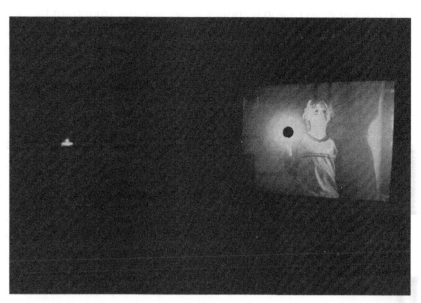

The example above shows a boy with a candle in his hand on the left-hand side, hardly noticeable by the film camera, while the CCTV camera sees it quite nicely as shown on the monitor right.

the real minimum illumination response of the sensor. When AGC is applied, it also amplifies the noise.

Let us say, for example, we have specifications that says 0.01 lx at the object with an F-1.4 lens, which presumes (but does not tell you) the AGC is switched to on. Another manufacturer may be very modest (or more realistic) in its specs, stating, for example, minimum illumination (where 50% of the video signal is obtained with the AGC set to off) of 0.1 lx with an F-1.4 lens. On paper, the first

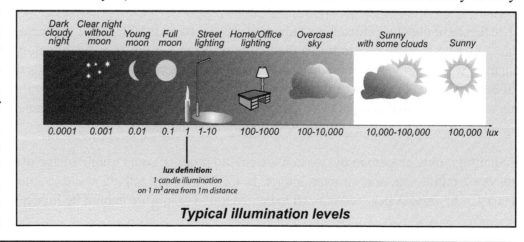

Typical illumination levels

case would seem to be a much more promising camera than the second one, although the second, in reality, would be much better.

Another matter for discussion is when some manufacturers state the minimum illumination at the object, while others may refer to the minimum illumination at the CCD chip itself. This is not the same; in fact, there is a big difference.

When the minimum illumination of a camera (with the illumination at the object) is stated, we should also read to what F-stop it applies. Also, another important factor to know is the reflectivity percentage of the object when the illumination is stated. If the minimum illumination is stated at the imaging sensor instead of at the object (which is very rare in CCTV), then not all the factors (such as reflectance and lens transmittance) have been taken into account. So, we have to compensate for all those factors when calculating the equivalent object illumination that is projected onto such sensor.

There is a rule of thumb (which I have elaborated on in the "Light onto an imaging device" section) that, with an F-1.4 lens, the minimum illumination at the chip is usually 10 times higher (represented by lower lux number) than the sensitivity at the object. For example, an illumination of 1 lx at the object with a reflectivity of 75% @ F-1.4 is equal to 0.1 lx illumination at the imaging sensor.

As can be concluded from the above, the real characteristics of a camera can be obscured quite easily by simply not stating all of the factors. Read the specs carefully or request better statements from manufacturers.

As mentioned and explained earlier, under "Spectral response of imaging sensors," cameras with infrared cut filter (IRC) removed would exhibit better lower light response.

It is true to say that the majority of color cameras these days, even without removing the infrared cut filter, have better low light response than the human eye.

You should also not believe manufacturers who specify their cameras minimum illumination performance with a few zeros after the full stop, like, for example, 0.000001 lux (3, 4 or 5 zeros after the point). Firstly, there is no instrument that can measure such a low illumination levels to verify such a statement. Secondly, any illumination levels below 0.01 lux are equivalent to night-time with only stars illumination, without moonlight. Lower than this is total darkness, with absolutely no light.

The theory of quantum efficiency of imaging sensors states that at 0.0001 lux there is hardly one photon-electron that may be penetrating a pixel area of 6 um x 6 um (a typical analog camera sensor pixel size). If we consider that under the room temperature there are already hundreds of electrons generated in each pixel (and even more at higher temperatures) there is no theory that will allow for that photon representing a signal at night to be detected or separated from the thermally generated electrons. Such specifications in CCTV are simply not supported by the theory.

In photography, or astronomy, we may be able to produce some usable image using long exposure, of maybe a minute, an hour or even longer. In CCTV, we are talking about usable live video, where at least 10~15 images per second need to be generated (exposure cannot be longer than 100 ms).

Camera resolution

Camera resolution is very simple but quite often misunderstood. It is also one of the most frequently quoted parameters of a camera or complete system. When talking about resolution of a complete system (camera-transmission-recording-monitor), the most important part is the input (i.e., the camera resolution). There are vertical and horizontal resolution, and they are measured using a test chart.

Vertical resolution is the maximum number of horizontal lines that a camera is capable of resolving. This number is limited to 625 horizontal lines by the CCIR/PAL standard, and to 525 by the EIA/NTSC recommendations. The real vertical resolution (in both cases), however, is less than these numbers. If we consider the active lines, the vertical sync pulses, equalization lines, and so on, the maximum for vertical resolution will be 576 lines for CCIR/PAL and 480 for EIA/NTSC. **This is the theoretical maximum for analog cameras.** In reality, this needs to be further corrected by the Kell factor of 0.7, to get the maximum realistic **vertical resolution** of 400 TV lines for CCIR/PAL (see "Resolution" in the "General characteristics of television systems" Chapter for a more in-depth study), similar deduction can be applied to the EIA/NTSC signal, where the maximum realistic vertical resolution is 330 TV lines.

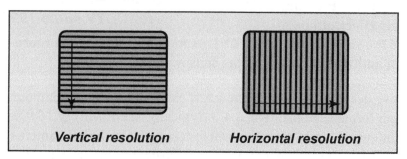

Vertical resolution   **Horizontal resolution**

Horizontal resolution is the maximum number of vertical lines that a camera is capable of resolving. This number is limited only by the sensor technology and the monitor quality. PAL analog cameras typically have a maximum (theoretical) horizontal resolution of 576 TV lines. This is because we are dealing with 576 active lines, and the aspect ratio of analog is 4:3, which yields 768 horizontal pixels, because 576 is 3/4 of 768. The above fits a camera with sensor that has "square pixels," and this is the case with most analog sensors. So, **the maximum horizontal resolution of analog CCD/CMOS cameras is usually 75% (4/3) of the number of horizontal pixels on the sensor and it is expressed in TV lines**. Clearly, if a lens is not of matching quality, and when the Kell factor is applied, this theoretical maximum is going to be lower. When counting vertical lines in order to determine horizontal resolution, we count only the horizontal width equivalent to the vertical height of the monitor. The idea behind this is to have ***square pixels***. So, when we describe analog camera resolution, **we always refer to horizontal resolution as TV lines (TVL) and not just lines**.

There is an exception to the above rule, and that is for imaging sensors that may not necessarily have square pixels. Such sensors have appeared a couple of years before the publishing of this book, mostly by the main sensor manufacturers such as Sony and Panasonic. Namely, they have produced sensors with over 970 pixels in horizontal direction and 576 in vertical (when PAL is used) or 480 (when NTSC is used). With such sensors it is possible to achieve higher horizontal resolution where pixels are no longer square, but rather rectangular (squashed in the horizontal direction). The theoretical horizontal resolution limit would be, again, 4/3 of the 976, which yields theoretical maximum horizontal resolution of around

730 TVL. In reality, manufacturers quote 650 TVL (and some more), which is still outstanding. This of course, will depend heavily on a good quality lens and good illumination. **Resolution should never be measured in low light as it will always show much lesser values due to noise masking the fine details.**

With the appearance of digital HD and Megapixel cameras, the concept of resolution changes slightly. We no longer have only 4:3 aspect ratio sensors, and we are no longer limited to the number of scanning line of the analogue system (576 for PAL and 480 for NTSC). So, in effect, although

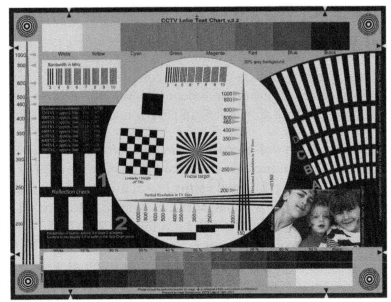

The CCTV Labs test chart was specially designed for CCTV since 1995.

it is possible to use the same definition of TVL as in analog video, it is more practical to simply state the horizontal and vertical pixels of the sensor view.

For this reason, I have designed a complete new test chart, first of it's kind in our industry, which can be used to check both formats, be that with a 4:3 (typically SD or MP) or 16:9 aspect ratio (typically HD). The only additional thing to consider when dealing with digital cameras is the compression artefacts. So, if stated that a sensor/camera is a 1080p high definition, for example, it doesn't necessarily mean that the resolution measured will be exactly 1920 x 1080 pixels. In fact, **it is always going to be lower** because there will be optical limitations if the lens is not adequate for such a sensor and, **more importantly, there will be compression artifacts which will reduce resolution appearance** (that is,

when IP cameras with compression are used, which will be most of the time). More details on what and how can be measured with this test chart will follow in the chapter near the end of this book.

If testing analog cameras with either the old test chart (which I created with the first edition of this book back in 1995) or the latest one mentioned above, the important thing to observe when measuring the resolution, is that the analog video signal must be properly terminated with 75 Ω and the image must be seen exactly to the tips of the black triangle When using analog CRT monitors, a high-resolution monitor

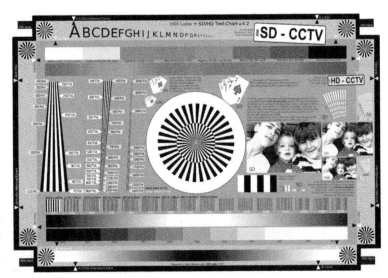

The new ViDi Labs test chart is a multi-format chart designed for both analog and digital CCTV.

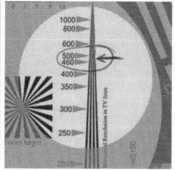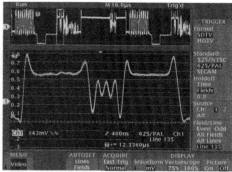

Visual detection of a horizontal resolution (in the middle) is not as precise as when using a proper oscilloscope with line selection and measuring 5% modulation.

(higher resolution than the camera being tested) with an under-scanning feature needs to be used. The camera is then set to the best focus possible (usually at a middle F-stop, 5.6 or 8), with the test chart fully in the field of view. Also, all internal camera correcting circuits (AGC, gamma, electronic-iris) need to be switched off.

Once a camera test chart setup is made like that, **the resolution can be visually checked by measuring where the resolution lines merge into each other**. For example, if the test chart shows four lines as in the example above, the point where these merge into three, or two, is the resolution limit. The ViDi Labs test chart introduced five converging lines, and in this case the resolution limit is the point when these five lines converge into four or three.

In broadcast TV, resolution is measured with studio lighting of 2000 lux and color temperature of 3100 K. In CCTV, there is no strict definition of the illumination required for testing resolution, so that less than 2000 lux would be acceptable, but common sense should be used, and resolution should not be measured in low light levels. The reason is very simple, noise levels at low illumination would reduce the resolution appearance of the camera.

For more accurate measurement of analog cameras, only the luminance signal should be analyzed, typically by turning the color completely down, or even better using Y/C connection on the analog camera (if such a camera is being tested). Since the merging of these lines does not have a clean cut, it represents only an approximate conclusion. The visual error of reading might be around 10%, which makes it very difficult to observe a difference between cameras with close resolution, for example 460 TVL and 480 TVL. Certainly, the new digital HD and MP cameras would not have an option to use Y/C output, as there is no composite analog video out of a digital signal.

For a more precise reading, a high-quality oscilloscope, with TV line selection feature, should be used. The measurement is then narrowed down to selecting **a line where the four lines modulation**

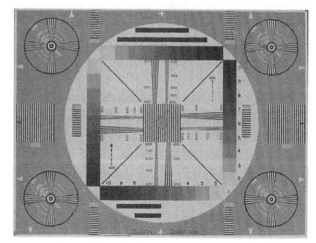

The (old) RETMA SD test chart

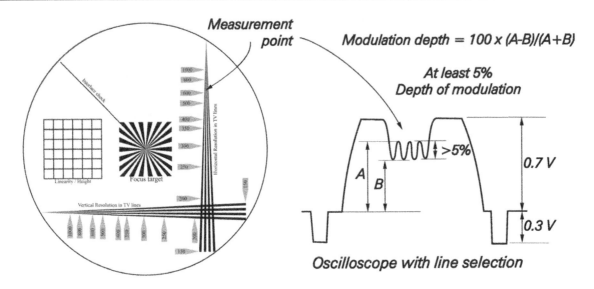

$$Modulation\ depth = 100 \times (A-B)/(A+B)$$

At least 5%
Depth of modulation

Oscilloscope with line selection

More accurate horizontal resolution measurement with 5% modulation depth

depth is equal to or better than 5%. How this is calculated is shown on the drawing above and is basically $100 \times (A - B)/(A + B)$, where A is the highest point and B the lowest of the measured lines. Using an oscilloscope in such a case enables us to not be affected by the monitor's resolution limits. In order to know which section of the test chart is measured (going vertically over the merging lines, when horizontal resolution is measured), there should be a way of telling which line is measured on the test chart. For this purpose, line selecting oscilloscopes should be used, such as the one shown in the photo (which can switch to a picture display). It shows the measured point precisely, with a line marker.

It should be repeated that only good optics need be used when measuring resolution; otherwise average lenses tend to have better center resolution and the corners of an image (center is always better than the corners). With such lenses resolution measurements are best if made in the central area of the test. Another important note is that wide angle lenses are not the best for testing camera resolution, as they typically introduce optical barrel distortions, making it difficult to align the black arrow-triangles for precise measurement. For this reason, **narrower view prime lenses are recommended if more accurate camera resolution testing is to be made.**

Analog camera resolution is closely related to the signal bandwidth such a camera is capable of reproducing. Their correlation was explained in the earlier section on resolution, but **a simple rule of thumb is that 80 TVL (TV lines) are equal to 1 MHz of bandwidth.**

Practical experience shows that the

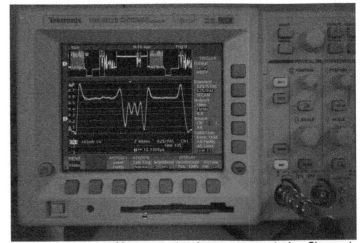

Measurement and screen capture by Les Simmonds
Highly recommended – the Tektronix TDS3012B

human eye can hardly distinguish a resolution difference of less than 50 lines. This is not to say that the resolution is not an important factor in determining camera quality, but small resolution differences are often hardly noticeable, especially if the resolution difference is smaller than 10% of the total number of lines.

Single-chip color cameras (as used in CCTV) have lower resolution than three-chip cameras used in broadcast, even if same sensor size is used. This is because of the way the single chip cameras process colors by the Bayer mosaic

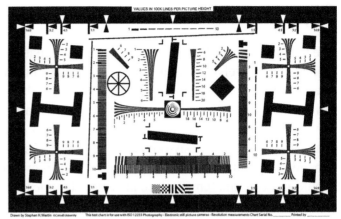
The (new) ISO-12233 HD test chart

filter. When analog signal is converted into digital other factors need to be taken into account, which we will explain in more details in the chapters discussing A/D digital conversion.

It should be noted that when measuring HD or MP cameras, the resolution is better expressed only as lines, rather than TV lines, or even more appropriate with just pixels. This allows to discuss and measure megapixel images of various aspect ratios and formats.

A number of test charts are available on the market for evaluating cameras resolution. The well-known one for standard definition (but now almost obsolete) was the EIA RETMA chart. For HD and MP cameras the ISO-12233 chart could be the choice. There are other charts which can be found on the Internet, but the newly designed ViDi Labs test chart should be used even more productively as it can measure not only the resolution, but also other important parameters for CCTV. Some of these, in addition to the resolution, include the face, playing cards and number-plate recognition, as well as compression artefacts. The ViDi Labs is the most complete chart for CCTV and it will keep evolving as the technologies and ideas develop. As with the previous editions of this book, we have enclosed a reproduction of this chart at the back of the book. This is the latest version (v.4.2), and with the evolution of versions over the years more measurement details are introduced.

For more accurate measurements we encourage the reader to obtain the larger format (A3+) from the *ViDi Labs* web site (*www.vidilabs.com*). Our book publisher would have taken maximum care to correctly reproduce the version available in this book, but exactitude is beyond our control. The procedure encompasses color inks and printing machinery that could not have been taken into account in our chart setup.

Measuring bandwidth is closely related to resolution

Signal/noise ratio (S/N)

The *signal to noise (S/N)* ratio is an expression that shows how good a camera signal can be, especially in lower light levels. **Noise cannot be avoided but only minimized. It depends mostly on the imaging sensor quality, the size of the sensor pixels, the electronics and the external electromagnetic influences, but also very much on the temperature.** The camera's metal enclosure (if, indeed, it is a metal one) offers significant protection from external electromagnetic influences. Internal noise sources include both passive and active components of the camera, their quality and circuit design; noise depends very much on the temperature. This is why, when stating the S/N ratio, a camera manufacturer should indicate the temperature at which this measurement is taken. If this is not done it is typically assumed to be room temperature.

The image noise is very similar to the noise in old audio tapes, only it is part of a video and not of an audio signal. On the screen, a noisy picture appears grainy or snowy, and if color signal is viewed, sparkles of colors may be noticeable. Extremely noisy signals may be difficult for equipment to synchronize these days, with the increased usage of digital video recorders. **More importantly, in digital CCTV, noisy pictures reduce the compression efficience and increase storage space used**, as most encoders see random noise as video detail.

The units for expressing ratios (including the S/N) are called *decibels* and are written as dB.

Decibels are only relative units. Instead of expressing the ratio as an absolute number, a logarithm is calculated. The reasoning behind this is simple: logarithms can show big ratios as only two- or three- digit numbers. Also, signal manipulation (as when calculating the attenuation of a medium or amplification of a system) is reduced to simple addition and subtraction. And perhaps one even more reason for using logarithmic decibels is due to the human senses for sound and vision. Namely, **the humans hear and see sound and light intensities (respectively) by obeying logarithmic laws.**

When a ratio of any two numbers with the same units is calculated, the units are expressed in dB only. If, however, a relative ratio is calculated – for example, a voltage level relative to 1 mV – the units are called **dBmV**. If the power value is shown relative to 1 μW, the units are called **dBμW**.

Seeing details in an image is affected not only by the resolution but by the noise

The general formula for voltage and current ratios is:

$$S/N = 20 \log(V_s/V_n) \tag{41}$$

where: V_s is the signal voltage and V_n is the noise voltage. Current values are used when a current ratio needs to be shown.

If a power ratio is the purpose of a comparison, the formula is a little bit different:

$$S/N = 10 \log(P_1/P_2) \tag{42}$$

We will not explain here why this is different (the factors 10 and 20 in front of the logarithm), but remember that it comes from the relation between the voltage, current, and power.

In CCTV we use decibels mostly for calculating voltage ratios, which means we would use the first formula (41).

The following table gives some dB values of voltage (current) and power ratios. Please note the difference between the two. While a 3 dB voltage difference means only a 41% higher value of the compared volts relative to the referred one, in terms of power this means 3 dB delivers twice as much power (100% increase) compared relative to the reference power.

dB	0	0.1	0.2	0.3	1	2	3	10	20	30	60
Voltage/current ratio	1	1.012	1.023	1.035	1.122	1.259	1.413	3.162	10	31.62	1000
Power ratio	1	1.023	1.047	1.072	1.259	1.585	1.995	10	100	1000	1,000,000

Decibels table

The S/N ratio of a CCTV camera video signal is measured differently from that of a transmitted signal.

In a broadcast TV signal the S/N ratio is the signal versus the accumulated noise from the transmission to the reception end. This is defined as the ratio (in dB) of the luminance bar amplitude to the RMS voltage of the superimposed random noise measured over a bandwidth of frequencies between 10 kHz and 5 MHz. There are special instruments that are designed to measure this value directly from the signal by using some of the video insertion test signal (VITS) lines.

The S/N ratio in a CCD camera is defined as the ratio between the signal and the noise produced by the sensor combined with the camera electronics. In order to get a realistic value for the S/N ratio of a camera, all internal circuits that modify the signal in one way or another need to be switched off or disabled. This includes gamma, AGC, electronic-iris, and backlight compensation circuitry. The temperature, as already mentioned, should be kept at room level. Of the few different methods used to measure camera video noise, the easiest one is to use a special instrument called a *video noise meter*. This unit selects the noise in the band between 100 kHz and 5 MHz and reads the S/N directly in decibels.

Practically, a S/N ratio of more than 48 dB is considered good for an analog CCTV camera.

A 3 dB higher S/N ratio means approximately 30% less noise, since the peak video level does not change. So, when comparing a 48 dB camera with a 51 dB camera, for example, the latter will show a considerably better picture, especially noticeable at lower light levels. It is assumed that the automatic gain control (AGC) is off when stating S/N ratios, otherwise noise gets amplified with the AGC and this will reduce the S/N ratio. For comparison purposes, let us just mention that broadcast cameras have a ratio of more than 56 dB, due to the fact they use separate sensors for each color, and also due to the better charge transfer.

The multi-functional EWM-40 Camera analyzer

Since the previous edition of the book, an instrument made by Elbex came on the CCTV market, called Camera Analyzer EWM-40. This instrument, unique in our industry, integrated a few instruments in one for measuring analog signals, such as oscilloscope, vectorscope, resolution meter, illumino-meter, spectrum analyzer, a test-pattern generator, and S/N meter.

Keeping any camera as cool as possible reduces the noise.

Lower temperatures, in any electronic device, produce less noise. This is especially important to consider when using modern HD and MP cameras where pixel size gets even smaller, thus increasing the apparent noise.

Photo courtesy of Spectral Instruments

Scientific astronomy cameras use cryogenic cooling to get to -60°C

In some industrial applications, especially astronomy, there are specially cooled cameras designed to keep the imaging sensor as cool as possible. This is important for capturing images of distant and faint stars and galaxies. Temperatures of below –60° C using cryogenic cooling are not uncommon. For such applications, cameras are available where the imaging block has provision for a coolant to be attached to it, as shown on the photo to the right. Some designs, like the one shown on the photo to the right here, uses Peltier cooling to keep the CCD chip always at 5° C, which reduces the noise to one-eighth of the normal room temperature.

In summary, in any CCTV system high temperature can play a significant role in lowering the picture quality, even more so if low quality cameras are used.

Photo courtesy of Cohu

A camera design with Peltier cooling operates at 5°C and reduces the noise by 85%.

Dynamic range of imaging sensors

Dynamic range (DR) is seldom mentioned in CCTV camera specification sheets. Nonetheless, it is a very important detail of the camera performance profile.

The dynamic range of an imaging sensor is defined as the maximum signal charge (saturation exposure) divided by the total RMS (root-mean-square) noise equivalent exposure. DR is similar to S/N ratio, but it only refers to the sensor dynamics when handling low to bright objects in one scene. While the S/N ratio refers to the complete signal including the camera electronics and is expressed in dB, the DR is a pure ratio number, but it can also be expressed in dB, i.e. as a logarithm of the ratio.

It is a known fact that **the dynamic range of a solid state sensor is pretty limited compared to human eye.** This refers to both CCD or CMOS sensors. **The dynamic range is limited by the pixel size (which is always getting smaller and smaller) and the noise levels (which can never be eliminated)**. There are always pixel imperfections, which are called *fixed pattern noise*, and, in addition, there are always thermally generated electrons in normal circumstances. This exist on any temperature above absolute zero (-273°C). A small pixel element can only contain so many electrons created by light photons. The DR actually shows the light range an imaging sensor can handle – only this light range is not expressed with the photometric units but with the generated electrical signal. It starts from the very low light levels, equal to the imaging sensor RMS noise, and goes up to the saturation levels. Since this is a normalized ratio of two values of electrons, it is a pure number, usually in the order of thousands. **The dynamic range of a good solid state imaging device at the time of writing this book is between 3,000~6,000:1**. Converted in engineering terminology, this is equivalent to 70~80 dB. This number is even smaller for sensors with smaller pixels. In comparison, **human eyes see anything from full moon-light levels (0.1 lux) up to full sunny beach (100,000 lux), and this is a ratio of 1,000,000 : 1, or, in engineering terminology, 120 dB.**

When saturation levels are reached during a sensor exposure (1/50 s for PAL, or 1/60 s for NTSC), the *blooming* effect may become apparent when excessive light saturates not only the picture elements (pixels) on which it falls but the adjacent ones as well. As a result, the camera reduces the resolution and detail information of the bright areas. To solve this problem, a special **anti-blooming** section is designed in most CCD chips. This section limits the amount of charges that can be collected in any pixel. When anti-blooming is designed properly, no pixel can accumulate more charges than what the shift registers can transfer. So, **even if the dynamic range of such a signal is limited, no details are lost in the bright areas of the image**. This may be extremely important in difficult lighting conditions such as looking at car headlights or looking at people in a hallway against light in the background.

On the left a camera with visible smear and on the right an almost invisible smear

There are some ingenious ways of increasing the apparent dynamic range even with limited DR by the sensor itself.

Some camera makers have introduced a special design that blocks the oversaturated areas during the digital signal processing stage. The AGC circuitry then does not see extremely bright areas as a white peak reference point so that much lower levels are taken as white peaks, thus making the details in the dark more recognizable.

Others are using new methods of imaging sensor operation where, instead of having one field exposure every field time, two exposures are done during this period. One at a very short

Courtesy of Axis

Wide Dynamic Range (WDR) effect: Top left short exposure, top right long and bottom combination of the two

time and the other at the normal time. Then, the two exposures are combined in one frame so that bright areas are exposed with short exposure duration giving details in the very bright areas, and the darker areas are exposed with the lower speed giving details in the dimmer part of the same picture. The overall effect is of the dynamic range of the camera being increased a number of times. Some manufacturers call this wide *"dynamic range" (WDR)*, others *"super-dynamic effect."*

Another interesting design was made possible with the introduction of the CMOS technology. Pixim made a sensor where electronic-iris was possible on a local pixel level in the sensor. When excessive light is projected in one area, the electronic-iris would shorten the exposure in that area only, leaving the rest of the pixels to be exposed at the longer electronic-iris times, allowing more details in the shadow areas of the picture. This was simply impossible with the CCD technology.

Photo courtesy of Pixim

Modern and sophisticated CMOS chip offers A/D conversion on the chip.

Special low-light-intensified cameras

Modern imaging sensors can see in as low light as human eyes can see. Certainly, this can be further improved by using extreme AGC (although this will increase the noise). Most important factor in how good a camera can see in low light, as already explained under the "Minimum illumination" heading, is the pixel size. But all of the above can be summarized and approximated as being able to see in as low light as human eyes.

The illumination levels visible to human eyes, from 100,000 (10^5) down to 0.1 (10^{-2})lux, as illustrated earlier, is called a *photopic vision* area.

Sometimes, for special purposes, there is a need for an even lower light level camera. The light range lower than 10^{-2} lx belongs to the *scotopic vision* area.

Although the human eye cannot see this low it is possible to get images from light levels much lower than 10^{-2} lx with the use of the integration function available on some cameras. This is a function where exposure time longer than 1/50 s (1/60 s for EIA) is used. Obviously, in such a case we lose the real-time effect and the camera actually becomes a kind of storage device. This might not be acceptable for viewing moving objects in low light levels, but it is a good alternative for viewing slow-moving objects in the dark. If we want to see real movement in the scotopic vision area, a special type of camera called *intensified,* or *low light level* (LLL), can be used.

An LLL camera

Intensified cameras have an additional element, called a *light intensifier*, that is usually installed between the lens and the camera. The light intensifier is basically a tube that converts the very low light, undetectable by the CCD chip, to a light level that can be seen by it. First, the lens projects the low light level image onto a special faceplate that acts as an electronic multiplying device, where literally every single photon of light information is amplified to a considerable signal size. The amplification is done by an *avalanche* effect of the electrons, which light photons produce when attracted to a high-voltage static field. The resultant electrons hit the phosphor coating at the end of the intensifier tube, causing the phosphor to glow, thus producing visible light (in the same manner as when an electron beam produces

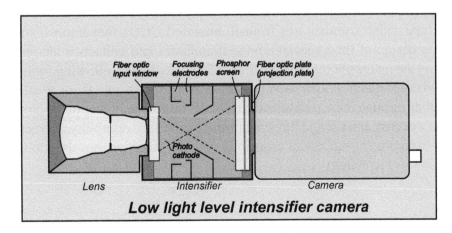

Low light level intensifier camera

light onto a B/W CRT). This now visible image is then projected onto the CCD chip, and that is how a very low light level object is seen by the camera. Because of the very specific infrared wavelengths of low light levels, as well as the monochrome phosphor coating of the intensifier, the LLL cameras will only display monochrome images.

It is to be expected that, having a phosphor coating inside the intensifier, the lifetime, or more correctly, the MTBF (*mean time between failure*) of an intensifier tube is short. It is usually in the vicinity of a couple of thousand hours.

Photo courtesy of e2Vtechnologies

A modern LLL camera

In order to prolong this lifetime, high F-stop lenses are necessary (with at least F-1200), especially if the camera is to be used day and night. Also, lenses with infrared light correction should be more adequate.

More advanced and purposely built LLL cameras have a fiber optic plate for coupling the phosphor screen of the intensifier tube to the CCD chip. This technique avoids any further light losses and improves picture sharpness.

Needless to say, the intensifier requires a power source in order to produce the high-voltage static field for the electrons' acceleration.

This type of intensifier can be bought separately and installed onto a camera, but specifically made integrated cameras have much better performance.

Another interesting and innovative design has been offered by PixelVision Inc. with its back-illuminated CCD camera that operates without an image intensifier. This camera, the manufacturer claims, is capable of acquiring quality images at low light levels previously attainable only with image intensifier tubes. Conventional video cameras use front-illuminated CCDs that impose some limitations on performance. The design of their special device illuminates and collects a charge through the back surface, permitting the image photons to enter the CCD unobstructed, allowing for high-efficiency light detection in the visible and ultraviolet wavelengths. The manufacturer claims greater resolution under low-light conditions through increased sensitivity, better target identification through superior contrast and resolution, lower cost, and a longer lifespan through increased reliability. As mentioned previously, this concept has been proven successful especially with the smart phones used today, whose imaging sensor is based on this technology.

Thermal imaging cameras

As discussed in the beginning of the imaging sensors chapter, the imaging sensors are very sensitive to the infrared spectrum, which is an unwanted phenomenon for a typical CCTV camera working with visible light. The infrared sensitivity, however, is very useful and promoted in thermal imaging cameras. These are cameras that have developed very fine resolution of determining temperature difference of the object with high precision, even lower than 0.1˚, and going as high as temperatures of 1200˚C. The highlighting of the temperature difference is the most important tool in the usage of the thermal imaging cameras. Typically hotter areas are shown in red, and colder in blue, with the temperature variations between these extremes changing through yellow and green.

Thermal imaging cameras can be used to detect intruders in total darkness for example, without having to use infrared illuminators. More importantly, thermal imaging cameras can see through night as through day, but also through fog and rain. Fire-figthers use thermal imaging cameras to see the possible source of fire. Also, detecting parts in an engine, for example, that heats up more than it should, can easily be picked by thermal imaging cameras. Electrical installation and fuse boxes can be checked for excessive current draw with thermal imaging cameras.

Thermal image of a house

The video output of thermal imaging cameras don't have to have very high resolution, and typical sensors would have 320 x 240 pixels, but 640 x 480 is not uncommon. The video output format can be in the form of analog PAL or NTSC, or digital stream as M-JPG, MPEG-4 or H-264. As such, thermal imaging signals can be recorded on any DVR as normal video signal.

Some manufacturers even put two cameras on one PT head, for example, one being a thermal imaging camera and the other a standard camera, so that when monitoring or surveying an area both types of video streams are available.

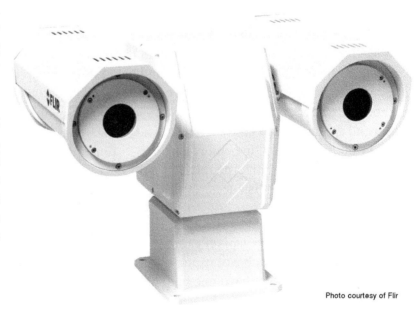

Photo courtesy of Flir

A typical thermal imaging camera(s) on a pan/tilt head

Multi-sensor panoramic camera

Just two years before writing this book, CCTV manufacturer Dallmeier produced a new type of camera, a multi-sensor ***Panomera***. This is a camera made up of multiple sensors, able to produce panoramic images of large areas with exceptional clarity. The most accurate description of this technology would be as a ***multi-sensor and multi-focal panoramic camera system with live views***.

The Panomera is, in essence, an array of megapixel cameras with optically and mechanically perfectly aligned views so that although they are separate sensors; they make a continuous panoramic view of a larger area.

There are some important differences between this concept and a typical single-sensor megapixel camera.

The first is that the Panomera can show incredible detail by digital zooming in, without pixelation. Also, because each sensor is using a very efficient H.264 video compression, Panomera is able to show live and playback images with minimum bandwidth requirements. The current maximum number of sensors is 17 sensors in one enclosure, making it effectively a 68 MP camera, since each sensor works in the 4 MP mode. The streaming bandwidth requirement is still less than 100 Mb/s of data traffic. The viewing experience is live motion in every section of the panoramic view, no matter how zoomed it is, and no matter if it is live or playback.

All camera sensors produce ***multicast*** streams, which means they are available to multiple operators simultaneously without increasing the streaming bandwidth or putting a load on the encoders. Each operator can concentrate on their own section of surveillance of the same panoramic view without losing or affecting the other operators.

The recording of the Panomera array of cameras are done on the Dallmeier recorders, with built-in RAID-5 or RAID-6 redundancy. Each of the camera sensors can be accessed individually if need be, but the actual seamless panoramic stitching is done on a powerful viewing station computer, which is then used to digitally pan and zoom around the panoramic view, both in live mode and playback. In actual fact, the playback after the event and being able to pan, tilt and zoom at any point is the most attractive feature to many surveillance operators. No need to worry about PTZ cameras being pointed and zoomed to a wrong location. There are various lenses with various focal lengths to suit different dis-

tances and applications. These are finely tuned in the factory, and the end result is the ability to recognize people or vehicles at very large distances, even exceeding 200 m.

Existing single sensor megapixel cameras, which might be capable of capturing 30 MP or even more, have one big limitation: they can only produce a very limited number of frames per second (1-2). Such a low rate might be sufficient

Panomera with 9 sensors

Multi-sensor Panomera is especially attractive for stadiums

for reading number-plates when triggered with a speeding radar, or perhaps a red-light loop, but it is insufficient for human action surveillance. In addition, even with such a low frame rate, the streaming bandwidth may exceed 100 Mb/s because only JPG can be used.

Installation costs of Panomera are very low when compared to installing individual cameras to cover the same area, because all it is required a power and a network connection. The netwok could be copper, fibre-optics, or even wireless.

Interestingly, Panomera is not limited to a standard 16:9 or 4:3 aspect ratio view, but it can have any combination of sensors to cover an area most efficiently. This could even be wide horizontal strip monitoring an airport tarmac, a stadium with thousands of spectators, or even a vertical strip monitoring tall buildings, elevators, opened lift shafts, or anything that is awkward to be seen with a horizontal 16:9 aspect ratio HD cameras.

Panomera viewing station

Camera power supplies and copper conductors

A typical CCD camera consumes between 3 and 4 W of energy. This means that a 12 V DC camera needs no more than 300 mA of current supply. A 24 V AC camera needs no more than 200 mA. As the technology improves, cameras will consume less current, although IP cameras that include power-hungry encoders as well as network circuitry may need a bit more.

When powering a number of cameras from a central power supply it is important to take the voltage drop into account and not to overload the supply.

Also, a very important factor to check with DC power supplies is whether or not they are regulated. For example, if a power supply of 12 V DC/2 A is used, it is advisable to have approximately 25% to 30% of spare capacity to minimize overheating. Be very critical when choosing a power supply and especially when you want to power multiple cameras from one source. When some manufacturers quote 12 V/2 A, the 2 A may only be a maximum rating. This is usually defined with short lengths of peak deliveries. In other words, you cannot count on a constant load of 2 A if the power supply is not designed for that. It really depends on the make and model. Very often, 12 V DC power supplies are actually made with a 13.8 V output used for charging batteries on security panels. Take this fact into account to minimize camera overheating, especially if there is only a short-run power cable between the camera and the power supply. Usually no intervention is required for a couple of hundred meters of power cable run because of the voltage drop, but if the camera is in the vicinity of the power supply, the excessive power must be dissipated somewhere, and this is usually in the camera itself. To put it simply, the 12 V DC camera gets hotter if it is powered from a 13.8 V rather than a 12 V power supply, and this influences the camera's S/N performance.

Unregulated DC power supplies (usually in the form of plug-packs) are not very healthy for CCTV cameras. First, there is a high probability of blowing the camera's fuse when the power is switched on, owing to voltage spikes created when turning the load on (the camera in this instance). Second, there is an extra power dissipation that occurs in the camera when more than 12 V DC are applied.

Finally, if the camera does not have any further voltage regulations inside (DC/DC conversion), or if the regulations are of a bad quality, the unregulated voltage ripples may get into the readout pulses, thus affecting the video signal.

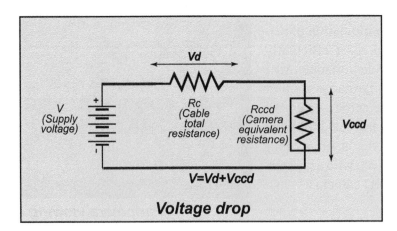

Voltage drop

On the other hand, in most of the good regulated power supplies there is a short-circuit protection. That means, even if the installer makes a mistake with the polarities or termination, the power supply will cut off the output, thus protecting the supply and the camera from further damage. Also, with regulated power supplies, the voltage can be adjusted to compensate for voltage drops.

This is not the case with unregulated supplies.

Voltage drop has to be taken into account when powering distant cameras. This is especially critical with 12 V DC cameras since the voltage drop at lower DC voltages is more evident. This is a result of the $P = V \cdot I$ formula, where for a certain camera power consumption level, the lower the voltage is, the higher the current will be, indirectly increasing the voltage drop through a long run power cable.

Very similar logic applies when using numerous 24 V AC cameras powered from a single source (transformer). When calculating the total amount of current required for all the cameras, always leave at least 25% to 30% of spare capacity.

When AC cameras are used, attention should be paid first of all to the voltage rating (24 V is what the majority of AC-powered cameras require). Very often, power transformers can be purchased that have secondary voltage stated with the transformer fully loaded, as with halogen lamps. This might be misleading, since with big and constant loads, transformers may show lower voltage than they would have if only one camera was connected to it.

An AC camera's current consumption is very minimal (200 to 300 mA), so you should look for transformers with an open circuit of 24 V AC rating. Not by any means least important is the sine wave appearance, which can be especially critical when uninterruptable power supplies (UPS) are used. If a step-sine wave UPS is used it may interfere with the camera electronics and phase adjustment. If a

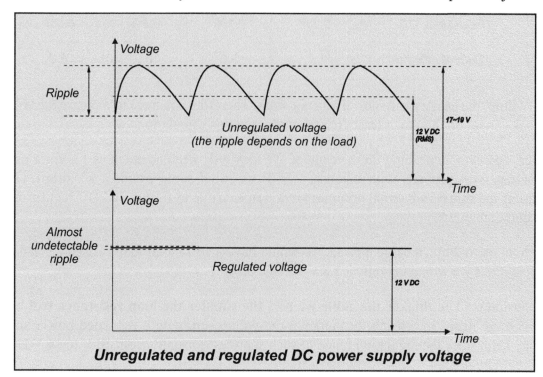

Unregulated and regulated DC power supply voltage

UPS is part of the CCTV system we should always intend to use a true sine wave.

We will see in the following a very basic calculation for the voltage drop which occurs in the so-called *figure-8* cable that powers a single 12 V DC camera.

The typical copper wire resistance, together with the cross section and the AWG (American Wire Gauge), is shown in the table below. This table could be very useful for various voltage drop calculations.

The popular figure-8 cable is, in most cases, a 14/0.20 type. The first number indicates the number of strands per conductor, and the second indicates the diameter of each strand in mm. The cross-sectional area of this cable is $14 \times (0.1)^2 \times 3.14 = 0.44$ mm^2. The resistance for a copper figure-8 wire, per meter, is approximately 0.04 Ω. A typical manufacturer's specification for the 14/0.20 states approximately 8 Ω/100 m DC loop resistance (loop, meaning 2 × 100 m). Using these numbers we can calculate the average voltage drop when powering a 12 V DC camera via a 300 m cable run, using the very simple Ohm's Law.

A realistic assumption would be that our 12 V CCD camera consumes 250 mA. This means that the camera is seen by the power supply as 12 V/0.25 A = 48 Ω resistor. For 300 m of 14/0.20 cable we will have a total loop resistance of 24 Ω. The supply voltage will now see a total resistance of 72 Ω. The 12 V will be divided between the R_c and R_{ccd} proportional to the resistance; that is, we will have a voltage divider. The calculation will show V_d to be 4 volts.

Copper wire size in AWG (cross section)	#24 (0.22 mm²)	#22 (0.33 mm²)	#20 (0.52 mm²)	#18 (0.83 mm²)
Resistance Ω/m	0.078	0.050	0.030	0.018
Resistance Ω/ft	0.257	0.165	0.099	0.059
Current rating (A)	1.5	2.0	3.0	6.0

With a 4 V drop, the camera will most likely not work. Therefore, we have to increase the voltage (and a plug-pack cannot do this) to at least 16 V, according to this calculation.

In practice, however, depending on the camera, we may only need as much as 13 V, for our camera under test may work properly with as low as 9 V (if we still assume around a 4 V drop). This would be the case if the camera's internal minimum requirement (due to further DC/DC regulations inside) were no higher than 9 V.

If we were to use a 24/0.20 cable instead, we would have a 15 Ω total loop resistance, and using the same calculations we would get only a 2.8 V voltage drop.

The conclusion is: **The thicker the cable we use, the smaller the loop resistance will be, thus a smaller voltage drop**. Increasing, or pumping-the-voltage-up, with a regulated power supply unit (PSU) may help, since the regulation range of such supplies is usually from 10 V to 16 V DC.

Nearest AWG	Stranding (No./diam. in mm)	Copper area (mm²)	Resistance (Ω/km)
10	65/0.30	4.59	4.0
12	41/0.30	2.90	6.0
14	26/0.30	1.84	9.4
14	50/0.25	2.45	7.0
16	7/0.50	1.37	13.0
16	16/0.30	1.13	15.3
16	30/0.25	1.47	12.0
17	32/0.20	1.00	20.0
18	16/0.25	0.78	23.5
18	24/0.20	0.75	26.0
19	1/0.90	0.65	27.0
20	1/0.80	0.50	35.0
20	7/0.30	0.49	35.0
20	9/0.30	0.64	28.0
20	10/0.25	0.49	35.0
20	16/0.20	0.50	39.0
21	1/0.70	0.40	46.0
21	14/0.20	0.44	44.0
22	1/0.64	0.32	54.8
22	7/0.25	0.34	54.5
24	1/0.50	0.20	89.2
24	7/0.20	0.22	84.3
26	1/0.40	0.13	136.0
26	7/0.16	0.14	139.4
28	7/0.127	0.08	221.5

A similar principle applies to 24 V AC cameras, only then we are talking about RMS voltages (root mean square); therefore, it may look as though there is a smaller voltage drop.

Ohm's Law is valid for both AC and DC voltages, so if we try to calculate the voltage drops for when the camera is powered with, let us say 24 V AC, we have to consider two things: the current consumption is lower (since the voltage is higher), and the 24 V AC we refer to are really RMS, that is, 24×1.41 = 33.84 V_{zp} (volts zero-to-peak). So, by applying Ohm's Law, a mathematical calculation will obtain a lower voltage drop compared to the 12 V DC power, but this is only due to the different current and voltage numbers. In other words, a lower voltage drop when 24 VAC is used is because of the lower current. This is, in fact, the same reason power used in households is not distributed from power stations at the level it is used in the household, but it is raised to tens of thousands of volts. Thus, the current and voltage drop, due to the power cables' resistance with long distances, becomes acceptable.

For the purposes of easy calculation and further reference, located on the previous page is a table of the typical copper wires found on the market, showing the relation between the nearest AWG number, the most common stranding technique, the area in mm^2, and the resistance in ohms.

Power over Ethernet (PoE)

Although this subject should be covered under the IT section where network switches are discussed, it is important to mention this here too, as it refers to power used by IP cameras, derived from the PoE network switches. During the previous edition of this book, and immediately after it, a new standard for powering IP cameras over the ethernet cable that is used to transmit streaming video over was developed.

This standard, known as Power over Ethernet (PoE), is actually an IEEE standard (Institute of Electrical and Electronic Engineers) known under the code of 802.3af and 802.3at.

These standards describe the technology that define how a system passes electrical power safely, along with data, on Ethernet cabling (Cat-5 or Cat-6). Basically, using the same network copper cable that transmits video streaming from the camera to the switch, the camera receives the power from the switch in first place. In the 802.3af and 802.3at standards there are certain procedures that define the initial handshaking and checking if the device being connected to is PoE and "expects" power from the switch. Clearly, for this to work the network switch has to be designed to be able to send power over Ethernet.

The easiest way to find out if a switch is PoE is to check what is written on the front panel itself, where PoE should be stated.

There are two main PoE standards (as at 2012) depending on the amount of power that can be delivered to the devices.

A typical 8-port PoE network switch

The key issue to consider with the PoE is the voltage drop over the given distance of copper wires (Cat-5 or Cat-6). So, it's not only the Ethernet limitation of the distance (100 m) relative to the Mb/s of data, but it is in addition to the voltage drop of powering an attached device that has to be considered. This depends on the current consumption of the peripheral device, and the copper resistance of the cable used. For this reasons, PoE switches use higher DC voltages at the switch source, with perhaps lower current, so that the electricity delivered to the PoE end device will achieve the desired power (power is nothing but voltage times current). The upper limit of what the DC voltage at the PoE switch is allowed to go is governed by the safety standard which consider voltages over 50 V DC to be dangerous to be handled by unqualified people.

The original *IEEE 802.3af*, from 2003, provides up to 15.4 W of DC power (minimum 44 V DC and 350 mA) to each device. Only 12.95 W is assured to be available at the powered device as some power is dissipated in the cable.

In 2009 an updated *IEEE 802.3at* PoE standard was introduced, also known as *PoE+*. This provides up to 25.5 W of power. The 2009 standard prohibits a powered device from using all four pairs for power. Some vendors have announced products that claim to be compatible with the 802.3 at standard and offer up to 51 W of power over a single cable by utilizing all four pairs in the Category 5 cable.

The idea behind this standards is that when a PoE camera is connected to the switch, and where both the camera and the PoE switch are compliant with the standards, then the camera should simply work. But, it is also important to understand that if the camera is not PoE capable, it should still work with such a PoE switch, although without receiving power from it, but, rather, be powered from a standard external power supply. This is all to do with the wires being used in the Cat-5 and Cat-6 wiring standard, where data is using certain wires, and the other pairs are used for the power over ethernet. If standards are adhered to, there should be no danger from applying 44 V DC for example on the terminals used for data transmission.

Beware that some PoE switches may only have certain ports PoE ready instead of all of them.

An important note is that PoE network switches must have the power source inside, which means they would usually consume more power than a non-PoE switch.

A final note: there are PoE inserters which may be used to deliver power to the camera without actually having a PoE switch. This is usually inserted between the camera and the Cat cable going to a non-PoE switch.

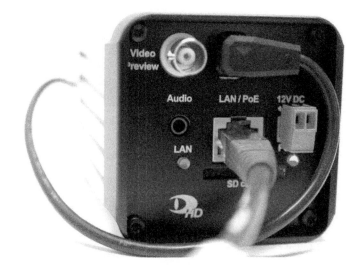

IP cameras that are PoE ready indicate it

V-phase adjustment

Analog AC-powered cameras are usually line-locked. This means that the vertical video frequency is synchronized with the mains frequency. If all cameras in a system are locked to the same power supply, that is, to the same phase (remembering that we can have three different phases, each of them displaced at 120° relative to the other two), then we will (indirectly) have synchronized cameras.

For the purpose of fine adjusting the vertical phase of each separate line-locked camera, a *V-phase adjustment* is available. V-phase adjustment can not only align the vertical sync of the cameras relative to their mains frequency zero crossing, but it can compensate even when different phase mains are used.

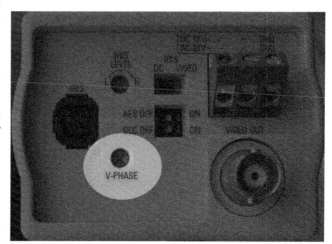

In order to do this an oscilloscope with two channels is required. One camera is then taken as a reference, to which a monitor's vertical adjustment is set so as to have no picture roll. The V-phase of the camera being adjusted is set so as to coincide with the V-phase of the referenced camera.

It should be noted that not all AC cameras are necessarily line-locked. That really depends on the camera design and provision in the electronics for such locking. If in doubt check with your supplier.

The majority of AC cameras have V-phase adjustment.

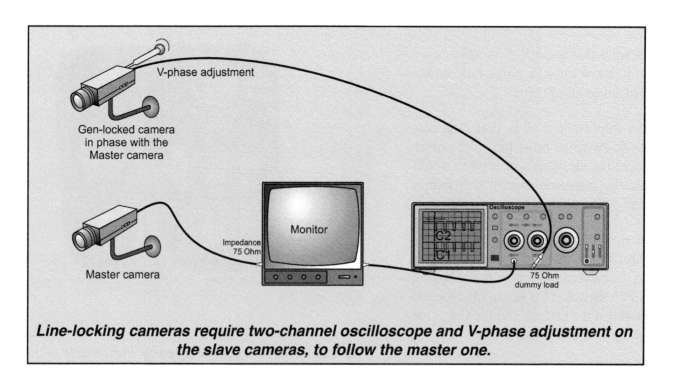

Line-locking cameras require two-channel oscilloscope and V-phase adjustment on the slave cameras, to follow the master one.

Camera checklist

In order to help people involved in installations, here is a list of things to be checked before the camera is installed in its position. Some may find this list very helpful, and others may even like to add a few more operations specific to their particular system. Many integral cameras (fixed and PTZ) made in recent years do not require all of the steps listed below since many things are now factory preset. However, there are still many camera setups that may require your thorough review.

The following is advisable to be checked before a camera is installed:

* Auto iris plug. This usually comes with the camera, not with the lens. Unfortunately, there is no standard among manufacturers, although lately it seems that a majority of them are compatible, but still it is better to check. AI connectors of all shapes and sizes are available, although the square ones are most common. Keep the connector with the camera. It might be very hard to find a spare if you lose it. Also, keep the AI pin-wiring diagram that usually comes with the camera instructions.

* If a DC camera is used, be sure to work out which is the positive and which is the negative end of the power plug. Sometimes the tip is positive, and sometimes it is negative. For some DC cameras there is no need for polarity to be known, as they are auto-sensing.

* Do the back-focus in the workshop, especially if a zoom lens is used. Doing the back-focus on site will be much harder and perhaps impractical. Follow the procedure described in the back-focus section, until you get more practice.

* Select a suitable lens for the angle of coverage required. For this purpose you can use focal-length viewfinders, hand calculators, tables, and so on. Take into account the imaging sensor size, as well as whether you have a C-mount or CS-mount camera/lens combination. In the last couple of years vari-focal lenses have been used instead and adjusted on site. Sometimes they may not have a wide enough or narrow enough angle of view to suit the application, so a fixed focal lens may be the answer.

* Further to the above point regarding the angle of coverage, additional care should be taken when IP and mega-pixel cameras are used. While in the past, with analog cameras, it was sufficient to know what sensor size the camera uses (1/3", 1/4" or similar) with the mega-pixel and multi-format sensors, the angle of view may change by just simply software selection of the camera mode, which can range anything from D1 size (4CIF), 720 HD, 1080 HD, 3 MP, 4 MP or 5 MP , all available on one chip. Clearly all of these modes will have different active areas, reflecting in different angles of view with the same lens. Consult the camera manufacturer brochure or specifications.

* Adjust the optimum picture for the estimated distance when the camera is installed. This is not so critical for a fixed lens, but installers tend to forget to adjust the camera focus on site, or unintentionally change the focus ring. If any out-of-focus problems appear, they will not be noticed during the daytime when the depth of field is big. They will become obvious and

problematic at nighttime, when the depth of field is minimal.

* Make sure that the level setting of the auto iris is good for day and night situations. ALC adjustment is important only if a very high-contrast scene needs to be monitored. The level may need some adjusting depending on the picture contrast.

* Get the mounting screws for the camera (if installed in a housing) and the bracket. These are 1/4" imperial thread screws, usually 10 to 15 mm in length. Sometimes trivial things like this will slow your installation.

* Make sure the camera/lens combination fits in the housing. If a zoom lens is used, take into account the focusing objective protrusion when focused to the minimum object distance (MOD). This should not add more than 10 mm to the lens length.

* Set the ID of the camera if such a model is used.

* If a camera with a electronic-iris is used, along with an auto iris, switch the electronic-iris off. Alternatively, use a manual iris or remote-controlled iris lens. Auto (lens) iris and electronic-iris do not go very well together.

* Set a higher shutter speed if the application requires. This is usually the case when high-speed traffic is observed and the signal is recorded on a DVR. Have in mind, however, that with higher shutter speeds you will need more light on the object, and the image smear may become more apparent.

* Set the power supply voltage value to what is required; that is, take into account voltage drop. Also, consider the current required by all cameras connected to the supply.

* If a 24 V AC camera is used and synchronization needs to be achieved, a V-phase adjustment may be necessary. You will need a two-channel oscilloscope and a reference camera for this purpose. Very often it is easier to make such adjustment in the workshop, and when such cameras are installed on site, just make sure they are powered from the same phase. Otherwise they will be displaced for 120°, as this is the phase difference in the mains three phase system.

* If a PoE IP camera is used, make sure you have PoE switch, and be aware of the longest distance you can run with Cat-5 or Cat-6 cable in PoE mode.

* If a color camera is used check the white balance setting. Some cameras have selectable indoor and outdoor white balance. Among the automatic white balance models you will find cameras with AWB (automatic white balance) and ATWB (automatic tracking white balance) selectable. In most situations ATW is the better choice.

* If an IP camera is used make sure you have a computer (notebook) and little switch with you in order to set up the camera focus, angle of coverage and compression and other parameters.

* If an IP camera is used, in addition to the computer, make sure you have the IP address of the camera so that you can connect to it. If the IP addressis unknown, make you have the IP address

discovery tool, which typically should be supplied by the IP camera manufacturer.

* Set the correct compression as required by the customer or system itself.

* If a PTZ camera is used set the camera ID and communication Baud rate data termination to the correct values.

* If a PTZ camera is used do not forget the mounting brackets, either wall or ceiling mount, with the suitable cable connectors, conduits, and sealants (especially for outdoor installation).

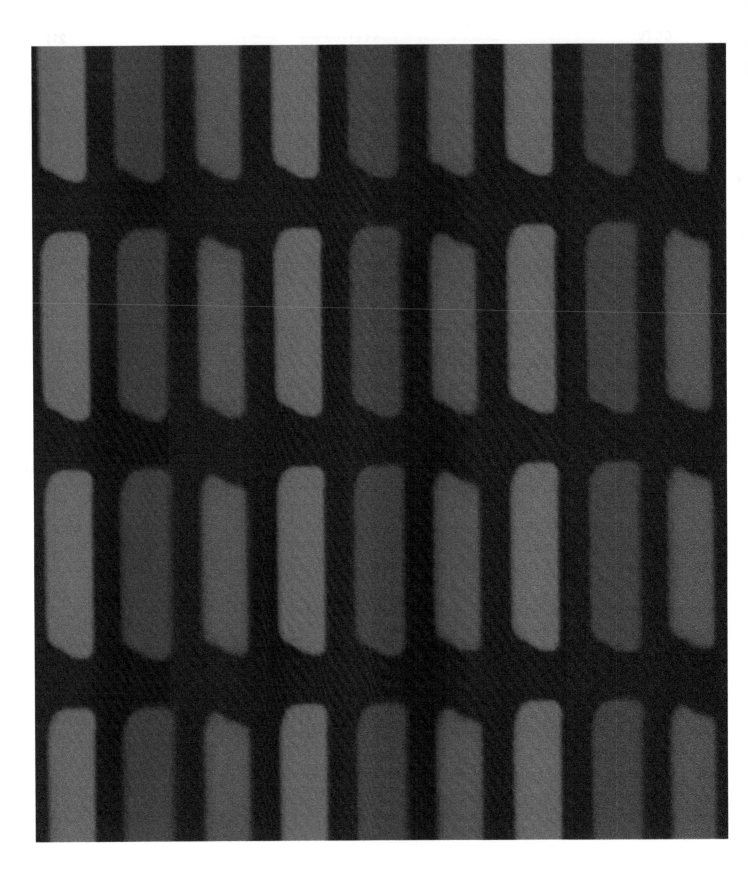

6. Displays

Displays in CCTV typically refer to electronic devices, often called monitors, but we will also discuss hard copy printed material and discuss its resolution details.

Monitors are often considered an unimportant investment compared to the other parts of a CCTV system. It is, however, very clear that if a monitor is not of matching or better quality than the camera's overall system quality will be diminished. Simple but worthwhile advice is: pay as much attention to your monitor as you do to your camera selection. Monitors have evolved in their quality and technology. From the good old Cathode Ray Tube (CRT), through Plasma and Liquid Crystal Displays (LCD) to the latest Organic Light Emitting Displays (OLED). In this chapter we will cover the most interesting current technologies for CCTV.

One of the most critical parameters of a monitor is the resolution. However, the design quality, contrast ratio, dynamic range, linearity of reproduction, power consumption, and size are also important and contribute to the overall picture quality.

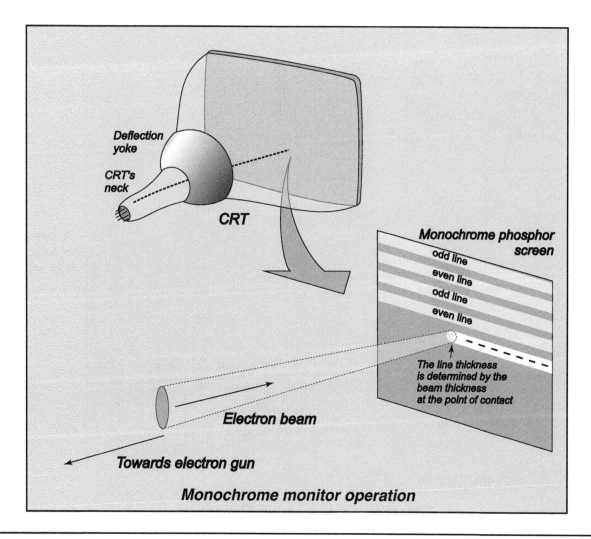

CRT monitors

In the past, in CCTV, the majority of monitor display units were CRTs. The CRT monitors used cathode ray tube technology, designed to convert the electrical information contained in the video signal into visual information. Basically, an electron beam produced by an electron gun is accelerated by high voltage tension at the screen end, and it emits light when hitting the phosphor coating. The intensity of electrons is proportional to the luminance information embedded in the video signal, which produces proportional phosphor luminance on the CRT screen.

On the inside the CRTs are coated with a phosphor layer that, when bombarded with electron beams, converts the kinetic energy of the electrons into light radiation. Different compositions of phosphor produce different colors. This is defined as the ***phosphor spectral characteristic***.

For a monochrome (or B/W) CCTV system, a phosphor layer that produces neutral color is used.

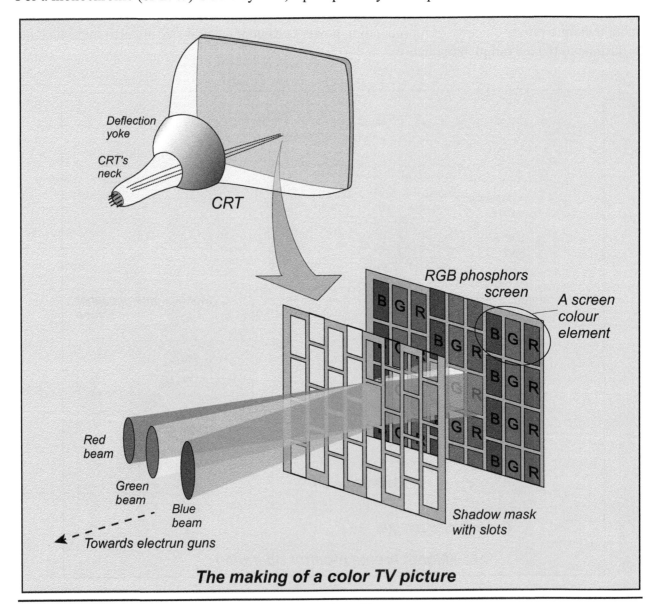

The making of a color TV picture

Color CRTs use a mosaic of three different phosphors that produce **red, green, and blue**, that are called *primary colors*. These are little pixels (limited by the physical size of the mask) that, when viewed from a distance, mix into a secondary (resultant) color in viewers' eyes.

It has been proven that with the red, green, and blue primaries the

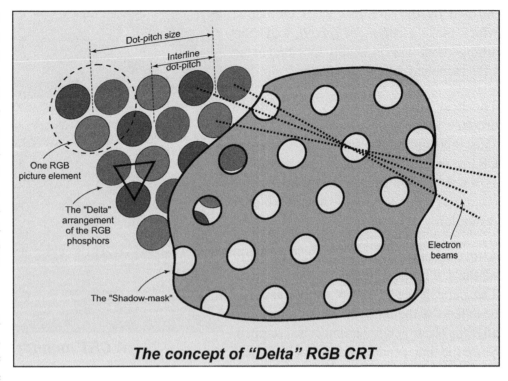

The concept of "Delta" RGB CRT

majority of natural colors can be simulated. This kind of color mixing is called *additive mixing* because light is added by each of the primary components to produce the resultant color. When all primary colors (red, green, and blue) in the additive color model have full intensity the resultant color is bright.

This is different to the *subtractive color mixing* as in painting and printing, where the term subtractive is used because these colors are produced by reflecting light. In that case we have an absorption of certain colors (depending on the color pigmentation), thus a passive method of producing the resultant color.

When all primary colors in the subtractive color model (such as printing with cyan, magenta, and yellow) are mixed together the resultant color is dark.

In terms of how the RGB micro-elements are arranged in the CRT screen, a few different technologies were possible in making color CRTs. The most common technology was "*In-line,*" as shown on the representation on the opposite page and the so-called "*Delta*" as shown above.

All of these technologies were used in CCTV CRT monitors, as well as in computers, at the end of the last and beginning of this century. The maximum resolution that can be reproduced is

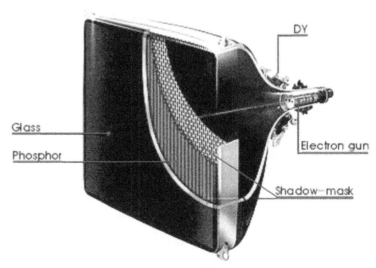

Cross section of a CRT

defined by the smallest RGB elements, which make a *color dot (pixel)*, and their arrangement. This is usually specified in the CRT technical data as *dot-pitch*.

The best CRT technology was able to produce a dot-pitch of around 0.21 mm. This then indirectly defined the smallest CRT screen size with a given resolution. This was one of the reasons why small color monitors (10" or less) could not offer higher resolution.

As in the human eye, an important property of the CRT phosphor is its *persistence*. The persistence of the phosphor layer is described as the duration of the luminance after the electron bombardment has stopped. Since the light produced does not disappear

A CRT monitor inside

abruptly, but decreases slowly, persistence is measured until the time when the luminance produced decreases to 1% of its initial value. Phosphor persistence is a useful feature because it helps minimize the flicker, but it should not be longer in duration than the TV frame duration (40 ms) as we want reproduction of dynamic images whose movements would be blurred if the persistence were too long. The persistence of the majority of CRTs used these days is around 5 ms. This is a bit more complicated with color CRT monitors since not all the phosphors have the same persistence (the blue phosphor has the shortest), but they are all around 5 ms.

Apart from the persistence, other important properties of the phosphor used in TV monitors are *efficiency* and *spectral characteristics*.

Efficiency is defined by the ratio between the produced light flux and the electron beam power.

The electron beam power depends on the acceleration produced by the CRT's high voltage and the electron beam itself. Different phosphors have different efficiencies; they can produce different luminance with the same amount of electrons and high voltage.

In color TV, the phosphor produces different intensities with different electron levels and hence they need to be modified to produce visually similar intensities. Hence, the following equation is applied to the RGB electron beams:

$$U_Y = 0.3\ U_R + 0.59\ U_G + 0.11\ U_B \qquad\qquad\qquad (46)$$

where Y signifies the luminance. This formula is based on the CIE (*Commission International De L'Eclairage* - the International Commission on Illumination) recommendation for color in electronic display devices.

Since CRTs work on the principles of electromagnetic deflection of electron beams, even a little stronger external magnetic field can affect this balance, which we sometimes see in the corners of monitors. To fix this degaussing coils are used, which produce a very strong electromagnetic pulse when turning on a monitor. Magnetic color distortion occurs frequently when loudspeakers are near a CRT monitor, or even if two such monitors sit next to each other. Their own magnetic field affects the other's precision of reproducing the colors. In order to minimize this we use metal-cased monitors.

Reproducing colors correctly on a color monitor is a delicate process. It all starts at the camera end, but calibrating the camera with its white balance and color temperature is only the beginning. The same process is repeated in the CRT. White color balance in monitors is one of the most delicate adjustments in the manufacturing of monitors and TVs; it is very difficult to do with the human eye, which is easily adaptable. Special color probes are used for accurate tuning.

In the past, the basic division of monitors in CCTV was made into B/W and color, but later on it was almost impossible to find a B/W CRT monitor. Today, while writing the edition of this book, even color CRTs are no longer manufactured, as they have become obsolete. B/W monitors used to have better resolution (since they have only one continuous phosphor coating) and were very useful in measuring resolution. The smallest dot element in B/W monitors was not defined by a dot-pitch (as there was no shadow mask with dot slots) but by the smallest electron beam cross section hitting the phosphor.

Despite the CRT technology now being obsolete, we will cover some details discussed in the previous edition as there still are many CCTV systems out there with CRTs. We will, however, expand a bit more on the current most common display technology, the LCDs.

CRT Monitor sizes

CRT Monitors are referred to by their diagonal screen size, which is usually expressed in inches, but sometimes in centimeters. B/W monitors have a variety of sizes; most often used are 9" (23 cm) and 12" (31 cm). Smaller sizes, such as 5" (13 cm) and 7" (18 cm), are not very practical apart from,

9" (23 cm) and 14" (36 cm) color monitors

perhaps, vehicle rear vision systems, video intercoms, and back-focus adjustments. Bigger ones are most often used where split-screen images are required, where sizes like 15" (38 cm), 17" (43 cm), and 19" (48 cm) are available.

The most popular color monitor size in CCTV was 14" (36 cm), around 10 years ago. This size was most suitable for the viewing distances typical in CCTV, where operator sits in front of the monitor at around 1.5 m. There are 9" monitors (some manufacturers were making 10" CRTs as well), which were more expensive than the 14" ones. This was due to the massive production of 14" CRTs for the household market, which brought the tube prices down. Larger color monitors, such as 17" or 20", were also available, but they were more expensive.

A lot of installers preferred to use a 14" TV receiver, around 10–15 years ago, instead of a proper monitor which would have offered some price advantage. CRT TV receivers were produced by the hundreds of thousands and, that made them low-cost. When a TV receiver is used, CCTV signal should be connected to the A/V input, since, as we said earlier, in CCTV we use basic bandwidth video signals, not RF modulated. In order to display the image on the screen the A/V channel has to be selected, that is, bypass the TV tuner.

The picture quality of a TV receiver, when compared to a CCTV CRT display, may or may not be of equal quality. This depends on the CRT type, the receiver quality, and the input bandwidth. Another important factor to consider is that TV receivers are usually housed in a plastic shell and are not protected against electromagnetic radiation from another set next to it.

CRT Monitor adjustments

CCTV monitors usually have four adjustments at the front of the unit: **horizontal hold, vertical hold, contrast, and brightness.** Not all of these adjustments are available on TV receiver monitors.

The *horizontal hold* circuit adjusts the phase of the horizontal sync of the monitor circuit relative to the camera signal. The effect of adjusting the horizontal hold is like shifting the picture left or right. When the horizontal phase goes too far to either end the picture becomes unstable and horizontal scanning lines break. A similar effect may appear when the horizontal sync pulses are too low or deformed; this usually happens with long coaxial cable runs (voltage drop due to significant resistance and high-frequency losses due to

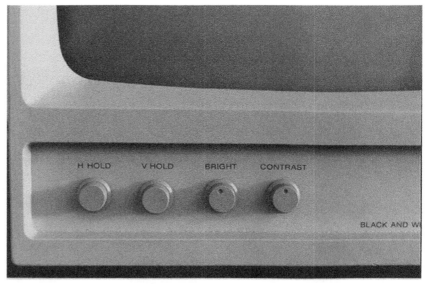

Adjustment potentiometers on a typical B/W monitor

significant capacitance). High frequency losses cannot be compensated for by the horizontal hold adjustments. By adjusting the horizontal hold the picture can only be centered.

The **vertical hold** adjusts the vertical sync phase. This has an effect of compensating for various cameras' vertical syncs. Usually, a monitor is adjusted for only one video signal, and so the picture stays stable. However, when more non-synchronized video signals are sequentially switched onto a monitor (which was a typical function of a matrix switcher), an unwanted effect called *picture roll* occurs. This is perhaps the most unwanted effect in CCTV. It occurs owing to the monitor's inability to quickly lock to the various signals as they are switched through a

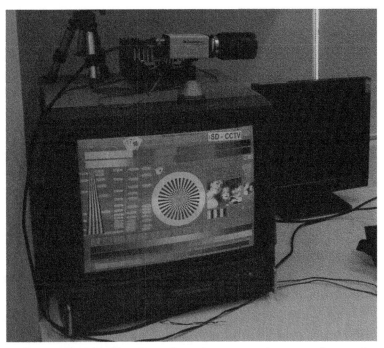

The gray scale of a test chart can be used to adjust brightness and contrast to optimum setting.

sequential or a matrix switcher (this is also discussed in the switcher section). This also means that various monitor designs have various locking times. Better monitors lock to vertical syncs quicker.

In the now older (analog) CCTV, switching numerous cameras onto one monitor was the most common system design. For this reason we will devote some more space to explain the synchronization techniques used in CCTV. Systems designed so that each camera goes onto its own monitor are possible, but done very rarely. Not only do the system costs become prohibitive in such a case, but the practicality

is not sustained. More space is required for more monitors, but, more importantly, no security operator can concentrate for long periods on so many different monitors.

Contrast adjusts the dynamic range of the electron beam, thus making the picture with higher or lower contrast (a difference from black to white). It is usually used when lighting conditions in the room where the monitors are change.

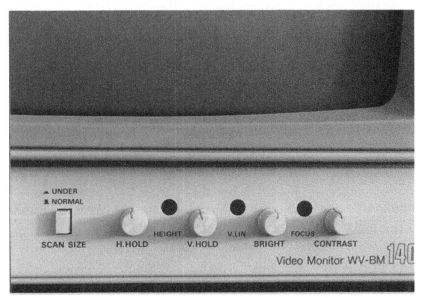

The front controls of a more advanced B/W monitor

Brightness is different from the contrast adjustment because it raises or lowers the DC level of the electron beam while preserving the same dynamic range. It is adjusted when the video signal tone reproduction is not natural.

A simple rule of thumb is to have the brightness and contrast adjusted so that the viewer can see **as many picture details as possible**. This is easily done with a test pattern generator or a good signal from a camera looking at a test chart. The less light in the monitor room the

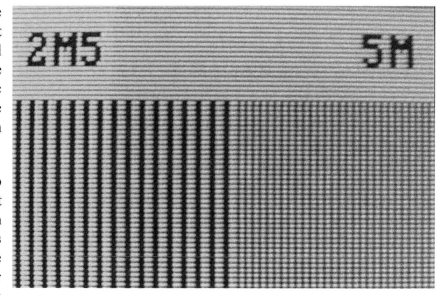

A close-up of a B/W monitor section showing 2.5 and 5 MHz test signal

lower the contrast setting can be. By reducing the contrast, picture sharpness improves (smaller electron beam cross section) and the CRT lifetime (phosphor burning) is prolonged.

Sometimes brightness and contrast are hard to adjust properly, especially when switching different cameras with different video signals. In order to have an objective setting for the brightness and contrast, a test pattern generator that produces an electronic gray scale should be used (i.e., where the gray levels are equally spaced). Then, the contrast and brightness are adjusted so as to distinguish all of the steps. After such an adjustment is made, the camera brightness and contrast can be judged more objectively. Consequently, we can decide whether a certain camera needs to have its iris level or ALC adjusted.

With time, the phosphor coating of a monitor's CRTs wears out. This is due to constant bombardment of the phosphor layer with electrons. The lifetime expectancy of a B/W CRT is around 20,000 to 30,000 hours. This means about a couple of years of constant operation. Worn-out CRT phosphor reproduces images with very poor contrast and sharpness. Color monitors should last a little longer because the smaller number of electrons (note that there are three separate beams for the three primary colors) are used to excite each of the three phosphors. In any case, after a few years of constant use contrast and brightness adjustment can no longer compensate for the CRT's aging, and that means the monitors need replacement.

Sometimes when a monitor is displaying one camera all the time, an imprinted image effect becomes noticeable (as was the case with tube cameras). If brightness and contrast adjustment are used carefully and in accordance with the ambient light, the monitor's life can be prolonged. The same applies to the domestic TV receivers.

Linearity and ***picture height*** are two other adjustments on most CRT monitors and are usually located at the back of the monitor.

Linearity adjusts the vertical scanning, which is reflected in the picture's vertical symmetry. If the linearity is not properly adjusted circles appear egg-shaped. In order to adjust monitor linearity a test pattern generator with a circular pattern is required. Sometimes a CCD camera can be used instead (CCDs do not have geometrical distortions) by positioning it to look perpendicularly at a perfectly circular object.

Picture height, as the name suggests, adjusts the height of the picture. With an improper picture height adjustment, the circles may appear elliptical. The scanning raster is also affected (increased or

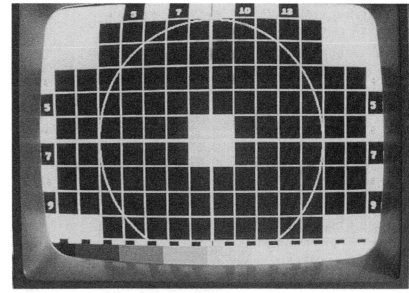

Linearity test signal

decreased), which indirectly changes the picture's vertical resolution.

Most monitors have **electron beam** **focus** adjustment, which is usually inside the monitor and close to the high-voltage unit. This adjustment controls the thickness of the electron beam when it hits the phosphor coating, indirectly affecting the sharpness of the picture. On some monitors this adjustment may be located at the front of the monitor and could also be called *aperture*.

Color monitors have **color adjustment** as well, which increases or decreases the amount of color in a color signal. This is different from brightness control. Color monitors are especially sensitive to static and other external magnetic fields because the color reproduction depends very much on the proper dynamic positioning of the three electron beams (red, green, and blue).

Even a slight presence of another magnetic field, such as a loudspeaker next to the CRT, may affect one of the beams more than the other two. This results in un-natural colored spots in certain areas of the screen

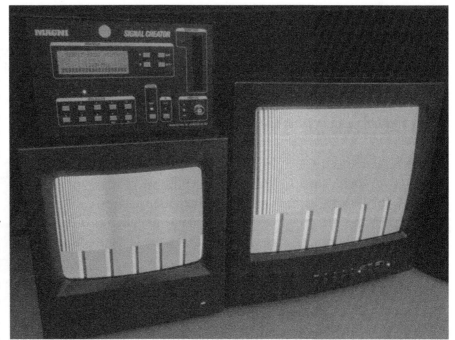

A special high-resolution sweep generator is used to check the bandwidth response of various monitors.

that are close to the magnetic field. In order to combat such effects color TV monitors have an additional element in their design called a ***degaussing coil***. The degaussing coil is a conductive loop around the CRT, through which every time the monitor is turned on a strong current pulse is injected. This creates a short but strong electromagnetic pulse that clears any residual magnetic fields. If the external field is very strong and permanent the degaussing coil might not be capable of clearing it.

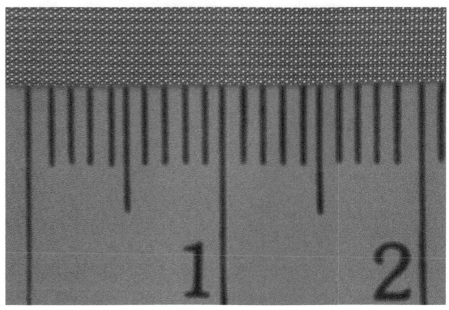

A close-up of the dot-pitch of a delta type color CRT, compared with the ruler below in millimeters.

Professional monitors (designed for the broadcast industry) are quite often used in bigger and better CCTV systems. They are equipped with sophisticated electronics and high-resolution CRTs whose horizontal resolution exceeds 600 TV lines. They quite often have adjustments in addition to the ones mentioned above. These may include ***hue*** (which is actually the color itself: red, green, orange, etc.); ***saturation*** (representing the purity of the color, i.e., how much white is mixed into it, where 100% saturated color has no white additives); ***H-V delay*** (a very useful feature where horizontal and vertical syncs are delayed so that the CRT will show the signal broken up into four areas, similar to a quad, so that horizontal and vertical syncs can be visually checked); and ***underscan*** (where the monitor shows 100% of the video signal, which is especially important when testing camera resolution).

Impedance switch

At the back of most CCTV monitors is an ***impedance switch*** next to two BNC connectors (In and Out). The purpose of the impedance switch is to allow for either terminating the video coaxial cable with 75 Ω (when the monitor is the last element) or leaving it to high position (impedance Off) if the monitor is not the last component in the video signal path.

As we have already discussed, the video sources used in CCTV are all designed so that their output stage has 75 Ω output impedance, which requires the same impedance from the signal receivers (monitors in this case).

The theory of electronic circuits says that only then, we will have 100% energy transfer and perfect picture reproduction. If however, the monitor is not the last element in the signal path, but perhaps another monitor is using the same signal, we then have to set the impedance of the first monitor to high (looping monitor) and set 75 Ω on the last one (terminating monitor).

Such a manual impedance switching is called *passive*. Most CCTV monitors have passive video inputs. There are some monitors, however, as well as various devices, such as VCRs, video printers, VDAs, where the video input is *active*. Active means the incoming video signal is going through an amplifier stage and the signal

The manual impedance switch is usually at the back of monitors.

is split into two or more components that are electronically matched with their impedances. In such cases, there are no manual switches to switch because there is no need for them. The 75 Ω impedance

is matched by the equal impedance of the electronic circuitry input and then distributed to two sections with correct active impedance matching. So, in other words, if no impedance switch is present at the monitor input, it means the video input is automatically terminated with 75 Ω, and the output of it should be treated as a new signal coming out of a camera.

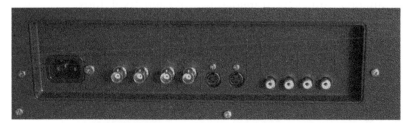

If a monitor does not have a manual switch, it means the electronics inside automatically termi-

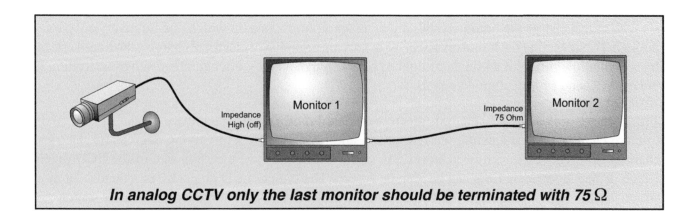

In analog CCTV only the last monitor should be terminated with 75 Ω

LCD monitors

LCD stands for *liquid crystal display*, which refers to organic substances that reflect light when voltage is applied. LCD technology was introduced back in 1970. The liquid crystal display consists of a liquid suspension between two glass or plastic panels. Crystals in this suspension are naturally aligned parallel with one another, allowing light to pass through the panel. When **electric current is applied the crystals change orientation and block light instead of allowing it to pass through**, turning the crystal region dark.

The concept of LCD operation is quite different to the CRT principles. Perhaps, the best description, or analogy, would be that LCD monitors compared to CRT monitors are what CCDs are to tube cameras. The image is not formed by electron beam scanning, but by addressing liquid crystal cells, which are polarized in different directions when voltage is applied to their electrodes. The amount of voltage determines the angle of polarization, which as a result determines the transparency of each pixel, thus forming an element of the video picture. The early version of liquid crystals was unstable and unsuitable for mass production. Today, things have changed. LCD monitors are also known as flat-panel, dual-scan, active matrix, and thin film transistor (TFT).

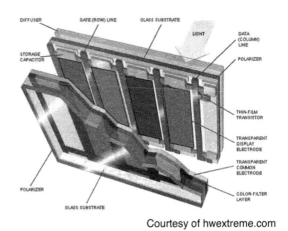

Courtesy of hwextreme.com

The LCD concept

LCD consists of several layers that are arranged in the following order: polarizing filter, glass, electrode, alignment layer, liquid crystals, alignment layer, electrode, glass, and polarizing filter. The cross section of the TFT LCD panel looks like a multilayer sandwich. At the outermost layer on either side are clear glass substrates. Between the substrates are the thin film transistor, color filter panel that provides the necessary red, blue, and green primary colors, and the liquid crystal layer. At the back of the LCD is a fluorescent backlight that illuminates the screen from behind. Under normal conditions when there is no electrical charge the liquid crystals are in an amorphous state, in which, the liquid crystal passes through. By subjecting the liquid crystal layer to varying amounts of electrical charges, the liquid crystal layer will allow different amounts of light to pass through as they orientate themselves according to the control center for the liquid crystals.

Just as in an ordinary CRT, the red, green, and blue liquid crystal "chambers" make up one pixel (picture element). By subjecting the red, green, and blue chambers to varying degrees of electrical charges, different colors can be achieved. As in the CRTs, we have a certain size of the LCD pixels which defines how many lines can be seen on screen. The typical LCD pixel size is around 0.28 mm, which is sufficient to produce a 1024 x 768 resolution screen on a 14" notebook computer screen.

Basically, LCD displays produce images by selectively filtering a white light used as a back-illumination. With the older LCD design the light is typically provided by a series of fluorescent lamps at the back of the screen. Most modern LCDs, however, use white LEDs as the back-light instead. Millions of individual LCD shutters, arranged in a grid, open and close to allow various amount of the white light through. Each shutter is paired with a colored filter to remove all but the red, green, or blue (RGB) portion of the light from the original white source. The shade of color is controlled by changing the relative intensity of the light passing through the sub-pixels. Each RGB group of sub-pixels forms a single pixel, the smallest element of a picture. The sub-pixels are so small that when the display is viewed from even a short distance the individual colors blend together in the viewer's eyes to produce any resultant color from the visible spectrum.

Photo courtesy of Apple

Apple's 27" displays with astounding 2560 x 1440 pixels

In the early days of the LCD technology the smallest pixels were around 0.3 mm (around 72 pixels per inch - PPI), but today resolution is much higher and retina displays, for example (as introduced by Apple), offer pixel resolutions of less than 0.1 mm (over 300 PPI). As we shall see later on, these pixel dimensions play important role in defining the optimum viewing distances. This could be very useful in planning and designing control rooms.

When compared to the CRT, there are many advantages of the LCD technology: no need for high-voltage elements; no phosphor coating wear (i.e., virtually unlimited lifetime of the screen); a flat and miniature appearance (compared to CRT); no geometrical distortions; low power consumption; and immune to external electromagnetic interferences.

The major disadvantages of the early LCD technologies were slow pixel response time, low dynamic range, and inability to reproduce real blacks or achieve high resolution. All of these shortcomings have actually been addressed successfully and resolved. In fact, at the time of writing this book, CRT monitors have been discontinued by all major manufacturers and LCD have become the primary choice for monitors. Their quality has increased dramatically while their costs have dropped considerably.

There are basically two kinds of LCD displays, based on how LCD pixels are driven: DSTN (*dual-scan twisted nematic*, also known as *passive*) and TFT (*thin film transistor*, also known as *active matrix*). The most common

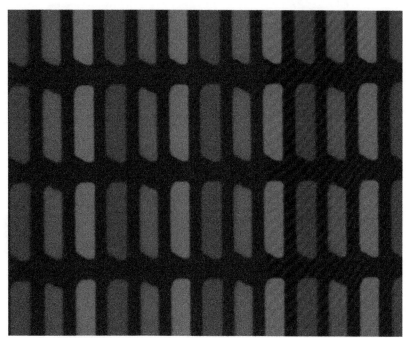

A magnified close up of an actual LCD display

today are the TFT based LCDs..

One other important parameters of the LCD screen is the ***pixel response time***. In the beginning of the LCD technology some monitors could only work with lower frequencies than what was possible with CRT monitors. Today, however, a pixel response time of less than 12 ms is achievable, which makes over 80 images per second possible. This is important for computer screens, although for CCTV displays 25 or 30 images per second are sufficient.

Also, the ***viewing angle*** is an important parameter of an LCD screen due to its liquid structure polarization. These days it is not uncommon to have one greater than 120°; some displays are coming close to 180°.

One of the weakest parameters in the early days of the LCDs (compared to CRTs) was the ***contrast ratio***. In CRTs it is easier to produce higher contrast, even though the total brightness of a CRT may be lower than in LCD screens, simply because the electron gun can be completely shut down if the image needs to display black. In LCD monitors, the backlight is always on, and, no matter how good the LCD pixels are at blocking light, when black needs to be represented there is a certain amount of light that still comes through.

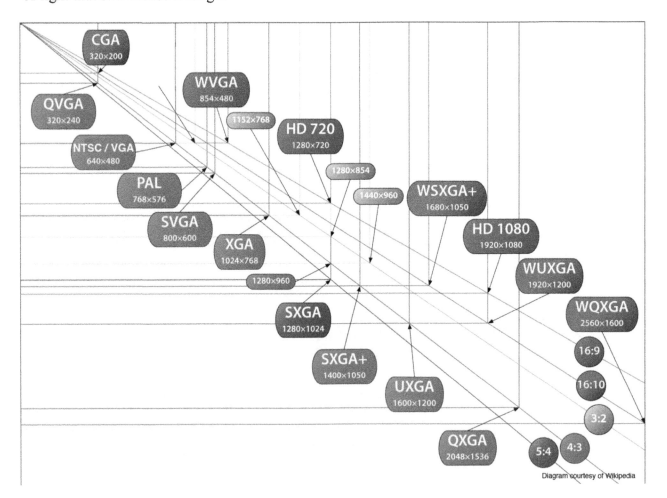

Various display formats based on their resolution

For a good LCD display, a typical contrast ratio should be at least 400 to 1, but contrast ratios of 1000:1 are possible today.

As opposed to contrast, LCD screens are also categorized by **brightness** parameter or **luminance**, expressed in Candellas per square meter (cd/m^2). Typical modern LCD display with LED back-light can have anything between 200 and 400 cd/m^2. For cinema experience and larger displays over 500 cd/m^2 are required.

LCDs do not have the CRT's geometric convergence, or focus problems, and their clarity makes it easier to view higher resolutions at smaller screen sizes. If you have an LCD TV/Monitor display, chances are you will have an RF antenna input which will be decoded by the in-built tuner/decoder.

If, however, a direct digital signal is displayed it would usually go via the HDMI (High Definition Multimedia Interface) or DVI (Digital Video Interface) input. The difference is that DVI is the digital video only interface, while HDMI is a digital connection that combines audio and video link. It is possible to convert DVI to HDMI by way of a simple adapter. Clearly the HDMI and DVI formats are purely digital and offer the best possible picture quality. Because such a signal is raw (uncompressed) digital signal the cable distances are limited to 15m of length. Going beyond this distance requires special media extenders.

Some LCD displays have computer (XGA) or composite (CVBS) video inputs, which means they can display analog CCTV or computer signals. Typically, composite video is in the form of RCA socket, but some monitors may have BNC.

It is important to note that when XGA or CVBS signal is connected to the LCD display there is a A/D conversion circuit in the LCD to allow such a view. The quality of such a conversion may vary, which means the quality of an analog signal shown on an HD LCD monitor will vary.

Because of the precise and discrete nature of each pixel element in LCD screens, the sharpest and best image appearance is achieved when the video signal pixel resolution matches the native resolution of the LCD screen. For this reason, if an HD signal of 1920 x 1080 pixels

Modern digital connections DVI and HDMI

needs to be shown on an LCD, the best picture quality would be if the LCD has the same resolution of 1920 x 1080 pixels. There are some manufacturers that overstate their monitors as HD, without disclosing that the display itself is only 1280x720 for example. In such a case the LCD monitor electronic performs down-sampling from 1920x1080 to 1280x720, and the clarity of the signal is not the same as if it was shown on a matching resolution display.

Video-walls and Projectors

Although LCD monitors are the most widely used, they can only be so big; their physical size is limited primarily by the available HD panel sizes. The largest LCD panels can be as large as 80", and some systems may require even larger than that (casinos, for example, or traffic monitoring control rooms). One method of having bigger displays is by adding intelligent Video-walls, whereby multiple displays are put close together to make a large uniform display. This might be a costly method, as it requires not only the number of display panels for the wall, but also dedicated and powerful computers capable of processing multiple video signals. Modern video-walls can also handle a variety of video

Video-wall used in traffic monitoring

compressions available with various IP CCTV cameras. Video-walls offer many advantages: brightness, resolution, and almost any display size can be achieved with either one video stream across multiple of display panels or muliple small split-screens (as the operators see fit).

Another method of having a larger display than what the LCD display panels can offer is by using LCD projectors. Some years ago, projection monitors were extremely big, expensive, and complicated for use and setup. They would usually consist of three separate optical systems, each projecting its own primary color. Today, video projectors are much smaller, cheaper, brighter, and

A typical small LCD projector

easier to use and set up. In most cases, they would accept a range of video inputs such as composite video, RGB (or component) video, computer and HDMI. Most of the projectors are single-lens color projectors that filter the light through an LCD film and these days even HD resolution is affordable. One of the biggest advantages of projectors is their ability to produce almost any image size required depending on the distance from the screen. Certainly, the further the projector is the lower the brightness will be, but typically LCD projectors are used inside dimmly lit rooms. Brightness is one of the main specifications of LCD projectors, and it is expressed in ***lumens***. Typical brightness of a small LCD projector can be anything from 1000 up to 4000 lumens.

The closest projection distance (and hence the smallest screen size) depends on the projector's lens ability. As with any lens, there is a minimum focus position. This defines the so-called ***throw distance***

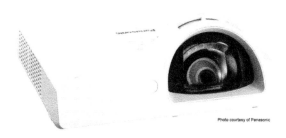

A close range projector

of a projector. Calculations may need to be made which define the position of the projector for the required projection screen size.

There are some new projector designs called **close range projectors** with a special anamorphic lens so that the projector can be installed very close above the screen itself, while preserving the screen's linear and rectangular geometry.

For larger and bigger venue projections, cinemas, and theaters, there are professional (and more costly) projectors producing even up to 40,000 lumens. Such projectors require special cooling for the lamps used; very often such large projectors have special sound cancellation design for the noisy cooling fans.

For the brighter projectors the most often used technology is the DLP (**Digital Light Processing**). The DLP idea was developed by Texas Instruments™ and it is based on the **micro-mirror device technology**. This is basically a memory chip with a matrix of millions of microminiature mirrors (similar size and appearance as CCD chips). A light source projects an image to the DLP chip so as to have the mirrors reflect the image onto virtually any size screen. The size of each mirror is 26 millionths of a

A professional 40,000 lumens projector

millimeter. The mirrors are so small that a grain of salt could obscure hundreds of them. Each mirror represents a screen pixel. All are controlled and switched on and off by the on-chip circuitry, and every

The heart of the DLP projectors, Texas Instruments' patented digital light processing chip

one of the hundreds of switches per second is performed with great precision and accuracy. The mirrors are programmed to remain at designated reflective angles for various time periods within a single frame of motion. This permits gray-scale projection or correct color presentation. For color projection the light is beamed through a condenser lens and then through a red, green, and blue color sequential filter. The filter switching is synchronized to the video information fed to the DLP chip at a rate three times that of the video (which results in 150 Hz switching for PAL and 180 Hz for NTSC signals). Filtered light is then projected onto the DLP integrated circuit, whose mirrors are switched on or off according to the digital video information written into the chip's memory circuits. Light shined on these mirrors is then reflected into a lens to project images from the DLP surface. Most important benefits from the DLP projectors (apart from the miniature physical size itself) include equal high resolution, brightness, and color fidelity, regardless of the screen size.

Plasma display monitors

When LCD monitors were first introduced, the competing technology was called *plasma*. Plasma monitors are still produced today, but it seems LCD improved its weaknesses and it is prevailing. Some scientists refer to *plasma* as the fourth state of matter (the first three being solid, gas, and liquid). Often plasma is defined as an ionized gas. The theory of plasma is beyond the scope of this book, but we should mention it for the sake of completeness.

Plasma monitors are made of an array of pixels, each composed of three phosphor sub-pixels – red, green, and blue. As opposed to CRTs where light radiation was caused by electron bombardment, in the plasma displays gas in a plasma state is used to react with phosphors in each sub-pixel. In plasma displays each subpixel is individually controlled in order to get all colors. Because each pixel is excited with the plasma process individually there is no geometrical distortion, as is the case in CRTs, and the picture sharpness and color richness are brought up to new heights.

One of the key advantages of plasma displays (at least in the early days) was the capability of producing deeper black, which increases the contrast ratio. This makes the picture contrast typically over 1000:1, making the plasma displays suitable for bright rooms and outdoor areas.

Since the plasma display does not require high voltage as is the case with CRT, much larger displays are possible. In fact, with plasma displays larger displays are made easier when compared to LCDs. Typical plasma display sizes can be anything from 75 cm (30") up to 380 cm (150"). The thickness of plasma displays is minimal, ranging from 5 to 15 cm (2–6").

One of the downsides of plasma displays is their power consumption, which is much higher than in LCDs.

The second is the wearing out of the posphor, causing display fading with time. Manufacturers usually claim 30,000 hours for the brightness to get reduced to 50% of its original quantity. This is equivalent to about three years of constant operation, which is more or less the same as what the CRT monitors are quoted.

A plasma display

OLED displays

A new type of display technology called OLED became popular at the time of writing this book. OLED stands for Organic Light Emitting Diode and is based on a process whereby electrical energy is converted into light. The term OLED (Organic Light-emitting Diode) is used because electric current generates light and because it features typical diode properties and relationships of voltage and electric current.

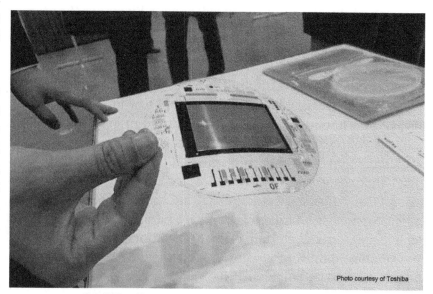

Solar cells, for example, absorb light and generate electricity, but with OLED the exact opposite occurs. OLED displays are based on component devices containing organic electroluminescent material that emits light when stimulated by electricity. An OLED component device consists of several layers of organic electroluminescent film with each layer measuring a mere 100nm. These layers are sandwiched

OLED displays are thin and flexible

between two electrodes. The electrodes create a flow of electrons and electron holes. These electrons and electron holes flow to the emissive layer (EML) inside the films and recombine. This activates the organic electroluminescent material in the EML to emit light.

When electrical current is applied, a bright light is emitted. The OLED emitted light does not require a backlight (unlike LCDs). In fact, each pixel is a small light-emitting diode. So, OLED displays are brighter, more efficient, thinner, flexible, and feature better refresh rates and contrast than either LCD or plasma.

These OLED component devices are marked by high contrast, rapid response, a fully sealed structure and wide viewing angle.

From the smallest to the largest OLED displays

About pixels and resolution

All of the digital video signals are composed of one tiny element. This is the "building block" of any digital image or video – the pixel.

Pixel is short for *picture elements*; pixels are sometimes referred to as *pels*. These are the smallest elements of any electronic (digitized) picture. **Pixels are the atoms of the image.** Understanding pixels is especially important in the digital photography, but it is also important for us in CCTV, especially when trying to identify faces and objects.

Pixel terminology is also used in printing processes (like glossy magazines), although they are different to electronic displays in how they are composed. In the offset printing industry with cyan, magenta, yellow and black (CMYK) colors are used, the picture elements are usually referred to not as pixels, but as **dots** and they have the same meaning. One cannot dissect it any more and get additional, meaningful information about the image, of which that pixel is a part. So, in very simple terms, **pixels contain elementary information about the smallest details of a picture, which is the information about the pixel's color and the brightness of that color.** Pixels can be associated with the image resolution, but understanding the differences between various kinds of pixels is very important.

CMYK Offset printing close-up

In television, we refer to the pixels' elementary attributes as chrominance and luminance of the picture element. Because of the need to represent a variety of colors and shades with only a limited number of primary colors, pixels are composed of smaller details, each representing a certain value of their primary color. So, in fact, pixels are not really the smallest elements of a picture, but only as a group of all the three primary elements they do represent a "complete" pixel. Interestingly, both the printing pixels as well as the electronic display pixels produce the resultant color when viewed from a certain distance, where the human eye cannot resolve the pixel's elementary details.

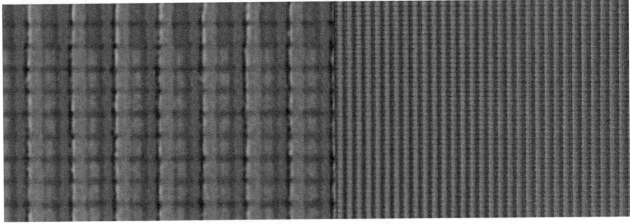

A close-up look at display pixels with RGB elements

So, the pixels used in digital photography, television, and printing are not of the same kind. The differences between various pixels are the source of many misunderstandings and misinterpretations in many imaging industries and applications, one of which is CCTV.

As we explained earlier in the book, we use red, green, and blue colors to "simulate" other colors. These three colors (RGB) are called ***primary colors***, and by varying their individual luminance it is possible to represent almost any other color. Almost any color and its saturation variations, from black to white, is possible, including skin colors. There are certain natural colors that may not be reproduced accurately with the RGB color space (like gold, for example), but they can be represented close enough.

The actual color mixing occurs in our eyes when viewing the pixels from a typical viewing distance. This distance is usually so large relative to the pixel size that we perceive the three elementary colors as one resultant – color dot. The resultant dot has the new color – a result from the additive mixing of the red, green, and blue phosphor in the TV screen pixels.

In digital television today, and, of course, in CCTV, pixels exist at both ends of the image chain – at the input (i.e., the camera end) and at the output (i.e., the monitor end). The pixels at the camera end are also using red, green, and blue primary colors, but there is a difference in how these are put together in single chip color cameras (as explained under the camera section) compared to displays. Using the so-called Bayer color filter array (CFA), a little square representing a pixel is divided into four quadrants, where green color appears twice diagonally opposite, and then red, and blue once. So, the "white light" is filtered into RGB spectrum components using organic or pigment dyes. These RGB values are then converted to colour space meaningful to human eye (YCb-Cr). The Y-Cb-Cr color space is also better during encoding, used in digital TV as recommended by ITU-R-BT 709. During the decoding the Y-Cb-Cr color space is converted again to RGB for additive color mixing on the display. So, **the display pixels are not of the same structure as the Bayer CFA**. The camera sensors have four RGB elements per pixel, while the displays have three per pixel. Most common RGB displays these days are in-line, where red, green, and blue pixel elements are next to each. When all pixel sub-colors are lit, they appear white from a distance (human eye acuity + additive color mixing).

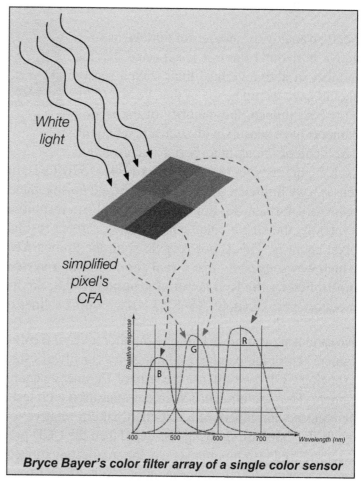

Bryce Bayer's color filter array of a single color sensor

The Bayer CFA applies to single-chip color cameras only, which is what we use in CCTV most of the time. In broadcast television, where three-chip color cameras are predominant, RGB is mixed with each of the imaging chips being dedicated to each of the three primary colors. In order to be able to represent the similar color system with only a single chip imaging sensor, Bryce Bayer has developed and patented a CFA where two green elements, one red, and one blue are combined to produce a color video signal. This is based on the fact that human vision is most sensitive to the green color and that the luminance of a scene is mostly influenced by the green spectrum. In addition, during the microlytography on a single sensor it is easier to divide the pixels into four elements, rather than three.

In analog CCTV, due to the known and fixed aspect ratio of the displays (4:3) the resolution was measured in TV lines (TVL). Basically, the maximum number of horizontal pixels across the screen width was divided by the 3/4, giving a number in TVL which is equal to the height of the monitors.

So, if an analog camera sensor for PAL has 576 active horizontal lines, it is expected that the number of active vertical lines such a sensor should have would be 4/3 of that, i.e. 4 x (576/ 3) = 768. Indeed, the majority of the analog cameras have sensors with such a pixel count. The theoretical maximum horizontal resolution

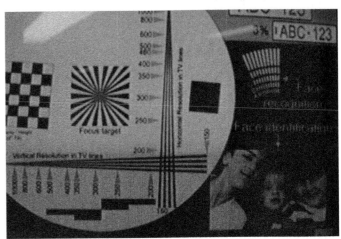

A snap-shot of a resolution test-chart

such a camera would have is 768 x (3/4) = 576 TVL. In practice, this is never achieved. In real life, lenses have limited resolution, low light and noise reduce it too, and on top of all of that the Bayer CFA reduces it further. So, in practice, the real life resolution is always below it's theoretical maximum. Typically, the single-color sensor analog camera resolution is approximately 70 – 80% of what the pixel count is. This is not very far from the known **Kell factor** used in the early days of television, which was suggesting that **a real resolution of a system is obtained by the theoretical maximum multiplied by the Kell factor of around 0.7~0.8**. So, for example, a 768 × 576 pixels CCD chip will produce approximately 768/4 × 3 × 0.8 = 460 TV lines of horizontal resolution.

We cannot show more details on a monitor (even if the monitor could display more) than what the actual camera sensor has captured. Although we can always state the number of pixels the camera sensor has, in analog CCTV we still use the term of TV lines as qualification of the quality of details we get from a camera. The resolution in TV lines is measured with test charts and in the real world there could never be a perfect alignment of a test chart pattern relative to the projected image on the CCD chip. As a result, TV lines are showing less detail than the CCD pixel count would indicate. When a video signal is reproduced on a monitor screen, the smallest picture element is clearly defined by the smallest of the two – camera CCD pixels or monitor pixels. If we have a very low-resolution monitor – for example, a small, 23 cm CRT with 330 TV lines specification – and our camera produces a high-resolution 480 TV lines signal, we can only see what the monitor shows – 330 TV lines. If we have, for example, a high-definition TV monitor capable of showing over 700 TV lines, and we put our 480 TV lines camera signal through, we can only see what the camera resolution shows.

The resolution in modern day HD and other digital cameras is no longer expressed in TVL, but rather simply in pixels. Although it is possible to introduce TVL as resolution parameter in HD as well, as this standard uses a known 16:9 aspect ratio, this is not being done in CCTV because of the many other variations in formats that are not necessarily HD standard. These include megapixel formats that might be using 3:2 aspect ratio for example, circular 360° fish-eye views, or even completely different and variable formats, such as those used by the Panomera from Dallmeier.

Suitable displays for the signal are very important for the overall quality of the image. As mentioned earlier, a true 1080 HD video signal displayed on a 720 HD monitor for example will not appear as good as it really is. Similarly, a megapixel image that exceeds the display resolution, for example, a 5 MP image shown on a 1080 HD display, will

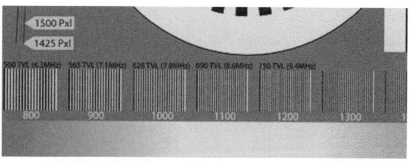

A resolution sweep bars in TVL and pixels

lose some details when shown in full screen. The display with less pixels has to show a "down-sampled" 5 MP image on it's 1920 x 1080 pixels display. For this reason, most of the front end software in CCTV has some kind of digital zooming function so that the user can get down to the real pixel resolution of such an image. To further complicate this discussion on pixels and resolution, compression artifacts must also be considered. They further reduce the actual resolution produced by a particular sensor.

It is possible to use some kind of correcting factor in digital CCTV (like the Kell factor in analog) to indicate the "real" resolution of a system, but unfortunately in modern digital CCTV there are too many variables to consider, of which the compression gives the most varying results. It makes no sense to use the same correcting factor as in analog CCTV. Because of this, if a video resolution in pixels is said to be HD (1920 x 1080) it does not necessarily mean that visually we can count 1920 pixels of a very fine pattern the camera sees. It is always going to be less than that due to the various post-processing and compression artifacts. Test charts are possibly one of the best methods to determine the real resolution.

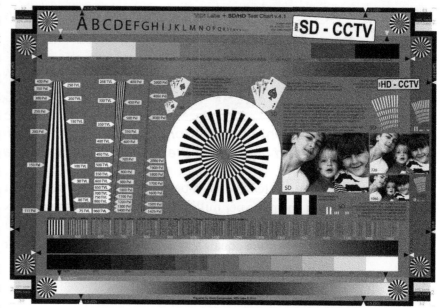

A multi-functional SD & HD CCTV test chart

Psycho-physiology of viewing details

In some CCTV systems the number of monitors can be quite large. It is very important to know how many monitors can be used in a place without exaggerating, as well as how to position them and what will be the correct viewing distance for the users. Even with one monitor in the system operators should be aware of certain facts and recommendations, especially when they are spending the majority of their time in front of the monitors.

The *ITU Recommendation 500* (used to be called CCIR R.500) states that the preferred viewing conditions are affected by the field frequency of the TV system, the size of the screen, and the distance relative to the screen size.

Designing an efficient control room depends on the available space, the number of operators, and the size of the system (number of cameras to be viewed).

Through experiments and testing it has been found that **the best a human eye can resolve is not more than 5 to 6 lp/mm** (line pairs per millimeter). This equates to approximately 0.1 mm line thickness at a distance between the object and the eyes of around 0.3 m (assuming 20/20 vision). In other words, the smallest detail we can see equates to a minimum angle of about one-sixtieth of a degree (1/60°).

This is called *the acuity of human vision*.

We can use this minimum angular vision for better understanding and optimizing the psycho-physiology of the viewing. Based on this parameter, we can draw conclusions about the optimal viewing distances for various displays used in CCTV. Viewing distance is an important factor for the psycho-physiological

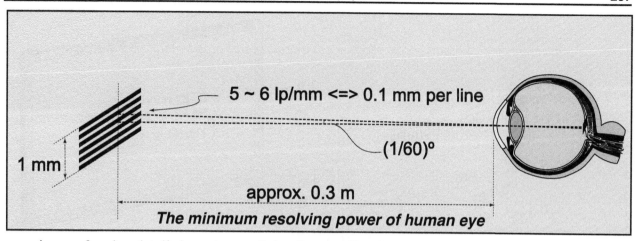

5 ~ 6 lp/mm <=> 0.1 mm per line

(1/60)°

1 mm

approx. 0.3 m

The minimum resolving power of human eye

experience of seeing details in an image. It is of no use if a viewer gets closer to the monitor, as it will start seeing the structure of the screen (RGB pixel), but it is also not going to get any better if he/she is positioned farther away from the monitor screen, as then there are details in the picture that can not be seen from further away.

Different resolutions on different display sizes will have different optimal viewing distances.

For example, if we take an analog NTSC CCTV 15" monitor, by knowing that there are 480 active TV lines in NTSC and that 15" monitor has a display height of approximately 230 mm we can deduce that one active line height should be around 0.48 mm. Let's round it to 0.5 mm to allow for the space in between lines and for easier calculations. If we have an operator with 20/20 vision, and if we assume his acuity of vision is 0.1 mm at around 30 cm distance, than smallest details of 0.5 mm (in this case a line width or pixel) will be 5 times bigger. So 30 cm times 5 is 1.5 m distance. This would be the optimal viewing distance for a 15" monitor showing an NTSC analog signal. This distance can be converted to approximately 6.5 times the 15" monitor height.

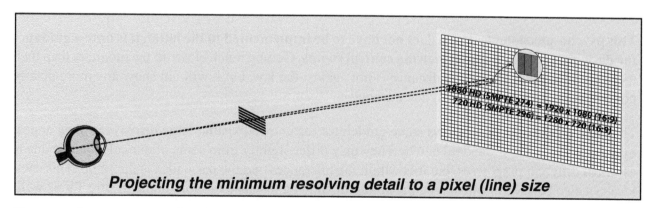

1080 HD (SMPTE274) = 1920 x 1080 (16:9)
720 HD (SMPTE296) = 1280 x 720 (16:9)

Projecting the minimum resolving detail to a pixel (line) size

If we use the same proportional calculation for analog PAL monitor, with 576 active lines, the optimal viewing distance for a 15" monitor with line thickness of around 0.4 mm will be around 1.2 m. Expressed in PAL monitor heights this will be approximately 5.5 times the 15" monitor height.

The same analogy can be used for all other display sizes, including the modern HD displays (1920 x 1080 pixels) or even the small tablets. In fact, the popular retina displays on Apple's smart phones and iPads has exactly that property: pixels so small that human eye cannot resolve them from a

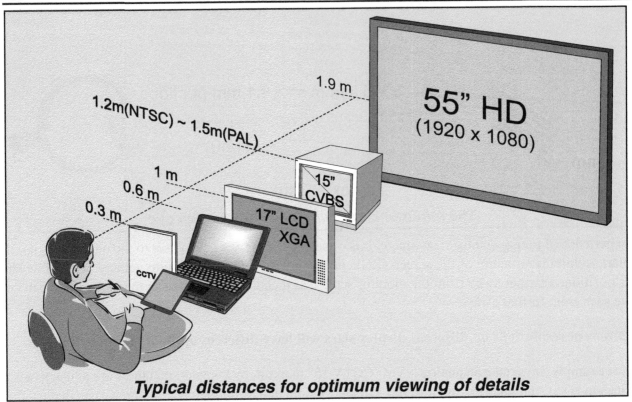

Typical distances for optimum viewing of details

normal viewing distance (which, in this case, would be around 30 cm). A 42" full HD screen for example has display panel dimensions of around 1210 x 680 mm. If we divide the 680 mm with 1080 active lines as for 1080 HD display standard, then a pixel height would be approximately 0.63 mm. This is about 6.3 times bigger than the minimum 0.1 mm resulting from the acuity limits. That means the optimal viewing distance for a 55" HD display would be around 6.3 times the 0.3 m reading distance, resulting in 1.9 m as an optimal viewing distance for a 1080 HD display of 55", which is around three times the HD screen height.

This psycho-physiological rule does not have to be implemented to the letter, it is only a guidance for designing proper optimal viewing control rooms. Getting much closer to the monitors than their recommended optimal viewing distance is not against the law, but it will not show any more details, nor will going further away be of any advantage.

The table on the next page shows some guidelines for known existing large displays as they appear in the new IEC CCTV standard 62676. They may differ slightly from some of our calculations due to assumed different pixel sizes, but it is within the tolerances and the basic idea is the same. This same idea can also be used in calculating the optimum viewing distance for your home theatre TV as well.

With larger CCTV systems, visual display management is of vital importance. The number of monitors, their size, position, and distance from the operators are all important considerations. For example, not all monitors need to display images all the time. It may be much more effective, for example, if the operator is concentrating on one or two active monitors (usually larger sized) and the rest of them are blank. In case of activity (i.e., an alarm activation, motion detection, or perhaps video fail notification), a blank monitor can be programmed to bring the image of a pre-programmed camera. In such a case, the operator's attention is immediately drawn to the new image and the system becomes more efficient.

In the modern digital CCTV increasing numbers of displays are nothing but computer displays, driven by software or hardware decoders. This is an additional consideration to what was the case with the analog system design where analog video was shown on various size displays, predominantly in full screen and sometimes in quad split-screens. With the digital CCTV we have much more options for split-screens and multi-screen displays, as almost any configuration can be

A hardware HD decoder

done by the computers graphic cards. It is, however, extremely important to understand that **because all signals are digital, the display computer power is used for decoding each channel.** The more video channels are shown on one computer display, the more processor's and graphic's power is needed. **It may come to a point where if there are too many streams showing on one display they no longer appear continuous and smooth, but interrupted and jerky.** This is especially the case with video compressions the which are processing power hungry, like H.264. This consideration for decoding capability of a display computer is in addition to the optimum viewing distances discussed here, but it does influence the overall system design efficiency. Some manufacturers resolve this issue by advising the maximum number of real-time streams that can be decoded by their system, others offer additional hardware decoding device, where only one single video channel is displayed, as decoded by a dedicated hardware decoder. With such displays it is easier to achieve the continuity of monitoring without depending on the computer power.

The optimum viewing distances as advised by the IEC 62676 CCTV standard

Screen size	Resolution	Pixel size mm	Distance m
20"(51cm)	SXGA+ (1 400x1 050)	0,29	1,00
50"(127cm)	SXGA+ (1 400x1 050)	0,71	2,50
70"(178cm)	SXGA+ (1 400x1 050)	1,00	3,40
80"(204cm)	SXGA+ (1 400x1 050)	1,14	3,90
50"(127cm)	Full HD (1 920x1 080)	0,57	1,98
70"(178cm)	Full HD (1 920x1 080)	0,80	2,75

Printed picture details

What we have covered up until now is the pixel as an electronic display picture element. As we discussed previously under "About pixels and resolution" these elements are composed of red, green, and blue (RGB) "sub-elements;" these are the smallest picture elements.

There are, however, pixels in the printed world as well, the brochures, books, and magazines. Those pixels also represent the smallest detail of a picture since they represent printed pixel of a digitally captured pixel, but they are different to the display pixels. The display pixels are made of RGB sub-elements and produce all colors with *additive color mixing*. The printed pixels are made of cyan, magenta, yellow, and black (CMYK) ink dots which are combined differently to produce resultant colors with *subtractive color mixing*. The actual "natural" colors of the ink jet printouts are obtained by a *dithering* process, which is actually spraying and mixing the ink jet dots with various sizes and combining them to produce a resultant color. In essence, the ink jet color printers are binary devices in which the cyan, magenta, yellow, and black dots are either "on" (printed) or "off" (not printed), with no intermediate levels possible. This is conceptually different from the RGB phosphor on a CRT or LCD, where each of the primary colors can have a variety of intensity. A "binary" CMYK printer can only

print five "solid" colors (cyan, magenta, yellow, black, and white). White is actually the non-printed paper background color, but that is also used. Clearly, this is not a big enough palette to deliver good color print quality, which is where *half toning* comes in. This is still the case even with the new photo quality ink jet printers that have additional two colors: light-cyan and light-magenta (for better human flash reproduction). Half toning algorithms divide a printer's native dot resolution into a grid of halftone cells and then turn on varying numbers of dots within these cells in order to mimic a variable dot size. By carefully combining cells containing different proportions of CMYK dots, a half toning printer can "fool" the human eye into seeing a palette of millions of colors rather than just a few. Depending on the type

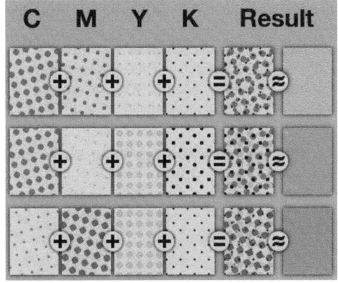

Half-toning - unfortunately colors can't be seen in BW print of this book

of printer used these dots can be smaller or bigger which depends on the printer's dots per inch (DPI) specification. Some good quality ink-jet printers may state 720, 1440, or even 2880 DPI, for example. Using these CMYK color dots with various sizes (droplets), of which the smallest are an inch divided by 1440 or 2880, the actual electronic pixels from a camera can be printed and shown as a continuous image when viewed from a certain distance. A simple rule of thumb by some imaging people, such as Adobe, is to divide the ink jet DPI as specified by the ink jet printer manufacturers with 4 to get the "real" color dots per inch. Practically, this means that a 720 DPI ink jet printer can reproduce 180 colored pixels per inch (PPI). To achieve highest quality, it is important to use high-quality printing paper. For best results, using the ink and photo paper of the corresponding printer manufacturer is recommended.

The electronic image can be described with pixels as an absolute pixel count, horizontal by vertical pixels (H x V). So for example if a screen is HD, then we know that HD display screen should have 1920 x 1080 pixels no matter what size it is. This means the RGB pixels on different size displays will have different pixel physical size to fit the screen. These pixels are fixed size for the display; they can't change. When using printers, however, the dots used to create the CMYK colors vary in size. These dots are the ones specified as Dots per Inch (DPI) in the printing industry. So when trying to print a CCTV exported image - it may produce different visual quality depending upon how many DPIs is the printer, what is the pixels count of the

An ink-jet printer

Photo courtesy of Epson

image, the PPI selected in the photo-editor and on what type of paper this is printed on. If we choose to print at 254 PPI resolution, for example, it practically means more than 10 dots per millimeter (since one inch is 25.4 mm). This is certainly a very tiny dimension, and the human eye cannot distinguish two different tiny color dots when they are very close to each other in a 254 PPI print.

The mixing of colors in the offset printing industry (not using domestic ink-jets but big press machines) is done in a completely different way. Instead of using small ink jets, the offset printing industry uses four plates each being a separate color of the CMYK. Black is added for additional dark tones, although theoretically, CMY are sufficient to produce other, resultant, colors. We use four different inks when printing color magazines or books, and in order to produce the resultant color the smallest picture elements in the printing industry are produced by having all these elementary color dots very close to each other. The difference compared to TV monitors is that the colors are not positioned next to each other in line (which is typically the case with most LCD and CRT phosphor these days), but rather they are displaced at various angles, such as 45° for black, 75° for magenta, 90° for yellow, and 105° for cyan. The resolution of the offset printing varies, but typical common standard for glossy magazines is 300 PPI. So when we read a

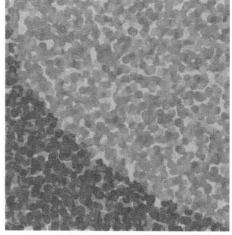

Magnified red and yellow ink-jet droplets

magazine from a normal reading distance (typically 0.5 m), the color pixels cannot be detected by a normal eye and we only see the resultant (subtractive) color mix.

Using the above, if we export a HD image with 1920 x 1080 pixels and print it out, the best visual quality would be when printed on quality glossy photo paper, with the native best resolution. If we choose this to be 300 PPI (this could be done in the photo-editing software), then the best size to print such an image would be 1920:300 x 1080 :300 = 6.4 x 3.6 inches = 162 x 91 mm, which is approximately a post-card size, convenient for viewing at an arm length distance.

Compression influence on picture details

All of the previously said and explained about the pixels per inch, the optimum viewing distance, and image quality **does not and cannot consider compression artifacts in an image.** As it is known, and will be explained later on in detail, most of the digital video processing, transmission, and storage is using some form of digital compression. Inevitably, any compression introduces artifacts in the stored and/or exported image, which may be visible when printed as a hard copy. **Such possible artifacts have nothing to do with the quality of printing resolution as they are a result of the compression, prior to exporting for print.**

It is however important to note that when a digitized and compressed image is exported and given as evidence to a third party (police, for example) it is suggested to be exported as uncompressed bitmap format (BMP or TIF), and printed as such, **so that no additional compression artifacts are introduced.** But artefacts already introduced by the original compression in the camera cannot be removed nor reverted, as most of the compression used in the CCTV industry are *lossy*. Lossless compression is possible, but only used in broadcast and cinematography for editing purposes and prior to mass distribution on DVDs, Blue-Ray discs, or streaming.

As mentioned in the previous chapter, often it can be found that high quality ink-jet printout of an exported image on a photographic paper visually shows more details than having the same image seen on a CCTV monitor. This is simply connected with the psycho-physiology of viewing various resolution details at various distances. If printing is done on a high quality paper with a

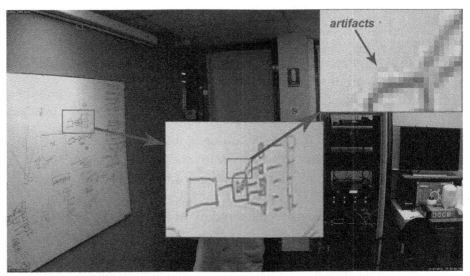

Compression artifacts on an HD image

quality settings on an ink-jet printer and at the same time let the pixel resolution of the exported image to match the ink-jet printing pixels, then the optimal viewing distance is holding the paper at reading distance. This is why it would appear as if there are more details on such a printout rather than viewing it on an LCD screen for example.

There is no simple formula one can use to indicate how much compression artefacts will appear in an image. This depends on many factors, the first and most important being the strength of the compression, but it also depends on the content of the image. Only if the picture content has significant transition in contrast (like from black to white, as in the example photo above) the artifacts might be noticeable.

Brightness, Contrast, and Gamma

There are certain monitor/display settings that might be changed by an operator which could affect the clarity and details of the displayed video. The first two such settings are the brightness and contrast of the display. **Brightness moves the average luminance level of the display high or low. Contrast increases or reduces the luminance difference between the brightest and the darkest part of the displayed image.** The simplest method of correct adjustment of brightness and contrast of a display is by use of an electronically generated test pattern, with preferrably a 11-step grey scale changing from black to white. The simple idea is to adjust the brightness and contrast so that all steps of the grey scale are visible and equally spaced from each other. Correct brightness and contrast are important for good reproduction of video, and even more important when multiple displays are used in one room. Same cameras should display the same on various monitors. If brightenss and contrast are not set correctly the same video signal will look different on different displays.

11-steps grey scale for adjusting brightness and contrast

The next important setting on a display is **gamma. Gamma is a process in encoding and decoding video to compensate for the non-linear properties of human vision, and to maximize the use of the display devices relative to how human eyes perceive light.**

Gamma setting may not necessarily be accessible on all displays, and, on some, there is a complete setup procedure available in the monitor advanced menu settings.

Human vision, like hearing, does not follow linear law, but rather logarithmic or exponential. If video is not corrected with this gamma, they allocate linear values to all luminance levels, from black to white, which human eye cannot differentiate properly, as in real life. The most simple illustration on how this works can be observed seeing at a sunny day, for example, at 50,000 luxes, half of that illumination of around 25,000 luxes will seem half of the original brightness; yet, in low light situations, when seeing a 50 lux scene for example, half of that brightness will of about 25 lux will appear to our eyes as half. Clearly, there is a big difference between 25,000 and 25 lux, but eyes see it as half relative to the illumination we compare it with.

Human vision is non-linear

This illustrates the non-linearity of human vision.

This is what gamma function intends to achieve.

So, in order to extract such details in the shadow, a non-linear curve is applied to the camera signal, and this curve can be described with a typical exponent of 2.2. Such a curve happens to be inverse to the curve that in the past tube cameras had, of 0.45, so that when both of these curves are put together, they resemble the mathematical symbol γ (*Gamma*), hence the name.

Modern solid state cameras sensors have linear response to illumination levels, but the intended reproduction of a scene is almost always nonlinearly related to the measured scene intensities at the camera end. In the process of rendering the camera linear raw data into conventional RGB video, a color space and rendering transformations is performed. **This represents the gamma correction at the camera end. This also improves the appearance of camera details in shadows, and indirectly increases the efficiency of compression.** All standard RGB color spaces and file formats use a non-linear luminance encoding (gamma conversion) of the intended intensities of the primary colors of the reproduction.

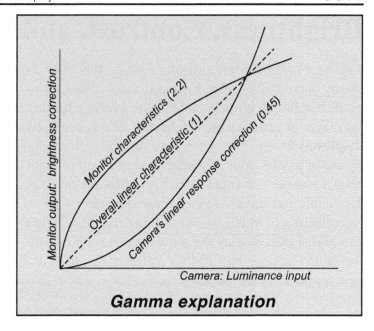

Gamma explanation

Output on the older CRT monitors does not usually require further gamma correction, since the standard video signals that are transmitted or stored incorporate gamma compression that provides a pleasant image after the gamma expansion of the CRT (although it is not the exact inverse function, but close). **For analog video signals, the actual gamma values are defined by the video standards (PAL or NTSC), and are always fixed and well known values.**

A common misconception is that gamma encoding was developed to compensate for the input–output characteristic of CRT displays. In CRT displays the electron-gun current, and thus light intensity, varies nonlinearly with the applied anode voltage. Altering the input signal by gamma compression can cancel this nonlinearity, such that the output picture has the intended luminance. However, the gamma characteristics of the display device

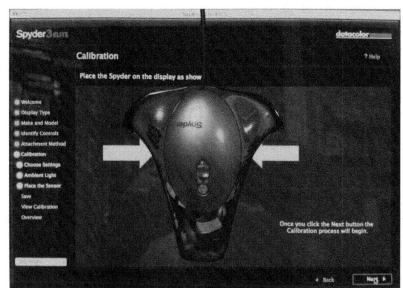

Color calibration tool

Y=2

Y=1 (original)

Y=1/2

Y=1/3

Y=1/4   Courtesy of Wikipedia

Various Gamma

do not play a factor in the gamma encoding of images and video — they need gamma encoding to maximize the visual quality of the signal, regardless of the gamma characteristics of the display device. The similarity of CRT physics to the inverse of gamma encoding needed for video transmission was a combination of coincidences which simplified the electronics in early television sets.

In modern computers with LCD monitor, images are encoded with a gamma of about 0.45 and decoded with a gamma of 2.2. **The values of gamma are embedded in the operating system itself.** A notable exception, until the release of Mac OS X 10.6 (Snow Leopard) were Macintosh computers: they used 0.55 and 1.8 respectively, but Snow Leopard Mac computers also use display gamma of 2.2.

All digital images and video signals have encoded gamma values. This is part of the various standards. Binary data in still image compressed files (JPEG, JPEG-2000) are explicitly encoded. **They carry gamma-encoded values, not linear intensities.** The same applies to video compressed streams, such as MPEG and H.264.

Sometimes a particular display or application may require different or more accurate gamma setting. This, for example, could be required in a desktop publishing system where display appearance has to match as close as possible the printed output. **The operating system can optionally further manage such cases, through color management, if a better match to the output device gamma is required.** This is usually done with hardware color calibration tools. The color calibration of a system (real life - camera - monitor - hard copy print) may be one of the most important processes for best color reproduction of real life images. This is of paramount importance for photographers, cinematographers, but it certainly can be used in CCTV. This is a very complex topics, especially because we have different color spaces (RGB vs. CMYK). We won't go in more details here, as it requires a book on its own, but it is important to know that such calibrations are possible.

It is important to highlight that with incorrect brightness, contrast, or gamma settings in the monitor, a nice video signal may be ruined and appear as if the source (camera or recorded footage) is of inferior quality.

Recognizing faces and details in CCTV

There are many interests in CCTV, but the most common can be categorized in the following groups:

- People identification

- Human activity

- Handling money and playing cards

- Vehicles and licence plates

- Traffic activity

Perhaps one of the most important functions is to be able to recognize or identify a face. Many standards have been developed with the recognition and identification as central points as the purpose of CCTV. So lets define first these terms in CCTV. To be able to identify a face from a recorded footage means that, based on our prior knowledge, we can say with high confidence that the person shown in the footage is a particular person we know or have seen before. We can positively confirm his/her identity. To be able to recognize a person, instead of identify, is less accurate in the jargon of CCTV. To recognize a person means we can determine if the person is male or female, whether it is wearing a blue shirt, or red skirt and similar. Understandably, to identify a person requires more details than to recognize.

So, how many details (pixels) are required to identify a person?

There are a number of standards around the world which may have similar or slightly different suggestions but they all come very close to each other's recommendations. For example, Australian Standards AS4806.2, after conducting independent testing and research, has concluded and suggested that when a camera angle of view is adjusted so that a 100% person's height is visible on a standard definition (SD) monitor, then the face should have sufficient details and it should be possible to *identify that person*. If we consider that Australia uses PAL analog system uses 576 active TV lines, and if we assume an average human head occupies approximately 15% of the total height (average used 1.7~1.8 mm), **then, based on this, 86~88 pixels would suffice to make a positive face identification**. The AS4806.2 standard clearly states that this applies to well focused lens, sufficient light to produce good image, and assumes the face details are not affected by video compression artifacts. Certainly, narrower views than suggested above may be even better for identifying a person, but, after extensive testing, the round figure of 100% of monitor height was proposed because of its simplicity when using it in practice. Any installer can easily position a camera and fit or adjust a vari-focal lens so that at a given distance the image of a person occupies 100% of the screen height. There is no need for special instruments or calculators, just a common sense of what is required.

The same AS 4806.2 standard suggests that to *recognize a person* it's height on the SD screen should occupy around 50% of the screen height. Larger than 50% of the monitor height people will give perhaps even more details, but at least 50% is required for an observer to identify if the person is male, female, what he/she wears, etc. but not necessarily being able to identify the person.

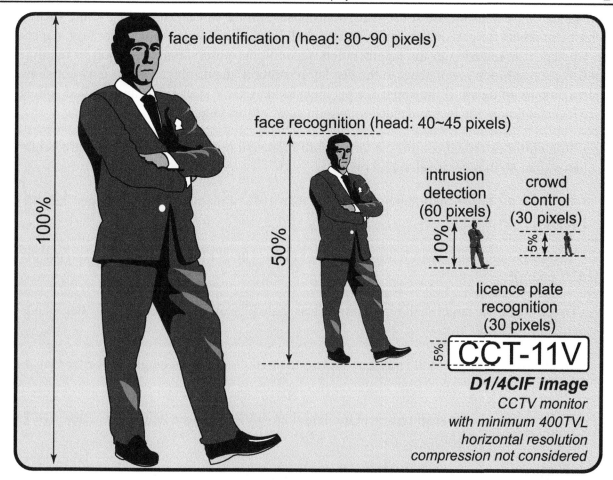

face identification (head: 80~90 pixels)

face recognition (head: 40~45 pixels)

intrusion
detection
(60 pixels)

crowd
control
(30 pixels)

licence plate
recognition
(30 pixels)

CCT-11V

D1/4CIF image
CCTV monitor
with minimum 400TVL
horizontal resolution
compression not considered

The AS-4806.2 recommendations

The next detail suggested by the AS4806.2 standard is a person's height on the screen for *intrusion detection*, which needs to be at least 10%.

And last, but not least, at least 5% of screen height needs to be occupied for *crowd control* purposes.

For visual *licence plate* recognition at least 5% of the SD monitor height needs to be occupied by the letters and numbers of a vehicle's licence plate. We highlight here "visual recognition," as the requirements for machine automatic licence plate recognition might be different and will depend on manufacturers' algorithms and level of software confidence.

Some experts, manufacturers, and consultants often use different metrics to indicate the same requirements, in *pixels/metre* (pix/m). Such numbers describe the density of pixels as seen on the screen, relative to the dimensions the camera is seeing at the point of the object. Using a simple mathematics, we can convert the AS4806.2 recommendations into pix/m. For example, if we assume that the 15% head-height of an average human is approximately 25cm (10 inches), and the AS4806.2 standard recommends 86~88 pixels for identification, then four times this number (because 1m / 25cm = 4) is it's pixels per meter density, i.e. 344 ~ 352 pix/m. Another way of expressing this could be approximately 3 mm per pixel (1,000mm / 344).

Other standards, such as the latest ISO-62676, proposes similar values. According to the ISO62676, the size of an object (target) on the display screen shall have a relation to the operator task, depending upon whether it is referring to the identification, recognition, observation, detection or monitoring. If the camera resolution is not equal to the display resolution, the displayed scene may not show the expected amount of detail. If the target is a person and the CCTV system has an installed equivalent PAL (576i) resolution, the recommended minimum sizes of this target are:

- To monitor or crowd control the target shall represent not less than 5% of picture height (or better than 80 mm per pixel = 12.5 pix/m);

- To detect the target shall represent not less than 10% of picture height (or better than 40 mm per pixel = 25 pix/m);

- To observe the target shall represent 25% of picture height (or better than 16 mm per pixel = 62.5 pix/m);

- To recognise the target shall represent not less than 50% of picture height (or better than 8 mm per pixel = 125 pix/m);

- To identify the target shall represent not less than 100% of screen height (or better than 4 mm per pixel = 250 pix/m);

- To inspect the target shall represent not less than 400% of screen height (or better than 1 mm per pixel = 1,000 pix/m).

Drawing courtesy of IEC - ISO 62676-4

The ISO-62676 recommendations

All of these recommendations can be applied to any resolution system, despite them being recommended for analog SD resolution signal. In other words, if a system needs to satisfy the identification requirement according to ISO-62676 for example, where 250 pixels per meter are required at the object (at the person that needs to be identified) plane, then it does not really matter if we are using SD camera, HD or mega-pixel, as long as the pixel density is 250 pix/m at the object plane.

We will now show a formula that may help in determining the lens required to achieve certain quality:

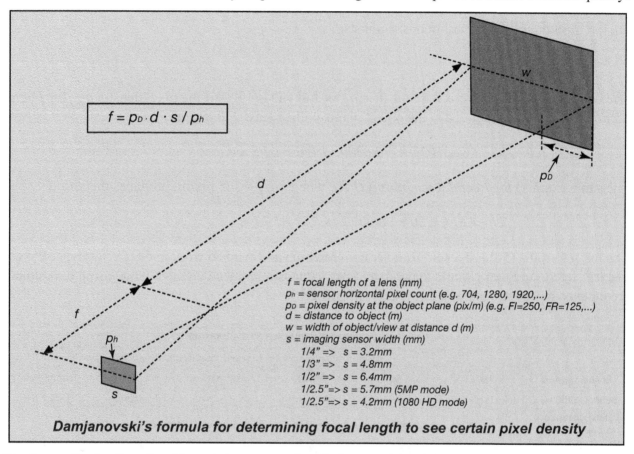

$$f = p_D \cdot d \cdot s \, / \, p_h$$

f = focal length of a lens (mm)
p_h = sensor horizontal pixel count (e.g. 704, 1280, 1920,...)
p_D = pixel density at the object plane (pix/m) (e.g. FI=250, FR=125,...)
d = distance to object (m)
w = width of object/view at distance d (m)
s = imaging sensor width (mm)
     1/4" => s = 3.2mm
     1/3" => s = 4.8mm
     1/2" => s = 6.4mm
     1/2.5"=> s = 5.7mm (5MP mode)
     1/2.5"=> s = 4.2mm (1080 HD mode)

Damjanovski's formula for determining focal length to see certain pixel density

Similar to the finding focal length lens or angle of view explained in the chapter *Optics in CCTV*, based on the drawing above, we can establish the proportion (which applies to all types of sensors):

$$f \, / \, s = d \, / \, w \tag{47}$$

From the above equation, the focal length of a lens can be found as:

$$f = (d \cdot s) \, / \, w \tag{48}$$

If we now know the pixel density that is required at the object plane, at distance d, we can express the w of the object as horizontal sensor pixel count p_h divided by the pixel density p_D:

$$w = p_h / p_D \tag{49}$$

Replacing (49) into (48) we get the minimum focal length of a lens (in mm) required to obtain details of certain pixel density p_D at the object which is at distance d, using a sensor with image width s and horizontal pixel count p_h :

$$f_D \geq (p_D \cdot d \cdot s) / p_h \tag{50}$$

So, let's work out an example, for a 1/3" sensor for full HD resolution. Let's assume we need to identify a person at 100m away, which would represent the variable d.

So, the formula (50) becomes more specific as:

$$f_{FI} \geq (p_{DFI} \cdot d \cdot s) / p_h \tag{51}$$

The width s of a 1/3" sensor is 4.8 mm. The p_h for full HD is 1920. If the pixel density p_{DFI} for **Facial Identification (FI)** is required at the object plane, with at least 250 pix/m.

$$f_{FI} \geq (250 \cdot 100 \cdot 4.8) / 1920 = 120{,}000 / 1920 = 62.5 \ mm \tag{52}$$

The same exercise for **Facial Recognition (FR)**, where p_{DFR} = 125 pix/m, produces this result:

$$f_{FR} \geq (125 \cdot 100 \cdot 4.8) / 1920 = 60{,}000 / 1920 = 31.25 \ mm \tag{53}$$

The main formula (50) is a result from my own research and original work in CCTV for over 30 years and it is free to use, but I would appreciate if my name is referenced/credited when using it. A simple credit, such as "Damjanovski's formula," will suffice.

	Face Identification (FI) (>250pix/m)					
	All calculations below are for a lens focal length in mm, to achieve FI					
Sensor type =>	1/4" SD	1/3" SD	1/3" HD	1/2" HD	1/2.5" HD	1/2.5" 5MP
sensor width =>	s = 3.2mm	s = 4.8mm	s = 4.8mm	s = 6.4mm	s = 4.2mm	s = 5.7mm
Below, distance to object in m	Below are calculated focal length of lenses which achieve FI at the distance on the left					
2	2.3	3.4	1.3	1.7	1.1	1.3
3	3.4	5.1	1.9	2.5	1.6	2.0
5	5.7	8.5	3.1	4.2	2.7	3.3
10	11.4	17.0	6.3	8.3	5.5	6.7
15	17.0	26	9	13	8	10
20	22.7	34	13	17	11	13
30	34.1	51	19	25	16	20
40	45.5	68	25	33	22	27
50	56.8	85	31	42	27	33
60	68.2	102	38	50	33	40
70	79.5	119	44	58	38	47
80	90.9	136	50	67	44	54
90	102.3	153	56	75	49	60
100	113.6	170	63	83	55	67
150	170.5	256	94	125	82	100
200	227.3	341	125	167	110	134
300	340.9	511	188	250	165	201
500	568.2	852	313	417	275	335

Face identification table, as per ISO 62676

The formula (50) is universal, and it can actually be further expanded and used for finding details on the screen of any object, providing the pixel density is known, the distance to the object (d), the sensor width (s) and the pixel count of the sensor (p_h).

The most common horizontal pixel counts in CCTV these days are 704 pixels for SD video, 1280 for 720 HD, and 1920 for 1080HD video.

The results in (52) means with a lens of at least 62.50 mm focal length it *should* be possible to identify a face at 100 m distance using a 1/3" full HD sensor. We need to highlight "should" as we are assuming a good quality lens (with a resolution at least as good as the sensor it is used on) and a good quality compression that does not noticeably reduce details.

Also, we have to assume that the display pixel quality matches the camera pixel quality. In other words, the display should be capable of showing all pixels that are encoded by the camera. For example, if we have an HD camera, delivering 1920 x 1080 pixels, and showing it on a 1280 x 720 HD monitor - it is obvious that the monitor will reduce the details (down-sample) due to lack of pixels. For such a system, a display with adequate resolution of 1920 x 1080 should be used. Similarly, if we have a mega-pixel camera with a resolution of 8000 x 4000 pixels, as an example, with intention to display it on an HD display with 1920 x 1080 pixels, we must be able to zoom in the image/video at least four times (8000 / 1920) in order to be able to see pixel for pixel resolution.

Similar calculations, using the formula (50), can be made for any format and any sensor size. The tables on this page show some calculations for the most common types of sensors, and most common formats (SD, HD and MP), for Face Identification and for Face Recognition as per ISO 62676.

Sensor type =>	Face Recognition (FR) (>125pix/m)					
	All calculations below are for a lens focal length in mm, to achieve FR					
Sensor type =>	1/4" SD	1/3" SD	1/3" 1080HD	1/2" 1080HD	1/2.5" 1080HD	1/2.5" 5MP
sensor width =>	s = 3.2mm	s = 4.8mm	s = 4.8mm	s = 6.4mm	s = 4.2mm	s = 5.7mm
Below, distance to object in m	Below are calculated focal length of lenses which achieve FR at the distance on the left					
2	1.1	1.7	0.6	0.8	0.5	0.7
3	1.7	2.6	0.9	1.3	0.8	1.1
5	2.8	4.3	1.6	2.1	1.4	1.9
10	5.7	8.5	3.1	4.2	2.7	3.7
15	8.5	13	5	6	4	5.6
20	11.4	17	6	8	5	7.4
30	17.1	26	9	13	8	11.1
40	22.7	34	13	17	11	14.8
50	28.4	43	16	21	14	18.6
60	34.1	51	19	25	16	22.3
70	39.8	60	22	29	19	26.0
80	45.5	68	25	33	22	29.7
90	51.2	77	28	38	25	33.4
100	56.8	85	31	42	27	37.1
150	85.3	128	47	63	41	55.7
200	113.7	171	63	83	55	74.2
300	170.5	256	94	125	82	111.3
500	284.2	426	156	208	137	185.5

Face Recognition table, as per ISO 62676

As you can see from the tables, face recognition or identification is possible with any sensor, be that SD or HD or MP, the only question is what focal length lens needs to be put on. So, most of the time the "trick" is to find a suitable lens and position for the camera so that it can see sufficient details for identifying people or license plates. But **the camera that is positioned to capture face for identification purposes, for example, cannot, at the same time, be used to see the whole wider area.** Customers are usually the ones who expect one camera to cover everything, see everything, and recognize everything. This topic has been discussed often, and it still is a stumbling block in designing various systems. If we all work under the pressure of a budget (and the budget is a very important consideration) the tendency is to have as small number of cameras as possible. Yet, when there is an incident and a positive identification is required, the CCTV system designer could be blamed for having a system that cannot recognize a face or a car, even if they are captured in the camera field of view. So, a simple recommendation is: **do not compromise your system design**; rather, **educate the customers** so that they understand why more cameras with certain coverage would be required. If necessary, have two cameras covering a foyer entrance, for example, one having a wide and global coverage, the other with a narrower angle of view picking up faces entering the foyer. Initially, this might seem to be an overkill, but when a suspected intruder is identified and captured, the system proves its existence. This **is** the purpose of a surveillance system.

In summary, all of the standards above suggest that for successful identification or recognition of a face, it practically doesn't matter what kind of camera is used. Whether it is SD, HD, or mega-pixel, all it matters is the number of pixels dedicated to identifying or recognizing a person. Understandably, with a lower resolution camera, like SD, for example, compared to HD, the angle of view needs to be narrower in order to see the same details.

All of the percentages and suggestions by the various standard described above refer to identification or recognition of people. It is possible however, to draw similar conclusions of minimum required pixels to identify money, playing cards, and anything else that might be needed by a CCTV system. Often, this is best done by testing and experiments.

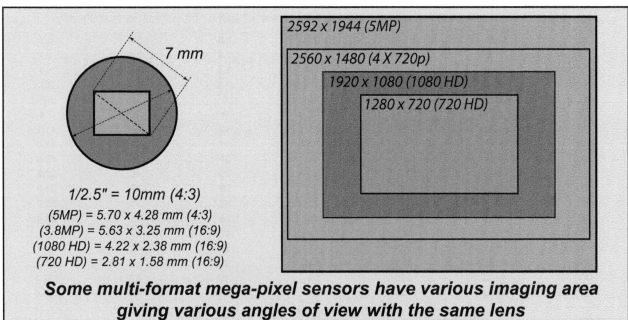

7 mm

2592 x 1944 (5MP)

2560 x 1480 (4 X 720p)

1920 x 1080 (1080 HD)

1280 x 720 (720 HD)

1/2.5" = 10mm (4:3)

(5MP) = 5.70 x 4.28 mm (4:3)
(3.8MP) = 5.63 x 3.25 mm (16:9)
(1080 HD) = 4.22 x 2.38 mm (16:9)
(720 HD) = 2.81 x 1.58 mm (16:9)

Some multi-format mega-pixel sensors have various imaging area giving various angles of view with the same lens

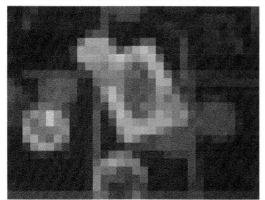

Playing card at 12 pixels

Playing card at 24 pixels

Playing card at 48 pixels

Playing card at 96 pixels

The sample images of playing cards on the left illustrate clearly how many pixels are required to recognise the cards and chips on the gaming table.

The first image with card being 12 pixels high has clearly insufficient information to make affirmative conclusion of the card and the chips.

The next image with 24 pixels card height is nearly convincing of what the card is, but may leave a little bit of doubt to a non-player. This means less than 24 pixels will be insufficient to be affirmative with the cards. The chips in the second photo occupy around 12 pixels in diameter and this too is insufficient for clear recognition.

The third image shows the playing card (queen of diamonds) with 48 pixels, which is now very clear and convincing. So, for playing cards, at least 40 pixels are required to make positive identification. The chips are now also recognizable, with their diameter now occupying 24 pixels, although not very convincingly.

In the last image on the left, the playing card has been already determined with even half of the 96 pixels shown here. So, having an image of playing cards with 96 pixels card height is certainly okay, albeit slightly unnecessary. But the chips with 48 pixels diameter are now certainly positively identified, which will make us conclude that at least somewhere around 40 pixels are needed for the chips identification too.

From the above, **a generic conclusion can be made that for objects with a reasonable amount of details, an approximate suggestion would be that they will require at least 40 pixels.** By knowing the playing cards physical dimensions of 9 cm x 6.35 cm, we can calculate the pixel density required at the card plane. This is simply:

(1 m / 9 cm) x 40 pix/m = 444 pix/m

Using the above pixel density, and the formula (50) we can also find out lenses required for card clear identification.

This is an experimental drawn suggestion but the theory and formula (50) satisfy the practical tests making it real and universal.

7. Analog video processing

Only very small CCTV systems use the simple camera-monitor concept. Most of the bigger ones, in one way or another, use video switching or processing equipment before the signal is displayed on a monitor. With the introduction of digital video, all switching and processing, even of the analog video signals (certainly after they are digitized), are done over network and network switchers. This chapter may well be outdated within the next couple of years, but there are still existing analog CCTV systems out there, and therefore there was a need to still cover this topic.

Analog switching equipment

The simplest and most common device found in small to medium-sized CCTV systems is the video sequential switcher. As the name suggests, they switch multiple video signals onto one or two video outputs sequentially, one after another.

Video sequential switchers

Since in the majority of CCTV systems we have more cameras than monitors on which to view them, there is a need for a device that will sequentially switch from one camera signal to another. This device is called a ***video sequential switcher***.

Sequential switchers come in all flavors. The simplest one is the 4-way switcher; then we have 6-way, 8-way, 12-way, 16-way, and sometimes 20-way switchers. Other numbers of inputs are not excluded, although they are rare.

The switcher's front panel usually features a set of buttons for each input. Besides the switch position for manual selection of cameras, there is a switch position for including a camera in the sequence or bypassing it. When a sequence is started, the dwell time can be changed, usually by a potentiometer. The most common and practical setting for a dwell time is 2 to 3 seconds. A shorter scanning time is too impractical and eye-disturbing for the operator, while a longer scanning time may result in the loss of information for the non-displayed cameras. So, in a way, sequential switchers are always a compromise.

Apart from the number of video inputs, sequential switchers can be divided into switchers with and without alarm inputs.

A simple, 8-channel video sequential switcher

When a sequential switcher has alarm inputs, it means that external normally opened (N/O) or normally closed (N/C) voltage-free contacts can halt the scanning and display the alarmed video signal. Various sources can be used as alarm devices. For indoor applications the choice of suitable sensors is often straightforward, but outdoor alarm sensors are more critical and harder to select. There is no perfect sensor for all applications. The range of site layouts and environmental conditions can vary enormously. The best help you can get in selecting a sensor is from a specialized supplier that has both the knowledge and the experience.

A more advanced switcher

Most common are the ***passive infrared*** (PIR) detectors, door reed switches, PE beams, and ***video motion detectors*** (VMDs). Care should be taken, when designing such systems, about the switcher activity after the alarm goes off, that is, how long the alarmed video input remains displayed, whether it requires manual or automatic reset, if the latter, how many seconds the automatic reset activates for, what happens when a number of alarms activate simultaneously, and so on. The answers to all of these questions are often decisive for the system's efficiency and operation. There is no common answer, and it should be checked with the manufacturer's specifications. Even better, test it yourself.

It is not a rule, but quite often simple sequential switchers (i.e., those without alarm inputs) have only one video output. The alarming sequential switchers on the other hand, quite often have two video outputs: one for video sequencing and the other for the alarmed picture. The first output is the one that scans through all the cameras, while the second one is often called the alarmed or spot output because it displays the alarmed picture (when the alarm activates).

Video sequential switchers (or just switchers for short) are the cheapest thing that comes between multiple cameras and a video monitor. This does not mean that more sophisticated sequential switchers are not available. There are models with text insertion (camera identification, time, and date) multiple configuration options via RS-232, RS-485, or RS-422 communications, and so on.

Some models like these either have the power down coaxial cable function, or they send synchronization pulses to the camera via the same cable that brings video signals to the switcher. All this is with the intention of synchronizing cameras, which will be discussed next. Most of these more sophisticated sequential switchers can easily be expanded to the size of a miniature matrix switcher.

Synchronization

One of the more important aspects of switchers, regardless of how many inputs they have, is the switching technique used. Namely, when more than one camera signal is brought to the switcher inputs, it is natural to have them with various video signal phases. This is a result of the fact that every camera is, in a way, a self-contained oscillator producing the line frequency of the corresponding TV system (i.e., for CCIR $625 \times 25 = 15,625$ Hz and for EIA $525 \times 30 = 15,750$ Hz), and it is hard to imagine that half a dozen cameras could have a coincidental phase. This is unlikely even for only two cameras. We call such random phase signals ***non-synchronized***. When non-synchronized signals are switched through a sequential switcher, an unwanted effect appears on the monitor screen: ***picture-roll***. A picture-roll appears owing to the discrepancies in the vertical synchronization pulses at various cameras that results in an eye-disturbing picture-roll when the switcher switches from one camera to another. The picture-roll is even more obvious when recording the switched output to a VCR. The roll is more visible with the VCR because the VCR's head needs to mechanically synchronize to the different cameras' vertical sync pulses, while the monitor does it electronically. The only way to successfully combat the rolling effect is to synchronize the sources (i.e., the cameras).

The most proper way of synchronizing cameras is by use of an external ***sync generator*** (***sync-gen***, for short). In such a case, cameras with an external sync input have to be used (please note: not every camera can accept external sync). Various cameras have various sync inputs, but the most common are:

- Horizontal sync pulses (usually known as horizontal drive pulses or HD)

- Vertical sync pulses (usually referred to as vertical drive pulses or VD)

- Composite sync pulses (which include both HD and VD in one signal, usually referred to as composite video sync or CVS)

In order to perform synchronization, an extra coaxial cable has to be used between the camera and the sync-gen (besides the one for video transmission), and the sync-gen has to have as many outputs as there are cameras in use.

This is clearly a very expensive exercise, although, theoretically, it is the most proper way to synchronize. Some camera manufacturers produce models where sync pulses are sent from the switcher to the camera via the same coax that sends the video signal back. The only problem here is the need to have all the equipment of the same make.

There are cheaper ways to resolve the picture-roll problems; one of the most accepted is through ***line-locked cameras***. Line-locked cameras are either 24 V AC or 240 V AC (110 V AC for the United States, Canada, and Japan) powered cameras. The 50 Hz (60 for the United States, Canada, and Japan) mains frequency is the same as the vertical sync rate, so these cameras (line-locked) are made to pick up the zero crossings of the mains sine

A line-locked camera (24 V AC) which also has an external vertical sync input terminal

wave and the vertical syncs are phased with the mains frequency. If all of the cameras in a system are powered from the same source (the same phase is required), then all of the cameras will be locked to the mains and thus synchronized to each other.

The above method is the cheapest one, although it sometimes offers instability of the mains phase owing to heavy industrial loads that are turned off and on at unpredictable intervals. Still, it is the easiest way. There is even a solution for different phases powering different cameras in the form of the so-called *V-phase* adjustment. This is a potentiometer on the camera body that will allow the camera electronics to cope with up to 120° phase difference. It should be noted that the low-voltage AC-powered cameras (i.e., the 24 V AC) are more popular and more practical than the high voltage ones primarily because they are safer.

Some cameras are designed to accept the video signal of the previous camera and lock to it. This is called *master-slave* camera synchronization. By daisy-chaining all of the cameras in such a system, synchronization can be achieved, where one is the master camera and the others are slave cameras. A coaxial cable is required between all of the cameras for this purpose, in addition to the coax for video transmission.

Still, not every sequential switcher can use the benefits of synchronized cameras. The switcher also needs to be a *vertical interval switcher*. Only vertical interval switchers can switch synchronized signals at the moment of the vertical sync pulse so that the switching is smooth and without roll. Nonvertical interval switchers switch on a random basis rather than at a specific moment relative to the video signals. With the vertical interval switcher, when a dwell time is adjusted to a particular value, the switcher switches with this specific dwell time, but only **when the vertical sync period occurs**. By doing so, **the switching is nice and clean and happens in the vertical blanking period**; that is, there is no picture break on the monitor screen.

Normal switchers, without this design, will switch anywhere in the picture duration; this means it could be in the middle of a picture field. So if the cameras are synchronized, there will be no picture-roll, but picture breaking will still be visible to the operator **owing to the abrupt transition from one signal to another in the middle of the visible picture field**.

The same concept of vertical interval switching applies to the sequential switcher's big brother, the video matrix switcher.

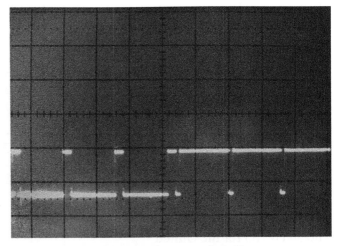

Vertical sync detail

Analog Video Matrix Switchers

The analog video matrix switcher, as we have noted, is the big brother of the sequential switcher. The bigger CCTV systems can only be designed with a video matrix switcher (VMS) as the brain of the system.

The name *"matrix switcher"* comes from the fact that the number of video inputs plotted against the number of video outputs makes a matrix, as it is known in mathematics. Quite often, video matrix switchers are called video cross-point switchers. These cross-points are actually electronic switches that select any video input onto any video output at any one time, preserving the video impedance matching. Thus, one video signal can simultaneously be selected on more than one output. Also, more video inputs can be selected on one output; only in this case we would have a sequential switching between more inputs, since it is not possible to have more than one video signal on one output at any single point in time.

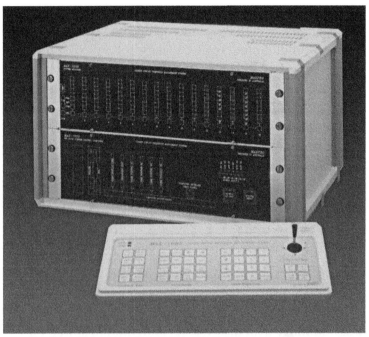

A sophisticated video matrix switcher

Video matrix switchers are, in essence, big sequential switchers with a number of advancements:

• A matrix switcher can have more than one operator. Remember that the sequential switchers usually have buttons at the front of the unit. Thus, only one operator can effectively control the system at any one time. Matrix switchers can have up to a dozen operators, sometimes even more, all of whom can concurrently control the system. In such a case, every operator controls (usually) one video output channel. A certain intelligent control can be achieved, depending on the matrix in use. Different operators may have equal or different priorities, depending on their position in the security structure of the system.

• The matrix switcher accepts many more video inputs and accommodates for more outputs, as already mentioned, and more importantly, these numbers can easily be expanded at a later date by just adding modules.

• The matrix has pan, tilt, and lens digital controllers (usually referred to as PTZ controllers). The keyboard usually has an integral joystick, or buttons, as control inputs and at the camera end, there is a so-called PTZ site driver within a box that is actually part of the matrix. The PTZ site driver talks and listens to the matrix in digital language and drives the pan/tilt head together with the zoom lens and perhaps some other auxiliary device (such as wash/wipe assembly).

• The matrix generates camera identification, time and date, operator(s) using the system, alarm messages, and similar on screen information, superimposed on the video signal.

• The matrix has plenty of alarm inputs and outputs and can be expanded to virtually any number required. Usually, any combination of alarms, such as N/O, N/C, and logical combinations of them (OR, NOR, AND, NAND), is possible.

• In order for the matrix switchers to perform the very complex task of managing the video and alarm signals, a microprocessor is used as the brain. With the ever-increasing demand for power and processing capacity, microprocessors are becoming cheaper and yet more powerful. These days, full-blown PCs perform these complex processes. As a consequence, a matrix setup becomes programming in itself, complex but with immense power and flexibility, offering password protection for high security, data logging, system testing, and re-configuring via modem or network. The latest trend is in the form of the graphical user interface (GUI), using popular operating systems, with touch-sensitive screens, graphical site layout representation that can be changed as the site changes, and much more.

• The matrix might be very complex for the system designer or commissioner, but they are very simple and user friendly for the operator and, more importantly, faster in emergency response.

There used to be a handful of manufacturers of matrix in the world, not many of which are left today. Many of them have stayed with the traditional concept of cross-point switching and a little bit of programmability, usually stored in a battery-backed EPROM. Earlier concepts with battery-backed EPROMs, without recharging, could only last a few weeks. But many have accepted clever and flexible programming, with the system configuration stored on floppy disks or hard drives, preventing loss of data even if the system is without power for more than a couple of months. The demand for compatibility has forced many systems to become PC-based, making the operation familiar to the majority of users and at the same time, offering compatibility with many other programs and operating systems that may work in conjunction.

The Maxpro video matrix at Sydney's Star City Casino handled over 1000 cameras and over 800 VCRs.

The new designs of matrix switchers take almost every practical detail into account. First of all, configuring a new system, or even re-configuring an old one, is as easy as entering details through a setup menu. This is, however, protected with high levels of security, which allows only authorized people who know the appropriate access code and procedures to play around with the setup.

The analog CCTV matrix at the Frankfurt Airport

Next, the matrix switchers became so intelligent and powerful that controlling other complex devices was not only possible, but various interfacing were demanded with each new project. These include lights in buildings, air conditioning, door access control, boom gates in car parks, power, and other regular operations performed at a certain time of the day or at certain detectable causes.

Unfortunately, there is no standard design or language for configuring and programming matrix switchers. Different manufacturers use different concepts and ideas, so it is very important to choose a proven expert for a particular system.

Matrix switchers usually come with their basic configurations of 16 or 32 video inputs and 2 or 4 video outputs. Other combinations of numbers are possible, but the above mentioned are the most common ones. Many of them come with a certain number of alarm inputs and outputs. Almost all of them, in their basic configuration, have a text insertion feature incorporated and a keyboard for control. A basic operator's manual and other technical information should be part of the switcher.

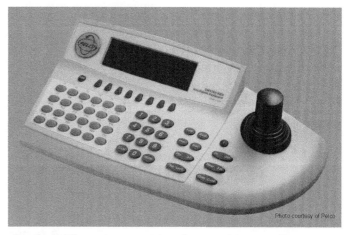

An intelligent, ergonomic, and reconfigurable matrix keyboard

Since the number of PTZs may vary from system to system, it is expected that the

Some larger matrices by Pacific Communications come neatly pre-wired

number will be specified when ordering. How many PTZ cameras can actually be used in a system depends on the make and model. In most cases, matrix switchers use digital control which has a limited number of sites on one output that it can address. This number depends on the controlling distances as well and it can be anything between 1 and 32 PTZ sites. This limit is mostly a result from the RS-485 communication limit. For a higher number of sites, additional PTZ control modules need to be used.

In the early days of matrix switchers, there was no compatibility between products of different manufacturers of PTZ site drivers. It is fair to say, however, that in the later years of mature matrix switchers, an increased number of matrix manufacturers have produced multifunctional driver boards so that you can control at least a couple of different brands. Furthermore, protocol converter boxes are now available that will allow the users, if they know the protocol of the PTZ camera and the matrix switcher, to have them talk to each other.

Small systems with up to 32 cameras can easily be configured, but when more inputs and outputs are required from a matrix switcher it is better to talk to the manufacturer's representative and work out exactly what modules are required. This selection

can make a big difference between an affordable and an expensive system, as well as between a functional and nonfunctional system.

Because of their capability and potential to do many things other than just video switching, video matrix switchers are often referred to as CCTV management systems.

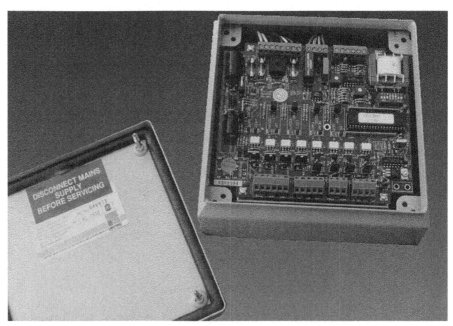

A typical PTZ site driver (receiver)

Switching and processing equipment

Quad compressors (splitters)

Because of the sequential switcher's inability to view all of the cameras simultaneously and other synchronization worries, the CCTV designers from the last century had to come out with a new device called a *quad compressor*, sometimes known as a *quad splitter*.

Quad compressors, as the name suggests, put up to four cameras on a single screen by dividing the screen into four quadrants (hence the name *"quad"*). In order to do that, video signals are first digitized and then compressed to corresponding quadrants. The quad's electronics does the time base correction, which means all of the signals are synchronized, so when the resultant video signal is produced all of the four quadrants are actually residing on one signal and **there is no need for external synchronization**.

Quad compressors in CCTV were one of the very first digital image processing devices with analog input and output.

As with any digital image processing device, we should know a few things that define the system quality: frame-store resolution expressed by the number of pixels (horizontal × vertical) and the image processing speed.

The typical frame-store capacities found in quads were 512 × 512 or 1024 × 1024 pixels. The first one is fine compared to the camera resolution, but do not forget that we split these 512 × 512 into four images; hence every quadrant will have 256 × 256 pixel resolution, which might only be acceptable for an average system. So, if you have a choice of quads, you should opt for the higher frame-store resolution. In those early days of digital products, there were still B/W cameras and color was only coming out, so there were quad splitters for B/W and quad splitters for color video cameras. A typical good quality B/W quad will have 256 levels of gray, although 64 levels are sufficient for some. However, 16 levels of gray are too little, and the image appears too digitized. Color quads of the highest quality will have over 16 million colors, which corresponds to 256 levels of each of the three primary colors (i.e., 256^3).

The next important thing about quads is the image processing time. In the early days of quads, the digital electronics were not fast. Quite often you would notice jerky movements, as the quad could only process a few images every second. Slow processing quads are still available today. In order to see smooth movements faster electronics that will process every image at the vertical frequency rate of the TV system (1/50 s or 1/60 s) was needed. Quads that were capable of such faster processing were called *real-time quads*. Real-time and high-resolution quads were more expensive. Color quads were more expensive than monochrome because there was a need for three frame stores for each

A typical quad compressor

channel (the three primary colors). If more than four cameras are in the system, the solution was the *dual quads,* where up to 8 cameras can be switched onto two quad images alternating one after the other. On most of these quads the dwell time between the two switching quads is adjustable.

Another very handy feature of most quads is the alarm input terminal. Upon receiving an alarm, the corresponding camera is switched from quad mode to a full screen. Usually, this is a live mode; that is, an analog signal is shown without being processed through the frame-store. This full-screen alarm activation is especially important when recording.

A typical quad view

No matter how good the quad video output may look on the monitor, when recorded onto a VHS VCR the resolution is reduced to the limits of the VCR. Admittedly, modern Digital CCTV systems would record all on DVRs rather than VCRs, but it is important to remind the readers that VCRs had a resolution limit. These limits were 240 TV lines for

a color signal and about 300 for B/W. When a quad picture is replayed from the VCR, it is very hard to compare details to what was originally seen in live mode. For this reason, CCTV systems employing quads were designed to activate alarms that will switch from quad to full-screen mode. The recorded footage details could then be examined much better. Various things could be used as an activation device, but most often they are passive infrared detectors (PIR), infrared beams, video motion sensors, duress buttons, and reed switches.

A dual-quad compressor

As with alarming sequential switchers, it should be determined what happens after alarm activation – that is, how long the quad stays with the full image and whether or not it requires manual acknowledgment. These minor details make a big difference in the system design and efficiency.

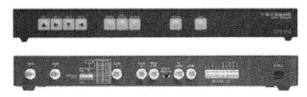

Sometimes a customer might be happy with quad images recorded as quad, in which case a plain quad, without alarm inputs, might suffice.

A quad compressor with alarm inputs

However, when full-screen recordings are required, care should be taken in choosing quad compressors that have zoom playback. They may appear the same as the alarming input quads, but in fact they do not record full-screen images as might be expected. Rather, they electronically blow up the recorded quadrants into a full screen. The resolution of such zoomed images is only a quarter (one-half vertical and one-half horizontal) of what it is after being recorded.

Multiplexers (MUX)

The natural evolution of digital image processing equipment produced video multiplexers as a better alternative to quads, especially when recording. Although now also obsolete products, MUXs can still be found in many existing CCTV installations. Multiplexers are devices that perform *time division multiplexing* with video signals on their inputs, and they produce two kinds of video outputs: one for viewing and one for recording. Over ten years ago, just before the introduction od digital video recorders, the only recording device in CCTV were VCRs, so the multiplexers recording was made on VCRs or Time Lapse VCRs. **In fact the MUX + VCR combination were predecesors of DVRs today.**

Photo courtesy of Dedicated Micros

Uniplex by DM was one of the first

The MUX live viewing output shows all of the cameras on a single screen simultaneously. This means, a 9-way multiplexer will show 9 cameras in a 3 × 3 mosaic of multi-images. The same concept also applies to 4-way multiplexers and 16-way multiplexers. Usually, with the majority of multiplexers, single full-screen cameras can also be selected. While the video output shows these images, the multiplexer's VCR output sends the time division multiplexed images of all the cameras selected for recording. This time division multiplexing looks like a very fast sequential switching, with the difference that all of these are now synchronized to be recorded on a VCR in a sequential manner. Some manufacturers produced multiplexers that only perform fast switching (for the purpose of recording) and full-screen images, but no mosaic display. These devices were called *frame switchers* and when recording is concerned they work like multiplexers.

In order to understand this process, we should mention a few things about the VCR recording concept. Unfortunately, VCRs are long ago obsolete and no longer manufactured, so I have removed the chapter from this edition of the book, but in order to explain the multiplexer operation, will remind the readers of the basic concept of VCR recording. The VCR video recording heads (usually two of them) are located on a 62 mm rotating drum, which performs a helical scanning of the videotape that passes around the drum. The rotation depends on the TV system: for PAL this is 25 revolutions per second and for NTSC it is 30. By using the two heads, positioned at 180° opposite each other on the video drum, the helical scanning can read or write 50 fields each second for PAL and 60 for NTSC. This means that every TV field (composed of 312.5 lines for PAL and 262.5 for NTSC) is recorded in slanted tracks on the videotape that are densely recorded next to each other. When the VCR plays back the

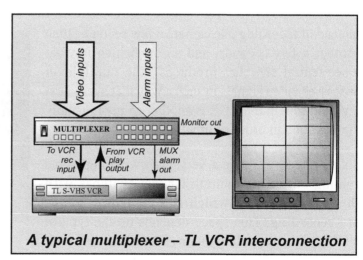

A typical multiplexer – TL VCR interconnection

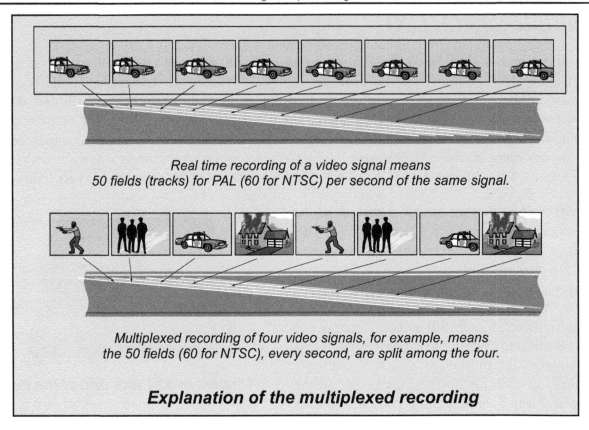

Real time recording of a video signal means
50 fields (tracks) for PAL (60 for NTSC) per second of the same signal.

Multiplexed recording of four video signals, for example, means
the 50 fields (60 for NTSC), every second, are split among the four.

Explanation of the multiplexed recording

recorded information, it does so with the same speed as the TV standard requires, so we once again reproduce motion pictures.

Clearly, however, because the VCR heads are electromechanical devices, the rotation speed precision is critical. Because of the electromechanical inertia, the VCRs have a longer vertical lock response time than monitors. This is the main reason for even bigger picture-roll problems when non-synchronized cameras are recorded through a sequential switcher.

With normal recordings and playback, the video heads are constantly recording or reading field after field after field. There are 50 (60 for NTSC) of them every second.

Instead of recording one camera a few seconds, then another a few seconds, and so on (which is what a sequential switcher produces), the multiplexer processes video signals in such a way that every next TV field sent to the VCR is another camera (usually the next one in order of inputs).

So, in effect, we have a very fast switching signal coming out of the multiplexer that **switches with the same speed at which the recording heads are recording**. This speed depends on the type of VCR and the recording mode (as was the case with

An 16-channel multiplexer

time-lapse VCRs). This is why it is very important to set the multiplexer to an output rate suitable for the particular VCR. This selection is available on all multiplexers in their setup menu. If the particular VCR model is not available on your multiplexer, you can either use the generic selection, or, if nothing else, use the method of trial and error to find an equivalent VCR. The major difference in TL VCRs is that some are field recorders and others are frame recorders.

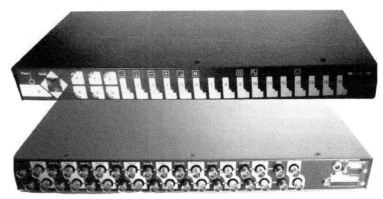

16-input looping multiplexer

Apart from this output synchronization (MUX-VCR), theoretically there is also the need for an input synchronization (cameras-MUX), but because multiplexers are digital image processing devices, this synchronization, that is, the ***time base correction*** (TBC) of the cameras, happens inside the multiplexer. This means that different cameras can be mixed onto a multiplexer and there is no need for them to be gen-locked (i.e., synchronized between themselves).

Some multiplexer models on the market, however, are made to synchronize the cameras by sending sync pulses via the same coaxial cable that brings the video signal back and then multiplex the synchronized cameras. These multiplexers do not waste time on TBC and, therefore, are supposed to be faster.

When playback is needed, the VCR video output goes first to the multiplexer, and then the multiplexer extracts the selected camera only and sends it to the monitor. The multiplexer can display any one camera in full screen, or play back all of the recorded cameras in the mosaic mode (multiple images on one screen).

Recording time gaps

The number of shots (images) taken from every camera during the recording depends on the total number of cameras connected to the multiplexer and the time-lapse mode of the VCR. Therefore, **it is not possible to record real-time images from all the cameras simultaneously** because, as the name suggests, this is a time division multiplexing.

There are, however, ways to improve performance by using external alarm triggers, usually with a built-in motion activity detector in the multiplexer. The best way, though, is to record in as short a time-lapse mode as is practically possible and also to keep the number of cameras as low as possible. Translated into plain language, if your customer can change tapes at least once a day, do not use more than a 24-hr time-lapse recording mode. If the system is unattended over weekends, then a 72-hr time-lapse mode should be selected. And, if the budget allows, instead of using a 16-way multiplexer for more than 9 cameras, it would be better to use two 9-way (some manufacturers have 8-way and some 10-way) multiplexers and two VCRs. The recording frequency will then be doubled, and two tapes will need to be used instead of one.

This is how you can calculate the time gaps between the subsequent shots of each camera. Let us say we have a time-lapse VCR that records in 24-hr time-lapse mode. Earlier we stated stated normal (real-time) recording VCRs make 50 shots every second in PAL and 60 in NTSC. If you open the TL VCR technical manual, you will find that when the VCR is in 24-hr mode it makes a shot every 0.16 s and even if you do not have a manual with the VCR, it is easy to calculate: When a PAL VCR records in real time, it makes a field recording every $1/50 = 0.02$ s. If the TL VCR is in

A VHS tape

24-hr time-lapse mode, it means $24 \div 3 = 8$ times slower recording frequency. If we multiply 0.02 with 8 we get 0.16 s. The same exercise for NTSC VCR will obtain a field recording every $1/60 = 0.0167$ s. For a 24-hr time-lapse mode, when using T120 tape, $24 \div 2 = 12$. This means that in 24-hr time-lapse mode in the NTSC format, the TL VCR moves 12 times slower to fit 24 hr on one 2-hr tape. Thus, the update rate of each recorded field in 24-hr mode is $12 \times 0.0167 = 0.2$ s.

All of these calculations refer to a single camera signal; therefore, if the multiplexer has only one camera, it will make a shot every 0.16 s in PAL and every 0.2 s in NTSC. If more cameras are in the system, in order to calculate the refresh rate of each camera, we need to multiply by the number of cameras, plus add a fraction of the time the multiplexer spends on time base correction due to non-synchronized cameras (which will usually be the case). So if we have, for example, 8 cameras to record, $8 \times 0.16 = 1.28$ s (PAL) and $8 \times 0.2 = 1.6$ s (NTSC). Adding to it the time spent on sync correction and the realistic time gaps between the subsequent shots of **each** camera should result in approximately 1.5 to 2 s. This is not a bad figure when considering that **all 8 cameras are recorded on a single tape**. If we have to identify an important event that happened at 3:00 P.M., for example, we can either view all of the cameras in a mosaic mode and see which cameras have important activity, or we can select each one of them separately in full screen.

For some applications, 2 s might be too long a time to waste; this is where the alarm input or the motion activity detection can be very handy. Most of the multiplexers have alarm input terminals and with this we can trigger the **priority encoding** mode. **The priority encoding mode is when the multiplexer encodes the alarmed camera on a priority basis**. Say we have an alarm associated with camera 3. Instead of the normal time division multiplexing of the 8 cameras in sequence 1, 2, 3, 4, 5, 6, 7, 8, 1, 2, it goes 1, 3, 2, 3, 4, 3, 5, 3, 6, 3, and so on. The time gap in such a case is prolonged for all cameras other than 3. But since number 3 is the important camera at that point in time, the

Typical monthly cycle of tapes with MUXs

priority encoding has made camera 3 appear with new shots every $2 \times 0.16 = 0.32$ s, or in practice almost 0.5 s (due to the time base correction). This is a much better response than the previously calculated 2 s for the plain multiplexed encoding. It should be noted, however, when more than one alarm is presented to the multiplexer inputs the time gaps between the subsequent camera shots are prolonged, and once we get through all of the alarmed camera inputs, we get plain multiplexed encoding.

In case a system cannot be designed to use external alarm triggers, it should be known that most of the multiplexers have a ***video motion activity detection*** built in. This is a very handy feature in which every channel of the multiplexer analyzes the changes in the video information in each of the framestore updates. When there is a change in them (i.e., something is moving in the field of view), they will set off an internal alarm, which in turn will start the priority encoding scheme. This feature can be of great assistance when replaying intrusions, or events, and determining the activity details.

Usually, the activity motion detection can be turned on or off. When turned on, on some MUX models, it will allow you to configure the shape of the detection area in order to suit various areas or objects.

When the "Real-time" time-lapse recorders (RTTL VCR) have appeared on the CCTV market they confused a little bit the calculating the refresh rate. The RTTL VCRs were faster recording machines where the TL VCR's mechanics is modified so as to record 16.7 fields per second in PAL (a field every 0.06 s) for 24-hr relative to E240 tape. In the case of NTSC, around 20 fields per second (a field every 0.05 s) can be recorded for 24 hrs on a T160 tape. Understandably, to calculate the refresh rate of multiplexed cameras on such a TL VCR, you would need to multiply the number of cameras with the above-mentioned field update. In reality, RTTL VCRs were actually not real-time recording as such, but they were definitely better (faster) than the ordinary time-lapse mode.

The NTSC system used VHS tapes at a higher recording speed (2 meters/minute) than PAL or SECAM (1.42 meters/minute). Usually, VHS tapes were marked in playing times as opposed to tape length. Therefore, a T120 (2-hr) tape bought in the United States was not the same as an E120 (2-hr) tape bought in the UK. The U.S. T120 tape was 246 meters in length and would provide 2 hours of play time on an NTSC VCR. This same tape used on a PAL VCR would give 2 hours and 49 minutes of play time. Conversely, a UK E120 tape was 173 meters in length and would provide 2 hours of play time on a PAL VCR. The same tape used on an NTSC VCR would provide only 1 hour and 26 minutes of play time. The chart on the right compares the recording times of each tape in NTSC and PAL.

Tape label	Tape length (m)	NTSC time (min)	PAL time (min)
E30	45	22	30
E60	88	44	60
E90	130	65	90
E120	173	86	120
E180	258	129	180
E240	346	173	240
T20	44	20	28
T30	64	30	42
T45	94	45	63
T60	125	60	84
T90	185	90	126
T120	246	120	169
T160	326	160	225

Simplex and duplex multiplexers

Most multiplexers will allow you to view images of any selected camera in a mosaic mode while they are encoding. When a recorded tape needs to be viewed, as we have already mentioned, the VCR output does not go directly to a monitor, but it has to go through the multiplexer again in order for the images to be de-multiplexed. While doing this, the multiplexer cannot be used for recording. So, if recording is very important and the playback needs to be used in the meantime, another multiplexer and VCR are required. Multiplexers that can do only one thing at a time are called ***simplex multiplexers***.

There are also ***duplex multiplexers***, which are actually two multiplexers in the one unit, one for recording and one for playback. Still, two VCRs will be required if both recording and playback are required at the same time.

Some manufacturers even made multiplexers, which were referred to as *triplex*. These were multiplexers with the same functionality as the duplex ones, with the addition of displaying a mixture of live and playback images on one monitor.

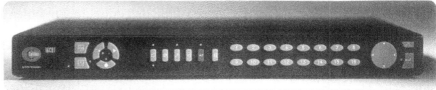

Photo courtesy by Calibur

A 16-channel triplex multiplexer

As with quad compressors, there were B/W and color multiplexers, with a limited resolution available. Needless to say, the bottleneck in the resolution reproduction was still the VCR itself. Newer systems were installed with Super VHS VCRs that offered an improved resolution of 400 TV lines as opposed to 240 with the ordinary VHS format.

Multiplexers were successfully used in applications other than just recording. This was especially useful in cases where more than one video signal needed to be transmitted over a microwave link, for example. By using two identical simplex multiplexers, one at each end of the link, it was possible to transmit more than one image in a time division multiplexed mode. In this instance, the speed of the refresh rate for each camera is identical to what it would be if we were to record those cameras in real (3-hr) mode on a VCR.

Video motion detectors (VMDs)

A ***video motion detector*** (VMD) is a device that analyzes the video signal at its input and determines whether its contents have changed. Consequently, it produces an alarm output, or prioritizes recording.

With the ever-evolving image processing technology, near the end of last century, it became possible to store and process images in a very short period of time. If this processing time is equal to or smaller than 1/50 (PAL) or 1/60 (NTSC), which as we know, is the live video refresh rate, we can process images without losing any fields and preserve the real-time motion appearance.

In the very beginning of the development of VMDs, only analog processing was possible. Those simple VMDs are still available and perhaps still very efficient relative to their price, although they are incapable of sophisticated analysis; therefore, high rates of false alarms are present. The principles of operation of the analog VMDs (sometimes called video motion sensors) are very simple: a video signal taken from a camera is fed into the VMD and then onto a monitor, or whatever switching device might be used. In the analyzed video picture, little square marks (usually four) are

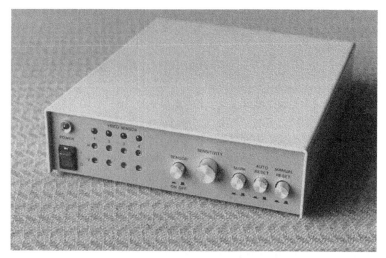

A simple single channel VMD

positioned by means of a few potentiometers on the front of the VMD unit. The square marks actually indicate the sensor areas of the picture and the video level is determined by the VMD's electronics. As soon as this level is changed to a lower or higher value, by means of someone or something entering the field of view at the marked area, an alarm is produced. The sensitivity is determined by the amount of video luminance level change required to raise an alarm (usually 10% or more of the peak video signal). The alarm is usually audible, and the VMD produces relay closure, which can be used to further trigger other devices. The alarm acknowledgment can be either automatic (after a few seconds) or manual. There is also a VMD sensitivity potentiometer on the front panel in such devices, and with the proper adjustment it can bring satisfactory results. There will always be false alarms activated by trees swaying in the wind, cats walking around, or light reflections, but at least the reason for the alarm can be seen when the VCR is played back (assuming the VMD is connected to a VCR).

A more sophisticated PC-based VMD

VMDs are often a better solution than passive infrared motion detectors (PIR), not only because the cause of the alarm can be seen, but also because it analyzes exactly what the camera sees, no less and no more. When using a PIR, its angle of coverage has to match the camera's angle of view if an efficient system is to be achieved.

When a number of cameras are used, we cannot switch signals through the VMD because it will cause constant alarms; therefore, one VMD is required per camera. In systems where further processing of the video signal is done, the sensing markers can be made invisible, but they are still active.

The next step up in VMD technology is the ***digital video motion detector*** (DVMD), which is becoming even more sophisticated and popular. This, of course, is associated with higher price, but the reliability is also much higher and the false alarm rate is lower.

One of the major differences between various DVMD manufacturers is the software algorithm and how motion is processed. These concepts have evolved to the stage where tree movement due to wind can be ignored, and car movement in the picture background can also be discriminated against and excluded from the process deciding about the alarm activity. In the last few years, DVMDs that take the perspective into account have been developed. This means that as the objects move away from the camera, thus getting smaller in size, the VMD sensitivity increases in order to compensate for the object's size reduction owing to the perspective effect. This effect, we should point out, also depends on the lens.

Many companies now produce a cheaper alternative to a full-blown stand-alone system in a form of PC card(s). The cards come with specialized software, and almost any PC can be used for VMD. Furthermore, image snapshots can be stored on a hard disk and transmitted over telephone lines connected to the PC. With many options available in the VMDs, a lot of time needs to be spent on a proper setup, but the reward will be a much more reliable operation with fewer false alarms.

A special method of recording, called ***pre-alarm history***, is becoming very standard in most of the VMD devices. The idea behind this is very simple but extremely useful in CCTV. When an alarm triggers the VMD, the device keeps a number of images recorded after the alarm occurrence, but also a few of them before. The result is a progressive sequence of images showing not only the alarm itself, but also what preceded it.

One of the interesting developments in this area, even many years after its introduction, is the concept of three-dimensional (3D) video motion detection. This concept offers extremely low rates of false alarms by using two or more cameras to view objects from different angles. Thus, a three dimensional volumetric protection area is defined, which like the other VMDs is invisible to the public, but it is quite distinguishable to the image processing electronics. With

PC-based 3-dimensional VMD

this concept, movement in front of any of the cameras will not trigger an alarm until the **protected volume area**, as seen by **both cameras**, is disturbed. Using this concept, valuable artworks in galleries, for example, can be monitored so that the alarm does not activate every time someone passes in front of the artwork, but only when the artwork is removed from its position.

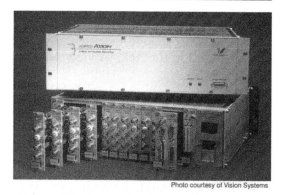

Photo courtesy of Vision Systems

A sophisticated multi-channel outdoor VMD system

Quite often, alarm detection is useful not when someone or something moves in the field of view, but, rather, when something fixed is removed from its location. This can be done with a *video non-motion detector* (VNMD). This unit is very similar to the VMD, except that additional information is collected for objects put in the field of view that are stationary for a longer period. Movements around the **selected** object cause alarms only when the protected object is removed from the stationary position.

In the last few years of analog camera manufacturing, many came with built-in VMD circuitry built inside the camera itself. This is practical in systems where recording and/or an alarm is initiated only when a person or object moves inside the camera's field of view. Today, almost all digital IP cameras have built-in VMD.

All of the above-mentioned VMDs, used as an alarm output, produce a closure of relay contacts that can trigger additional devices in the CCTV chain, like VCRs, or matrix switchers, frame-stores, sirens, or similar. If you decide to use one, make sure you clarify the type of alarm output with the supplier, for it can be anything from voltage-free N/O contacts to a logic level voltage (5 V) N/C output.

One of the early adaptation of intelligent VMD system, apart from the motion detection recording, was also to dial remote receiving stations and send images via telephone lines. With such devices, remote monitoring was possible from virtually anywhere in the world, long before the implementation of IP systems.

Video Analytics

With the introduction of the IP CCTV and more intelligent software video analytics has become more powerful and more promising. Starting from the ubiquitous automatic face recognition, vehicle number plate recognition, and counting objects up to detecting loitering in an area and speed and direction of vehicles in motion.

We are not going to go in more details of Video Analytics in this book, as it will require a book in itself, but it is important to acknowledge that more and more intelligent analysis is requested by various project and customers. Not everything is possible, or, at least, not every analysis can be made with 100% confidence (which, unfortunately, some customers expect to happen) but there are products and companies that invest heavily in this and there is serious progress made there.

Some of the analysis may require more powerful and dedicated software and hardware, but one of the simplest and most common analysis is the so-called "Smart find" or "Smart search," where an object can be found in the field of view when it's approximate location is known, but not the time when it was left there.

The simplest idea behind video analytics is to get a computer to find an event or instance which requires human attention, or perhaps warn an operator of a possible incident. The incredible amount of CCTV cameras that are being installed every day throughout the world, can not be viewed or analyzed by a couple of operators sitting 8 hours a day in the control room. Very often, incidents or events are discovered after being reported, or after a bomb had exploded. Then the CCTV recorded footage may help finding some clues of who did that, but it may take days and days to go through all the cameras (if there have been any cameras near the incident place indeed) by a human operator. This is where computer video analysis may be very helpful and may cut the search time dramatically.

Some of the many video analysis that can be made with Video Analytics software

Frame-stores

Frame-stores were important in the days of analog video switching and recording. These were basically simple electronic devices used to temporarily store images. Two important parts of a frame-store device are the analog-to-digital (A/D) conversion (ADC) section and the random access memory (RAM) section. The ADC section converts the analog video signal into digital, which is then stored in the RAM memory, for as long as they were powered.

The main advantage of the frame-stores, in the times when VCRs were the main recording device, was their response time. Since they do not have any mechanically moving parts, the storage of the alarmed picture was instant on activation. This was then fed back, usually, to a video printer or a monitor for viewing or verification purposes.

A framestore device

More sophisticated frame-stores were usually designed to have a few frame-store pages that constantly store and discard a series of images using the first in first out (FIFO) principle until an alarm activates. When that happened, it was possible to view not only the alarmed moment itself, but also a few frames taken before the alarm event took place, thereby giving a short event history. This was the same concept as the "***pre-alarm history***" used in VMDs.

Another important application of the frame-stores was as the ***frame locking device***. This device constantly processes video signals present at its input and also does the time base correction to be in sync with the master clock inside. Since this processing takes place at a very fast real-time rate and the frame-store has a high resolution, there was no perceptible degradation of the video signal. **This was especially practical and useful for showing (switching) non-synchronized cameras on a single monitor without the dreaded picture roll.** In such cases, the frame lock device acts as a synchronizer (i.e., it eliminates picture-roll while cameras are scanning).

Video printers

Long before the digital CCTV and high resolution ink-jet printers, in CCTV, *Video printers* were used. **These were practically printers specially designed to print (hard-copy) video image straight from a video source.** This was often required as an evidence of the recorded intruder or an event for analysis or evaluation. Since at the time there were B/W and color cameras, there were also two types of video printers: monochrome and color. The monochrome video printers usually used thermal paper as an output medium, but some more expensive ones could print on plain paper. The thermal-paper video printers, used for monochrome signals, were similar in operation to the facsimile machine, and they could print images with a size and resolution dependent on the printer's resolution. With thermal printers the hard-copy was not as durable and stable (due to thermal paper aging).

Color video printers printed on special paper, and the process of printing was similar to dye-sublimation printing, using CMYK filters. The printing quality produced by such technology was excellent, but the number of copies that can be produced was limited and cartridges needed to be replaced with every new set of paper.

The more sophisticated video printers had a number of controls, including titling, sharpening, duplication into more copies, and storing the images in their frame-stores until a printout is necessary. In many instances, CCTV users did not want to invest in video printers, so often they used specialized services of some bureaus. The videotape was taken to such services and the certain event(s) extracted and printed out.

Photo courtesy of Panasonic

A color video printer

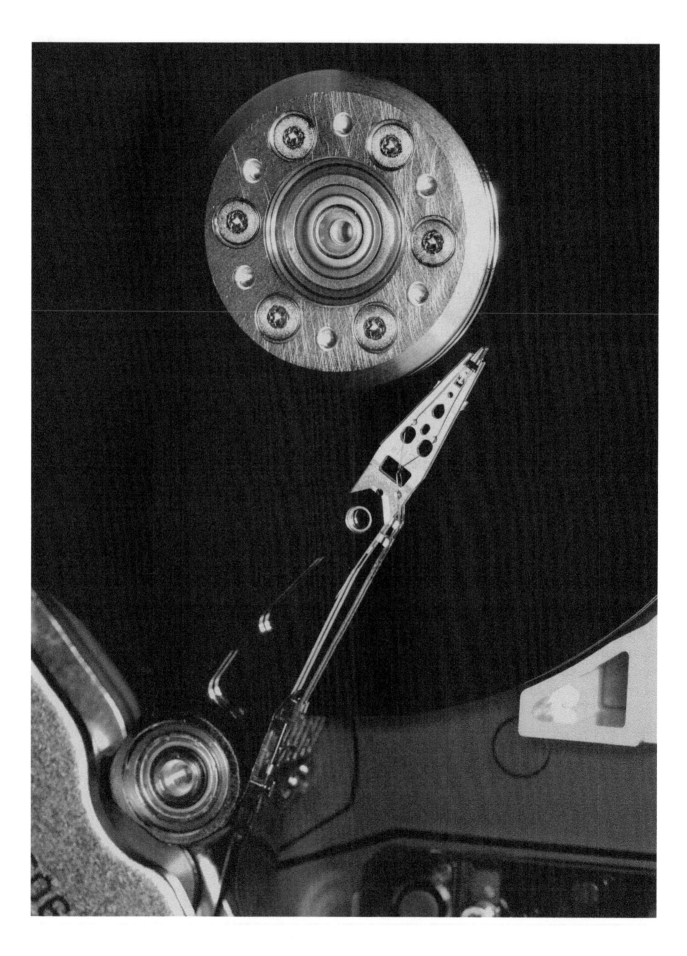

8. Digital CCTV

Most of the previous discussions in this book have involved PAL and NTSC television standards, which refer to analog video. Although the majority of existing CCTV systems would still have analog cameras, all new projects and tenders today call for digital video cameras, digital storage, and IP networks.

The very few components in CCTV that, only a decade ago, used digital video were the frame-stores, quad compressors, multiplexers, and the internal circuits of the digital signal processing (DSP) cameras. Today, there is no new system or upgrade that doesn't call for complete digital system. Camera quality is an important starting point in the CCTV system video chain, but the quality of the recorded images and its intelligent processing have become equally important.

In the last two decades there have been revolutionary developments in electronics in general, in TV, communications, photography, and CCTV. All of these developments are based on digital technology. One of the locomotives of the real new boom in CCTV has been the switch to digital video processing, transmission, and recording. This development gathered a real momentum in CCTV in the last ten years, hence the reason for yet another updated edition of this book with extended discussions on digital, video compression, networking, and IP technology.

There are technical reasons why digital is better than analog, and we will discuss this next, but the commercial aspect is important too. **Out of many human activities and industries electronics is probably the only one where newer, faster, and better products are cheaper.** Only a few years ago, the price of high-speed digital electronics capable of live video processing was unaffordable and uneconomical. With the ever increasing performance and capacity of memory chips, processors, and hard disks, as well as their decrease in price, digital video signal processing in real time is not only possible and more affordable, but it has become the only way to process a large number of high-quality video signals.

Digital video was first introduced in the broadcasting industry in the early 1990s. As with any new technology it was initially very expensive and used for special projects. Today digital video is the new standard, replacing the nearly half a century old analog television. It comes basically in two flavors – *Standard Definition (SDTV)* with the aspect ratio of 4:3 and the quality as we know it, and *High Definition (HDTV)* with the aspect ratio of 16:9 and around 5 times the number of pixels of SDTV. Many countries around the world are already broadcasting digital video, usually in both formats (SDTV and HDTV). Not surprisingly, the HDTV is the current new broadcast standard, owing to its much higher resolution and the theatrical experience one has watching movies. In CCTV, there are still many analog standard definition cameras in the existing systems, but most of the new cameras are installed as high definition cameras. To help such a transition, many manufacturers offer hybrid systems, where analog cameras can be mixed up with HD and IP cameras and all of them recorded in digital format.

Digital video recorders (DVRs) and IP cameras have now become the main reason for the new CCTV growth, a source of higher revenue and inspiration for new and intelligent system design solutions that have blurred the line between computers, IT technology, networking, and CCTV.

Why digital video?

Analog signals are defined as signals that can have any value in a predefined range. Such analog signals could be audio, but also video. As we know, the predefined range for video signal is anything from 0 volts (corresponding to black) to 0.7 volts (corresponding to white).

The main problem with analog signals is the noise. Noise is defined as an unwanted signal that masks the real signal. Like an audio noise from traffic in a busy street in real life that prevents you from understanding what the person next to you tells you, **in real life noise cannot be avoided.** It is accumulated at every stage of the signal path. Starting from the thermal noise at the camera imaging chip, the camera electronics itself, it adds in the transmission media (cables) and at the receiving end (recorders, monitors, etc.). The longer this path is, and with more stages, the more noise it will get induced. This is where digital signals can make a big difference. Once a signal is converted to digital, it becomes virtually immune to noise. It is simply

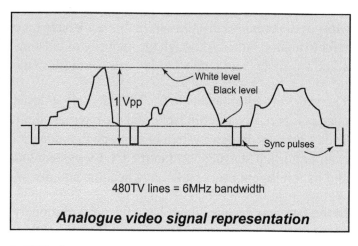

Analogue video signal representation

480TV lines = 6MHz bandwidth

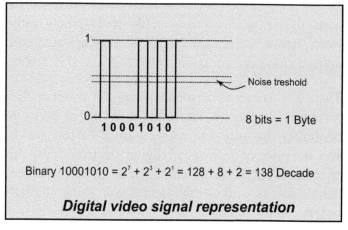

Binary $10001010 = 2^7 + 2^3 + 2^1 = 128 + 8 + 2 = 138$ Decade

Digital video signal representation

detected as either one or zero, as long as the noise does not exceed the noise threshold. So, one of the most important differences between an analog and a digital signal (apart from the form itself) is how noise is avoided.

The immunity to noise is the first and most important reason why most communications in the modern times are in digital format. It doesn't mean that noise can't affect the digital signals at all, but it is the possibility of having only two values a signal can have (zero and one) that makes it almost immune to noise. As mentioned previously, noise may be introduced in various stages in the video chain, starting from the thermally generated electrons in a CCD or CMOS chip, external induction from power lines, etc. So excessive noise may affect the digital signals as it may interfere with the digital circuit threshold deciding if a signal is zero or one. If the noise exceeds the noise threshold values then the digital signal may completely loose it's content. **So digital signal is either perfect when within it's limits, or disappears completely if out of them.** This ultimately depends on it's usage within the maximum allowable parameters, such as the maximum distance of 100m for Cat-5 cable for example, specified for 100 Mb/s in the Fast Ethernet standard 802.3.u. **If we stick to the predefined limits of digital signals, then they (digital signals) appear as if they are not affected by the noise at all.** This is why we say **digital signals are virtually immune to noise**.

The second important advantage of digital video signals is the possibility for digital processing and storage. This includes video or image enhancement, compression, transmission, storage, and, very important for security, encryption.

The third important feature is that there is no difference in quality between the copies and the original. Digital signals can be reproduced multiple times with the exact same quality. Whether we make one, two, or ten copies of the image captured in a digital format, the quality is exactly the same as the original no matter what generation copy it is.

The fourth reason why digital is more attractive than analog is the possibility to introduce new system functionality without changing the hardware. By just changing the software, new functionality can be introduced to the same physical hardware. For example, a computer, which is in itself a digital device, using the same hardware and only updating the operating system software may introduce new functionality. Similarly, a digital (IP) camera can be enhanced with it's functionality by just updating the firmware.

And the last, but not the least important reason for digital, is the possibility of verifying the originality of a copy. Typically, this is done by some smart algorithm that can verify if a video footage is identical copy of the original. This feature is often referred to as *"watermarking."* It enables the protection of digital signals against deliberate tampering. This is very important feature in legal proceedings where CCTV footage is used.

Modern day digital CCTV matrix rack

Types of compressions

CCTV has it all: JPEG, M-JPEG, Wavelet, H.263, MPEG-1, MPEG-2, JPEG-2000, MPEG-4, H.264, and now the latest H.265. There are too many different image compression techniques.

How do you know which one is best for you?

The answer is, without any doubt, not easy to find. One has to understand the concept of digitized images and the limitation of the TV standards, on top of which digitized video and compression limitations have to be added.

In general, there are two basic types of image/video compressions based on losses: *loss-less* and *lossy*.

Loss-less is a compression (encoding) that when it is decoded - it is bit-for-bit exactly as the original uncompressed video. **Loss-less compression offers very low compression ratios (usually not more than 3 to 4 times) and because of it - is generally not used in CCTV, but in broadcast and cinematography, before the video content is edited and (lossy) compressed for video distribution.**

So, the compression types we will concentrate on in this book are the lossy ones. **Lossy compression means that certain details of the image (video) are lost and cannot be retrieved, no matter what we do after they are compressed. Good compression in CCTV is** not the one that offers the highest squeeze, but **the one that offers the best compromise between quality and small file size.**

The second and important division of compressions used in CCTV is based on the type of codecs used, and can be divided in *image compressions* and *video compression*.

The **image compressions use two dimensions** of image only, horizontal x vertical pixels, and compresses with known, simple and fast image compression.

The **video compressions use three dimensions**, and this is the horizontal x vertical pixels of a signal, plus time. The video compressions are more efficient, but require specialised codecs and need more processing power. Many DVRs, encoders and IP cameras use encoder chips with a choice between one image compression (typically JPEG) and a video compression (typically H.264).

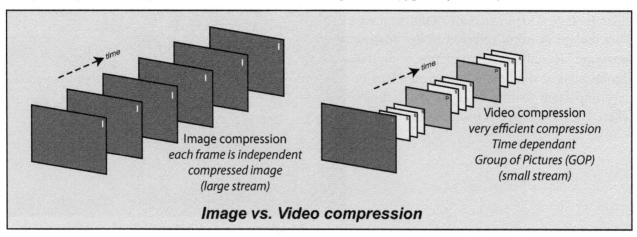

Image vs. Video compression

Digital video recorders (DVRs)

In CCTV today, there is hardly any recording made on VCR tapes. Nearly all recordings today are made on digital video recorders or computer servers. Most, if not all, of these recordings are made with compressed video signal. Digital raw (un-compressed) data recording is possible, but not practical due to excessive bandwidth required.

So what are the real benefits of using DVRs in CCTV, as opposed to the old VCRs? Although this was a topic worth a discussion around ten years ago, this is no longer the case. There is hardly any new installation where VCRs will be considered today, but for the sake of old times, and maybe some outdated system out there, we will elaborate the obvious advantages of digital video recording.

First, with the VCR's analog method there is no direct and quick access to the desired camera, except when using a reasonably quick Alarm Search mode (available on most TL VCRs). In VCRs the information is stored in an analog format and cannot be further processed. The VCR recorded video quality is always lower than the actual original source, although it would be fair to say that all DVRs reduce the visual quality slightly (some more than others, depending on the settings), simply because we use lossy compression. With an educated usage, DVRs however will almost always produce better video quality than VCRs.

VCRs were (and for some still are) the main method of recording

Furthermore, DVRs footage can be exported and saved in long term storage with the same quality as the original video. This was not the case with VCR tapes. They are very much dependent on the mechanical and magnetic integrity of the tape itself. It is not difficult to accidently ruin an important recording that has been stored on tape.

Initially, attempts were made to implement digital video recording on a digital audio tape (DAT) format. Though digital, such recorded material still required a sequential search mode, which is not as efficient as the random access used on a hard disk. Hard disks have a much higher through-output than other digital storage media and higher capacity; better than S-VHS quality images are achievable with appropriate video compressions. What was a problem only a few years ago – the length of recording – these days is no longer such. Hard disks with capacities of 3 TB are readily available, and DVRs with internal capacities of 12 TB or more are not a rarity. Servers with even more internal RAID capable storage are available.

Many months of digital recording of a number of cameras is no longer a problem too. This was unthinkable with VCRs and tapes, where at best, a multiplexed video of 16 cameras could be fitted on a 24 hrs tape, recording 1 image per second per camera. The only way to extend such recordings were a library of video tapes for 30 days for example, recycling each month.

Hard disk drives (HDD) now have very fast access time, and by using good compression it is possible to record and play back multiple images from one hard disk – in real time (meaning "*live*" video rate).

The hard disk prices are falling daily and it is interesting to note that at the time of writing the previous edition of this book (2005), a single 3.5 inch HDD with the capacity of around 300 GB was becoming available. For the same price, today (2013), we have a nearly tenfold increase in capacity, with 3 TB available. Because of the importance of the hard drives, the need arose for a complete new chapter discussing all the important aspects of it.

How many days or weeks of video recording can be stored on a single 3 TB disk, for example, depends first on the video type (SD, HD or mega-pixel) and, secondly, on the type of compression. Also, an important factor would be if the recording were made permanent or if it was based on video motion detection. The latter has become very popular in CCTV, as it extends the recording capacity at least two or threefold.

Certainly, extending the hard drive storage is also possible, but providing redundancy (safety) as well, is an important factor. Because hard disks in digital CCTV are working continuously, 24/7, there is a new saying "it's not a question if a hard disk will fail, but when." Offering redundancy has become a very important thing to do. Luckily, with the mature computer technology this is easily done using the redundant arrays of independent disks (RAID) configurations. More on these details are discussed later in the book.

Another important additional variable possible today is the mix of analog signals and digital IP streams in one DVR for example. This is typically referred to as "hybrid DVR." Hybrid DVRs are great for seamless transition from an old analog CCTV system, to a modern with HD cameras for example, without necessarily removing the old cameras immediately, but gradually. In a hybrid system there is variety of video formats, compressions, and pixels counts. There is a mix of analog signals, brought via coaxial cables, and IP streams brought in via Cat-5 or Cat-6 Ethernet cables and network switches. Some hybrid DVRs even offer additional feature of RAID configuration in the same box. Very convenient when switching from an old analog system to a modern digital one.

All of the above leads us to various considerations we have to have in mind when selecting digital compression, storage media, and data transfer rate. In a hybrid system for example, anything can be expected from the point of view of compression - MPEG-4, M-JPEG, H.264 and any of these may work with various pixel counts and motion activities, producing different storage requirements. This is why we have to understand the theory of digital video and image representation with various compression techniques. The following few headings will try to explain some of the basics.

A hybrid DVR working with analog and IP cameras

Image and Video compressions

A few international bodies engage in various standards for digitized video. The most well known is the ***International Telecommunication Union*** (**ITU**), which is the United Nations specialized agency in the field of telecommunications. This was in the past called CCIR (***Consultative Committee on International Radio***). The ITU Telecommunication Standardization Sector (ITU-T) is a permanent organ of ITU. ITU-T is responsible for studying technical, operating, and tariff questions and for issuing recommendations on them with a view to standardizing telecommunications on a worldwide basis. The World Telecommunication Standardization Assembly (WTSA), which meets every four years, establishes the topics for study by the ITU-T study groups which, in turn, produce recommendations on these topics. The approval of ITU-T Recommendations is covered by the procedure laid down in WTSA Resolution 1. In some areas of information technology that fall within ITU-T's review, the necessary standards are prepared on a collaborative basis with ISO and IEC.

ISO (the ***International Organization for Standardization***) and **IEC** (the ***International Electrotechnical Commission***) form the specialized system for worldwide standardization. National bodies that are members of ISO and IEC participate in the development of International Standards through technical committees established by the respective organization to deal with particular fields of technical activity. ISO and IEC technical committees collaborate in fields of mutual interest. Other international organizations, governmental and nongovernmental, in liaison with ISO and IEC, also take part in the work. In the field of information technology, ISO and IEC have established a joint technical committee, ISO/IEC JTC1.

IEC's working group TC-79 has recently completed a series of standard recommendation, under the code 62676, which deal specifically with CCTV, and I highly recommend for everybody involved in our industry to obtain copies of it.

Some recommendations, such as the now popular H.264, are prepared jointly by ITU-T SG16 Q.6, also known as

The clarity of an image depends on the pixels used

VCEG (*Video Coding Experts Group*), and by ISO/IEC JTC1/SC29/WG11, also known as MPEG (*Moving Picture Experts Group*). VCEG was formed in 1997 to maintain prior ITU-T video coding standards and to develop new video coding standard(s) appropriate for a wide range of services. MPEG was formed in 1988 to establish standards for coding moving pictures and associated audio for various applications such as digital storage media, distribution, and communication.

As we already mentioned earlier under the "Types of compressions" heading, it should be made clear that even though in CCTV we use video signals rather than static images, we still use image compressions. Often encoder

H.264 is one of the most common video compressions used today

chips can be set to either. In order to make a clear distinction between these two, we refer to them as *image compression* and *video compression*.

Image compressions use only two dimensions: the horizontal and vertical dimension of the image. Typical *image compressions* representatives are JPEG and Wavelet (JPEG-2000).

Video compressions use three dimensions when compressing: horizontal and vertical picture dimensions, as well as time. As a result, such compressions are often referred to as *temporal compressions* (*tempus* in Latin means time). Typical representatives of temporal (video) compressions are MPEG-1, MPEG-2, MPEG-4, H.263, and H.264.

The difficult challenge we face in CCTV is deciding which compression is the best for a particular product or project. There is no simple or single answer. Often it depends on how much we understand about the differences between compressions, but more importantly on what is the intended usage.

For example, if a digital CCTV system is designed to monitor human activity, protect a cash teller in a bank, or a card dealer in a casino, a high image rate would be preferred. This is best done with a video compression, like H.264. Often live rates should be used (25 for PAL or 30 for NTSC), although in some instances 10 images/second might be sufficient. Tests are often the best indicator. Video compression are stated in Mb/s.

ITU was the body that approved the latest H.265 (HEVC)

If, for example, we have a traffic red-light camera, or speed camera, then having only one high quality image capturing the off-ending vehicle might be sufficient, as long as it is captured at the right moment. This is typically done with a ground loop trigger (or speed detector trigger). In cases like this, image compression is sufficient and perhaps the only one that can be used, since we only have one image to compress, and not a stream of motion video. Furthermore, when using cameras for specific applications like this, typically mega-pixel cameras are used, where the image sensor produces very high quality images, and may not be compressed with anything else but image compressor. Image compression is stated in kB per image.

Casinos prefer playing cards to be covered with "live" recording

Ground loops in the road help trigger cameras to take a snap-shot at the right moment

It is wrong to ask "what is the image size" of a particular H.264 encoder, since H.264 is a video compression, and as such can only be specified in Mb/s.

Very often, the number of images required by a tender specification could be a mine-field. On one hand, some insist nothing short than live images at 25 or 30 i/s, on the other some try to stream-line it to the required bandwidth by reducing the images/second to the minimum allowed where human activity can be monitored, be that 1, 2, 6, 10 or 12 images per second.

The reality is, **in video compression the only parameter that defines how much storage is taken by a video stream is the stream data itself in Mb/s, not the images per second within the stream**. If a stream is said to be 4 Mb/s (a typical good quality HD stream using H.264 encoding), then this is what defines the storage taken by that stream. If every second 4 Mb are produced, then this is 4 mega-bits each second, irrespective how many images per second are captured. Since this is a video compression, the stream is continuous with the appropriate number of pictures in groups of pictures that repeat, called Group Of Pictures, or GOPs (this will be explained in detail later on). These GOPs may vary in size, but the goal is to allow for live streaming with certain visual quality depending on the encoder settings.

The end result is data streaming, defined by the Mb/s, irrespective of how many pictures each second are contained in the signal. Certainly visual quality will be different for different sized GOPs in each stream, but once this data stream is set or selected in the encoder, **this is what defines the storage needed, not the number of images inside this stream.** This is why this is called video compression and it is measured in Mb/s.

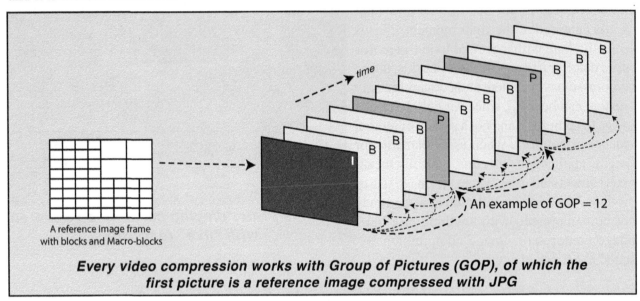

A reference image frame with blocks and Macro-blocks

An example of GOP = 12

Every video compression works with Group of Pictures (GOP), of which the first picture is a reference image compressed with JPG

The video compression uses a "trick" where there are many reconstructed images (in addition to real reference JPG images) based on the past and future GOPs. These are not actual pixels data but rather encoder instructions how the decoder should reconstruct the details in the GOP frames without transmitting all details. **This is how the encoder efficiency and saving of bandwidth is achieved.** To calculate how many megabytes (MB) are recorded on a hard disk every hour, when using video compression of 4 Mb/s, for example, we divide the 4 mega-bits with 8 (since there are 8 bits in a byte), which makes 0.5 MB/s. Multiplied by 60 seconds we get the 30 MB per minute, and then this multiplied by 60 minutes gives us 1,800 MB per hour (this is approximately 1.8 GB per hour).

If compressed images are streamed from an IP camera (rather than using a video compression), then, certainly, each compressed image data packet adds to the overall data traffic. If an average image is about 2 MB for example (which is a good quality JPG produced by an HD camera with 1920 x 1080 pixels), then, one such image per second means 2 MB/s, which averages to 16 Mb/s since 2 MB needs to be multiplied by 8 to get bits per second (one Byte equals 8 bits). So, even with one image per second, using JPG image compression this is already 4 times larger than an H.264 video streaming of the same quality signal with 25 or 30 i/s.

For us in CCTV there is no question, **video compression is more efficient**, saves a lot of network bandwidth, and achieves higher storage capacity then doing it with image compression. Also, all video compressions include audio channel, if needed. **The downside with the video compression is that requires a lot of complex calculations by the codecs, a lot of number-crunching.** When this is done by a dedicated hardware encoder, like it is the case in CCTV IP cameras, there is no problem. But when the decoding is done on a typical computer connecting to variety of cameras, the decoders are then typically in the software (video management system, VMS) it could create problems if the CPU power is insufficient. Depending on this processing power there will only be a limited number of streams that can be decoded simultaneously.

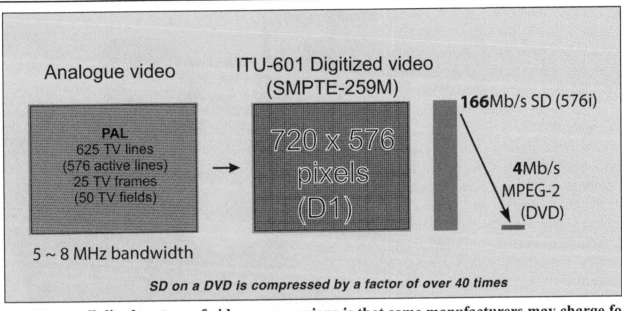

SD on a DVD is compressed by a factor of over 40 times

Another small disadvantage of video compressions is that some manufacturers may charge for the Intellectual Property of the encoder/decoder they are using.

Last, but not least, important downside of video compression is the introduction of **latency (delay)**, which is due to the need for prediction calculations and the GOP structure. Latency is present in all modern digital television broadcasting. Some times there is a delay of a few seconds after the real event being televised, but it doesn't represent a problem for the viewers as there is no need for interaction with the televised scene. In CCTV, however, when real-time camera control is required by operators, it is expected that camera reaction is instant (pan and tilt for example). If there is a delay, it is difficult and confusing to follow a real event, especially if the latency is in order of seconds. However, with careful design, it is possible to find a good compromise between decoding processing required, low latency and great bandwidth savings, while still achieving excellent image quality. In working with video compression latency the important consideration is the speed of human reflex, i.e. the latency between

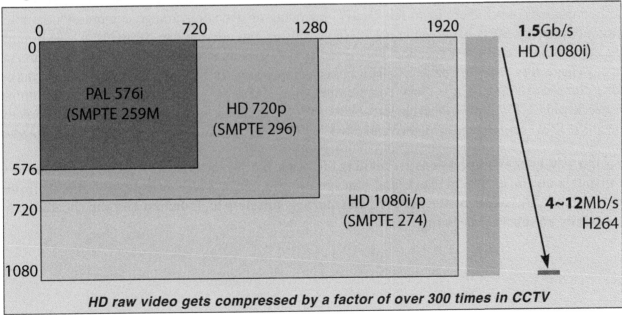

HD raw video gets compressed by a factor of over 300 times in CCTV

the eye and brain processing the visual stimuli. This is in the order of 200 ms ~ 300 ms. This means if the digital system latency is reduce to be not greater than 200 ms, then operators don't have an issue following objects with pan and tilt cameras. This is possible, and this is how large digital CCTV systems have been designed to work. It is important to note that network latency (switchers, routers) as well as decoding latency (which depends on the computer performance) add to the initial encoding latency, but typically they are much lower than 200 ms. Some systems allow for dual streaming, where encoding with lower visual quality, but lesser latency too, are sent to the operators for live PTZ controls, while higher quality encoding with greater latency is used

Photo courtesy of Dallmeier

Camera response to PTZ keyboard control is the key in working with latency

for recording. Live viewing and monitoring of static cameras doesn't impose any problems, as there is no direct camera control needed.

The advantages of image compression on the other hand are: it is very simple to implement and quick and easy to decode. It does not require powerful computers for decoding. **JPG image compression for example, is a universal standard implemented in almost all software and hardware devices dealing with images.** All web browsers, irrespective of the type and operating system, are able to decode JPG compressed images easily and quickly, without a need for a special plug-in software to be installed (many video compressions in order to work with a particular web-browser need a special plug-in to be installed). Also, there is no fee for JPG encoding or decoding. Image compression is much easily applied on mega-pixel cameras with large sensors and odd number of pixels, where it might be impossible to use video encoder. This is the reason why most of digital photographic cameras always have JPG compression as a default image compression. In fact, certain details are lost during JPG compression (since it is lossy compression), but if chosen well, these losses are negligible, especially in CCTV. Professional photographers will always choose raw file format, as they know some details if JPEG compressed are irreversibly lost, but the price paid for that is much larger picture files.

So, in a way, **image compressions are making life easier for the encoders and decoders (computers) but put a strain on the network and storage requirements. Video compression on the other hand makes life easier for the network and storage, but puts high demand on the encoding and especially on software decoding.**

The SD digital video

In the beginning of digital CCTV, before "pure" IP cameras were produced, the very first type of digital video was created when composite video was converted to digital by *analog to digital conversion (A/D) circuitry*. Such a circuit could exist on it's own (as an *encoder*), or inside a standard definition (SD) IP camera, or perhaps inside a digital video recorder (DVR) designed for analog video signals. It is fair to mention at this point that all camera sensors, irrespective whether they are analog, digital, HD, or mega-pixels, all first produce an analog electronic representation of the optical image in front of the lens. So, in other words, **the very first stage of any camera is analog, as explained in the Camera chapter.** Only the output produced may be analog (analog camera) or digital (IP camera).

The A/D is a stage where the analog signal is sampled and quantized (broken into discrete values) in order to be converted to digital format. The sampling rate and levels of quantization depend on the quality and speed of the electronics, and they define the resolution (image quality) and the speed of the digital frame grabbing device. It is important to understand here that, although theoretically a variety of A/D quality conversions might be used in terms of sampling rates and quantizing levels, a television digitization standard called *ITU-R BT.601* has been established and the majority of CCTV products use it.

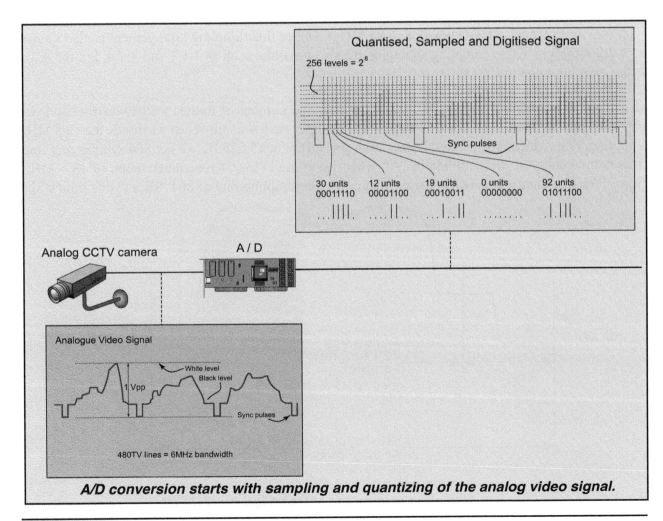

A/D conversion starts with sampling and quantizing of the analog video signal.

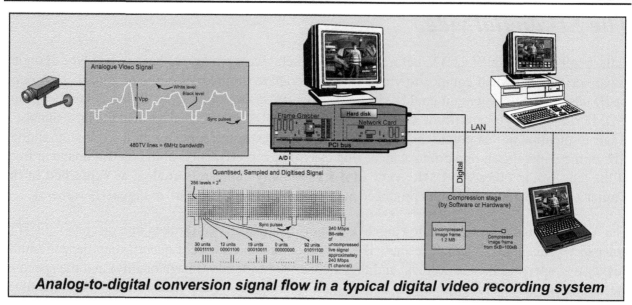

Analog-to-digital conversion signal flow in a typical digital video recording system

The ***ITU-R BT.601*** recommendation specifies the digitization of analog video signal comprised of luminance Y, red color-difference component, and blue color-difference component with a ***sampling base frequency*** of 3.375 MHz, common for both PAL and NTSC. The luminance Y is sampled with four times of this "base" frequency (i.e., 3.375 × 4 = 13.5 MHz), and the color difference components with two times the base frequency (i.e., 6.75 MHz). Hence this sampling arrangement is also known as 4:2:2 sampling. Other sampling strategies are also possible, such as 4:1:1 and 4:4:4, but the 4:2:2 is the most common in CCTV.

If we refresh our memory about PAL scanning lines and number of frames we get each second, we can calculate that there are 625 × 25 frames = 15,625 lines each second. When we divide the 13.5 MHz sampling rate (which is 13,500,000 times each second) with 15,625 Hz we get 864 samples per line. This is the quality of the sampling in PAL when using the ITU-601 recommendation of 13.5 MHz. Since PAL line duration is 64 μs (see the diagram), the sampling rate of 864 "slices" this time width

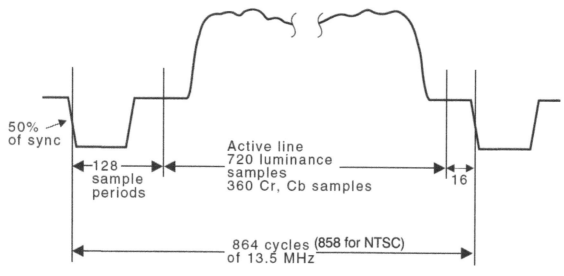

The sampling rate as recommended by ITU-601

in pretty fine slices. It should be noted that this "slicing" includes the sync pulses as well.

The same type of calculation for NTSC, using 525 scanning lines at 59.94 Hz field rate (the accurate field frequency is 59.94, not 60) obtains 525 × 29.97 Hz = 15,734.25 lines each second. Dividing 13.5 MHz by 15,734.25 Hz gives 858 samples per line, including, again, the sync pulses.

So, just to recap, **using the ITU-601 recommendation in PAL luminance sampling, we get 864 samples/line, and in NTSC we get 858 samples/line. In both cases a sampling frequency of 13.5 MHz is used**.

Based on the above, we can reach a very important conclusion about the ITU-R BT.601: **the ITU-601 is the first international recommendation that tries to merge the two incompatible analog composite television standards (NTSC with 525/59.94 and PAL with 625/50) to a common component digital sampling concept. The major achievement of Rec 601 is choosing a set of sampling frequencies of 13.5 MHz which is common to both standards.**

Out of the 864 samples for PAL and 858 for NTSC, **the active line in both cases is given to have 720 samples**. This is the maximum horizontal resolution a digitized signal using ITU-601 sampling recommendation can have. The term *resolution* should be used loosely here because it has a slightly different meaning than the analog video signal resolution expressed in TVL. We shall explain this in more detail further in the text.

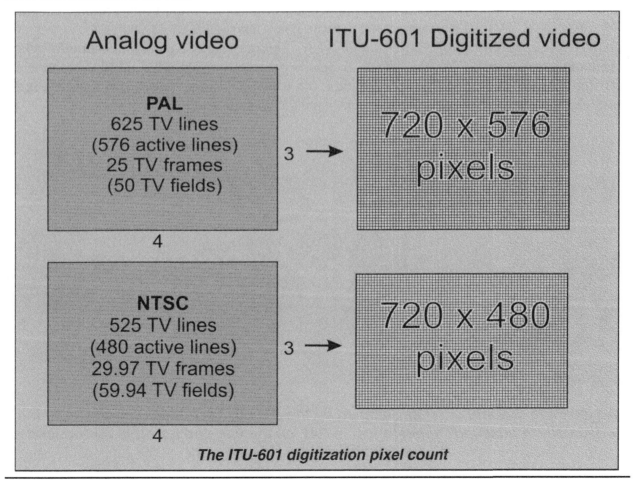

The ITU-601 digitization pixel count

Some of you may ask, "Why 720, and not less or more than that?" This is because 720 is a number, divisible by 8 (i.e., 2^3), which is very useful for most of the video compressions using *discrete cosine transformation* (such as JPG, MPEG and H series) where video images are subdivided in blocks of 8×8 pixels. Often, you will find that some digital processing equipment will narrow the active video signal by 8 samples to the left and 8 to the right of the main video signal contents of 720 samples, making an active line consist of 704 pixels instead of 720. This is to allow for the various camera active signal fluctuations (tolerances).

The vertical sampling recommendation by ITU-601 is equal to the number of active lines, which is 288 per TV field (or 576 for a full TV frame) in PAL and 240 per TV field (or 480 for a full TV frame) in NTSC.

Therefore, **the digitized TV frame according to the ITU-601 recommendation is 720 × 576 for PAL and 720 × 480 for NTSC. Often, this is referred to as D1 resolution. If the active lines are considered, i.e. 704 x 576 and 704 x 480, then often, in CCTV, this is referred to as 4CIF.**

This fact also indicates that the ITU-601 considers the interlaced scanning effect and in many digital recorders a choice can be made if the playback is to be in field or frame mode.

An observant reader would notice something in these numbers that may make digital CCTV confusing; because of that, it is worth clarifying now. This is the aspect ratio of the standard definition TV and the aspect ratio of the images produced when sampled with ITU-601 recommendation. As we all know, all TVs and monitors in CCTV use an aspect ratio of 4:3 = 1.33, and yet the aspect ratio of 720:576 = 1.25 for PAL and 720:480 = 1.5 for NTSC. This introduces so-called "*non-square*" pixels in both of these standards. The PAL gets "horizontally squashed" pixels, which need to be stretched out before reproduction onto a 4:3 aspect monitor, while the NTSC gets "vertically squashed" pixels, which need to be expanded vertically before displaying onto a monitor.

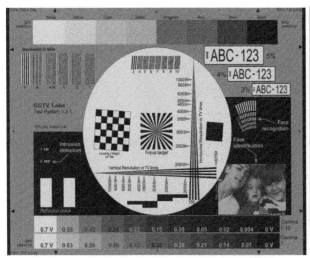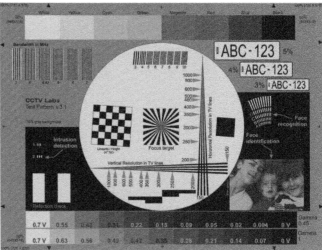

An example of a test chart as sampled by the ITU-601 recommendations (note the horizontally "squashed" appearance) in PAL on the left, and how it is reproduced at the analog video output on the right

This expansion/stretching correction is usually done (or at least it should be done) in the last stage of the decoding before it gets displayed. It may seem as if this is an unnecessary stage in digital, but in actual fact it makes the decoding electronics cheaper and more universal, since they are used in both PAL and NTSC.

The resolution of the ITU-601 digitized (SD) video

Based on the Nyquist theory, **an analog and continuous signal can be reproduced from its discrete samples, if the sampling frequency is at least twice the highest bandwidth frequency.** Higher frequencies than the highest bandwidth are not wanted and in actual fact if they do exist they cause aliasing (like the well-known Moiré patterning). In order for the aliasing to be minimized, the sampled signal has to pass through a low-pass filter where frequencies higher than the upper frequency (equal to half the sampling frequency) are being deliberately eliminated. An ideal brick-wall low-pass filter doesn't exist in practice, so the actual filter cutoff frequency is slightly lower than what the theory needs it to be. This fact has a direct bearing on the frequency response and the number of horizontal picture elements (pixels) that a digitized system can handle.

In ideal conditions, if no additional filtering was done, given the Nyquist frequency of 6.75 MHz (i.e., a sampling rate 13.5 MHz), the 720 pixels per active line would be equivalent to a horizontal resolution of 3/4 × 720 = 540 TVL, as defined by the analog TV.

The ITU-601 recommendation, however, specifies an anti-aliasing and reconstruction filter cutoff of 5.75 MHz, which reduces the luminance analog horizontal resolution to around 460 TVL.

Further reduction of the resolution is introduced by the video compression artifacts themselves, so in practice it is fair to say that **no video signal in digitized CCTV can have any higher horizontal resolution than around 450 TVL**. It now becomes very clear that choosing a video compression that has as little losses as possible is of paramount importance. This contradicts the "CCTV typical" requirement for long recording storage. We will discuss the various video compressions further in this chapter, but **it is important to highlight again that the above resolution limit applies to the digitized analog video signal, before it undergoes the compression.**

The human eye is less sensitive to color resolution, and because of this in CCTV we do accept 4:2:2 sampling (four units luminance, two for color difference of red and two for color difference of blue) strategy as good enough. So, the chrominance signals are sub-sampled by a factor of two, at 6.75

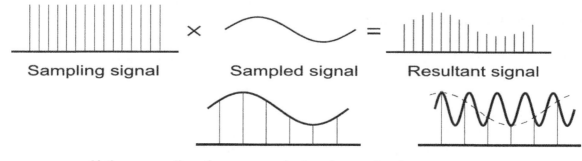

Sampling signal   ×   Sampled signal   =   Resultant signal

If the sampling frequency is too low, aliasing may occur.

MHz (only half the luminance sampling of 13.5 MHz). This results in 432 total pixels for PAL and 429 pixels for the NTSC scanning standard (includes the sync pulses period). So, the digital active line accommodates the 360 red color-difference samples and 360 blue color-difference samples in both standards. Under ideal conditions, given the Nyquist frequency of 3.375 MHz, 360 pixels per active line is equivalent to 3/4 × 360 = 270 TVL of color resolution. Rec 601 specifies an anti-aliasing and reconstruction filter cutoff of 2.75 MHz, resulting in a color differences signal resolution on the order of 220 TVL.

All of the above represent a very important conclusion when discussing resolution in digitized video. It should be noted that this is the ITU-601 digitization recommendation, and it is in use in the majority of digitization products in the CCTV dealing with analog video (typically called *encoders*, or *streamers*). There is no advantage of using cameras with much higher resolution than 450 TVL when the same is to be recorded on ITU-601 compliant recorders. This is the same argument as when we had high-resolution cameras (460 TVL for example) being recorded on VHS VCRs (which were limited to 240 TVL by the low-pass filter design).

The difference here is not so dramatic, as some CCTV manufacturers lately have come up with color cameras offering 520 TVL, for example, or lately even 650 TVL. Practically, this means one cannot see any difference between a 460 TVL or 520 TVL or even 650 TVL camera, when these are recorded on ITU-601 compliant DVRs. More attention should be directed

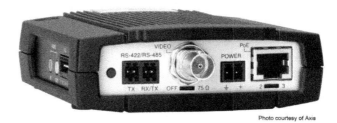

A typical analog video encoder (streamer)

to choosing a camera with a better signal/noise ratio, less smear, or better dynamic range than to slight differences in horizontal resolution that can't be seen.

If, however, a system is designed and used for just live monitoring on high-quality CCTV monitors, such a difference in resolution might be visible, but only for live viewing, not playback.

The above is all true for horizontal resolution, but let us now talk about vertical resolution. In some system designs, vertical resolution is as important, especially when detecting and recognizing license plates, or faces at a distance.

The number of quantized levels in ITU-601 is chosen to be represented with 8 bits, that is, making a total of 256 levels ($2^8 = 256$). The reason for such a choice is very practical from an engineering point of view: **CCTV monitor can reproduce a very good video signal without the need of more than around 250 shades of luminance, so there is no need to sample the analog video signal with more levels than this**. The 256 is chosen because it is a binary number, and as we know, in the digital world everything is represented with zeros and ones (i.e., with the binary numbering system).

So, ITU-601 suggests that out of the 256 combinations of 8 bits, the 0 and 255 are used for representing the syncs, which leaves 1 to 254 values to be used for video. The luminance level of black is given a value of 16 (binary 00010000) and the white level is given a value of 235 (binary 11101011). The value 128 is used to indicate that there is no chrominance in the signal.

As we mentioned earlier, **the number of vertical pixels in a TV frame offered by the PAL system is 576, while in the NTSC this is 480, which corresponds to the actual number of active lines in each standard.** It is important to remind the reader that each analog camera in CCTV generates interlaced video (50 fields/s and 29.97 fields/s). The interlaced video consists of TV fields displaced in time (1/50 s for PAL and 1/29.97 s for NTSC). As a result, when digitizing video with moving objects the **interlaced effect** may show up if the recording is made in frame mode. **This is a "natural" analog television effect – a result of the interlaced scanning. It is not an error on behalf of the digitization, as some may think.** The objects may seem blurred in the direction of movement, and the faster the object moves the more obvious this effect is.

Certain techniques called ***de-interlacing*** can minimize or completely eliminate this effect. Such functions are available in various video and photo editing programs (such as PhotoShop, Premiere, VLC,...), but many DVRs and applications in CCTV can also perform de-interlacing on the fly (while playing back).

The result of playing back in frame mode, as opposed to field, is twice the vertical resolution, making object edges smoother and showing more details in an exported image. When playing a footage in frame mode, the artifacts of playing alternating fields becomes apparent and this is the jumping up and down of each next field by one line. This is, again, a natural result of how interlace television works, and it is not an error in the playback as some may think. Basically, the digitized fields are displaced by one line (since they are coming from cameras complying with the 2:1 interlaced PAL or NTSC TV standard). When playing back in frame mode, basically the DVR/software plays two fields at the same time.

It is interesting, then, to ask the following question: how does a digitized video recorded in field mode (720 × 288 for PAL, or 720 × 240 for NTSC) get reproduced to appear with 720 × 576, that is, 720 ×

Left: TV field exported; Center: TV frame interlaced effect; Right: De-interlaced frame. Note the jagged edges on the car when field recording is used (left)

The difference between field recorded image on the left and the frame recorded

480 on a screen or when exported? **This is done simply by duplicating each line**. Such duplicating produces another prominent effect of ***jagged edges***. The human eye is more sensitive to resolution in the horizontal direction than in the vertical, and this is the reason perhaps why the majority of DVR systems in practice are set to record this way. In some systems, however, vertical resolution might be more important, so frame recording should be used. For other DVRs, it is simply not possible to record in any other than field mode as that could be the only option they have.

The interlaced effect explained above appears only in analog cameras with image compressions, which, as explained, are compressions that work with static TV fields and will be treated as still images. In temporal video compressions, such as MPEGs and H.26x series, the interlaced effect is compensated by the process of motion prediction vectors; therefore this effect is not as noticeable. In digital cameras (HD or MP) all signals are captured in progressive, non-interlaced mode, so there are no such issues.

All of the discussion we have had so far refers to the so-called ***full TV frame*** resolution. There are a number of compression techniques that use one-quarter of TV frame pixels. This size is usually referred to as the ***Common Interchange Format*** (CIF) and was initially used by MPEG-1 and H.261 video compressions (early video conferencing format or movies on CDs). This makes the CIF video pixels real estate ***one-quarter*** of the full frame as defined by the ITU-601 recommendation (one-half of the horizontal number of pixels and one-half of the vertical number of pixels). The purpose was to reduce the digitized streaming to an acceptable size, with resolution comparable to the early VHS quality. The CIF size was originally designed for video conferencing, and it was very suitable for live video with talking heads movement. The reduced resolution, when compared to 4CIF, needs to be considered when using it in a system where face identification or license plates recognition are required. Also, a QCIF size is available, which refers to Quarter CIF size (i.e., 176 × 144 pixels). All of the calculations and picture resolution we have made earlier are applicable by halving these numbers.

Thus, an **equivalent analog resolution that a CIF size image would have is around 230 TVL, which is almost the same as VHS resolution**. For some applications this might be of sufficient quality, especially for CCTV with remote access, since smaller size of frames takes less time and streams less data to encode and transmit.

It is understandable that we want the best possible picture quality. But no matter what steps we take, **in lossy compression the image cannot be of a better quality than the original, uncompressed image**. The pixel count of a digitally recorded image of any analog CCTV camera, even if it is a full frame size, is only just over 410,000 picture elements for PAL and just over 340,000 for NTSC. One can certainly appreciate the difference between a 400,000 and say a still image of a digital photo-camera with, for example, 14,000,000 pixels. Using D1 (4CIF) resolution as a starting point, we can draw conclusions about angles of view and pixel density required when using HD signals, or beyond. Essentially, everything comes down to how many pixels a human head or licence plate occupies. With appropriate lens selection, even CIF size image may give sufficient details to recognise or identify a person. Certainly, in such a case the angle of view will be much narrower than when using a HD camera for example looking at the same person and producing the same number of pixels of his/her head shot. This was already explained in the *Displays* chapter.

In a nutshell, for face identification as per the analog standards (where 100% of a person's height was required at D1 or 4CIF), twice the person's height is required when using only CIF size (a person is viewed to his waist). When switching to 1080 HD, a full person's height has to occupy at least half (50%) of the 1080 monitor height, or 80% of 720 HD monitor height. The above does not consider compression artifacts, so it is expected that a compression with minimal visual losses is used.

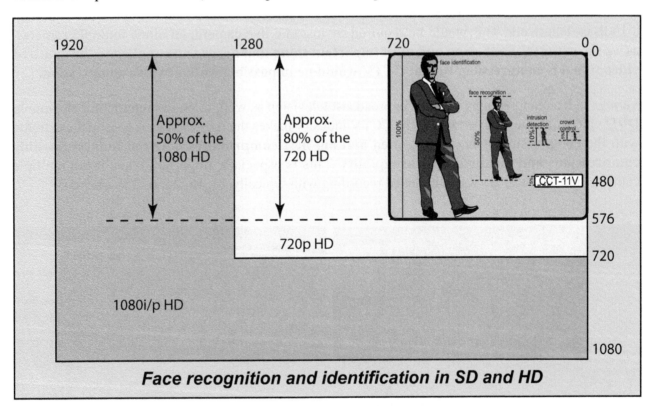

Face recognition and identification in SD and HD

The need for compression and networking

The following simplified calculations will help understand the amount of streaming data when analog signal is converted to digital using the ITU-601 recommendations, before compression is applied. Multiplying the samples of each line (864 for PAL and 858 for NTSC) with the number of lines in the system (625 and 525), then with the number of TV frames of the system (25 and 30) and assuming 8 bits representation of luminance and 8 of the color differences (4 for Cr and 4 for Cb), we get approximately **the same bit rate for both (NTSC and PAL) digitized TV systems**.

For **PAL**: $864 \times 625 \times 25 \times (8 + 8) = 216$ Mb/s, of which the **active video streaming is $720 \times 576 \times 25 \times 16 = 166$ Mb/s.**

For **NTSC**: $858 \times 525 \times 29.97 \times (8 + 8) = 216$ Mb/s, of which, similarly, the **active video streaming is $720 \times 480 \times 29.97 \times 16 = 166$ Mb/s.**

This is a bit rate for a digitized uncompressed live video streaming, according to ITU-601 with 4:2:2 sampling. If the 4:4:4 sampling strategy is used, or even 10 bits instead of 8 bits sampling (which is done in the broadcast TV for video editing and processing), this number of over 166 Mb/s streaming becomes almost twice as large. Streaming 166 Mb/s of just one video signal is impossible even with a 100BaseT network. This would be required by just one live camera, let alone multiple cameras, as we use in CCTV. **So the first and most important thing that needs to be applied to the digitized video signal is compression. Digital CCTV would be impossible without video compression.**

Various video compressions are used in broadcast television as well, or on the Internet for streaming, DVD video, or Blue-Ray. However, the CCTV industry makes the most of it as it goes to the extreme with the compression technologies, often **making best compromises between highest possible compressions and maximum picture quality.** This is especially important when using multiple cameras into a typical DVR (multiplexed recording with typically 16, 18, 24, or 32 cameras).

Noncompressed image (720 x 576 pixels) on the left (digitized and noncompressed around 1.2 MB) and the same on the right compressed 100X with JPG compression

*2X enlargement of a section
100X JPEG compression*

*2X enlargement of a section
100X wavelet compression*

A typical uncompressed full frame video image might be over 1.244 MB in size in PAL (720 × 576 × 3 = 1.2 MB) where we assume 3 colors and 8-bit sampling. Hence 8 bits can be converted to 1 byte. In CCTV we work with a compressed image size (typically JPG) of less than 100 kB, and often even lower than 10 kB.

If we have 25 fps (frames per second) at good quality image compression, at 100 kB each frame, we would get 2,500 kB/s for "live streaming." This number when converted to kb/s produces around 20 Mb/s (2.5 MB/s x 8). **The Video compression is more efficient than this due to the temporal factor, so that for the same visual quality we could use MPEG-2 at 4 Mb/s for example, which is at least 5 times less bandwidth when compared to the image compression.** With the modern H.264 encoders, we could produce a similar visual quality with only 1 Mb/s, which makes it a **20 times less streaming bandwidth compared to a M-JPG or 20 Mb/s as above.**

When video compressions is used it is not expressed in field or frame size in kB, but rather as a streaming in kb/s or Mb/s. So, for example, a high DVD quality video compression using MPEG-2 works with around 4 Mb/s. A decent quality of MPEG-4 streaming over the Internet can be around 256 to 512 kb/s. How far you can go in increasing the compression depends on how much detail one is willing to sacrifice from the image and the type of compression used, but, again, there is no doubt: **compression must be applied**. The only question is: which one?

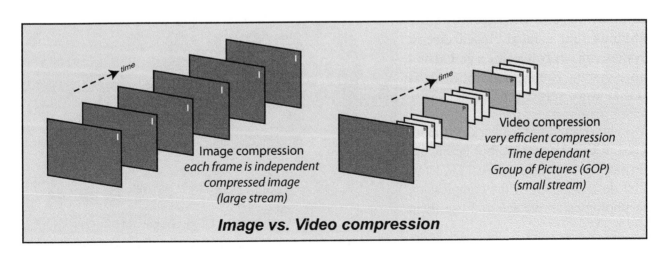

Image vs. Video compression

Certain image processing can be done on a digitized signal, before or after the compression. Some processing makes a simple division and recalculation in order to put the images in smaller screens (as is the case with split screen compressors or multiplexers), other processing may perform sharpening (which is actually an algorithm where every pixel value of the image is changed on the basis of the values of the pixels around it) or video motion detection, and still others may reduce the noise in the signal, and so on.

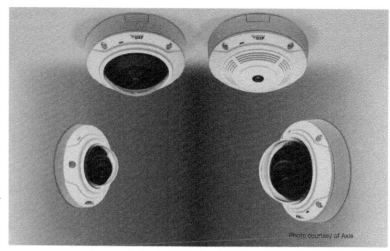

A variety of IP dome cameras

Digital networks today are used not only for sending e-mails and web browsing, but they have become the main transport media for digital CCTV.

One of the biggest advantages of digital video was the possibility to transmit the video over almost any network. Whether this is copper Cat-6 cable, fiber optics Ethernet, or Wi-Fi it is almost irrelevant, as digital video is data like any other digital signal. Certainly, special considerations have to be made in regards to the bandwidth and speed, but video can be sent over almost any modern network.

Although a typical advice for a new system would be to run the digital CCTV network on it's own, there are projects where using an existing network will prove beneficial and will save a lot of money. Certainly, an agreement from the responsible IT people is needed, but digital CCTV system can easily be retrofitted in such environments. Separation of traffic is possible with the so-called Virtual LANs (V-LANs), which offer extra security and guaranteed bandwidth.

Despite the distance limitations of local (copper) networks, longer distances are easily achieved by using fiber, wireless, and combined networks with repeaters (this is usually done by network switches and routers).

In today's age of Internet it is quite obvious that a local closed circuit system can easily become a global and open system, connecting continents as if they were across the street.

We will discuss the networking theory used in CCTV later on, but before we go onto the networking in CCTV, let us first describe each of the compression technologies as we see and use them in CCTV.

HD PoE cameras work with only one cable

Raw and compressed digital video

When digital video was introduced near the end of the last century, the current High Definition (HD) digital standard didn't exist. So the first form of digital video was basically analog video signal converted to digital. We refer to this as ***Standard Definition (SD)*** digital signal. As explained earlier, the actual pixel count of this standard for PAL is 720 x 576, and for NTSC it is 720 x 480. The actual standard that defines the parameters of SD video is known as ***SMPTE-259-B*** (SMPTE stands for **Society of Motion Pictures and Television Engineers**) for PAL, and ***SMPTE-259-A*** for the NTSC. **Since the SMPTE-259-B refers to a PAL signal with 576 active interlaced lines, this is sometimes referred to as 576i. Similarly, the SMPTE-259-A, which refers to the NTSC digital version with 480 active interlaced lines, sometimes it is referred to as 480i digital format.**

It is important to highlight that these standards deal with raw (uncompressed) digital streams. So, the digital stream of uncompressed SD video, using 8-bits sampling, is approximately ***166 Mb/s*** for the SD 576i, as well for the SD 480i. This refers to a stream of one live video channel, where live refers to 25 frames (50 fields) per second in PAL and 30 frames (60 fields) for NTSC with the corresponding pixel counts. It is quite obvious from these numbers that in CCTV, when using multiple cameras, it is almost impossible to work raw digital video. **With the amount of storage required we have no choice but to compress the video data.**

The important recent advancement in the broadcast industry, and hence also in CCTV happened in the time between the previous two editions of this book: the official acceptance of the ***High Definition*** television standard ***(HD)***. In fact, one of the main reasons for this revision is exactly this — the introduction of the digital and high definition technology. So, although analog TV and CCTV are still covered in this book, the main standard we now discuss is the HD and beyond.

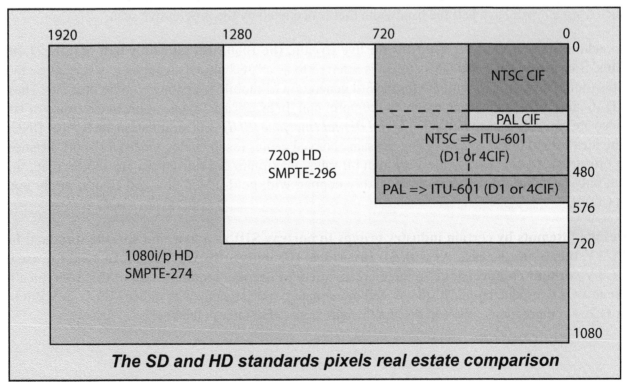

The SD and HD standards pixels real estate comparison

The HD video standard is now a world-wide standard, and not a country or continent based.

The key points of HD are: the signal is digital from the source itself (camera), it contains five times more details (resolution) than the analog SD and it's aspect ratio is 16:9, as opposed to 4:3 in SD. There are, admittedly, some minor differences between some countries in the number of TV frames per second (25, 30, 50 or 60), but the pixel count format is internationally compatible, meaning a TV from US will work in Europe or Australia and other way around. This was not the case with analog TV.

Digital video formats	Standard	Signal mVpp	Active pixels	Live uncompressed bandwidth	Compressed with CCTV neglegible visual losses
PAL SD 576i	SMPTE-259M-B	800	720 x 576	166 Mb/s	1~4 Mb/s
NTSC SD 480i	SMPTE-259M-A	800	720 x 480	166 Mb/s	1~4 Mb/s
HD 720p	SMPTE-296	800	1280 x 720p	1.485 Gb/s	2~10 Mb/s
HD 1080i	SMPTE-274	800	1920 x 1080i	1.485 Gb/s	2~10 Mb/s
HD 1080p	SMPTE-274	800	1920 x 1080p	2.970 Gb/s	4~20 Mb/s

The SMPTE-296 comes as 720p, meaning it defines 1280 x 720 pixels in "progressive" mode only, while SMPTE-274 comes as 1080i and 1080p, meaning it can either be "interlaced" or "progressive" scanned image.

The idea behind the interlaced (denoted with "i") is similar as in analog, to reduce the bandwidth of data produced by such a signal. Interlaced stream alternates the lines between the fields so that it still produces the 1080 visual experience, but this is composed of 540 alternating lines (odd and even). Such a stream will have half the bandwidth that is occupied by the progressive scan.

So, while 1080p occupies 2.970 Gb/s for live stream, the 1080i will take only half of it, i.e. 1.485 Gb/s. Sometimes the 2.970 Gb/s stream is referred to as a "3G" digital signal as it is very close to 3 Gb/s (not to be confused with 3G for the 3rd generation of mobile telephony communication). These HD streams refer to raw data, before the compression. In broadcast TV uncompressed streams of HD video are typically transmitted using *Serial Digital Interface (SDI)* with terminals in the form of BNCs. The idea behind this is to keep using the same coaxial cables (as in analog video) for short distances in broadcast studios. Due to the very high bit-rates, maximum coax distances are shorter than; this can only be achieved with very high quality copper, with gold-plated pins and BNCs, rarely used in CCTV.

Despite attempts by certain industry groups to portray SDI as a new and specific standard for CCTV, this is not the case. As with SD raw video, HD native raw video is impractical for use in CCTV without compression. So there are a variety of encoding (compressions) that we typically use in CCTV, producing much smaller and more manageable streams with either MPEG-2, MPEG-4 or H.264 compressions. We will discuss them in more detail later in the book.

The most common compressions in CCTV

As explained previously, compressions can be divided in two main categories: compression applied to still images, which we call *image compressions*, and compressions applied to a continuous streaming video signal, and therefore called *video compressions*. The image compressions use static images, while the video compressions use the time as an important variable when reducing image redundancy. This is why video compressions are often referred to as *temporal compressions*.

Some authors make a compression division based on what standard group has proposed them, such as ITU-T or ISO, JPEG, or MPEG. There are also proprietary compressions that some manufacturers offer as their own, and this is especially typical for the CCTV industry. Some of these proprietary compression may do an excellent job of compressing, even better than some standard compressions. The only problem with proprietary compressions is that there are no standard off-the-shelf codec for them, but typically they would use purpose built encoding chips with their own firmware. Also, the receiving side has to have the manufacturer's proprietary decoding software. Despite some of them being very advanced, the actual problem is their compatibility with the rest of the world. Not being able to playback (decode) such encoded video on a third party computers without specialised software makes it a difficult choice. So, later in this chapter we are going to list the most popular image and video compressions.

The following are the most common **image compressions** used in CCTV today, in the order of time appearance:

> • **JPEG** – A widely spread standard, over 20 years in existence. The original Joint Photographic expert Group (JPEG) group was organized in 1986, issuing the first JPEG standard in 1992, which was approved in 1994 as ***ISO/IEC 10918-1***. It offers simple and quick encoding and decoding, and no royalties are attached. It uses the mathematical Discrete Cosine Transformation (DCT) to achieve it's redundancy, i.e. to effectively remove details from a picture that human eye will not notice. It is incorporated and used by many programs, such as image editing and web browsers. All digital photographic cameras have this codec, and almost all CCTV IP cameras have JPG (or M-JPG) as one of the options in the encoder settings. Typical file extension is *.jpg*.

> • **M-JPEG** – Motion JPEG, a variation on JPEG, and not really a standard, but offers a stream of multiple successive JPG files. Each image is an independently compressed TV frame using JPEG compression, and may allow for audio as well. M-JPEG is less efficient than the MPEG-2 and H.264, but requires less processing power and it is easily implemented. Often mega-pixel cameras, for which there is no off-the-shelf video codec (due to unusual, or large number of pixels), have M-JPG as the only available compression. File formats vary, but *.avi, .amv, .mov* are the most common.

> • **Wavelet** – Used to be popular image compression in CCTV, but chipsets are superseded with JPEG-2000. Started with Haar's work in the early 20th century, used in various scientific researches, such as studying the reaction of the ear to sound, astronomy, and similar subjects. Offered better details than JPEG and it did not divide the image in blocks of 8 × 8 pixels, but rather followed the outline of objects. Typical file extensions were *.wav*.

• **JPEG-2000** – A standardized version of the Wavelet compression. It was created by the Joint Photographic Experts Group committee in the year 2000, with the intention of superseding their original discrete cosine transform-based JPEG standard with the wavelet-based method. It is also known as *ISO/IEC 15444*. The standardized filename extension is *.jp2*. Plug-ins are available for JPEG-2000 for a variety of image editing programs and web browsers. Some DVRs in CCTV use JPEG-2000 codecs.

• **M-JPEG-2000** – Similar to the M-JPEG, but uses JPEG-2000 and it is standardised under *ISO/IEC 15444-3*. It specifies the use of the JPEG-2000 format for timed sequences of images (motion sequences), possibly combined with audio, and composed into an overall stream. Some high end broadcast and cinematographic cameras offer loss-less version of M-JPEG-2000 for audio/video recording, which is then very accurate to edit, as each frame is a key frame. This may be not so attractive for CCTV as the streaming produced by even lossy M-JPEG-2000 is still much larger than any video compression. Filename extensions for M-JPEG-2000 video files are *.mj2* and *.mjp2* according to RFC 3745.

The following is a list of the evolution of **video compressions**:

• **H.261** – Video compression introduced in 1984 by the ITU-T, intended for video conferencing over then new digital ISDN lines. It was ratified in November 1988 and it is the first member of the H.26x family of video coding standards. H.261 was originally designed for low bit-rates transmission over the first digital communication lines introduced in the last century — the ISDN lines. The coding algorithm was designed to be able to operate at video bit rates between 40 kbit/s and 2 Mbit/s. H.261 supports two video frame sizes: the Common Interchange Format known as CIF (352x288 luma with 176x144 chroma) and Quarter CIF, known as QCIF (176x144 with 88x72 chroma) using a 4:2:0 sampling scheme.

• **MPEG-1** – While H.261 was primarily intended for video conferencing, the MPEG-1 was created at the time when computers and data CDs became the latest technology, around 1993. The intention was to be able to compress and put a full digitized movie on a CD. It was designed to produce VHS quality video onto a CD with low bit rates of around 1.5 Mb/s. MPEG-1 is the most widely compatible lossy video compression in the world, perhaps not as well known for its video compression as much as for its audio component, the popularly known as MP3 audio format. The MPEG-1 standard is published as *ISO/IEC 11172*. When saved as movie file, typical extension is *.mpg.*

• **H.263** – H.263 is, as H.261, a video compression standard originally intended for video conferencing, building on H.261. It was developed by the ITU-T around 1995. H.263 was developed as an evolutionary improvement from H.261, and MPEG-1 and MPEG-2 standards. Its first version was completed in 1995 and provided a suitable replacement for H.261 at all bit-rates. The next enhanced codec developed after H.263 was the H.264 standard, which provided significant improvement in capability beyond H.263, so that H.263 standard is now considered a legacy design. Most new video conferencing products now include H.264 as well as H.263 and H.261 capabilities.

MPEG-2 – Introduced after the MPEG-1, around 1996, but intended for broadcast quality video. Initially was known as H.262, but got more popular as MPEG-2, although the official standard was released under the name of *ISO/IEC 13818*. It also belongs to the group of lossy audio and video compressions. MPEG-2 is widely used for broadcasting digital television signals over terrestrial antennas, cables, and satellites. It is also used for distribution of movies on DVD disks. Although MPEG-2 was initially applied and used (and still is) on standard definition (SD) analog video, it also works with high definition (HD) video signals. The audio compression under MPEG-2, also known as Advanced Audio Coding (AAC) was standardized as an adjunct to MPEG-2 (as Part 7). Typical good DVD quality SD video streams with MPEG-2 are around 4Mb/s. When HD video is compressed with MPEG-2 it may go up to 40Mb/s. When saved as movie file, typical extensions are *.mpg, .mpeg, .m2v.*

• **MPEG-4** – Introduced in the late 1998, in the time when computers and internet became important for businesses, entertainment and communications. It was published as an official standard under the name of *ISO/IEC 14496*, although it is more popular as MPEG-4, based on the main contributing group name of Motion Pictures Expert Group. MPEG-4 was intended to be more encompassing then all preceding video compression standard and more scalable. It included compression for web streaming media, CD and DVD distribution, video conferencing, broadcast television and certainly CCTV. MPEG-4 absorbs many of the features of MPEG-1 and MPEG-2 and other related standards, adding new features such as support for 3D computer rendering, support for Digital Rights Management (DRM) and various types of interactivity. MPEG-4 is still a developing standard and is divided into a number of parts. Companies promoting MPEG-4 compatibility do not always clearly state which "part" level compatibility they are referring to. To deal with this, the standard includes the concept of "profiles" and "levels," allowing a specific set of capabilities to be defined in a manner appropriate for a subset of applications. It is interesting to mention here that the now most popular H.264 video compression, is actually MPEG-4 Part 10. MPEG-4 is efficient across a variety of bit-rates ranging from a few kilo-bits per second to tens of megabits per second. Most common file formats when video is saved using MPEG-4 are *.mpg, m4v.*

• **H.264** – This is one of the most popular and most efficient standards used in broadcast television, internet and certainly CCTV today. It is also known as *Advanced Video Codec (AVC)* and also as *MPEG-4 part 10*. The final standard came out somewhere in early 2003. The actual standard code is also known as *ISO/IEC 14496-10*. H.264 is perhaps best known as being one of the codec standards for Blu-ray Discs. It is also widely used for streaming on the internet, such as videos from Vimeo, YouTube, and the iTunes Store. Most of the digital television that used MPEG-2 for broadcasting, recording, and transmission are now switching to H.264. Like the initial MPEG-4, the goal of H.264 was to provide enough flexibility to be applied to a wide variety of applications from mobile phones to broadcast TV and cinemas. The H.264 standard can be viewed as a "family of standards" composed of the profiles. A specific decoder decodes at least one, but not necessarily all profiles. The decoder specification describes which profiles can be decoded. In CCTV for achieving similar visual quality as with MPEG-2, the H.264 requires only a third, or quarter, of the streaming bandwidth. This is huge savings on network bandwidth and in storage capacities, which is why almost all CCTV manufacturers these days have at least one of the H.264 profiles in their cameras or recorders.

Some new compressions

At the time of preparing this book, a new and improved video compression on H.264 was adopted, called H.265. Since this has only came out, there are no products on the market using this new video compression, but it's worth noting the key features.

> • **H.265** – This is a successor of H.264, and it is also referred to as ***High Efficiency Video Coding (HEVC)*** video compression. The formal name under which ITU-T published it as a standard this year is ***ISO/IEC 23008-2***. HEVC is said to improve video quality, producing visually similar quality as H.264 at half of the bit-rate. This means a good quality 1080HD stream might be squeezed down to 2 Mb/s and still have acceptable quality. It is also said that it can support the next step in resolution after HD - the 8K Ultra-High Definition (UHD) with resolutions of up to 8192×4320 pixels. At this stage this is brand new video compression technology. It remains to be seen how soon and how much our industry will implement it in CCTV equipment. However, there is no doubt it will help further reduce network bandwidth and increase (double) the storage.

> • **Dirac** – Despite not being used in CCTV yet, Dirac is an open and royalty-free video compression format, specification, and system developed by BBC Research at the BBC. Dirac format aims to provide high-quality video compression for Ultra HDTV and beyond and, as such, competes with existing formats such as H.264 and now H.265. The specification was finalized in January 2008, and further developments are only bug fixes and constraints. Dirac Pro was used internally by the BBC to transmit HDTV pictures at the Beijing Olympics in 2008. The format implementations are named in honour of the theoretical physicists Paul Dirac and Erwin Schrödinger, who shared the 1933 Nobel Prize in Physics.

DCT as a basis

One of the most common mathematical transformations used on the compression of two-dimensional images is the ***Discrete Cosine Transformation*** (DCT). This is the basis for almost all compression techniques used in CCTV, with the exception of the Wavelet based. So JPEG, MPEGs, and H series compressions all use DCT in one form or another. In fact, the reference frame in all video compressions is in actual fact a JPEG frame compressed with a DCT. Because of this, it is important to say a few words about it.

The DCT is based on the ***Fourier Transformation***. The Fourier Transformation is a very good method for analyzing signals in frequency domain. The only "problem" is that it always works with an assumption of signals being periodical and infinite. This is never the case in reality, and this is why an alternative to the Fourier Transformation, the ***Fast Fourier Transformation*** (FFT), was introduced in the 1960s. **The DCT is based on FFT.**

So how does the Discrete Cosine Transformation work? Spatial redundancy is found in all video material, be that CCTV or broadcast television. If there is a sizable object in the picture (TV field), all of the pixels representing that object will have quite similar values. This is redundancy, that is, it is possible to reduce the amount of information of each pixel with a value of the one giving the average and by

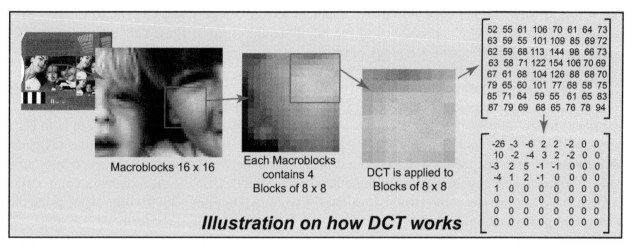

Illustration on how DCT works

defining the area only. Large objects produce low spatial frequencies, whereas small objects produce high spatial frequencies. Generally, these frequencies will not be present at a high level at the same time. Digitized video has to be able to transmit the whole range of spatial frequencies, but if a frequency analysis is performed, only those frequencies actually present need be transmitted. Consequently, an important step in compression is to perform a spatial frequency analysis of the image.

The illustration here shows how the two-dimensional DCT works. The image is converted a block at a time. A typical block is 8 × 8 pixels. The DCT converts the block into a block of 64 coefficients. A coefficient is a number that describes the amount of a particular spatial frequency that is present. In the figure the pixel blocks that result from each coefficient are shown. The top left coefficient represents the average brightness of the block and so is the arithmetic mean of all the pixels or the DC component. Going across to the right, the coefficients represent increasing horizontal spatial frequency. Going downwards, the coefficients represent increasing vertical spatial frequency. Now the DCT itself does not achieve any compression. In fact, the word length of the coefficients will be longer than that of the source pixels. **What the DCT does is convert the source pixels into a form in which redundancy can be identified**. Because not all spatial frequencies are simultaneously present, the DCT will output a set of coefficients where some will have substantial values, but many will have values that are almost or actually zero. If a coefficient is zero, it makes no difference whether it is sent. If a coefficient is almost zero, omitting it will have the same effect as adding the same spatial frequency to the image but in the opposite phase. The decision to omit a coefficient is based on how visible that small unwanted signal would be, which is defined by the compression scale. If a coefficient is too large to omit, compression can also be achieved by reducing the number of bits used to carry the coefficient. This has the

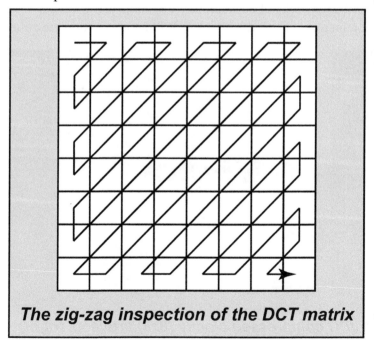

The zig-zag inspection of the DCT matrix

same effect as when a small noise is added to the picture. Some typical, and unwanted, artifacts when using DCT are the "blocky" appearance of the highly compressed images. This is due to the DCT function being applied to each 8 × 8 block of pixels.

Readers should note that Wavelet compression is different compared to JPG in that Wavelet compression "looks at" the whole image, not blocks of 8 × 8, and hence Wavelet artifacts do not have such "blocky," but rather "foggy" appearance. Wavelet uses Discrete Wavelet Transformation (DWT). In both DCT-based and DWT-based compressions, there are losses, and this is why these compressions are called lossy compressions. The idea is to find the best compromise between a high compression in order to reduce file size and the best image quality without too much visible loss. The modern image compression standard, based on wavelet compression, is called JPEG-2000. The JPEG-2000 works on the principles of wavelet transformation.

JPEG

JPEG stands for ***Joint Photographic Experts Group*** of the ISO, which is the original name of the committee that prepares the digital photographic standard.

JPEG is also named the standardized image compression mechanism which uses DCT in order to reduce the image redundancy. It works only with still digital images, and resolution is not specified.

Although it is widely used in digital photography and web-based technology, we use it in CCTV as well, where compression is applied to the digitized video (TV fields or TV frames), treating them as independent still snapshots.

JPEG has a subgroup recommendation for loss-less compression (of about 2:1), but, as we mentioned earlier, in CCTV we are more interested in the lossy compression of JPEG, where compression factors of over 10× are possible. JPEG works by transforming blocks of 8 × 8 picture elements using the discrete cosine transformation (DCT). The compression factors achieved with lossy JPEG compression are quite

 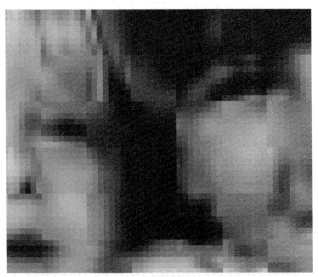

ViDi Labs test chart detail (left) and JPG compressed blocky (8x8) artefacts (right)

high (over 10 times), and the picture quality loss appears insignificant to the human eye.

JPEG is designed to exploit the known limitations of the human eye, like the fact that fine chrominance details are not perceived as well as fine luminance details in a given picture. For each separate color component, the image is broken into 8 × 8 blocks that cover the entire image. These blocks form the input to the DCT. Typically, in the 8 × 8 blocks, the pixel values vary slowly. Therefore, the energy is of low spatial frequency. A transformation that can be used to concentrate the energy into a few coefficients is the two-dimensional DCT (8 × 8). This transformation, studied extensively for image compression, is extremely efficient for highly correlated data.

JPEG stores full-color information: 24 bits/pixel (8 bits per color, 16 million colors) compared to the ***graphics interchange format*** (GIF), for example (another popular compression technique among PC users), which can store only 8 bits/pixel (256 or fewer colors). Gray-scale images do not compress by such large factors with JPEG because the human eye is much more sensitive to brightness variations than to hue variations and JPEG can compress hue data more heavily than brightness data. An interesting observation is that a gray-scale JPEG file is generally only about 10 to 25% smaller than a full-color JPEG file of similar visual quality. Also, it should be noted that JPEG is not suitable for line art (text or drawings), as the DCT is not suitable for very sharp B/W edges.

JPEG can be used to compress data from different color spaces such as RGB (video signal), YCbCr (converted video signal), and CMYK (images for the printing industry) as it handles colors as separate components. The best compression results are achieved if the color components are independent (non-correlated), such as in YCbCr, where most of the information is concentrated in the luminance and less in the chrominance. This is why such a color space conversion is used when encoding RGB raw signal coming from an image sensor.

Since JPEG files are independent of each other, when used in CCTV recording they can be easily and quickly be played back in reverse direction. JPEG codecs are royalty free and fast. The only downside is, when encoding and streaming live image sequences, JPEG is not as bandwidth efficient as the video compression counterparts, such as MPEG-2 and H.264.

M-JPEG

Motion JPEG (or M-JPEG) is a JPEG derivative typically used in CCTV only. M-JPEG does not exist as a separate standard but rather it is a rapid flow of JPEG images that can be played back at a sufficiently high rate to produce an illusion of motion. M-JPEG may allow for audio channel, although strictly speaking, image compression doesn't care about audio. As mentioned above, in M-JPEG there is no differential analysis and treatment between various frames, each one is independent of each other. For that reason the M-JPEG doesn't achieve the level of efficiency of the temporal compressions, such as the H.26x or MPEG described later. However, M-JPEG is used by some DVR manufacturers where multiple cameras are used. Also, many IP cameras, in addition to a video compression in their encoder settings, typically they offer M-JPEG as a second choice.

Wavelet

For many decades scientists have wanted more appropriate functions than the sines and cosines that comprise the bases of DCT's Fourier analysis to approximate choppy signals. By their definition, sines and cosines are nonlocal functions (they are periodical and stretch out to infinity). This is the main reason they do a very poor job of approximating sharp changes, such as high-resolution details in a finite, two-dimensional picture. This is the type of picture we most often have in surveillance time-lapse multiplexed recording, as opposed to a continuous stream of motion images in broadcast television. Wavelet analysis is one that works differently, and **it is more efficient in preserving the small details**.

The wavelet mathematical apparatus was first explicitly introduced by Morlet and Grossman in their works on geophysics during the mid-1980s. As a result, wavelet compression was first used in scientific data compression such as astronomy and seismic research. It was soon discovered that it would be extremely useful in CCTV, when Analog Devices,

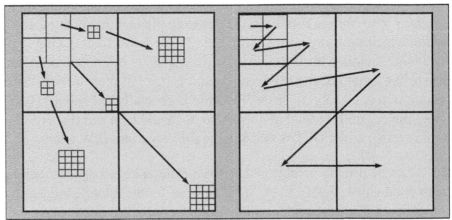

One of the clever wavelet ways of coding a picture and reducing the redundancy by a zig-zag method

wavelet compression chip 601 was introduced. Wavelet compression transforms the entire image, as opposed to 8 × 8 sections in JPEG, and is more natural, as it follows the shape of the objects in a picture. This is why wavelet has become especially attractive for CCTV.

With wavelet we can use approximating functions that are contained in finite domains. Wavelets are functions that satisfy certain mathematical requirements and are used in representing data or other

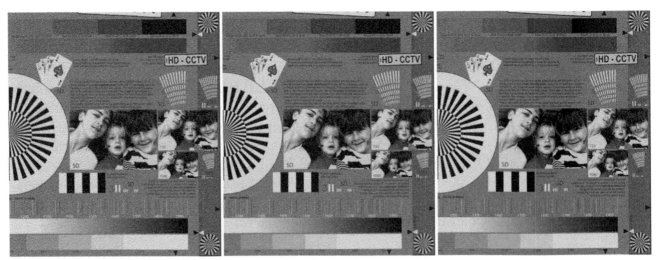

500x450 pixels TIF file on left (800kB), JPG file (150kB) and JPG-2000 (150kB)

functions in wavelet analysis. The main difference compared to the FFT (DCT) analysis is that the **wavelets analyze the signal at different frequencies with different resolutions**, i.e., many small groups of waves, hence the name wavelet. The wavelet algorithms process data at different scales or resolutions and try to see **details and the global picture**, or as some wavelet authors have said, "see the forest through the trees," as opposed to Fourier analysis which "sees just the forest."

Wavelets are well-suited for approximating data with sharp discontinuities. The wavelet analysis procedure is to adopt a wavelet prototype function, called an analyzing wavelet or mother wavelet. Time analysis is performed with a contracted, high-frequency version of the prototype wavelet, whereas frequency analysis is performed with a dilated, low-frequency version of the prototype wavelet. Because the original signal or function can be represented in terms of a wavelet expansion (using coefficients in a linear combination of the wavelet functions), data operations can be performed using just the corresponding wavelet coefficients.

Another interesting feature of wavelet is the "Area of Interest" or "Quality Box" function, where an image presence can be detected based on motion, for example, and have that area compressed with better quality relative to the rest of the same image. By using such an intelligent selection the file size is extremely small, yet it offers the best details related to where the important object is.

JPEG-2000

JPEG-2000 (ISO 15444) is basically a standardized version of the wavelet compression, produced by the JPEG group. The JPEG group realized the superiority of wavelet and released a new standard in year 2000, hence the name JPEG-2000.

The main advantage of JPEG 2000 (in addition to the better efficiency of the compression) is the significant codestream flexibility. The codestream obtained after JPEG-2000 compression is scalable in nature, meaning that it can be decoded in a number of ways. For example, by truncating the codestream at any point, one may obtain a representation of the image at a lower resolution. This is very useful for CCTV remote applications, where a complete image, although slightly blurred, can be delivered over very narrow bandwidth connections. By ordering the codestream in various ways, applications can achieve significant performance increases. However, as a consequence of this flexibility, JPEG 2000 requires encoders/decoders that are complex and computationally more demanding then JPEG. Another difference, in comparison with JPEG, is that JPEG 2000 produces ringing artifacts, manifested as blur and rings near edges in the image.

The ViDi Labs SD faces details with TIF (800kB), JPG (150kB) and JPG-2000 (150kB)

A few years ago JPEG-2000 was only available as an optional plug-in; today, it is a standard included in many photo editing applications. Many CCTV manufacturers today have JPEG-2000 hardware codec for their cameras and DVRs.

Furthermore, the JPEG-2000 standard defines usage of embedded information about the author or source of the image, or, what is interesting for us in CCTV, the originality of the image. There are some variations of JPEG-2000, one of which refers to motion video and is called Motion JPEG-2000.

Motion JPEG-2000

Motion JPEG-2000 is defined in ISO/IEC 15444-3 and in ITU-T T.802. Similarly to Motion-JPEG, it specifies the use of the JPEG 2000 format for timed sequences of images (motion sequences), possibly combined with audio. This standard also defines a file format, based on ISO base media file format (ISO 15444-12). Filename extensions for Motion JPEG 2000 video files are .mj2 and .mjp2 according to RFC 3745.

It is an open ISO standard and an advanced update to MJPEG (or MJ), which was based on the legacy JPEG format. Unlike common video compressions, Motion JPEG-2000 does not employ temporal or inter-frame compression. Each frame is a key frame, an independent entity encoded by either a lossy or loss-less variant of JPEG 2000. Key frame compression provides extremely accurate frame-by-frame time stamps needed for surveillance and evidentially procedures. This is important for multiplexed recording in CCTV, but also for video editing. For audio, it supports LPCM encoding and various MPEG-4 variants as "raw" or complement data.

MPEG-1

MPEG-1 (ISO 11172) is one of the first video compression standards, proposed by the ISO's *Motion Picture Experts Group* soon after the introduction of the H.261. It was intended for digitizing movies and not for video conferencing, as was the H.261. It belongs to the video compression group and it works with continuous digitized video signal and includes two channels of audio. **The output visual quality at typical bit rates (as used in VCDs, for example) is comparable to that of an analog VHS VCR. The audio layer of MPEG-1 is the actual, now popular, audio format MP3.**

MPEG-1 is defined to work with CIF size (352 × 288 pixels for PAL; 352 × 240 pixels for NTSC) video sequence. The color information is sampled with half of that resolution: 176 × 144 (i.e., 176 × 120). Typical video rates MPEG-1 works with are between 1 Mb/s and 3 Mb/s. Around 1.5 Mb/s was the achievable data speed of majority CD players in the time when MPEG-1 was introduced, and this was one of the major intentions of the MPEG-1 video compression. Up to one hour can be stored on a CD of 700 MB, which is the reason two CDs were needed for VCD movies.

MPEG does not define compression algorithms (although it is based on DCT), but rather the compressed bit stream – the organization of digital data for recording, playback, and transmission. The actual compression algorithms are up to the individual manufacturers, and their quality may vary.

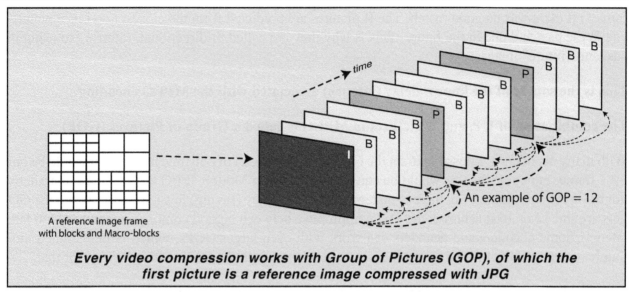

A reference image frame
with blocks and Macro-blocks

An example of GOP = 12

*Every video compression works with Group of Pictures (GOP), of which the
first picture is a reference image compressed with JPG*

The basic idea in all temporal compressions is to predict motion from frame to frame in the temporal direction and then to use the DCT to organize the redundancy in the spatial directions. The DCTs are done on 8 × 8 blocks, and the motion prediction is done in the luminance (Y) channel on 16 × 16 blocks. In other words, the block of 16 × 16 pixels in the current frame is coded with a close match to the same pixel block in a previous or future frame. This describes the backward prediction mode, where frames coming later in time are sent first to allow interpolating between frames. The DCT coefficients (of either the actual data or the difference between this block and the close match) are quantized, which means that they are divided by some value to drop bits off the bottom end. Hopefully, many of the coefficients will then end up being zero. The quantization can change for every "macro-block" (a macro-block is 16 × 16 of Y and the corresponding 8 × 8's in both U and V). The results of all of this, which include the DCT coefficients, the motion vectors, and the quantization parameters (and other stuff), are encoded using the so-called Huffman code, using fixed tables.

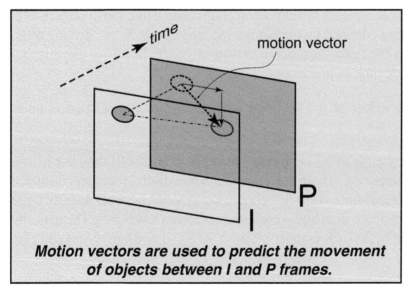

*Motion vectors are used to predict the movement
of objects between I and P frames.*

There are three types of coded frames (pictures) in MPEG-1 (the same applies to MPEG-2): the *intra (or key) frames* (**I**), the *predicted frames* (**P**), and the *bi-directional frames* (**B**).

The I pictures are basically still pictures compressed as JPG, and they are used as reference pictures (key frames). The P pictures are predicted from the most recently reconstructed I or P frame. Each macro-block in a P frame can either come with a vector and difference DCT coefficients for a close match in the last I or P, or it can just be "intra" coded (as in the I frames) if there was no good match. The B pictures are predicted from the closest two I or P pictures, one in the past and one in the future. This is why they are called bi-directional, referring to using the past and "future" images.

This is the source of the known delay (latency) associated with the MPEG encoding.

The combination of I, P, and B pictures in MPEG is called a Group of Pictures (GOP).

With many encoders it is possible to set the GOP to be composed of only one image, that would be only the I frame. Functionally, this would be equivalent to having Motion-JPEG as there are no predictive nor bi-directional frames. So, there is no temporal redundancy (saving) in such a case. When the GOP gets around 12 or 15 it achieves the best compromise between a good compression and not too large latency. Some encoders and decoders can work with even larger GOPs, which requires much more number-crunching by the codec, but the image gets better with the bandwidth allocated for that.

A typical GOP sequence of 12, which always repeats itself, would look like this:

IBBBPBBBPBBB - IBBBPBBBPBBB - IBBBPBBBPBBB...

Latency is a new side effect of the MPEG efforts to reduce redundancy by motion prediction. This is the price MPEG is paying for getting better picture quality at lower data rates. Most of the MPEG machines offer a choice of bit rates and GOP sizes a combination of which can be selected to reduce the latency to below noticeable levels by choosing higher bandwidth and smaller GOP sizes. Basically, the number of pictures in a GOP define the latency. So if, for example, we have a GOP size of 12, in PAL this makes a half a second in time, which becomes such a delay. If we add to this delay the network latency, it becomes clear why a latency of nearly a second or even more is sometimes noticeable in MPEG coding.

The latency may not even be noticed in a fixed camera CCTV system, but clearly this could be a problem with PTZ camera control. So what is the acceptable latency when controlling a live video camera over LAN and a video streaming? This really is defined by the human reaction speed. When driving a car, for example, around 200 ms is taken as the fastest reaction a person can have. So, if we use this as a guide, it will make an acceptable latency time in practice.

Another interesting, but positive, side effect of the bidirectional macro-block prediction is noise reduction due to averaging.

The practical application of MPEG-1 is most often in storing video clips on CD-ROMs, but is also used in cable television and video conferencing. There are, however, some digital recorders designed especially for CCTV applications where real-time video is recorded using MPEG-1 technique. In these applications, real-time video is more important than high-resolution video at a lower rate. The majority of higher quality MPEG-2 recorders are backwards compatible with MPEG-1 and can record and play back video streaming made in MPEG-1.

MPEG-2

MPEG-2 (ISO 13818) is not a next generation MPEG-1 but rather another **standard targeted for higher quality digital video with audio**. It was proposed by the ISO's MPEG group in 1993, and, like MPEG-1, MPEG-2 is also an Emmy Award-winning standard. **The MPEG-2 standard specifies the coding formats for multiplexing high-quality digital video, audio, and other data into a form suitable for transmission or storage.**

MPEG-2, like MPEG-1, does not limit its recommendations to video only, but also includes audio. It should be highlighted again that **MPEG-2 is not a compression scheme or technique**, but rather a standardization of handling and processing digital data in the fastest and most optimized way. MPEG-2 encoding can produce data rates well above 18 Mb/s, although in most practical CCTV applications there is hardly any visible difference between a live camera signal and its 4 Mb/s encoded video.

MPEG-2 is designed to support a wide range of applications and services of varying bit rate, resolution, and quality. The MPEG-2 standard defines four profiles and four levels for ensuring the interoperability of these applications. The profile defines the color space resolution and scalability of the bit stream. The levels define the maximum and minimum for image resolution, and Y (Luminance) samples per second, the number of video and audio layers supported for scalable profiles, and the maximum bit rate per profile.

Common MPEG-2 Profiles and Levels combinations (source: wikipedia.com)					
Profile @ Level	**Resolution (px)**	**Framerate max. (Hz)**	**Sampling**	**Bitrate (Mbit/s)**	**Example Application**
Simple Profile @ Low Level	176 × 144	15	4:2:0	0.096	Wireless handsets
Simple Profile @ Main Level	352 × 288	15	4:2:0	0.384	PDAs
	320 × 240	24			
Main Profile @ Low Level	352 × 288	30	4:2:0	4	Set-top boxes (STB)
Main Profile @ Main Level	720 × 480	30	4:2:0	15 (DVD: 9.8)	DVD, SD-DVB
	720 × 576	25			
Main Profile @ H-14	1440 × 1080	30	4:2:0	60 (HDV: 25)	HDV
	1280 × 720	30			
Main Profile @ High Level	1920 × 1080	30	4:2:0	80	ATSC 1080i, 720p60, HD-DVB (HDTV). (Bitrate for terrestrial transmission is limited to 19.39Mbit/s)
	1280 × 720	60			

As a compatible extension, MPEG-2 video builds on the MPEG-1 video standard by supporting interlaced video formats and a number of other advanced features. MPEG-2 today is used in almost all broadcast television services, such as DBS (*Direct Broadcast Satellite*), CATV (*Cable Television*), HDTV (*High Definition Television*), and of course the now popular movie format – DVD. A single-layer, single-sided DVD has enough capacity to hold two hours and 13 minutes of high-quality video, surround sound, and subtitles.

Like MPEG-1, the MPEG-2 is based on GOPs made up of I, P, and B pictures. The I pictures are **intra-coded**, that is, they can be reconstructed without any reference to other pictures. The P pictures are forward predicted from the last I picture or P picture, that is, it is impossible to reconstruct them without the data of another picture (I or P). The B pictures are both forward predicted and backward predicted from the last and next I pictures or P pictures; that is, there are two other pictures necessary to reconstruct them. Because of this, P pictures and B pictures are referred to as **inter-coded** pictures.

In its prediction algorithm, MPEG-2 works with motion vectors. Imagine an I frame showing a circle on white background. A following P frame shows the same circle but at another position. Prediction means to supply a motion vector, which declares how to move the circle on an I frame to obtain the circle in a P frame. This motion vector is part of the MPEG stream and it is divided in a horizontal and a vertical part. These parts can be positive or negative. A positive value means motion to the right or motion downwards, respectively. A

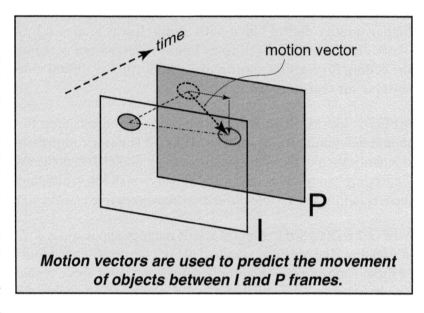

Motion vectors are used to predict the movement of objects between I and P frames.

negative value means motion to the left or motion upwards. But this model assumes that every change between frames can be expressed as a simple displacement of pixels. There is also a prediction error matrix in MPEG stream, which helps in accurate reconstruction of the motion.

In the beginning of the CCTV digital era, only a very few DVR manufacturers were using MPEG-2. Today many more see the benefits of high-quality digital video recording, and many more unexplored MPEG-2 functionality are used in CCTV. For example reverse playback, slow play forward or in reverse direction, as well as the incredible fast forward or rewind speed of up to 1024 times the normal speed and even video motion detection triggered recording.

MPEG-2 (and all other temporal compressions) cannot multiplex various cameras in one recording partition, as was the case with image compression. Temporal compression, as the name suggests, works best when one and the same camera stream is present over time. So, a DVR or NVR in CCTV, designed to work with video compression will usually stream individual cameras to its own channel (partition) to a hard disk. They cannot be multiplexed in the CCTV sense, as we used to do with multiplexers. But, multiple channels can be recorded concurrently. Depending on the overall data speed of the network, computer hardware, hard disk, processor, and

MPEG-2 encoders

motherboard — it is possible to combine a number of channels per recorder. Knowing the video data rate for a good quality video, for example, it can easily be calculated how many channels can be recorded in one box, allowing for simultaneous playback while recording on the same drive.

Although MPEG-2 encoding can be done in the software, with reasonably fast processors, in CCTV it is always preferable to have this done by dedicated hardware compression chips, so at least the encoding (recording) is not compromised and there are no gaps in it. Decoding (playing back) can be made via software decoders, and there are quite a few around, since MPEG-2 is a standard. Windows Media Player, Apple Quick Time, VLC, and so on are all examples of software players capable of playing the MPEG-2. With some MPEG-2 DVRs it is possible to burn a CD or DVD with MPEG-2 footage, which can be directly played onto a commercial DVD player.

Many high-end DVR manufacturers offer hardware encoding (recording) while doing hardware decoding on a composite or Y/C monitor, as well as software decoding of the same machine from another point in time in the past (referred to as Time Shift technology) for playback or backup over network. This multiple functionality sometimes is referred to as triplex, quad-plex, or penta-plex operation. In the latter case the following five functions would be performed concurrently: continuous recording, playback a certain instance from the past on a composite monitor, export a footage from the past on local external drive or CD, for example, playback another instance from the past via network on a PC, and back up to another PC or network storage via network, all at the same time. If all these processes are coming out of, or being written to, the same hard disk (which would usually be the case), then the hard drive data transfer rate needs to be able to sustain such a speed. This limits the number of channels that can be recorded on one physical machine.

Considerations have to be made when PTZ cameras need to be controlled over LANs with MPEG-2 and latency taken into account, but as mentioned earlier, this can be reduced to around 200 ms, or less, with the bit rate and GOP size smart adjustments.

Although MPEG-2 is still used in many CCTV projects, it is fair to say that H.264 has overtaken MPEG-2. Not only H.264 is more efficient, but it also offers scalability not possible with MPEG-2.

MPEG-4

MPEG-4 (ISO 14496) is another MPEG standard developed not long ago, but after MPEG-2. MPEG-4 is the result of another international effort involving hundreds of researchers and engineers from all over the world. MPEG-4 was finalized in October 1998 and became an International Standard in the first months of 1999. It adds to the many features of MPEG-1 and MPEG-2 and other related standards, such as VRML support for 3D rendering, object-oriented composite files (including audio, video and VRML objects), support for externally specified Digital Rights Management and various types of interactivity. The audio encoding AAC (*Advanced Audio Coding*) was standardized as an adjunct to MPEG-2 (as Part 7) before MPEG-4 was issued.

The MPEG-4 visual standard is developed to provide users a new level of interaction with visual contents. It provides technologies to view, access, and manipulate objects rather than pixels, with great error robustness at a large range of bit rates. **MPEG-4 application areas range from digital television and streaming video to mobile multimedia, games, and, of course, CCTV and surveillance. A major**

difference between MPEG-4 and the previous audiovisual standards is the object-based audiovisual representation model. An object-based scene is built using individual objects that have relationships in space and time, offering a number of advantages. The MPEG-4 standard opens new frontiers in the way users play with, create, reuse, access, and consume audiovisual content. The MPEG-4 object-based

Photo courtesy of Axis

MPEG-4 encoders

representation approach where a scene is modeled as a composition of objects, both natural and synthetic, with which the user may interact, is at the heart of the MPEG-4 technology. The handling of objects (especially the synthetic ones) and interactivity are the heart of MPEG-4, which, unfortunately, those of us working in CCTV cannot make any use of.

MPEG-4 is still a developing standard and is divided into a number of parts. Companies promoting MPEG-4 compatibility do not always clearly state which "part" level compatibility they are referring to. The key parts to be aware of are MPEG-4 part 2 (including Advanced Simple Profile, used by codecs such as DivX, Xvid, Nero Digital and 3ivx and by QuickTime 6) and MPEG-4 part 10 (MPEG-4 AVC/H.264 or Advanced Video Coding, used by the x264 encoder, by Nero Digital AVC, by QuickTime 7, and by high-definition video media like Blu-ray Disc).

Most of the features included in MPEG-4 are left to individual developers to decide whether to implement them. This means that there are probably no complete implementations of the entire MPEG-4 set of standards. To deal with this, the standard includes the concept of "profiles" and "levels," allowing a specific set of capabilities to be defined in a manner appropriate for a subset of applications. The MPEG-4 conformance points are defined at the Simple Profile, the Core Profile, and the Main Profile. The Simple Profile and Core Profile address typical scene sizes of QCIF and CIF size, with bit rates of 64 kb/sec, 128 kb/s, 384 kb/s, and 2 Mb/s. The Main Profile addresses a typical scene sizes of CIF (352 × 288), full standard definition ITU-R 601 (720 × 576), and High Definition (1920 × 1080), with bit rates at 2 Mb/s, 15 Mbit/s, and 38.4 Mb/s.

Initially, MPEG-4 was aimed primarily at low bit-rate video communications; however, its scope as a multimedia coding standard was later expanded. MPEG-4 is efficient across a variety of bit-rates ranging from a few kilobits per second to tens of megabits per second.

Motion compensation is block-based, with appropriate modifications for object boundaries. The block size can be 16 × 16, or 8 × 8, with half-pixel resolution. MPEG-4 also provides a mode for overlapped motion compensation. Texture coding is based in 8 × 8 DCT, with appropriate modifications for object boundary blocks. Coefficient prediction is possible to improve coding efficiency. Static textures can be encoded using a wavelet transform. Error resilience is provided by re-synchronization markers, data partitioning, header extension codes, and reversible variable-length codes. Scalability is provided for both spatial and temporal resolution enhancement. MPEG-4 provides scalability on an object basis, with the restriction that the object shape has to be rectangular. This is perhaps one of the most useful features for us in CCTV, for it allows scalable streaming over narrow bandwidths.

MPEG-4 Video offers technology that covers a large range of existing applications as well as new ones. The low bit rate and error-resilient coding allows for robust communication over limited rate wireless channels, useful for mobile videophones, space communication, and certainly, CCTV. At high bit rates, tools are available to allow the transmission and storage of high-quality video suitable even for studio and other very demanding content creation applications. The standard has evolved through many versions, and support data rates beyond those of MPEG-2.

A major application area, outside our industry, is interactive web-based video. Software that provides live MPEG-4 video on a web page are very common.

MPEG-4 provides support for both interlaced and progressive scan (although progressive scan is rarely used in CCTV) video material. The chrominance format that is supported is 4:2:0. In this format, the number of Cb and Cr samples are half the number of samples of the luminance samples in both horizontal and vertical directions. Each component can be represented by a number of bits ranging from 4 to 12 bits.

MPEG-4 is perhaps best known for its latest successor, the H.264, which, in fact, is often referred to as MPEG-4 part 10, or AVC. This is currently one of the most used video compressions in CCTV.

H.264 (MPEG-4 part 10, AVC)

H.264, MPEG-4 Part 10, or **AVC** (Advanced Video Coding) is a standard for video compression and is currently one of the most commonly used formats for the recording, compression, and distribution of high definition video. In CCTV, many DVRs are using H.264 for converting analog standard definition cameras to digital. The final drafting work on the first version of the standard was completed in May 2003.

H.264/MPEG-4 AVC is a block-oriented motion-compensation-based codec standard developed by the ITU-T Video Coding Experts Group (VCEG) together with the ISO/IEC JTC1 Moving Picture Experts Group (MPEG). The project partnership effort is known as the Joint Video Team (JVT). The ITU-T H.264 standard and the ISO/IEC MPEG-4 AVC standard (formally, ISO/IEC 14496-10 – MPEG-4 Part 10, Advanced Video Coding) are jointly maintained so that they have identical technical content.

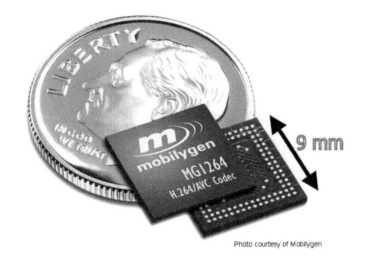

Photo courtesy of Mobilygen

An H.264 codec

The intent of the H.264/AVC project was to create a standard capable of providing good video quality at substantially lower bit rates than previous standards (i.e., half or less the bit rate of MPEG-2, H.263, or MPEG-4 Part 2), without increasing the complexity of design so much that it would be impractical or excessively expensive to implement. An additional goal was to provide enough flexibility to allow the standard to be applied to a wide variety of applications on a wide variety of networks and systems, including low and high bit rates, low and high resolution video, broadcast, DVD storage, RTP/IP packet networks, and ITU-T multimedia telephony systems.

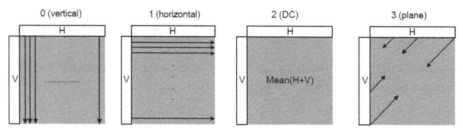

H.26L Intra 16x16 prediction modes (all predicted from pixels H and V)

H.264 uses sophisticated prediction of macroblocks.

The H.264 standard can be viewed as a "family of standards" composed of various profiles. A specific decoder decodes at least one, but not necessarily all, profiles. The decoder specification describes which profiles can be decoded. The H.264 name follows the ITU-T naming convention, where the standard is a member of the H.26x line of VCEG video coding standards; the MPEG-4 AVC name relates to the naming convention in ISO/IEC MPEG, where the standard is part 10 of ISO/IEC 14496, which is the suite of standards known as MPEG-4. The standard was developed jointly in a partnership of VCEG and MPEG, after earlier development work in the ITU-T as a VCEG project called H.26L. It is thus common to refer to the standard with names such as H.264/AVC, AVC/H.264, H.264/MPEG-4 AVC, or MPEG-4/H.264 AVC, to emphasize the common heritage.

Level	Max decoding speed		Max frame size		Max video bit rate for video coding layer (VCL) kbit/s			Examples for high resolution @ highest frame rate (max stored frames)
	Luma samples/s	Macroblocks/s	Luma samples	Macroblocks	Baseline, Extended and Main Profiles	High Profile	High 10 Profile	
1.1	768,000	3,000	101,376	396	192	240	576	176×144@30.3 (9) 320×240@10.0 (3) 352×288@7.5 (2)
2.2	5,184,000	20,250	414,720	1,620	4,000	5,000	12,000	352×480@30.7(10) 352×576@25.6 (7) 720×480@15.0 (6) 720×576@12.5 (5)
4	62,914,560	245,760	2,097,152	8,192	20,000	25,000	60,000	1,280×720@68.3 (9) 1,920×1,080@30.1 (4) 2,048×1,024@30.0 (4)
5	150,994,944	589,824	5,652,480	22,080	135,000	168,750	405,000	1,920×1,080@72.3 (13) 2,048×1,024@72.0 (13) 2,048×1,080@67.8 (12) 2,560×1,920@30.7 (5) 3,672×1,536@26.7 (5)

In the table above, some of the H.264 levels are shown, with the corresponding bit-rates.

One of the key advantages of H.264 over the predecessors is the higher efficiency of the compression.

What was possible with MPEG-2 in the past, converting an SD raw stream to a pleasant 4 Mb/s DVD quality video, the H.264 standard has managed to produce similarly pleasant quality of a HD raw stream with only 4 Mb/s stream.

Perhaps the latest video compression standard update of H.264, called H.265, will become as common in the CCTV industry as H.264 is now.

H.265 (HEVC)

The **H.265** is also known as **High Efficiency Video Coding (HEVC)**, a successor to H.264/MPEG-4 AVC (Advanced Video Coding). It is officially published as ISO/IEC 23008-2. This is a very new compression, only published during the writing of this edition of the book. HEVC is said to improve video quality, double the data compression ratio compared to H.264/MPEG-4 AVC, and can support up to 8K Ultra High Definition resolutions of up to 8192×4320.

This would mean, a HD stream which would take 4 Mb/s with H.264, would look similar with H.265 at only 2 Mb/s.

The H.265 is very new so we don't have any experience with it yet in CCTV, but, judging by the preliminary specification, it would further improve the storage and network capability of HD and beyond HD video streaming. It is expected that the H.265 codec will be more complex than the predecessor H.264, which will mean computer processing power for software decoding will need to be even higher.

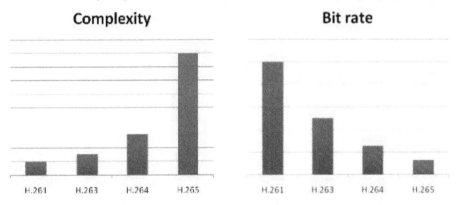

The complexity of codecs versus the bit-rate

9. Video Management Systems

Video Management System (VMS) is the most common reference given to computers, hardware, and software used in modern digital IP video systems. The operator of a VMS is the controller and manager of multiple digital signals, alarms, and PTZ controls.

VMS is a modern digital replica of analog matrix switchers, dealing with the recording, live viewing, and playback of camera streams, be that analog cameras converted to digital or pure digital streams coming directly of IP cameras. All of the communication in a VMS system goes over Ethernet, be that copper, fiber, or wireless.

The key hardware control component in a VMS system is a computer, or computers, most often running Windows operating system and a VMS application. The VMS software is a dedicated application installed on top of the operating system, which is capable of selecting cameras, decoding video streams and playback from recording hardware. The recording hardware may be a dedicated and purposely made hardware, often referred to as *digital video recorder* (DVR), or *network video recorder* (NVR). The DVRs and NVRs are treated as self-sufficient recording devices, and, as such, they rarely run on Windows OS, but rather on Linux, which is proven to be a more stable OS.

VMS by Axxon

If a digital system has multiple concurrent users, then a computer from which an operator works with the VMS would usually be referred to as a *VMS Workstation*. In such a case, in order for the system and cameras to be accessible by multiple operators concurrently, the digital system has to have a central computer (often referred to as server) with a data-base that contains information about all the devices in the system, their IP addresses, location, level of priority, and functionality. Understandably, all components in a digital system are interconnected in a network with network switches and routers.

There is no rule on how a VMS Workstation software should look like, but since most of VMS is written for Windows OS, they most often follow a rule of a folder tree structure. On the left hand side is

A typical VMS screen by Axis

A 3D representation of a system layout by Axxon

typically a listing of all cameras, grouped in a folders, and the split screen display on the right.

Another common module in a VMS system is graphical representation of the system layout, so that operators don't have to remember which camera is located where and what is their number or name. The graphical user interface (GUI) is typically an interactive map, whereby pointing and clicking on a camera icon the video from that camera is selected on the main screen.

Another important function customers expect from VMS is its integration with access control, building management and point of sales systems. This may not necessarily be a part of every VMS, but many do have some form of integration built-in. Others customize their systems as required by the project(s).

With the increased CPU power of computers, VMS can not only handle video streams, but may also allow for some intelligent video analysis. Most modern CCTV systems, cameras and DVRs, offer built-in video motion detection (VMD) in order to optimize the recording and extend the length of recording. But others can also offer intelligence such as face recognition, number-plates recognition, and even human activity recognition, such as running, direction of moving, loitering, or even fight starting.

VMS can be as simple as one operator controlling half a dozen cameras, but it can also grow up to hundreds of operators controlling a system with a couple thousand cameras.

Currently one of the largest IP CCTV systems in one physical location is by Dallmeier in casinos in Macau, some of which are with over 5,000 IP cameras in one system. Soon, integrating a few casinos in one will increase this number to 20,000 cameras.

One of the largest distributed digital systems today is the CCTV system of Moscow, done by Axxon, with over 124,000 cameras, running on 8,000 servers, with 120 local monitoring centers, and 2300 operators.

A multi-screen control center by Dallmeier

Operating Systems (OS)

In order for a computer to work it needs appropriate hardware and software that understands the environment in which it works. When a computer starts up, the first thing that happens is that it loads the BIOS table (***Basic Input Output Set***), which has all the hardware configuration details, hard disks, video cards, keyboard, mouse, communication ports, parallel ports, and so on. Once the BIOS defines all these details, it goes to a special section of the hard drive called ***boot record area*** and looks for an ***operating system*** (OS).

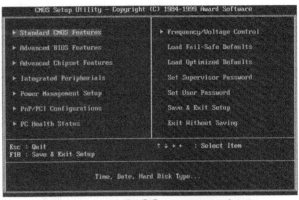

An award BIOS screen shot

The OS is software that usually resides on the hard disk, but lately there are computers with solid state drives instead of electromagnetic hard disk drives, although the principles are the same. When the OS loads, it brings up a graphical user interface (GUI) and connects the whole system in a meaningful interactive environment, loading various drivers, displaying certain quality images, and accepting and executing commands as defined by the computer user or application. **The OS is what the name suggests: it is a system to operate with the computer**, and it is the basis for various applications and specialized programs, such as spreadsheet program, word processing, photo editing, or video editing.

An actual BIOS chip

Many DVRs in CCTV fall into this category, since they use one of the few popular OSs and add video processing as specialized application to it. Most common OSs used in CCTV are Windows and Linux. There are other OSs as well, such as Unix, Solaris, and Mac OS X, but none of these is used in CCTV; this is why we will not be comparing and analyzing them in-depth.

A firmware chip

Some DVRs do not load their OS from a hard disk, but, rather, from a chip (usually flash memory or EPROM, which stands for Erasable Programmable Read Only Memory). Such a chip which has instructions on how the hardware should perform is typically referred to as ***firmware***. Some DVR manufacturers refer to this as a ***Real Time Operating System*** (RTOS) or ***Embedded OS***.

Running a DVR with an embedded OS simplifies things, as the OS is then smaller and it is faster to load. Also, if a hard disk fails, there is no need to install the OS, just swap the faulty drive with a new one.

With the increasing size of flash memories and solid state drives many DVRs and NVRs today embed a complete OS on a flash drive, which is effectively the same concept as the Embedded OS or firmware-based OS. The key outcome is beneficial, not only to the end users, but to the installers as well. When the hard drive that records the video streams fails there is no need to reinstall the OS, only replace the faulty hard disk with a new one.

The trademarks of the three main operating systems: Mac, Linux, and Windows

One of the most important requirements in security and surveillance is the stability of the OS. Long-term operation in CCTV is sometimes more demanding than a busy web server, and even more demanding than a typical office or home computer usage. A web server can go down for a few minutes, or maybe even hours, for maintenance purposes, but in security a DVR or NVR is expected to run without interruption for months, if not years. This is a big task. The intensity of writing and reading data to and from hard disks is usually higher than web servers, as video data is much larger than handling web pages or e-mails, for example. Not all operating systems and hardware are suitable for such a long and uninterrupted operation. One reason why the majority of web servers on the Internet these days are running on Linux is exactly that – the **long-term stability**. This is not to say that the popular and wide spread Windows OS is not suitable at all, but readers should be made aware that the current statistic shows that identical hardware with Intel processors (which makes the majority of PCs) will perform faster and more reliably with Linux than with Windows.

Linux is an operating system written by a Finnish student by the name of Linus Torvald, and it is based on Unix, one of the oldest and most robust industry standards. Linux has picked up so much in only 10 years of its existence because, unlike Unix (which is licensed), the source code for **Linux is freely available to everyone** (developed under GNU – *General Public License*). When the first version of Linux was written and made freely available to everyone the author's only requirement was to have any additional improvements or drivers written by others to be made available to everyone.

Thousands of software developers, students, and enthusiasts wholeheartedly accepted the idea of a free, open source OS. This is why Linux not only became more popular, but, with more hardware drivers, it also offered more applications and was continually improving. Stability is only one part of the Unix concept, but it is further developed and improved with new kernels and file systems. Linux comes in many "flavors," called distributions, but all use the same kernel (the core of the operating system) and have a variety of additions, programs, tools, and GUIs, all of which are license-free. So, when Linux is used in a DVR in CCTV, it has not only a cost benefit, but, maybe more importantly, long-term license independence. If a hard disk inside a DVR with Linux OS fails (and drives will fail regardless of the OS), installing a new version does not require that any fees or license numbers be entered when re-installing. This is not the case when Windows is used.

Some DVR manufacturers use certain versions of Windows and have gone an extra step forward by having their own software engineers "tweak" the Windows engine to suit their hardware better than when it comes from Microsoft themselves, hence achieving higher stability and reliability.

Today a typical OS with the basic functions would hardly use more than 1 GB gigabyte of storage space. This will easily fit on a 1 GB flash chip, which can be used to boot computer from.

Digital video recorders (DVRs) in CCTV are designed to use the maximum hard disk space available. With a typical large size hard disk available these days of 4 TB, the internal DVR hard drive capacity (without using a RAID array) can get extended to over 16 TB using up to four such drives. Larger systems, requiring higher redundancy, may even include RAID storage controllers and expand the storage capacity to many more TB. A typical DVR, as used in CCTV, would be working re-

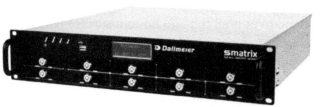

Some modern DVR/NVR include RAID

ally hard, day and night, 24 hours a day, 7 days a week without (ideally) being shut down. DVRs are without any doubt a symbiosis of software and hard disk technology. If any of the two fails, you will have a failed DVR and loss of important recordings. This is why lately a lot of DVRs and NVRs with built-in redundancy in the form of RAID-1, RAID-5 and RAID-6 are becoming as a standard choice.

The need for a better understanding of hard disks and their limitations is greater than ever, especially for those in CCTV. **Even the most stable OS depends on the hardware reliability**. If the hardware fails, the OS can no longer run, even if, technically, the OS has not failed. **The most vulnerable hardware parts of any computer are the moving parts, notably the cooling fans and the spinning hard disks**. These parts most commonly fail simply because of wear and tear, increased temperature, dust, moisture, and mechanical shocks.

Educating customers about the importance of clean and cool air-conditioned equipment rooms may only extend the life expectancy of the DVR.

The hard disk drives are the most important hardware component in an overall digital CCTV IP system.

Solid State Drives (SSD) are becoming more and more common, and some mobile CCTV applications already use them, but they are still very expensive compared to the electro-magnetic drives. No doubt, as their prices are coming down, they will become more common in CCTV too, but at this stage, over 99% of CCTV uses the electro-magnetic drives. For that reason we will devote more space to these further in this chapter.

Hard disk drives

Hard disk drives are an essential part of any modern computer device, and this includes the DVRs in digital CCTV. It is therefore very important to understand how they work and to learn what are their performances and limitations. A hard disk or hard drive is responsible for long-term storage of information. Unlike volatile memory (often referred to as RAM), which loses its stored information once its power supply is shut off, a **hard disk stores information permanently**, allowing you to save programs, files, and other data. Hard disks also have much greater storage capacities than RAM; in fact, current single hard disks may contain over 4 TB of storage space.

A hard disk is comprised of four basic parts: **platters, a spindle, read/write heads, and integrated electronics**. Platters are rigid disks made of metal or plastic. Both sides of each platter are covered with a thin layer of iron oxide or other magnetizable material. The platters are mounted on a central axle or spindle, which rotates all the platters at the same speed. Read/write heads are mounted on arms that extend over both top and bottom surfaces of each disk. There is at least one read/write head for each side of each platter. The arms jointly move back and forth between the platters' centers and outside edges. This movement, along with the platters' rotation, allows the read/write heads to access all areas of the platters. The integrated electronics translate commands from the computer and move the read/write heads to specific areas of the platters, thus reading and/or writing the needed data.

Computers record data on hard disks as a series of binary bits. Each bit is stored as a magnetic polarization (north or south) on the oxide coating of a disk platter. When a computer saves data, it sends the data to the hard disk as a series of bits. As the disk receives the bits, it uses the read/write heads to magnetically record or "write" the bits on the platters. Data bits are not necessarily stored in succession; for example, the data in one file may be written to several different areas on different platters. When the computer requests data stored on the disk, the platters rotate and the read/write heads move back and forth to the specified data area(s). The read/write heads read the data by determining the magnetic field of each bit, positive or negative, and then relay that information back to the computer. The read/write heads can access any area of the platters at any time, allowing data to be accessed **randomly** (rather than sequentially, as with a magnetic tape). Because hard disks are capable of random access, they can typically access any data within a few millionths of a second (milliseconds).

An open SATA hard disk drive

In order for a computer operating system to know where to look for the information on the hard disk, hard disks are organized into discrete, identifiable divisions, thus allowing the computer to easily find any particular sequence of bits. The most basic form of disk organization is called **formatting**. Formatting prepares the hard disk so that files can be written to the platters and then quickly retrieved when needed.

Before a brand new hard drive is used it needs to be formatted. **Formatting is a method of organizing what is saved to the disk** and it depends on the OS.

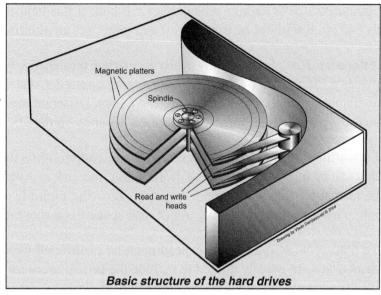

Basic structure of the hard drives

Hard disks must be formatted in two ways: *physically and logically.* Physical formatting is done before a logical one.

Formatting is made in *sectors, clusters (a group of sectors), and tracks* according to the operating system used. Tracks are concentric circular paths written on each side of a platter, like those on a record or compact disc. The tracks are identified by number, starting with track zero at the outer edge. Tracks are divided into smaller areas or sectors, which are used to store a fixed amount of data. Sectors are usually formatted to contain 512 bytes of data (there are 8 bits in a byte). A cylinder is comprised of a set of tracks that lie at the same distance from the spindle on all sides of all the platters. For example, track three on every side of every platter is located at the same distance from the spindle. If you imagine these tracks as being vertically connected, the set forms the shape of a cylinder. Computer hardware and software frequently work using cylinders. When data is written to a disk in cylinders it can be fully accessed without having to move the read/write heads. Because head movement is slow compared to disk rotation and switching between heads, cylinders greatly reduce data access time.

After a hard disk is physically formatted, **the magnetic properties of the platter coating may gradually deteriorate**. Consequently, it becomes more and more difficult for the read/write heads to read data from or write data to the affected platter sectors. **The sectors that can no longer be used to hold data are called bad sectors**. Fortunately, the quality of modern disks is such that bad sectors are rare. Furthermore, most modern

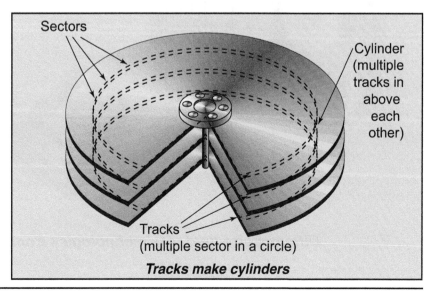

Tracks make cylinders

computers can determine when a sector is bad; if this happens, the computer simply marks the sector as bad (so it will not be used again) and then uses an alternate sector.

After a hard disk has been physically formatted, it must also be logically formatted. *Logical formatting* places a file system on the disk, allowing an operating system (such as Windows or Linux) to use the available disk space to store and retrieve files. Different operating systems use different file systems, so the type of logical formatting applied depends on the OS installed.

Formatting the entire hard disk with one file system limits the number and types of operating systems that can be installed on the disk. If, however, the disk is divided into partitions, each partition can then be formatted with a different file system, allowing multiple operating systems. Dividing the hard disk into partitions also allows the use of disk space to be more efficient.

To read or write data the disk head must be positioned over the correct track on the rotating media. *Seek times* are usually quoted to include the time it takes for the head to stop vibrating after the move ("settling time"). Then a delay occurs until the correct data sector rotates under the head ("rotational latency"). Modern disks use accelerated track positioning, so that the head moves faster and faster until about the halfway point and then is decelerated to a stop at the target track. This is why the **average seek** is only a few times the **minimum seek**. The **maximum seek** time is usually about twice the average seek time because the head reaches its maximum speed before the middle track of the disk. The minimum track-seek time is the time it takes to move the heads from one track to the next adjoining track. For reading large blocks of data, such as our DVR recorded footage, this is the most significant seek performance value. The **average track seek time is more important** for random access of small amounts of data such as traversing a directory path.

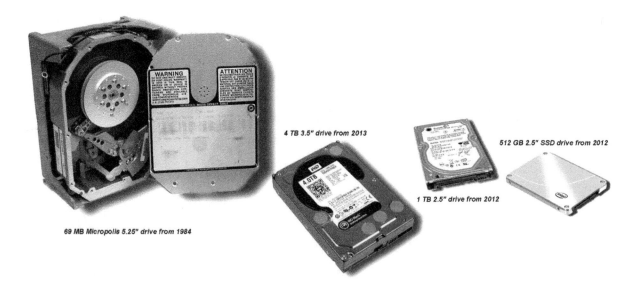

69 MB Micropolis 5.25" drive from 1984

4 TB 3.5" drive from 2013

1 TB 2.5" drive from 2012

512 GB 2.5" SSD drive from 2012

Different size hard disk technologies from different times

Access Time is equal to the time to switch heads + time to seek the data track + time for a sector to rotate under the head + repeat for the next sector. More heads reduce the need to mechanically seek a new track.

Faster rotational speed (spindle speed) increases the maximum data transfer rate and reduces the rotational latency. The rotational latency is the additional delay in seeking a particular data sector while waiting for that sector to come under the read head. The following table illustrates typical differences between various rotational speed hard drives and their maximum transfer rates (discussed later in the book), which are the most important indicator of how much data we can put through the magnetic plates of the hard disks.

Rotational Speed	Rotational Latency	Maximum (Burst) Transfer Rate
3600 rpm	16.7 ms	60 MB/s
4500 rpm	13.3 ms	80 MB/s
5400 rpm	11.1 ms	100 MB/s
7200 rpm	8.3 ms	140 MB/s
10,000 rpm*	6.0 ms	200 MB/s
12,000 rpm*	5.0 ms	250 MB/s
15,000 rpm*	4.0 ms	300 MB/s

*The higher speeds require better cooling of the drive.

Every hard disk is specified with its rotational speed or *spindle speed*. Expressed in *revolutions per minute* (rpm), this specification gives a very good indication of the drive performance. Desktop drives generally come in 5400 rpm and 7200 rpm varieties, with 7200 rpm drives averaging 10% faster (and 10 to 30% more expensive) than 5400 rpm models. High-end 10,000 rpm and 15,000 rpm hard drives offer only marginally better performance than 7200 rpm drives and cost much more, in part because they are typically SCSI drives with added reliability features. Also, higher spindle hard drives need more current; as a result, they get **hotter. Cooling is very important for all hard drives**, and more so for the faster ones. For a typical DVR, hard disks with 5400 rpm or 7200 rpm are a good compromise between sufficient speed, reasonable cost, and being relatively "cool" drives.

If two drives have the same spindle speed, the seek time shows which one is better. Differences in seek times, which range from 3.9 milliseconds (ms) for ultra-fast SCSI drives to 12 ms for slower IDE drives, may be noticeable in database or search applications where the head scoots all over the platter, but also when doing a VMD or time/date search in a digital recorder footage.

Cache is another term used in hard drives, and it refers to the **amount of memory built into the drive.** Designed to reduce disk reads, the cache holds a combination of the data most recently and most frequently read from disk. Large caches tend to produce greater performance benefits when multiple users access the same drive at once. Although small differences in cache size may have little bearing on performance, a smaller cache may be a sign of an older, slower drive. Operating systems try to maximize performance by minimizing the effect of mechanical activity. Keeping the most recently used data in memory reduces the need to go to the disk drive, move the disk heads, and so on. The writing of new data may also be cached and written to disk at a later, more efficient time. Other strategies include track buffering where data sectors are read into memory while waiting for the correct sector to rotate under the head. This can eliminate the delay of rotational latency because later sectors have already been read after reading of the sought sector is complete. For modern disks, this track buffering is usually handled by a memory cache on the disk drive's built-in controller. Modern disk drives usually have cache that ranges from 8 MB to 16 MB of cache memory to buffer track reads and hence eliminate rotational latency. Some high-end drives, like the latest 4 TB, have 64 MB. However, the rotational speed still limits the maximum transfer rate.

The latest 4 TB drive by WD for example, scored very good write and read time (as shown in the diagrams to the right) with an average data transfer of around 140 MB/s, which is consistent with the table in the previous page. 140 MB/s is equivalent to around 1 Gb/s.

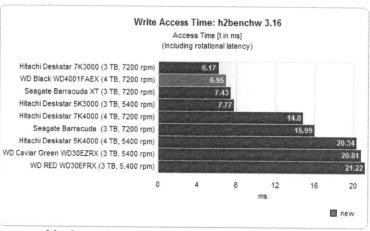

Various hard disk write access times

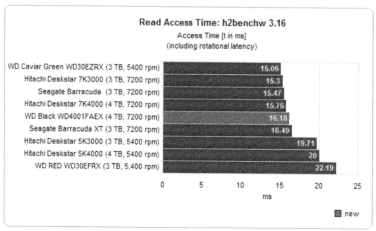

Various hard disk read access times

Despite the electronic methods of improving the hard disk's performance, the performance is determined primarily by the mechanical characteristics of the drive. This is the reason external factors affecting the mechanical performances of the hard disks also affect their reliability and lifetime expectancy. **Exposure to high temperatures, dust, moisture, shocks, and vibrations are external factors that can cause hard disk failures.**

High temperatures and dust are the two most common causes of hard disk failures we experience in practice in CCTV.

It is realistic to say that hard drives in some DVRs work even harder than hard drives on many Internet web servers. Unfortunately, CCTV customers don't have the same culture as IT customers, where all servers are well protected, air-conditioned, and put in nice, clean rooms. DVRs are very often mistreated and installed in places with minimum air conditioning, often with plenty of dust, and various evaporations. It is up to us, CCTV suppliers, integrators, and installers, to insist on a better treatment of these hard-working machines and protect hard disks the same way as when they are running inside the company servers.

In the current race of making more and better DVRs, most CCTV manufacturers concentrate on having higher compressions or recording more frames per second, and making it cheaper by cutting corners in manufacturing, using inferior materials and avoid good quality control. Few among them have gone in the direction of taking care of the actual DVR environment, including filtering the cooling air and measuring the fan's revolutions and external and internal temperature. In addition to using a stable OS, these are extremely important factors influenc-

A Linux-on-flash based DVR with air filtering and temperature sensing

ing the longevity of the DVRs. At the end of the day, it is no good having even the highest frame rate if data cannot be saved on a healthy hard disk.

In not too distant past there were a variety of hard disk interface standards between the computer and the hard disks, such as ATA, SCSI, RAID, and SATA. Today, there is no longer Parallel ATA available, nor SCSI. The most common hard disk drive standards used in CCTV are the SATA and RAID.

Today, in CCTV, there are at least a couple of hundred different DVR models.

The hard disk sustained data transfer rate ultimately defines how many cameras or images per second a DVR can have. This is the bottleneck of data transfer as it depends on mechanically moving parts. The sustained transfer rate is always less than the burst transfer rate. Generally ranging from 30 MB/s to 160 MB/s (megabytes per second!), it indicates how fast data can be read from the outermost track of a hard drive's platter into the cache. **The sustained transfer rate is an important parameter of the DVR's hard drives, which ultimately defines the upper limit of how many cameras your system can record and play back.**

This performance also depends on the operating system, processor, compression stream, file sizes, and the like, but, ultimately, if the hard drive cannot cope with such a through output the DVR cannot achieve what it is (theoretically) capable of.

File system	Creator	Year introduced	Original operating system
DECtape	DEC	1964	PDP-6 Monitor
V6FS	Bell Labs	1972	Version 6 Unix
DOS (GEC)	GEC	1973	Core Operating System
CP/M file system	Gary Kildall	1974	CP/M
FAT (8-bit)	Marc McDonald, Microsoft	1977	Microsoft Disk BASIC
CBM DOS	Commodore	1978	Microsoft BASIC (for CBM PET)
DOS 3.x	Apple Computer	1978	Apple DOS
Pascal	Apple Computer	1978	Apple Pascal
AFS	Carnegie Mellon University	1982	Multiplatform MultoOS
ProDOS	Apple Computer	1983	ProDOS 8
FAT16	Microsoft	1984	MS-DOS 3.0
MFS	Apple Computer	1984	Mac OS
Amiga OFS[15]	Metacomco for Commodore	1985	Amiga OS
HFS	Apple Computer	1985	Mac OS
HPFS	IBM & Microsoft	1988	OS/2
ISO 9660:1988	Ecma International, Microsoft	1988	MS-DOS, Mac OS, and AmigaOS
ext2	Rémy Card	1993	Linux, Hurd
NTFS Version 1.0	Microsoft, Tom Miller, Gary Kimura	1993	Windows NT 3.1
Joliet ("CDFS")	Microsoft	1995	Microsoft Windows, Linux, Mac OS, and FreeBSD
Be File System	Be Inc., D. Giampaolo, C. Meurillon	1996	BeOS
FAT32	Microsoft	1996	Windows 95b[3]
GPFS	IBM	1996	AIX, Linux, Windows
QFS	LSC Inc, Sun Microsystems	1996	Solaris
HFS Plus	Apple Computer	1998	Mac OS 8.1
NSS	Novell	1998	NetWare 5
ext3	Stephen Tweedie	1999	Linux
GFS	Sistina (Red Hat)	2000	Linux
NTFS Version 3.1	Microsoft	2001	Windows XP
ReiserFS	Namesys	2001	Linux
FATX	Microsoft	2002	Xbox
OCFS	Oracle Corporation	2002	Linux
Google File System	Google	2003	Linux
Non-Volatile File System	Palm, Inc.	2004	Palm OS Garnet
Reiser4	Namesys	2004	Linux
VxCFS	VERITAS, (now Symantec)	2004	AIX, HP-UX, Solaris, Linux
ZFS	Sun Microsystems	2004	Solaris, FreeBSD, PC-BSD, FreeNAS
OCFS2	Oracle Corporation	2005	Linux
ext4	Various	2006	Linux
GFS2	Red Hat	2006	Linux
Oracle ACFS	Oracle Corporation	2009	Linux - Red Hat Enterprise Linux 5 and Oracle Enterprise Linux
LTFS	IBM	2010	Linux, Mac OS X, planned Microsoft Windows,
ReFS	Microsoft	2012, 2013	Windows 2012 Server

The history of some file systems (source: www.wikipedia.com)

In today's digital world a typical server dedicated to recording multiple IP cameras will be limited to a number of cameras which can be handled by the total system throughput. This primarily depends on the hard disk performances. If we assume we use HD cameras with, for example, 4 Mb/s per camera stream, then, if we have for example 32 cameras, all recorded on the same hard disk, the total amount of streaming created by these cameras will be no less than 128 Mb/s. If we now assume the worst-

A typical computer server with built in RAID

case scenario, where some operators are trying to playback the recorded streams as well, then we have to double this and allow for such an additional bandwidth. The worst case scenario, with the cameras, then will be close to 250 Mb/s (which is around 30 MB/s). If we add some overheads for various applications and programs that need to read from or write to the hard disk, as well as allow for possible back-up of recorded footage, it is realistic to assume that over 50 MB/s (400 Mb/s) would be more comfortable to work with. So, if the hard disks are not capable offering at least this throughput there could be no recording possible of 32 cameras on such a server/computer. This analysis suggests that there is a limit on how many streams can be accepted and recorded by a recording server, and this limit is primarily imposed by the hard disk transfer rate.

The different file systems

Each different operating system uses some kind of file system in order to write data on hard drives and removable media, so that later on the user is able to find it and read it. Inherently, this is a fundamental and important concept that defines the flexibility, capacity, and security of various systems. This is why we will mention the most common ones here.

All file systems consist of structures necessary for storing and managing data. These structures typically include an operating system boot record, directories, and files.

A file system performs three main functions: it tracks the allocated and unused space; it maintains directories and filenames; and it tracks the physical coordinates where each file is stored on the disk.

Different file systems are used by different operating systems. Some operating systems (such as Windows) can recognize only some of its own file systems, while others (such as Linux and Mac OSX) can

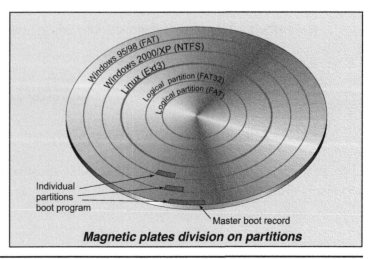

Magnetic plates division on partitions

recognize several, including file systems from another OS.

One important example of this interchangeability is the USB keys, or flash-disks, as they are sometimes called. Typically, a USB storage device is formatted with Microsoft's FAT32, but it can be read by almost all OS. But, the FAT32 comes with its own limitations, such as the maximum file size in one block, or the maximum capacity it can address. This may eventually affect some exporting capability from some machines, so installers and integrators should be aware of such limits.

USB keys use FAT32

It is also important to mention that some manufacturers use raw data writing on hard disks without formatting them. This allows for quick installation of new drives when the old ones fail, as it does not require time-consuming formatting. Other manufacturers have their own formatting method where video data is written sequentially, thus making the USB storage device quicker in searching and playback.

Some of the most common file systems in use today are:

Ext/ext2/Ext3/Ext4 – Extended file system, designed for Linux systems

FAT – Used on DOS and Microsoft Windows, working with 12 and 16 bits

FAT32 – FAT with 32 bits

HFS/HFS+ – Hierarchical File System, used on Mac OS systems

HPFS – High Performance File system, used on IBM's OS/2

ISO 9660 – Used on CD and DVD-ROM disks on a variety of platforms

GPFS – IBM Journaling File system, provided in Linux, Windows and AIX

NTFS – Used on Windows NT-based systems (Windows 2000, XP, 7)

ReiserFS/4 – File system that uses journaling, used in Linux and Unix

FATX – Used in Microsoft's X-Box

LTFS – IBM's file system for Linux, Mac OSX and planned for Microsoft Windows

ReFS – Microsoft's new 2012 Server file system

Later in this chapter we are going to address the most common file systems in CCTV workstations, servers, DVRs, and NVRs more in-depth.

FAT (File Allocation Table)

Introduced by Microsoft in 1983, the *File Allocation Table* (FAT) is a file system that was developed for MS-DOS and used in consumer versions of Microsoft Windows up to and including Windows ME. Even with 512-byte clusters, this could give up to 32 MB of space – enough for the 10 MB or 20 MB XT hard drives that were typical at the time. As hard drives larger than 32 MB were released, large cluster sizes were used. The use of 8192-byte clusters allowed for file system sizes up to 512 MB. However, this increased the problem of internal fragmentation where small files could result in a great deal of wasted space; for example, a 1-byte file stored in a 8192-byte cluster results in 8191-bytes of wasted space.

The FAT file system is considered relatively uncomplicated, and because of that, it is a popular format for floppy disks. Moreover, it is supported by virtually all existing operating systems for personal computers, and because of that it is often used to share data between several operating systems booting on the same computer (a multi-boot environment). It is also used on solid-state memory sticks and other similar devices.

The FAT file system also uses a root directory. This directory has a maximum allowable number of entries and must be located at a specific place on the disk or partition.

Although it is one of the oldest file formats, FAT is likely to remain in use for a long time because it is an ideal file system for small drives. It is also used on other removable storage for non-computer devices, such as flash memory cards for digital cameras, USB flash drives, and the like.

FAT32 (File Allocation Table 32-bit)

In 1997, Microsoft created FAT32 as an extension to the FAT concept because the cluster growth possibility was exhausted. The FAT32 was an enhancement of the FAT file system and was based on 32-bit file allocation table entries, rather than the 16-bit entries used by the previous FAT system. As a result, FAT32 supports much larger disk or partition sizes (up to 2 TB). This file system can be used by Windows 95 SP2 and Windows 98/2000/XP. Previous versions of DOS or Windows cannot recognize FAT32 and are thus unable to boot from or use files on a FAT32 disk or partition. The FAT32 file system uses smaller clusters than the FAT file system, has duplicate boot records, and features a root directory that can be of any size and can be located anywhere on the disk or partition. The maximum possible size for a file on a FAT32 volume is 4 GB. Video applications, large databases, and some other software easily exceed this limit which is the reason some DVRs/NVRs exporting to USB keys break up the files to such sizes.

The FAT32 cluster values are represented by 32-bit numbers, of which 28 bits are used to hold the cluster number. The boot sector uses a 32-bit field for the sector count, limiting the FAT32 volume size to 2 TB for a sector size of 512 bytes and 16 TB for a sector size of 4,096 bytes. FAT32 was introduced with Windows 95 in 1996, although reformatting was needed to use it, and DriveSpace 3 (the version that came with Windows 95 OSR2 and Windows 98) never supported it. Windows 98 introduced a utility to convert existing hard disks from FAT16 to FAT32 without loss of data. In the Windows NT line, native support for FAT32 arrived in Windows 2000. In theory, this should support a total of approximately 268,435,438 clusters, allowing for drive sizes in the multi-terabyte range. However, due to limitations in Microsoft's ScanDisk utility, the FAT is not allowed to grow beyond 4,177,920 clusters, placing the volume limit at 124.55 GB. The open FAT+ specification proposes how to store larger files up to 256 GB on slightly modified and otherwise backwards compatible FAT32 volumes, but imposes a risk that disk tools or FAT32 implementations not aware of this extension may truncate or delete files exceeding the normal FAT32 file size limit.

NTFS (New Technology File System)

NTFS or *New Technology File System* is the standard file system of Microsoft Windows NT and its descendants, Windows 2000, Windows XP, Windows 7 and Windows Servers. NTFS is a descendant of HPFS, the file system designed by Microsoft and IBM for OS/2 as a replacement for the older FAT file system of MS-DOS. The improvements over FAT was support for meta-data and the use of advanced data structures in order to improve performance, reliability, and disk space utilization. NTFS incorporates these plus additional extensions such as security access control lists and file system journaling. In NTFS everything that has anything to do with a file (name, creation date, access permissions, and even contents) is written down as meta-data. Internally, NTFS uses binary trees in order to store the file system data; although complex to implement, this allows fast access times and decreases fragmentation. A file system journal is used in order to guarantee the integrity of the file system itself (but not of each individual file). Systems using NTFS are known to have improved reliability, a particularly important requirement considering the unstable nature of the older versions of Windows NT.

NTFS has gone through an evolution of versions, starting from v.1.0 in mid-1993 with Windows NT 3.1, then v.1.1 in 1994 for NT 3.5, then v.1.2 for NT4 in 1996 (some times referred to as NTFS 4.0), v.3.0 for Windows 2000 in year 2000 (some times referred to as NTFS 5.0), up to v.3.1 for Windows XP in 2001 (some times referred to as NTFS 5.1).

The central system structure of the NTFS file system is the **master file table** (MFT). NTFS keeps multiple copies of the critical portion of the MFT to protect against corruption and data loss. Like FAT and FAT32, NTFS uses clusters to store data files. However, the size of the clusters is not dependent on the size of the disk or partition. A cluster size as small as 512 bytes can be specified, regardless of whether a partition is 6 GB or 60 GB. Using small clusters not only **reduces the amount of wasted disk space**, but **also reduces file fragmentation**, a condition where files are broken up over many non-contiguous clusters, resulting in slower file access. **Because of its ability to use small clusters, NTFS provides good performance on large drives.** Finally, the NTFS file system supports hot fixing, a process through which bad sectors are automatically detected and marked so that they will not be used.

In theory, the maximum NTFS volume size is $2^{64}-1$ clusters. However, the maximum NTFS volume size as implemented in Windows XP Professional is $2^{32}-1$ clusters partly due to partition table limitations. For example, using 64 kB clusters, the maximum Windows XP NTFS volume size is 256 TBs. Using the default cluster size of 4 kB, the maximum NTFS volume size is 16 TB. Because partition tables on master boot record (MBR) disks only support partition sizes up to 2 TB, dynamic volumes must be used to create NTFS volumes over 2 TB. As designed, the maximum NTFS file size is 16 EB (16×1024^6 or 2^{64} bytes). As implemented, the maximum NTFS file size is 16 TB. With Windows 8, the maximum NTFS file size is 256 TB.

Ext2/3/4

The *ext2* or *second extended file system* was the standard file system used on the Linux operating system for a number of years and remains in wide use. It was initially designed by Rémy Card based on concepts from the extended file system. It is quite fast, enough so that it is used as the standard against which to measure many benchmarks. Its main drawback is that it is not a journaling file system. The Ext2 file system supports a maximum disk or partition size of 4 terabytes. Its successor, Ext3, has a journal and is compatible with Ext2.

The *ext3* or *third extended file system* is a journaled file system that is coming into increasing use among users of the Linux operating system. Although its performance and scalability are less attractive than those of many of its competitors such as ReiserFS and XFS, it does have the significant advantage that users can upgrade from the popular Ext2 file system without having to back up and restore data. The ext3 file system adds a journal without which the file system is a valid Ext2 file system. An Ext3 file system can be mounted and used as an Ext2 file system. All of the file system maintenance utilities for maintaining and repairing the Ext2 file system can also be used with the Ext3 file system, which means Ext3 has a much more mature and well-tested set of maintenance utilities available than its rivals.

The *ext4* is the fourth extended, journaling, file system for Linux, developed as the successor to ext3. **The journal allows the file system to quickly return to a consistent state after an unscheduled system shutdown caused by a power outage or a system crash**. This feature greatly **reduces the risk of file system corruption** (and the need for lengthy file system checks). ReiserFS also handles directories containing huge numbers of small files very efficiently. Unfortunately, converting a system to ReiserFS requires users of Ext2 to completely reformat their disks, which is a disadvantage not shared by its main competitor Ext3. Because of its advantages many Linux distributions have made it the default file system.

The ext4 file system can support volumes with sizes up to 1 Exabyte (EB) (1,000 Terabytes = 10^{18} Bytes) and files with sizes up to 16 Terabytes (TB). Ext4 is backward compatible with ext3 and ext2, making it possible to mount ext3 and ext2 as ext4. Ext3 is partially forward compatible with ext4.

Ext4 does not yet have as much support as ext2 and ext3 on non-Linux operating systems.

File systems of various OS handle addressing of hard disk magnetic areas

Hard disk connectivity standards (SATA)

The type of connection between the hard drive and the system (motherboard and CPU) is defined by one of a few standards.

The most popular up until 2010 was the ***Enhanced Integrated Drive Electronics*** drives (EIDE), which was also known as ***Advanced Technology Attachment*** (ATA). Some people called it Parallel ATA (PATA), because it had multiple parallel cables connecting the hard disk drive with the computer motherboard. This is now obsolete technology, giving way to the ***Serial ATA*** (SATA) standard. Serial offers

several advantages over the PATA: reduced cable size and cost (seven conductors instead of 40), native hot swapping, faster data transfer through higher signalling rates, and more efficient data transfer. The SATA bus interface connects not only the host bus adapters to hard disk drives, but also to optical drives as well. Physically, the cables in SATA are the largest change. The data is carried by a light, flexible seven-conductor wire with 8 mm wide wafer connectors on each end. It can be anywhere up to 1 meter long. Compared to the short (45 cm), ungainly 40 or 80 conductor ribbon cables of Parallel ATA, this is a great relief to system builders. In addition, airflow and, therefore, cooling in equipment is improved due to smaller cables. The concept of a master-slave relationship between devices has been dropped. SATA has only one device per cable. The connectors are keyed so that it should no longer be possible to install cable connectors upside down, which often is a problem with PATA types.

PATA and SATA 3.5" drives

PATA and SATA 2.5" drives

The other popular standard of communicating with hard disk in the past was the ***Small Computer System Interface*** (SCSI). This has also become obsolete technology when using PATA drives as part of the SCSI.

So the current and most popular standard of connecting computer motherboard to hard disk(s) is the ***Serial ATA*** (SATA). The SATA drives dominate the PC industry today, and this is the case with the DVRs as well.

SATA host adapters and devices communicate via a high-speed serial cable over two pairs of conductors. In contrast, PATA used a 16-bit wide data bus with many additional support and control signals, all operating at much lower frequency. To ensure backward compatibility with legacy ATA software and applications, SATA uses the same basic ATA and ATAPI command-set as legacy ATA devices. SATA industry compatibility specifications originate from The Serial ATA International Organization (SATA-IO). The SATA-IO group corroboratively creates, reviews, ratifies, and publishes the interoperability specifications and test cases.

There are a number of SATA standard revisions dealing with data transfer speed:

    - SATA v.1.0, speed up to 150 MB/s (1.5 Gb/s)

    - SATA v.2.0, speed up to 300 MB/s (3 Gb/s)

    - SATA v.3.0, speed up to 600 MB/s (6 Gb/s)

SATA v.1.0 was released in 2003. First-generation SATA interfaces, now known as SATA 1.5 Gb/s, **communicate at a rate of 1.5 Gb/s**, and do not support Native Command Queuing (NCQ). Due to some encoding overheads, the SATA v.1.0 has an actual transfer rate of 1.2 Gb/s (150 MB/s). The theoretical burst throughput of SATA 1.5 Gb/s is similar to that of PATA/133, but newer SATA devices offer enhancements such as NCQ, which improve performance in a multitasking environment. During the initial period after SATA 1.5 Gb/s finalization, adapter and drive manufacturers used a "bridge chip" to convert existing PATA designs for use with the SATA interface. Bridged drives have a SATA connector, may include either or both kinds of power connectors, and, in gen-

SATA drive connections (data pins on the left, power on the right)

eral, perform identically to their PATA equivalents. Most lack support for some SATA-specific features such as NCQ. Native SATA products quickly eclipsed bridged products with the introduction of the second generation of SATA drives. As of April 2010 the fastest 10,000 RPM SATA mechanical hard disk drives could transfer data at maximum (not average) rates of up to 157 MB/s,which is beyond the capabilities of the older PATA/133 specification and also exceeds a SATA 1.5 Gb/s link.

The second generation *SATA v.2.0* interfaces **run with a native transfer rate of 3.0 Gb/s**, and the maximum un-coded transfer rate is 2.4 Gb/s (300 MB/s). The theoretical burst throughput of SATA 3.0 Gb/s is double that of SATA revision 1.0.

All SATA data cables meeting the SATA spec are rated for 3.0 Gb/s and handle current mechanical drives without any loss of sustained and burst data transfer performance. However, high-performance flash drives can exceed the SATA 3 Gb/s transfer rate, which was addressed with the next, SATA 6 Gb/s, interoperability standard. SATA 3 Gb/s is backward compatible with SATA 1.5 Gb/s.

The *SATA v.3.0* was proposed in 2008, and **deals with speeds of up to 6 Gb/s**. In actual fact, after the encoding overheads, achieves maximum un-coded transfer rate of 4.8 Gb/s, which is equivalent to 600 MB/s. The theoretical burst throughput of SATA 6.0 Gbit/s is double that of SATA revision 2.0. There were some minor improvements on the previous SATA standards, such as improved power management capabilities, that are aimed at improving quality of service for video streaming and high-priority interrupts. In addition, the standard continues to support distances up to one meter. SATA v.3.0 is backward compatible with SATA v.2.0.

The SATA hard disks require a different power connector as part of the standard. Fifteen pins are used to supply three different voltages if necessary – 3.3 V, 5 V, and 12 V. The same physical connections are used on 3.5" and 2.5" (notebook) hard disks.

In 2004, there was a proposal for an external SATA, called *eSATA.* This was a variant of SATA meant for external connectivity. It uses a more robust connector, longer shielded cables (up to 2 m), and stricter (but backward-compatible) electrical standards. The protocol and logical signaling (link/transport layers and above) are identical to internal SATA.

SATA (left) and eSATA (right) connectors

The eSATA connector is mechanically different to prevent unshielded internal cables from being used externally. The eSATA connector discards the "L"-shaped key and changes the position and size of the guides. The eSATA insertion depth is deeper: 6.6 mm instead of 5 mm. The contact positions are also changed. The eSATA cable has an extra shield to reduce EMI to FCC and CE requirements. Internal cables do not need the extra shield to satisfy EMI requirements because they are inside a shielded case. The eSATA connector uses metal springs for shield contact and mechanical retention. The eSATA connector has a design-life of 5,000 connections, the ordinary SATA connector is only specified for 50.

Multiple SATA drives can be connected in a serial connection, which is called *Serial Attached SCSI* standard, or short *SAS*. The SAS is a point-to-point serial protocol that moves data to and from computer storage devices such as hard drives and tape drives. SAS replaces the older Parallel SCSI bus technology that first appeared in the mid-1980s. SAS, like its predecessor, uses the standard SCSI command set. SAS offers backward compatibility with second-generation SATA drives. SATA 3 Gb/s drives may be connected to SAS backplanes, but SAS drives cannot connect to SATA backplanes. The SAS has no termination issues, like the Parallel SCSI.

The maximum number of drives that SAS allows to be connected is 65,535 through the use of expanders, while Parallel SCSI has a limit of 8 or 16 devices on a single channel. SAS allows a higher transfer speed (3 or 6 Gb/s) than most parallel SCSI standards. SAS achieves these speeds on each initiator-target connection, hence getting higher throughput, whereas parallel SCSI shares the speed across the entire multidrop bus.

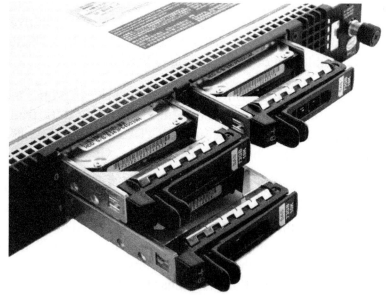

Serial Attached SCSI arrangement

DAS

Direct-Attached Storage (DAS) **refers to a digital storage system directly attached to a server or workstation, without a network in between.** A typical DAS system is made of a data storage device (for example enclosures holding a number of hard disk drives) connected directly to a computer through a host bus adapter, which in the past used to be SCSI, but these days more often eSATA, SAS or Fiber Channel. The most important differentiation between DAS and NAS is that between the computer and DAS there is no network device (like a hub, switch, or router).

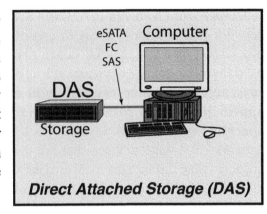

Direct Attached Storage (DAS)

NAS

Network-Attached Storage (NAS) is a computer data storage connected to a network, providing data access to various group of clients. NAS not only operates as a file server, but it is also specialized for this task either by its hardware, software, or configuration of those elements.

NAS systems are networked appliances which contain one or more hard drives, often arranged into logical, redundant storage containers, or RAID. Network-attached storage removes the responsibility of file serving from other servers on the network. They typically provide access to files using network file sharing protocols. In CCTV, NAS devices have gained popularity, as a convenient method of sharing files among multiple computers. **The benefits of network-attached storage, com-**

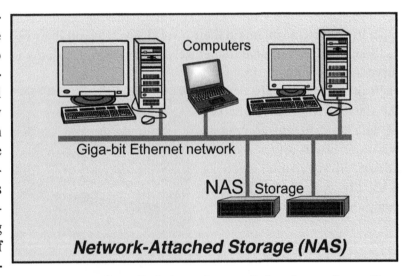

Network-Attached Storage (NAS)

pared to file servers, include faster data access, easier administration, and simple configuration.

Although it may technically be possible to run other software on a NAS unit, it is not designed to be a general purpose server. For example, NAS units usually do not have a keyboard or display, and are controlled and configured over the network, often using a browser. A full-featured operating system is not needed on a NAS device, so often a stripped-down operating system is used, like FreeNAS, a simplified version of FreeBSD.

SAN

A ***Storage Area Network*** (SAN) is a dedicated network that provides access to consolidated, block level data storage. **SANs are primarily used to make storage devices, such as disk arrays, accessible to servers so that the devices appear like locally attached devices to the operating system.** A SAN typically has its own network of storage devices that are generally not accessible through the local area network by other devices. The cost and complexity of SANs has dropped to levels allowing wider adoption across both enterprise and small to medium sized business environments. Their usage in CCTV is also increasing, allowing for more efficient long-term storage.

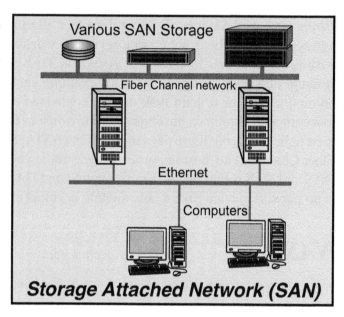

Storage Attached Network (SAN)

A SAN does not provide file abstraction, only block-level operations. However, file systems built on top of SANs do provide file-level access and are known as SAN file systems, or shared disk file systems.

Operating systems maintain their own file systems on their own dedicated, non-shared virtual drives, as though they were local to themselves. If multiple systems were simply to attempt to share a virtual drive these would interfere with each other and quickly corrupt the data. Any planned sharing of data on different computers within a virtual drives requires advanced solutions, such as SAN file systems or clustered computing.

SANs help to increase storage capacity utilization, since multiple servers consolidate their private storage space onto the disk arrays. Common uses of a SAN include provision of data that require high-speed, block-level access to the hard drives such as e-mail servers, databases, and high-usage file servers.

Multiple servers might be used in a SAN configuration

RAID

RAID stands for ***Redundant Arrays of Independent Disks,*** and the name describes its concept. It combines multiple small, independent (and inexpensive, a term that was originally used instead of independent) disk into an array of drives that yields **performance and/or redundancy**. Such an array of multiple disks appears as one logical unit to the operating system. In the past, when PATA drives were used for RAID arrays, by using RAID they could be further improved on speed. Today, when only SATA drives are used in RAID configurations, the speed is no longer an issue because SATA drives do come with speed in excess of 1.5 Gb/s.

The redundancy offered with RAID is especially useful for those of us in CCTV. If one drive fails, data is not lost and there is usually a way to remove and replace the faulty drive by hot-swapping (while running) it with a new one.

RAID usually requires interface electronics, like RAID controller, although it can also be done through the software, but not as efficiently as when it is done via the appropriate hardware. The RAID array of drives appears to the computer as a single logical storage unit or drive and can principally be used with any hard drive. Many RAID controllers are these days built-in the computer motherboards.

There are several different types of RAID, called "RAID levels," depending on the level of redundancy and performance required. The different levels or architectures are named by the word RAID followed by a number (e.g. RAID-0, RAID-1). Each level provides a different balance between the key goals: reliability and availability, performance and capacity. RAID levels greater than RAID-0 provide protection against unrecoverable (sector) read errors, as well as whole disk failure.

Photo courtesy of Synology

A RAID box with 5 drives

Originally, there were five RAID levels, but more variations have evolved and some non-standard levels (mostly proprietary). RAID levels and their associated data formats are standardized by the Storage Networking Industry Association (SNIA) in the Common RAID Disk Drive Format (DDF) standard.

RAID-0

***RAID-0* (block-level striping without parity or mirroring) has no redundancy,** so technically this is not of interest to CCTV, but it should be mentioned here, as one may come across the term RAID-0. **It provides improved performance and additional storage but no fault tolerance.** Any drive failure destroys the array, and the likelihood of failure increases with more drives in the array.

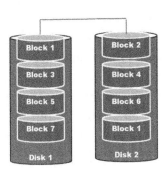

RAID-0

RAID-1

In *RAID-1* (**mirroring without parity or striping**), **data is written identically to two drives**, thereby producing a "mirrored set." The read request is serviced by either of the two drives containing the requested data, whichever one involves least seek time plus rotational latency. Similarly, a write request updates the stripes of both drives. The write performance depends on the slower of the two writes (i.e. the one that involves larger seek time and rotational latency). **At least two drives are required to constitute such an array.** While more constituent drives may be employed, many implementations deal with a maximum of only two. The array continues to operate as long as at least one drive is functioning.

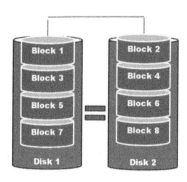

RAID-1 (mirroring)

RAID-2

In *RAID-2* (**bit-level striping with dedicated code parity**), **all disk spindle rotation is synchronized, and data is striped such that each sequential bit is on a different drive.** Code parity is calculated across corresponding bits and stored on at least one parity drive. **This theoretical RAID level is not used in practice**.

RAID-3

In *RAID-3* (**byte-level striping with dedicated parity**) **all disk spindle rotation is synchronized, and data are striped so each sequential byte is on a different drive.** Parity is calculated across corresponding bytes and stored on a dedicated parity drive. **Although implementations exist, RAID 3 is not commonly used in practice.**

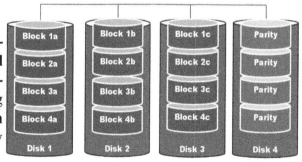

RAID-3 (parity on separate disks)

RAID-4

RAID-4 (**block-level striping with dedicated parity**) **is equivalent to RAID-5 (see next) except that all parity data are stored on a single drive.** In this arrangement files may be distributed among multiple drives. Each drive operates independently, allowing I/O requests to be performed in parallel. RAID-4 was previously used primarily by NetApp, but has now been largely replaced by an implementation of RAID-6.

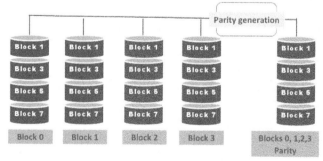

RAID-4 (all parity on one drive)

RAID-5

***RAID-5* (block-level striping with distributed parity) distributes parity along with the data and requires all drives but one to be present to operate; the array is not destroyed by a single drive failure.** This is one of the most common RAIDs used in CCTV recording servers. Upon drive failure, any subsequent reads can be calculated from the distributed parity such that the drive failure is masked from the end user. RAID-5 requires at least three disks.

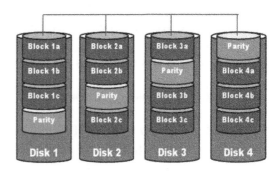

RAID-5 (parity across disks)

RAID-6

***RAID-6* (block-level striping with double distributed parity) provides fault tolerance up to two failed drives. This makes larger RAID groups more practical, especially for high-availability systems.** This becomes increasingly important as large-capacity drives lengthen the time needed to recover from the failure of a single drive. Like RAID-5, a single drive failure results in reduced performance of the entire array until the failed drive has been replaced and the associated data rebuilt. This RAID is also used in some CCTV recorders.

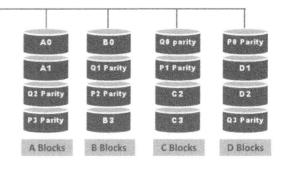

RAID-6 (two drives can fail)

RAID-10

In ***RAID-10***, often referred to as RAID 1+0 **(mirroring and striping), data is written in stripes across primary disks that have been mirrored to the secondary disks.** This array achieves high speed and redundancy for each drive, but it is very expensive and unlikely to be used in CCTV, as it needs a lot of drives without extending the recording time.

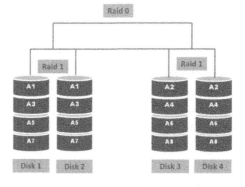

RAID-10 (mirroring+striping)

Clearly, CCTV has become very interested in RAID arrays, since offers protection of the most important purpose of CCTV - the recorded footage. It also offers an ease way of extending recording time by just adding hard disks. The most commonly used arrays are RAID-5, RAID-6, and RAID-1, which is typical for some of the largest casino IP systems, where mirroring per camera channel is achieved with RAID-1.

It should be noted that when a drive fails in one of the configurations above, **it takes a considerable time when rebuilding a faulty drive with a new one.** This time is proportional to the size of the drives. Based on some empirical studies, **it takes approximately 2 hours per terabyte of the RAID array.**

Hard disk failures

Since a hard disk drive is an electromechanical device, wear and tear will cause it to eventually fail.

It's not a question if a drive will fail, the question is when it will fail?

When analyzing hard disks life expectancy, there are three common parameters manufacturers give.

The *Annualized Failure Rate* (AFR), which is the **percentage failure share of a certain amount of hard disks (extrapolated to one year based on expectation values).**

The *Mean Time Between Failures* (MTBF) **specifies the expected operating time between two consecutive failures of a device type in hours** (definition according to IEC 60050 (191)). The MTBF considers the life cycle of a device that fails repeatedly, then is repaired and returned to service again. Because repair of hard drives rarely happens, we don't really have mean time between failures. We only have mean time to a failure, after which the drive is discarded. Therefore, *Mean Time To Failure* (MTTF) **is used, which specifies the expected operating time until the failure of a device type in hours** (definition according to IEC 60050 (191)).

The acronyms MTBF and MTTF are often used synonymously in terms of drives. Some manufacturers, for example, estimate the MTBF as the number of operating hours per year divided by the projected failure rate. This view is based on a failure without repair. As such , the MTTF would be more practical parameter, but it will still show unrealistic high number of hours of life expectancy, something that is not the case with a standard electronic definition of MTBF.

An extreme case of a broken hard drive

For example, a typical hard disk MTBF's numbers range between 300,000 and 1,000,000 hours. This is quite a high number, equivalent to 34 to 114 years. These numbers are way too high, and practical experience shows that the posted lifetime of a drive is more likely one tenth of it, typically 3~5 years. In addition, technology progress and new standards and increased capacity does not allow hard drives to be used for more than a couple of years. Moore's law can easily be applied to hard disk capacity too, they double almost every year or two.

The stated MTBF/MTTF high numbers are a result of a specific definition by hard disk manufacturers, which refers to a testing a number of drive, the testing period selected (in hours) and the percentage of drives failed in that period. This can be written as the following formula:

$$MTTF = ([test\ period] \times [number\ of\ drives]) / [number\ of\ failed\ drives] \quad (47)$$

For example, if testing has been conducted over one month (720 hrs), and out of 1,000 drives, two have failed, the MTTF will be:

$$MTTF = (720 \times 1,000 / 2) = 360,000\ hrs.$$

In reality, this does not mean any drive may fail on and around 360,000 hrs of operation. The better interpretation of our example is that out of 1,000 drives tested for one month, two have failed, which is equivalent to having failure of one drive out of 500, each month (1,000 / 2). This is a 0.2% for one month. Over a period of one year, as the annual failure rate (AFR) defines it, this would be equivalent to having probability of 2.4% drives to fail. So, following our example, in a system with 1,000 drives with MTTF of 360,000 hrs, statistically, there will be 24 failed hard disk drives. If the MTTF was 1,000,000 hrs, which is typically quoted for enterprise drives, this statistics will mean 1 failed rive among 1,000 over a period of 1 year.

This same calculation can be generalized using the common ***exponential model of distribution***:

$$\text{Failure Probability} = R(t) = 1 - F(t) = 1 - e^{(-t/M)} = 1 - e^{-\lambda t} \qquad (48)$$

where e is the natural base number $e = 2.71$, t is the time for which this probability is calculated, and M is the MTBF.

So, if we do the same calculation for the previous example, for a drive with 360,000 hour MTBF drive ($M = 360,000$), we could calculate the failure probability for 1 year ($t = 8,760$ hrs) to be:

$$R(t) = 1 - e^{(-8,760/360,000)} = 0.024 = 2.4\%.$$

Clearly, the above numbers are statistical only and very much depend on the environment in which the tests have been conducted. Most notable are certainly temperature, humidity, mechanical shocks, and static electricity during handling and installing. It is quite understandable that manufacturers would try and conduct such tests in as ideal conditions as possible. This means, in practice, we can only expect higher numbers than the above, drawn as a statistical calculation from manufacturers' tests.

An interesting study was made by Carnegie Mellon University, which confirmed the empirical knowledge in mass-produced consumer electronics: **if a new product doesn't fail immediately after its inital usage, it will serve its purpose until approximately its MTBF time.**

This study evaluated the data of about 100,000 hard drives that were used in several large-scale systems and found a large deviation of the manufacturer's information. The average failure rate of all hard drives was six times higher for systems with an operating time of less than three years and even 30 times higher for systems with an operating time of 5-8 years. This statistic took them to a conclusion that in the early "in-

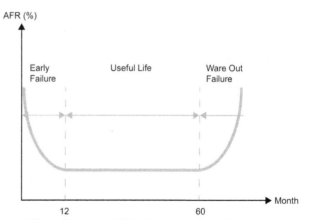

The Carnegi Mellon bath-tub curve

fancy" usage time there is much higher rate of failure, then it settles for a longer period, which is the expected useful working life. After that, it starts failing due to age, wear, and tear, which coincides with practical experience of around five years (60 months) before increased failure rate is experienced.

Processors (CPUs)

A *Central Processing Unit*(CPU), often *microprocessor* or short *processor*, is the brain of a computer. It is an electronic integrated circuit, composed of hundreds of thousands, and since the last few years, over millions of transistors. A processor (or CPU) is the main electronic hardware onto which software runs. Without CPUs, computers would be just a piece of electronic hardware that does repetitive thing without ability to change the outcome of a process. CPU can produce new outcome each time the software input changes. CPUs can perform hundreds and thousands of instructions very quickly. They are multipurpose devices, that can be re-programmed each new cycle to execute different process. These executions are based on the software instructions, using logical loops, calculations with various input parameters, and producing results that can be shown on a screen, printed, saved on permanent storage, or taken further to another CPU for more processing.

The most important part in this process is the ability of a CPU to make logical decisions, based on the set of rules contained in the software, combined with various external input and their logical interaction, which could be a human input or another external process.

Typically, CPUs are understood to be the brains of modern day computers, but many more processors are part of embedded systems used in our daily lives, which provide digital control of a myriad of objects from appliances to automobiles to mobile phones and industrial process control.

It all started in the 1960s, after the discovery of *Transistors* and the introduction of mathematical logic topology in array of transistors. Later on that decade, the development of micro-electronic enabled integrating miniature transistors into circuits, called *Integrated Circuits* (IC). Computer processors were constructed out of small and medium-scale ICs—each containing from tens to a few hundred transistors. These were placed and soldered onto printed circuit boards.

The first mass usage of microprocessors were *electronic calculators*, replacing the awkward *slide rulers*, *logarithmic*, and *trigonometric tables* used in all engineering practices. In the 1970s, microcomputers capable of executing simple programs, and even simple games were introduced. This was the decade when personal computers became possible.

The integration of a whole CPU onto a single chip or on a few chips greatly reduced the power consumption, it reduced and the complexity of circuits with discrete transistors, but it increased processing power. Single-chip processors increase reliability as there are many fewer electrical connections to fail. Because CPUs were produced by highly automated processes large numbers could be produced but to keep the unit cost low.

Early use of microprocessors

Perhaps electronics is one of the very few industries where newer products using processors are getting faster and more capable, and yet the prices go lower, not higher. The speed of progress, capacity, and CPU capability has never stopped growing. In fact, there is now an empirically proven law, advised by Intel's co-founder, Gordon E.Moore, in 1965, that the increase in capacity of microprocessors follows a law (now called **Moore's law**): **the number of components that can be fitted onto a chip doubles every year.** With present technology, it is actually every two years; as such, Moore later changed the period to two years.

The real pioneers of computer CPUs as we know them today are Intel's 8080 processor and Motorola's 6800 series.

Intel's original and popular 8080 design evolved into the 16-bit Intel 8086, the first member of the x86 family, which powers most modern PC type computers. Intel then released the 80186 and 80188, the 80286 and, in 1985, the 32-bit 80386, cementing their PC market dominance with the processor

The Motorola 6800 processor

family's backwards compatibility. The 8086 and successors had an innovative but limited method of memory segmentation, while the 80286 introduced a full-featured segmented memory management unit. The 80386 introduced a flat 32-bit memory model with paged memory management.

Intel's first 32-bit microprocessor was the iAPX 432, which was introduced in 1981 but was not a commercial success. It had an advanced capability-based object-oriented architecture, but poor performance compared to contemporary architectures such as Intel's own 80286 (introduced 1982), which was almost four times as fast on typical benchmark tests.

From 1985 to 2003, the 32-bit x86 architectures became increasingly dominant in desktop, laptop, and server markets, and these microprocessors became faster and more capable. Intel had licensed early versions of the architecture to other companies, but declined to license the Pentium, so AMD and Cyrix built later versions of the architecture based on their own designs. During this span, these processors increased in complexity (transistor count) and capability (instructions/second) by at least three orders of magnitude. Intel's Pentium line is probably the most famous and recognizable 32-bit processor model, at least with the public at broad.

Intel's 80286 processor

After the 32-bit microprocessors, the 64-bit were the natural evolution and such designs have been in use in several markets since the early 1990s (like the Nintendo 64 gaming console), their real introduction in the PC market was in the early 2000. First, it was AMD with their introduction of a 64-bit architecture, backwards-compatible with x86, x86-64 in September 2003. Then, it followed by Intel's near fully compatible 64-bit extensions. Both versions can run 32-bit legacy applications without any performance penalty as well as new 64-bit software. With operating systems Windows XP, Windows Vista, Windows 7, Linux, and Mac OS X that run 64-bit native, the software is also geared to fully utilize the capabilities of such processors.

Multi-core

The complexity of processors increases every year, following the Moore's Law. For example, the Intel 80286 processor had 134,000 transistors, the 80386 around 275,000, and 80486 already over a million. Pentium III had close to 10 million transistors, Pentium 4 over 40 million transistors.

It has become increasingly challenging as chip-making technologies approach their physical limits. A different approach to improving a computer's performance is to add extra processors, as in symmetric multiprocessing designs, which have been popular in servers and workstations since the early 1990s. This is how processor manufacturers have introduced multi-core processing in order to increase the processing power of CPUs.

Processor	Transistor count	Date of introduction	Manufacturer	Process	Area
Intel 8080	4,500	1974	Intel	6 µm	20 mm²
Motorola 6800	4,100	1974	Motorola	6 µm	16 mm²
Zilog Z80	8,500	1976	Zilog	4 µm	18 mm²
Motorola 68000	68,000	1979	Motorola	4 µm	44 mm²
Intel 8088	29,000	1979	Intel	3 µm	33 mm²
Intel 80186	55,000	1982	Intel		
Intel 80286	134,000	1982	Intel	1.5 µm	49 mm²
Intel 80386	275,000	1985	Intel	1.5 µm	104 mm
Intel 80486	1,180,000	1989	Intel	1 µm	160 mm
Pentium	3,100,000	1993	Intel	0.8 µm	294 mm
Pentium Pro	5,500,000[1]	1995	Intel	0.5 µm	307 mm
AMD K5	4,300,000	1996	AMD	0.5 µm	251 mm
AMD K6	8,800,000	1997	AMD	0.35 µm	162 mm
Pentium II	7,500,000	1997	Intel	0.35 µm	195 mm
Pentium III	9,500,000	1999	Intel	0.25 µm	128 mm
AMD K7	22,000,000	1999	AMD	0.25 µm	184 mm
AMD K6-III	21,300,000	1999	AMD	0.25 µm	118 mm²
Pentium 4	42,000,000	2000	Intel	180 nm	217 mm
AMD K8	105,900,000	2003	AMD	130 nm	193 mm
Core 2 Duo	291,000,000	2006	Intel	65 nm	143 mm
Dual-Core Itanium 2	1,700,000,000[6]	2006	Intel	90 nm	596 mm
POWER6	789,000,000	2007	IBM	65 nm	341 mm
AMD K10 quad-core 2M L3	463,000,000[2]	2007	AMD	65 nm	283 mm
AMD K10 quad-core 6M L3	758,000,000[2]	2008	AMD	45 nm	258 mm
Core i7 (Quad)	731,000,000	2008	Intel	45 nm	263 mm
Atom	47,000,000	2008	Intel	45 nm	24 mm²
Six-Core Xeon 7400	1,900,000,000	2008	Intel	45 nm	503 mm
Six-Core Opteron 2400	904,000,000	2009	AMD	45 nm	346 mm
8-Core Xeon Nehalem-EX	2,300,000,000[9]	2010	Intel	45 nm	684 mm
Six-Core Core i7 (Gulftown)	1,170,000,000	2010	Intel	32 nm	240 mm
10-Core Xeon Westmere-EX	2,600,000,000	2011	Intel	32 nm	512 mm
Six-Core Core i7/8-Core Xeon E5	2,270,000,000 [8]	2011	Intel	32 nm	434 mm
Quad-Core + GPU Core i7	1,160,000,000	2011	Intel	32 nm	216 mm
8-Core Itanium Poulson	3,100,000,000	2012	Intel	32 nm	544 mm

Processors evolution *(source: www.wikipedia.com)*

A multi-core processor is simply a single chip that contains more than one microprocessor core. This effectively multiplies the processor's potential performance by the number of cores. Certainly, the operating system and software is supposed to be designed to take advantage of more than one processor core. Some components, such as bus interface and cache, may be shared between cores. **Because the cores are physically very close to each other they can communicate with each other much faster than separate processors in a multiprocessor system, thus improving overall system performance.**

In 2005, the first personal computer dual-core processors were announced. As of 2012, dual-core and quad-core processors are widely used in home PCs and laptops while quad, six, eight, ten, twelve, and sixteen-core processors are common in the professional and enterprise markets with workstations and servers.

The modern desktop sockets do not support systems with multiple CPUs but very few applications outside of the professional market can make good use of more than four cores and both Intel and AMD currently offer fast quad- and six-core desktop CPUs so this is generally a moot point anyway. AMD also offers the first and still currently the only eight core desktop CPUs with the FX-8xxx line but anything with more than four cores is generally not very useful in home desktops. As of January 24, 2012, these FX processors are generally inferior to similarly priced (sometimes cheaper) Intel quad-core Sandy Bridge models.

The desktop market has been in a transition towards quad-core CPUs since Intel's Core 2 Quads were released and now are quite common although dual-core CPUs are still more prevalent. This is largely because of people using older or mobile computers, both of which have a much lower chance of having more than two cores than newer desktops and because of how most computer users are not heavy users.

Historically, AMD and Intel have switched places as the company with the fastest CPU several times. Intel currently win on the desktop side of the computer CPU market, with their Sandy Bridge and Ivy Bridge series. In servers, AMD's new Opterons seem to have superior performance for their price points. This means that AMD are currently more competitive in low- to mid-end servers and workstations that more effectively use fewer cores and threads.

In CCTV, the CPU processing power and number of cores is important for practical reasons. Most VMS workstations perform software decoding, which is basically done by the software decoders running on the CPU. The more powerful the CPU, the more concurrent streams can be decoded.

Some manufacturers recommend a simple rule of thumb: each HD stream encoded with H.264 requires one processor's core for smooth decoding. So, for viewing uninterrupted quad split-screen with 4 HD streams, a 4-core processor would be required. This is not to say that more streams cannot be placed on the screen, it's just they won't be decoded easily and smoothly.

Intel's i7 4-core processor

Software decoders and players

When a CCTV Video Management System decodes a stream from an IP camera, chances are it will use some of the embedded software decoders it has in itself. **There are many IP cameras on the market with various compression encoders. Many are standard or using standard compression hardware chips (after all, there are only so many standard hardware encoders). However, there are many differences in compression profiles, how they are applied, and how the stream is processed and encapsulated by various manufacturers.**

There are many VMSs that have various decoders built-in, borrowed from the codecs library various camera manufacturers have for their own IP cameras. The way how this works is as follows—a VMS manufacturer requests camera decoder (driver) modules from a camera manufacturer, and they insert them in the VMS, with full or limited functionality. It really depends on the VMS agreement with a particular hardware manufacturer, and clearly, some (like competitors) may not agree to share such details. The variety of software decoders in a VMS is the "multi-decoder software functionality." Some VMS manufacturers like to call their software "open platform," although, strictly speaking, this is not correct, as every new IP camera model, with even slightest variation in processing of the video stream needs to be (yet again) processed and included in the never-ending VMS compatibility list.

This is a sore point in today's digital IP CCTV. Because of such a dissonant range of image formats and encoders, and not having any global digital CCTV standards yet, it is difficult to ask for a universal VMS that will work smoothly with any IP camera. We have not come to the stage where a camera from one manufacturer will simply work with any DVR, NVR, or VMS, by simply plug and play, although efforts by some alliances and organizations, like ONVIF and PSIA are being made to standardize communications between various brands and protocols (a little bit more on this in the next heading).

One important expectation in digital CCTV, irrespective of the brand of the VMS, is the **possibility to export a footage of any incident in a self-contained, secured, and universal format** so that authorities (police or others) can play-back the footage directly from the media supplied, without the need to install any programs on the computer from which the footage is played. This is a typical request from highly secure and closed computer systems, like police or other government agencies, where installing external and unsecured executable programs is simply not allowed. The only option such agencies will allow is to either run the VMS's proprietary format

Some USB key can be security locked

from the media itself (USB key or DVD disk) without installing anything on the computer, or export the footage in one of the common formats that can be played on some of the common video players. **In addition, in security, it is always required to be able to verify the originality of the footage.** This is a huge ask, but very important in security and surveillance, which is why an original footage is rarely stored in a format that can be decoded by common format players, edited, and its originality questioned. It usually is possible to export the footage in a common format, like AVI or MPG, but it should always be possible to verify originality before or during the exporting process. This is extremely important in legal proceedings using CCTV footage.

Some of the most popular video players, that are capable of decoding quite a few different compressions are Windows Media Player, Apple's QuickTime, Real Player, DivX, and VLC.

Windows Media, QuickTime, Real, DivX, and VLC players

There are different versions of each one of the above, and every new update may embed some more decoders, but among this variety, there is one highly portable, *free and open-source,* cross-platform media player and streaming media server, called *VLC*. VLC is produced by the French VideoLAN project group and it is universal in regards to which operating systems it can run on. This means **VLC is available on Windows platform, Linux and Mac.** Very often, when using and testing IP cameras with various compressions and various streaming capabilities, VLC can be used as a tool to discover the connectivity and properties of an IP camera.

VLC media player supports many audio and video compression methods and file formats, including MPEG-2, MPEG-4 and certainly H264 compressions. It is able to stream over computer network and to trans-code multimedia files from one compression to another. The default distribution of VLC includes a large number of free decoding and encoding libraries, avoiding the need for finding/calibrating proprietary plug-ins.

VLC, like most multimedia frameworks, has a very modular design which makes it easier to include modules/plugins for new file formats, codec, or streaming methods. The VLC version 1.0.0 had more than 380 modules.

Because VLC is a packet-based media player, it can play the video content of some damaged, incomplete, or unfinished videos. It also plays MPEG transport streams (.TS) files while they are still being digitized from an HDV camera via a FireWire

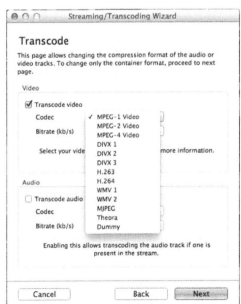

With VLC transcoding is also possible

cable, making it possible to monitor the video as it is being played. The player can also use to access *.iso* files so that users can play files on a disk image, even if the user's operating system cannot work directly with *.iso* images.

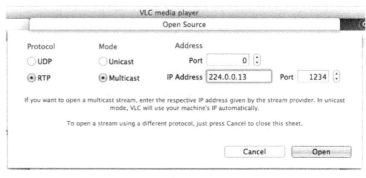

Using a FireWire or LAN connection from cable boxes to computers, VLC can stream live, un-encrypted content to a monitor or HDTV. VLC media player can display

VLC can be used for testing various streaming

the playing video as the desktop wallpaper, by using DirectX, only available on Windows operating systems. VLC media player can create screen-casts and record the desktop. **VLC can be installed or run directly from a USB flash drive or other external drive.**

It needs to be re-iterated that all VMS decoding, such as the media players above, are done by the computer's CPU using the software code contained within the player/decoder. We call this software decoding, as opposed to the hardware one, where dedicated decoder chips do nothing but decode one stream. **Software decoding may be very processing-intensive, especially when more complex video compressions are used, like H.264, and even more the new H.265.**

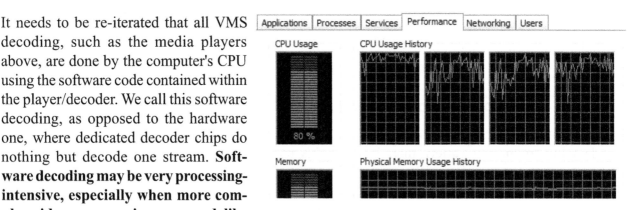

A very busy 4-core CPU usage diagram

This processing takes a large chunk of the CPU cycles, and the more streams/cameras are used, the more cycles are used by the CPU. There is a limit on the maximum number of streams that can be decoded by one computer. As already mentioned earlier under CPU heading, a rough rule of thumb today is that a HD stream encoded with H.264 requires one CPU core for smooth decoding. From this simple rule, it can easily be seen that having more than 4 HD split-screen streams may be more damaging then useful. A very busy CPU gets hot, and then becomes unstable running concurrent applications in the background, which, considering the weaknesses of Windows OS, may eventually lead to crashes.

A simple and useful suggestion for system designers is not to have too many cameras stream onto one workstation PC. Not only it would be difficult for the CPU to cope with large number of streams, but operators themselves cannot focus on more than a couple of cameras. When more cameras need to be monitored a better solution would be to split the streams to multiple workstations.

Even more efficient solution is to have a workstation with full control and minimal number of streams (up to 4) on the PC, but the rest of the cameras of interest to be decoded by dedicated hardware decoders, and have them configured as independent monitors, the same way we did in the good old analog matrix days.

ONVIF and PSIA

The variety of IP cameras and VMS on the CCTV market today and the difficulty to integrate various manufacturers into one system, which was extremely easy at the time of analog standardised video, has prompted a few reputable manufacturers to get together and start working on the interoperability standards.

There are basically two alliances today:

- *Open Network Video Interface Forum* (ONVIF, more info on their website, *www.onvif.org*)

- *Physical Security Interoperability Alliance*(PSIA, more info on their website, *www.psialliance.org*).

Driving network video through global standardization

When these movements started in 2008 there were only a handful of member companies, mostly the initiators. Five years later, there is almost no CCTV company that is not a member of one or both.

Both of these alliances are proposing standardized network interfaces and communications protocols between different CCTV IP network video devices. Most often this refers to cameras and recording equipment, but also control platforms and VMSs. **ONVIF and PSIA do not, and cannot, propose universal encoding and decoding compressions, as this is typically up to the codec chip manufacturers whose products are used in various cameras and encoders.** ONVIF and PSIA's role is like a translator and mediator between two different language speakers who may have different talents and interests.

Today, it seems that ONVIF has more members and a stronger voice in CCTV, but PSIA deserves credit for their efforts too. It is important to note that these proposals for standards in the communication and interoperability of CCTV IP devices are still developing and not finished at the time of writing this book.

Many products compliant with ONVIF and/or PSIA have been developed already and tested their compatibility with others. Typically, if a camera for example complies with one or the other alliance recommendations, it would be said, for example, "ONVIF compliant."

The key purpose of ONVIF, as stated in their original document, is that it aims to standardize the network interface (on the network layer) of network video products.

The ONVIF core specification ver. 1.0 defines a network video communication framework based on relevant IETF and Web Services standards including security and IP configuration requirements.

The following areas are covered by the core specification version 1.0:

- IP configuration

- Device discovery

- Device management

- Media configuration

- Real-time viewing

- Event handling

- PTZ control

- Video analytics

- Security

- Search for recorded information

- Replay and export of recordings

- Network video receiver

Overall, Web Services offer the most suitable technology to guarantee interoperability. The technology also allows easy and fast integration thanks to source code generation through the standardized WSDL. In addition, several rich and well-tested frameworks already exist in Web Services. ONVIF does not see any other technology on the market providing such a broad device support capabilities as Web Services.

Web Services relieves manufacturers and developers from interpreting the interface as it is based on source code generation. The generated code provides built-in data type conformance and guarantees interoperability to the interface, thus ensuring conformance to the interface specification. The generated code for the interface is always the same, thus eliminating the risk for misinterpretations.

ONVIF has chosen Web Services for the interface and utilizes other well-established standards when they are more appropriate, e.g. RTP/RTSP for streaming, motion detection and audio. ONVIF also relies on compression standards like H.264.

In version 2.0 of the ONVIF specifications, the latest iteration, even PTZ commands are being worked on, with the idea of being able to control a camera from manufacturer A and software from manufacturer B.

The PSIA alliance suggests working with the similar principles of Web Services, and this what they say in their original document:

Security and/or network management applications require the ability to change configurations and control the behaviors of IP media devices – cameras, encoders, decoders, recorders, etc. This functionality can be achieved by sending a standard HTTP(S) request to the unit. The scope of this specification is to define all HTTP(S) application programming interfaces (APIs) for media devices and their functionality; namely, for setting/retrieving various configurations, and controlling device behaviors.

In the latest version of PSIA specifications, the Video Analysis is also taken care of, and PSIA suggests protocols to make use of the video analysis coming from a camera or recording device.

It may take some time until all of these suggestions, both by ONVIF and/or PSIA, are implemented in all devices, but there is a momentum in the industry to have this accepted and implemented, as it makes system design and usage much easier and much more practical.

To make life easier for users, ONVIF has introduced *profile* categorization to their specifications.

For example, **an ONVIF device compliant to the Profile S is an ONVIF device that sends video data over an IP network to a client. The Profile S also includes support for PTZ, audio and metadata streaming, and relay outputs if those features are present on the device. So, a device compliant to the Profile S may be an IP network camera or an encoder device.**

An ONVIF client compliant to the Profile S is an ONVIF client that can configure, request, and control streaming of video data over an IP network from an ONVIF device compliant to the Profile S. The Profile S also includes support for control of PTZ, receiving audio and meta-data stream, and relay outputs if those features are supported by the client.

An ONVIF profile is described by a fixed set of functionalities through a number of services that are provided by the ONVIF standard. A number of services and functionalities are mandatory for each type of ONVIF profile. An ONVIF device and client may support any combination of profiles and other optional services and functionalities.

Profile C deals with access control, and the new Profile G deals with recording and storage functionality.

It is interesting to mention that in both ONVIF and PSIA, the video streaming protocol of choice is the RTSP (Real Time Streaming Protocol).

10. Transmission media

Once the image has been captured by a lens and a camera and then converted into an electrical signal, it is further taken to a switcher, a decoder, a recording device, or a monitor display. In order for the video signal, audio and control data to get from point A to point B, it has to go through some kind of transmission medium. This applies to both analog and digital signals.

The most common media for video and data transmission in CCTV are as follows:

- Coaxial cable (analog and digital)

- Twisted pair cable (analog and digital)

- Microwave link (analog and digital)

- RF open-air transmission (analog and digital)

- Infrared link (analog and digital)

- Telephone line (analog and digital)

- Fiber optics cable (analog and digital)

- Network (digital)

For analog video transmission, a coaxial cable is most often used, but twisted pair also, and fiber optics is becoming increasingly popular. Mixed means of transmission are also possible, such as video via microwave and PTZ control data via twisted pair, for example.

We will go through all of them separately, but we will pay special attention in this chapter to the coaxial cable and fiber optics transmission. Although the original chapter on transmission in the very first edition of this book (nearly 18 years ago now) was referring to analog signals only, all of these media can be used for digital (IP) transmission as well.

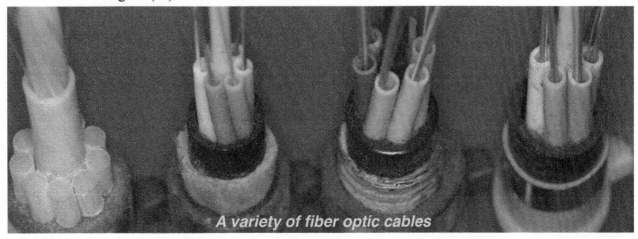
A variety of fiber optic cables

Coaxial cables

The concept

The ***coaxial cable*** is the most common medium for transmission of video signals and sometimes video and PTZ data together. It is also known as ***unbalanced*** transmission, which comes from the concept of the coaxial cable (sometimes called "***coax***" for short).

A cross section of a coax is shown to the right. It is of a symmetrical and coaxial construction. The video signal travels through the center core, while the shield is used to common the ground potential of the end devices – the camera and the monitor, for example. It not only commons the ground potential, but also serves to protect the center core from external and unwanted ***electromagnetic interference*** (EMI).

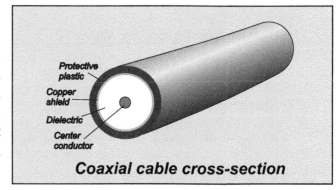

Coaxial cable cross-section

The idea behind the coaxial concept is to have all the unwanted EMI induced in the shield only. When this is properly grounded, it will discharge the induced noise through the grounds at the camera and monitor ends. Electrically, the coaxial cable closes the circuit between the source and the receiver, where the coax core is the signal wire, while the shield is the grounding one. This is why it is called an unbalanced transmission.

Most of the analog CCTV system would use coaxial cables for distances of up to a couple of hundred meters, before using an amplifier. Coaxial cables are used today in the broadcast industry for digital serial signals (SDI) as well, although maximum distances are shorter for HD signals (1.5 Gb/s or 3 BGb/s).

In the early days of networking, coax was used for the first 10Base2 Ethernet 10 Mb/s networks.

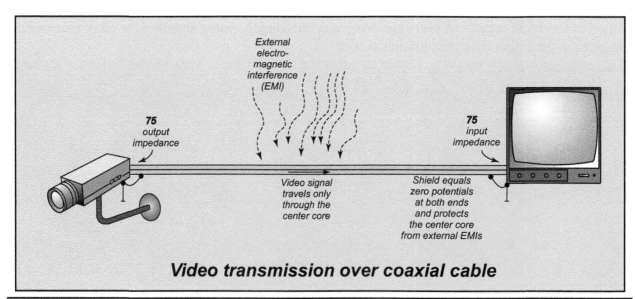

Video transmission over coaxial cable

Noise and electromagnetic interference

How well the coax shield protects the center core from noise and EMI depends on the percentage of the screening. Typically, numbers between 90 and 99% can be found in the cable manufacturer's specifications. Have in mind, however, even if the screening is 100%, that it is not possible to have 100% protection from external interference. The penetration of EMI inside the coax depends on the frequency.

Theoretically, only frequencies above 50 kHz are successfully suppressed, and this is due mostly to the skin-effect attenuation. All frequencies below this will induce current in smaller or bigger form. The strength of this current depends on the strength of the magnetic field. Our major concern would be, obviously, the mains frequency (50 or 60 Hz) radiation, which is present around almost all man-made objects.

This is why we could have problems running a coaxial cable parallel to the mains. The amount of induced electromagnetic voltage in the center core depends first on the amount of current flowing through the mains cable, which obviously depends on the current consumption on that line. Second, it depends on how far the coax is from the mains cable. And last, it depends on how long the cables run together. Sometimes 100 m might have no influence, but if strong current is flowing through the mains cable, even a 50 m run could have a major influence. When installing, try (whenever possible) not to have the power cables and the coaxial cables very close to each other; at least 30 cm (1 ft) would be sufficient to notably reduce the EMI.

The appearance of a "Ground loop"

The visual appearance of the induced (unwanted) mains frequency on analog video signal is a few thick horizontal bars slowly scrolling either up or down. The scrolling frequency is determined by the difference between the video field frequency and the mains frequency and can be anything from 0 to 1 Hz. This results in stationary or very slow-moving bars on the screen.

Other frequencies will be seen as various noise patterns, depending on the source. A rule of thumb is that the higher the frequency of the induced unwanted signal, the finer the pattern on the monitor will be. Intermittent inducting, like lightning or cars passing by, will be shown as an irregular noise pattern.

Characteristic impedance

Short wires and cables used in an average electronic piece of equipment have negligible resistance, inductance, and capacitance, and they do not affect the signal distribution. If a signal, however, needs to be transmitted for a longer distance, many factors add up and contribute to the complex picture of such transmission media. This especially influences high-frequency signals. Then, the resistance, inductance, and capacitance play a considerable role and visibly affect the transmission.

A simple medium like the coaxial cable, when analyzed by the electromagnetic theory, is approximated with a network of resistors (R), inductors (L), capacitors (C), and conductors (G) per unit length (as shown on the diagram on the previous page). For short cable runs this network has a negligible influence on the signal, but for longer runs it becomes noticeable. In such a case the network of R, L, and C elements becomes so significant that it acts as a crude low-pass filter that, in turn, affects the amplitude and phase of the various components in the video signal. The higher the frequencies of the signal, the more they are affected by these non-ideal cable properties.

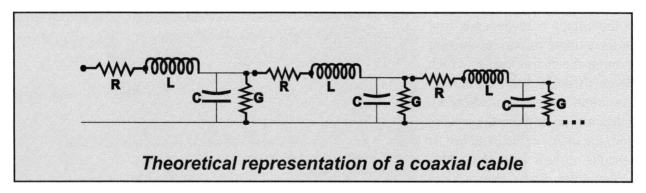

Theoretical representation of a coaxial cable

Each cable is **uniformly** built and has its own *characteristic impedance*, which is defined by the R, L, C, and G per unit length.

The main advantage of the unbalanced video transmission (which will be shown a little bit later) is based on the fact that **the characteristic impedance of the medium is independent of the frequency** (refers mainly to the mid and high frequencies), while the phase shift is proportional to the frequency.

The amplitude and phase characteristics of the coax at low frequencies is very dependent on the frequency itself, but since the cable length in such cases is reasonably short compared to the signal wavelength, it results in negligible influence on the signal transmission.

When the characteristic impedance of the coaxial cable is matched to the video source output impedance and the receiving unit input impedance, it **allows for a maximum energy transfer between the source and the receiver.**

For high-frequency signals, as is the video, impedance matching is of paramount importance. **When the impedance is not matched, the whole or part of the video signal is reflected back to the source, affecting not only the output stage itself, but also the picture quality**. A 100% reflection of the signal occurs when the end of the cable is either short circuited or left open. The total (100%) energy of the signal (voltage × current) is transferred only when there is a match between the source, transmission media, and the receiver. This is why we insist that **the last element in the video signal chain should always be terminated with 75 Ω** (the symbol Ω stands for ohms).

In CCTV, 75 Ω is taken as a characteristic impedance for all the equipment producing or receiving video signals. This is why the coaxial cable is meant to be used with 75 Ω impedance. This does not exclude manufacturers producing, say, 50 Ω equipment (which used to be the case with some RF and Ethernet standards), but then impedance converters (passive or active) need to be used between such sources and 75 Ω recipients.

Impedance matching is also done with the twisted pair transmitters and receivers, which will be discussed later in this chapter.

The 75 Ω of the coax **is a complex impedance, defined by the voltage/ current ratio at each point of the cable. It is not a pure resistance, and therefore it cannot be measured with an ordinary multi-meter.**

To calculate the characteristic impedance, we will make use of the electromagnetic theory as mentioned earlier and we will

Coaxial cable braiding machine

represent the cable with its equivalent network, composed of R, L, C, and G per unit length. This network, as shown on the schematic diagram previously, has an impedance of:

$$Zc = \sqrt{\frac{R + j\omega L}{G + j\omega C}}$$

(48)

where, as already explained, R is the resistance, L is the inductance, G is the conductance, and C is the capacitance between the center core and the shield, per unit length. The symbol j represents the imaginary unit (square root of -1), which is used when representing complex impedance, $\omega = 2\pi f$, where f is the frequency.

If the coaxial cable is of a reasonably short length (less than a couple of hundred meters), R and G can be ignored, which brings us to the simplified formula for the coax impedance:

$$Zc = \sqrt{\frac{L}{C}}$$

(49)

This formula simply means that the **characteristic impedance does not depend on the cable length and frequency but on the capacitance and inductance per unit length**. This is not true, however, when the length of a cable like RG-59/U exceeds a couple of hundred meters. The resistance and the capacitance then become significant, and they **do affect** the video signal. **For reasonably short lengths, however, the above approximation is pretty good.**

The cable limitations we have are mainly a result of the accumulated resistance and capacitance, which are so high that the approximation (49) is no longer valid and the signal is distorted considerably. This is basically in the form of voltage drop, high-frequency loss, and group delay.

The most commonly used coaxial cable in CCTV is the RG-59/U, which can successfully and without in-line correctors, transfer analog B/W signals up to 300 m and analog color up to 200 m.

This difference in maximum distance between B/W and color signals are due to the way Composite Video Burst Signal (CVBS) embeds the color carrier at higher frequencies (4.43 MHz for PAL and 3.58 MHz for NTSC). Since a long coaxial cable acts as a low-pass filter, the color information will obviously be affected sooner than the lower frequencies, so the loss of color information will happen before the loss of details in the lower frequencies. Eventually color is lost before the signal itself, which is the reason for stating shorter maximum cable lengths for color signals.

The other popular cable is the RG-11/U, which is thicker and more expensive. Its maximum recommended lengths are up to 600 m for a B/W signal and 400 m for a color signal. There are also thinner coaxial cables with 75 Ω impedance, like RG-179, with only 2.5 mm diameter or even coax ribbon cables. They are very practical for crowded areas with many video signals, such as matrix switchers with many inputs. Their maximum cable run is much shorter than the thicker representatives, but sufficient for links and patches. Note that these numbers may vary with different manufacturers and signal quality expectations.

Miniature coaxial cable can save a lot of space and improve accessibility.

If longer runs are required, additional devices

can be used to equalize and amplify the video spectrum. Such devices are known as **in-line amplifiers, cable equalizers, or cable correctors**. Depending on the amplifier (and cable) quality, double or even triple lengths are possible.

In-line amplifiers are best if they are used in the middle of the cable run because of the more acceptable S/N ratio, but this is quite often impossible or impractical

Physical size comparison between RG-59 and RG-11 coax

due to the need for power supply and storage. So, the majority of in-line amplifiers available in CCTV are designed to be used at the camera end, in which case we actually have ***pre-equalization*** and ***pre-amplification*** of the video signal. There are, however, devices that are used at the monitor end, and they have $1V_{pp}$ output with ***post-equalization*** of the video bandwidth.

Starting from the above theoretical explanation of the impedance, it can be seen that the cable uniformity along its length is of great importance for fulfilling the characteristic impedance requirements. **The cable quality depends on the precision and uniformity of the center core, the dielectric, and the shield. These factors define the C and L values of the cable, per unit length. This is why careful attention should be paid to the running of the cable itself and its termination.** Sharp loops and bends affect the cable uniformity and consequently the cable impedance. This results in high-frequency losses (i.e., fine picture detail loss), as well as double images due to

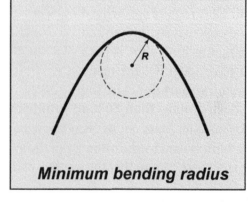

Minimum bending radius

signal reflections. So, if a short and good-quality cable is improperly run, with sharp bends and kicks, the picture quality will still be far from perfect.

Bends no smaller than 10 times the diameter of the coax are suggested for best performance. This is the equivalent of saying **"bending radius should not be smaller than 5 times the diameter, or 10 times the radius of the cable."** This means an RG-59/U cable should not be bent in a loop with a diameter smaller than 6 cm (2.5"), and an RG-11/U should not be bent in a loop smaller than 10 cm (4") in diameter.

Good quality coaxial cable

Copper is one of the best conductors for a coaxial cable. Only gold and silver will show a better performance (related to resistance and corrosion), but these are too expensive to be used for cable manufacturing. A lot of people believe that copper-plated steel makes a better cable, but this is not correct. Copper-plated steel can only be cheaper and perhaps stiffer, but for longer lengths, in CCTV, copper would be the better choice. Copper-plated steel coaxial cables are acceptable for master antenna (MATV) installations, where the transmitted signals are RF modulated (VHF or UHF). Namely, with higher frequencies the so-called skin effect becomes more apparent where the actual signal escapes on the copper-plated surface of the conductor (not the shield, but the center conductor). CCTV signals are, as explained, in the basic bandwidth, and this is why a copper-plated steel coaxial cable might be okay for RF signals but not necessarily for CCTV. So one should always look for a copper coaxial cable.

BNC connectors

A widely accepted coaxial cable termination, in CCTV, is the BNC termination. BNC stands for *Bayonet-Neil-Concelman* connector, named after its designers. There are three types: screwing, soldering, and crimping.

Crimping BNCs are proven to be the most reliable of all. They require specialized and expensive stripping and crimping tools, but it pays to have them. Of the many installations done in the

Male and female crimping BNC elements

industry, **more than 50% of problems are proven to be a result of bad or incorrect termination**. An installer does not have to know or understand all the equipment used in a system (which will be commissioned by the designer or the supplier), but if he or she does proper cable runs and terminations, it is almost certain that the system will perform at its best.

There are various BNC products available on the market, of which the male plug is the most common. Female plugs are also available, as well as right angle adaptors, BNC-to-BNC adaptors (often called "barrels"), 75 Ω terminators (or "dummy loads"), BNC-to-other-type of video connection, and so on.

Breaking the cable in the middle of its length and terminating it will add to the losses of the signal, especially if the termination and/or BNCs are of a bad quality. A good termination can result in as small as 0.3 to 0.5 dB losses. If there are not too many of them, this is an insignificant loss.

Various BNC connectors and adaptors

There are silver-plated and even gold-plated BNC connectors designed to minimize the contact resistance and protect the connector from oxidation, which is especially critical near the coast (salt water and air) or heavily industrialized areas.

A good BNC connector kit should include a gold-plated or silver-plated center tip, a BNC shell body, a ring for crimping the shield, and a rubber sleeve (sometimes called a "strain relief boot") to protect the connector's end from sharp bends and oxidation.

Coaxial cables and proper BNC termination

Never terminate a coaxial cable with electrical cutters or pliers. Stripping the coaxial cable to the required length using electrical cutters is very risky. First, small pieces of copper fall around the center core, and one can never be sure that a short circuit will not occur. Also, the impedance changes even if they do not short circuit the core and the shield. Second, using normal pliers for fixing the BNC to the coaxial cable is never reliable. All in all, these are very risky tools to terminate crimping BNCs, and they should only be used when no other tools are available (remember to always take utmost care when using them).

If you are an installer, or a CCTV technician who regularly terminates coaxial cables, get yourself a proper set of tools. These are precise cutters, a stripping tool, and a crimping tool.

Make sure you have the crimping and stripping tools

Samples of bad BNC terminations

for the right cable. If you are using RG-59/U (overall diameter 6.15 mm) do not get it confused with RG-58/U (overall diameter 5 mm) even though they look similar. For starters, they have a different impedance, i.e., RG-59/U is 75 Ω, compared to RG-58/U which is 50 Ω. Next, RG-59/U is slightly thicker, both in the center core and the shield. There are BNC connectors for the RG-58/U which look

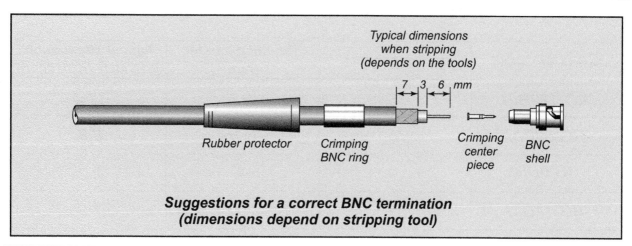

Typical dimensions
when stripping
(depends on the tools)

7 3 6 mm

Rubber protector   Crimping
                        BNC ring

Crimping
center
piece

BNC
shell

**Suggestions for a correct BNC termination
(dimensions depend on stripping tool)**

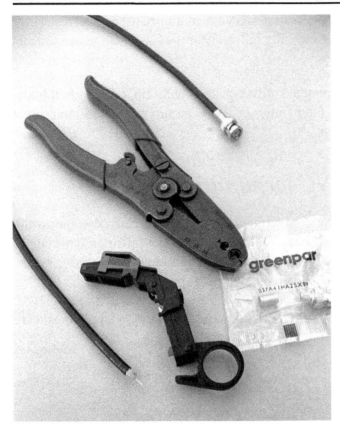

Good stripping and crimping tools

identical externally, but they are thinner on the inside.

The best thing to do is to waste one and try terminating it before proceeding with the installation. Sometimes a small difference in the cable's dimensions, even if it is RG-59/U, may cause a lot of problems fitting the connectors properly.

Technically, a solid center core coaxial cable is better, both from the impedance point of view (the cable is stiffer and preserves the "straightness") and from the termination point of view. Namely, when terminating the solid core cable, it is easier to crimp the center tip, compared to the stranded core cable which is too flexible. Some people may prefer a stranded center core coax, mainly because of its flexibility, in which case care should be taken when terminating as it is very easy to short circuit the center core and the shield because of its flexibility.

If there are no other tools available, it is best to get the soldering-type BNC connectors and to terminate the cable by soldering. Care should be taken with the soldering iron's temperature, as well as the quality of the soldering, since it can easily damage the insulation and affect the impedance. In this instance, a multistranded core coax would be better.

If you have a choice of crimping connectors look for the ones that are likely to last longer in respect to physical use and corrosion, like silver-plated or gold-plated BNCs. A good practice would be to use "rubber sleeves" (sometimes called "protective sleeves") for further protection of the interior of the BNC from corrosion and to minimize bending stress from plugging and unplugging.

A new type of BNC crimping technique became popular in the last few years, offering very quick and

Coaxial cable	Impedance (Ω)	Overall diameter (mm)	Typical attenuation @10 MHz (dB/100 m)
RG-179B/U	75	2.5	17.4
RG-59B/U	75	6.15	3.3
RG-6B/U	75	6	2.2
RG-11B/U	75	10	1.3

easy to implement crimping, which at the same time is also very durable (difficult to brake it apart by mechanical force). This is the so-called *pressure crimping* BNC termination.

In special cases, as with pan/tilt domes, there might be a need for a very thin and flexible 75 Ω coaxial cable (due to constant panning and tilting of the camera). Such cables are available from specialized cable manufacturers, but do not forget that you need special BNCs and tools for them.

Pressure crimping BNC

Even if such a cable could be as thin as 2.5 mm, as is the case with the RG-179 B/U cable, the impedance would still be 75 Ω, which is achieved by the special dielectric and center core thickness. The attenuation of such a cable is high, but when used in short runs it is negligible.

For installations where much longer runs are needed, other 75 Ω cables are used, such as RG-11B/U with an overall diameter of more than 9 mm. Needless to say, an RG-11 cable also needs special tools and BNCs for termination. Some installers use machines purposely built to strip or label coaxial cables. Although these machines are expensive and hard to find, they do exist and if you are involved in very large installations they are a worthwhile investment.

In the accompanying table below, typical attenuation figures are shown for various coax cables. Please note that the attenuation is shown in decibels and that it refers to the voltage amplitude of the video signal. If we use the decibel table shown under the section of S/N for cameras, it can be worked out that 10 dB is equivalent to attenuation of the signal to 30%, that is, 0.3 V_{PP}. The RG-59 will attenuate the signal for 10 dB after 300 m. Such low-signal amplitude may be insufficient for a monitor or VCR to lock onto. This is the point of attenuation where we would usually require an amplifier to boost the signal.

Installation techniques

Prior to installation, it should be checked what cable length can be obtained from the supplier. Rolls of approximately 300 m (1000 ft) are common, but 100 m and 500 m can also be found. Naturally, it is better to run the cable in one piece whenever possible. If for some reason the installers need a longer run than what they have, the cable can be extended by terminating both the installed and the extension cable. In such a case, although it is common practice to have a BNC plug connected to another BNC

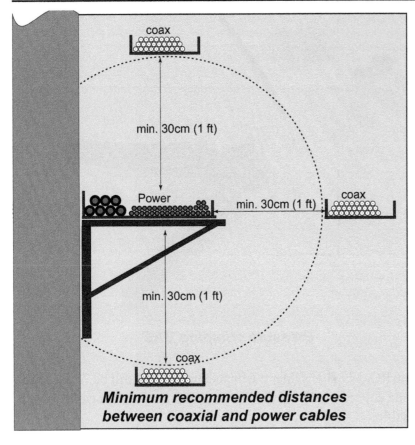

min. 30cm (1 ft)

coax

Power

min. 30cm (1 ft)

coax

min. 30cm (1 ft)

coax

***Minimum recommended distances
between coaxial and power cables***

plug with a so-called ***barrel adaptor***, it is better to minimize joining points by using one BNC plug and one socket (i.e., "male" and "female" crimping BNCs).

Before cable laying commences, the route should be inspected for possible problems such as feed-through, sharp corners, and clogged ducts. Once a viable route has been established, the cable lengths should be arranged so that any joints, or possible in-line amplifier installation, will occur at accessible positions.

At the location of a joint it is important to leave an adequate overlap of the cables so that sufficient material is available for the termination operation. Generally, the overlap required does not need to be more than 1 m.

Whenever possible, the cable should be run inside a conduit of an adequate size. Conduits are available in various lengths and diameters, depending on the number of cables and their diameters. For external cable runs, a special conduit with better UV protection is needed. In special environments, such as railway stations, special metal conduits need to be used. These are required because of the extremely high electromagnetic radiation that occurs when electric trains pass.

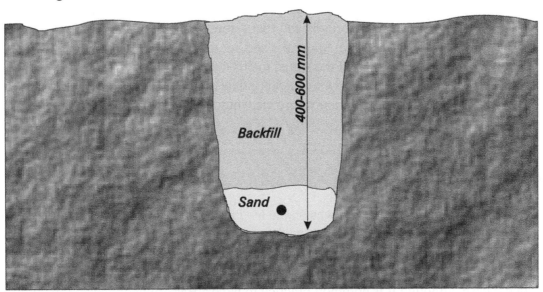

400-600 mm

Backfill

Sand

Trenching and burying recommendations

Similar treatment should be applied when a coaxial cable is run underground. When burying a cable, foremost consideration should be given to the prevention of damage due to excessive local loading points. Such loading may occur when backfill material or an uneven trench profile digs into the cable. The damage may not be obvious instantly but the picture will get distorted due to the impedance change at the point of the cable's distortion. No matter what, the cost of digging up the cable and repairing it makes the expenditure of extra effort during laying well worthwhile.

The best protection against cable damage is laying the cable on a bed of sand approximately 50 to 150 mm deep and backfilling with another 50 to 150 mm of sand. Due care needs to be exercised in the cutting of the trench so that the bottom of the trench is fairly even and free of protrusions. Similarly, when backfilling, do not allow soil with a high rock content to fall unchecked onto the sand and possibly put a rock through the cable, unless your conduit is extremely tough.

The trench depth is dependent on the type of ground being traversed as well as the load that is expected to be applied

Cabling and termination by Wegtech Services. Courtesy of Pacific Communications.

A brilliant example of very neat cabling practices

to the ground above the cable. A cable in solid rock may need a trench of only 300 mm or so, whereas a trench in soft soil crossing a road should be taken down to about 1 m. A general-purpose trench, in undemanding situations, should be 400 to 600 mm deep with 100 to 300 mm total sand bedding.

Placing a coaxial cable on cable trays and bending it around corners requires observing the same major rule: **minimum bending radius.** As mentioned, the minimum bending radius depends on the coaxial cable size, but the general rule is that **the bending radius should not be smaller than 5 times the diameter of the cable (or 10 times the radius)**. The minimum bending radius must be observed even when the cable tray does not facilitate this. The tendency to keep it neat and bend the coaxial cable to match power and data cables on the tray must be avoided. Remember, bending coax more than the minimum bending radius affects the impedance of the cable and causes a video signal quality loss.

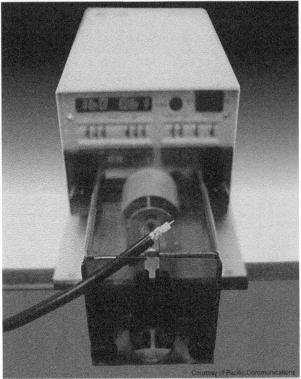

Automatic coax cable-stripping machine

The pulling of coaxial cables through ducts is performed by using a steel or plastic leader and then joining and securing all the cables that need to go through. Some new, tough plastic materials, called "snakes," are becoming more popular.

The types of cable ties normally used to tie the cables together are generally satisfactory, but remember that excessive force should not be applied, as it squashes the coax and therefore changes the impedance again.

Should a particular duct require the use of a lubricant, it is best to obtain a recommendation from the cable manufacturer. Talcum powder and bean-bag-type polystyrene beans can also be quite useful in reducing friction.

In some conditions the cable may already be terminated by connectors. These must be heavily protected while drawing the cable. The holes in such a case need to be bigger.

Between the secured points of a cable it is wise to allow a little slack rather than leaving a tightly stretched length that may respond poorly to temperature variations or vibration.

If the cable is in some way damaged during installation, then leave enough extra cable length around the damaged area so that additional BNC joiners can be inserted.

Time domain reflectometer (TDR)

When a complex and long coaxial cable installation is in question, it would be very useful to get a time domain reflectometer to help determine the location of bad cable spots.

The TDR works on a basic principle in that it inserts short and strong pulses and measures the reflected energy. By determining the delay between the injected and the reflected signals, a pretty accurate localization of bad termination points and/or sharp bends can be made.

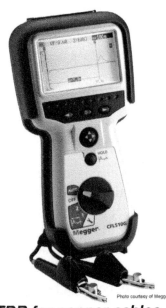

This can be especially important if the cable goes through inaccessible places.

TDR for copper cables

Twisted pair video transmission

Twisted pair cable is an alternative to the coaxial cable. It is useful in situations where runs longer than a couple of hundred meters have to be made. It is especially beneficial when only two wires have already been installed between two points.

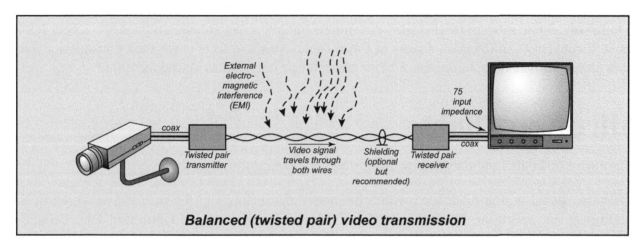

Balanced (twisted pair) video transmission

Twisted pair cable is reasonably cheap when used with normal wires, but if a proper cable (as per the recommendations by the manufacturers) is used, with at least 10 to 20 twists per meter and with shielding, the price becomes much higher.

Twisted pair transmission is also called **balanced video transmission**.

Unlike coaxial cable transmission, the twisted pair doesn't have common ground at both ends. The signal is converted into balanced mode where the signal is only the difference between the two twisted wires, not relative to ground. Unwanted electromagnetic interference and noise will eventually induce an equal amount of current in both of these wires. So, with sufficient twists per cable length, both wires will be exposed to external interference equally. When the signal arrives at the twisted pair receiver end, it comes across a differential amplifier input, with a well-balanced and good common mode rejection ratio (CMRR) factor. This differential amplifier reads the **differential signal between the two wires. The unwanted interference will be cancelled out because it will induce equal signal in both wires. The most important outcome of the twisted pair concept is that it virtually eliminates ground loops** because the ground potentials between a camera and a monitor (or DVR) cannot close a circuit (as is the case with the coaxial cable transmission).

When the CCTV industry switched to digital and networking, it was found out that Cat-6 cables are perfect for twisted pair analog video transmission. In fact, Ethernet over copper uses the same concept of common mode rejection as in analog, which is why Cat-6 cable especially is very suitable even for analog video. For many new installations this fact makes a very easy decision when running cables irrespective of whether the customer wants an analog

Twisted pair video over Cat cable

or IP system. Simply using Cat-6 makes it future proof.

Although the maximum distances with Cat-6 for 100 Mb/s Ethernet are quoted as 100 m, when used for analog video twisted pair transmission, much longer distances can be achieved than what is possible with an RG-59 or even an RG-11 cable. Manufacturers usually quote over a kilometer (1000 m) when active transmitters (powered with signal boost) are used, without any in-line repeaters. Termination of the twisted pair cable does not require special tools and connectors, as they are typically with screw-in terminals. Some modern twisted pair transmitters and receivers are already made to accept RJ-45 network connectors utilizing all 4 pairs of Cat-6 cable. All these facts make such transmission even more attractive. The output impedance of the twisted pair transmitters is usually 100 Ω.

Microwave links

Microwave links are used for good-quality wireless video transmission.

The video signal is first modulated with a frequency that belongs in the microwave region of the electromagnetic spectrum. **The wavelengths of this region are between 1 mm and 1 m**. Using the known equation between wavelength (λ) and the frequency (*f*) :

$$\lambda = c / f \qquad [\text{m}] \qquad\qquad (50)$$

where *c* is the speed of light 300,000,000 m/s, we can find out that the microwave region is between 300 MHz and 30 GHz. The upper region actually overlaps with the infrared frequencies that are defined as up to 100 GHz. Therefore, the lower part of the infrared frequency spectrum is also in the microwave region. In practice, however, the typical frequencies used for microwave video transmission are between 1 GHz and 10 GHz.

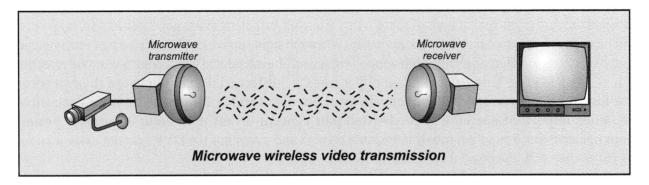

Microwave wireless video transmission

When considering other frequencies in the electromagnetic spectrum, microwaves approximately occupy the middle region of the spectrum. Typically the electromagnetic spectrum below the microwave region is occupied by TV and FM broadcasters and various UHF (ultra high frequency) satellite and communication systems. The electromagnetic spectrum above the microwave region is used for infrared systems and then gradually merges into the visible light region.

Since many services, such as the military, the police, ambulances, couriers, and aircraft radars use artificial frequencies, there is a need for some regulation of frequency. This is done on an international level by the International Communications Union (ITU) and by the local authorities in your respective country. For Australia this was the Department of Transport and Communications, which was recently renamed the ***Spectrum Management Agency***. Thus, a very important fact to consider when using microwave links in CCTV is that each frequency and microwave power needs to be approved by the local authority in order to minimize interference with the other services using the same spectrum. This is to protect the registered users from new frequencies, but it is also a downfall (at least in CCTV) for using microwaves and one of the reasons why a lot of CCTV designers turn to microwaves only as a last resort.

Microwave links transmit a very wide bandwidth of video signals as well as other data if necessary (including audio and/or PTZ control). The transmission bandwidth depends on the manufacturer's model. For a well-built unit, a 7 MHz bandwidth is typical and sufficient to send high-quality analog video signals without any visible degradation. Microwaves are usually ***unidirectional*** when a CCTV video signal is sent from point A to point B, but they can also be ***bidirectional*** when a video signal needs to be sent in both directions, or video in one and data in the other. The latter is very important if PTZ cameras are to be controlled.

The encoding technique in video transmission is usually frequency modulation (FM), but amplitude modulation (AM) can also be used. If audio and video are transmitted simultaneously, usually the video signal is AM modulated and the audio FM, as is the case with broadcast TV signals.

A line of sight is needed between the transmitter and the receiver. In most cases, the transmitting and receiving antennas are parabolic dishes, similar to those used for satellite TV reception.

The distances achievable with this technology depend on the transmitter output power and on the diameter of the antenna that contributes to the gain of the transmitter and the sensitivity of the receiver.

Like with the coax and twisted pair, microwave were initially for analog signals only, but there are now digital, Ethernet-ready, microwaves offering high bit-rates of over gigabit per second, and achieving distances of over a few kilometers.

Obviously, atmospheric conditions will affect the signal quality. The same microwave link that has an excellent picture during a nice day may have considerable signal loss in heavy rain if it is not designed properly. Fog and snow also affect the signal. If the parabolic antenna is not anchored properly, wind may affect the links indirectly by shaking it, causing an intermittent loss of line of sight.

Modern digital microwave of over 1 Gb/s

Many parabolic antennas come with a plastic or leather cover that protects the actual inner parabola. This protector simultaneously breaks the wind force and protects the sensitive parts from rain and snow.

The fitting and stability of a microwave antenna are of paramount importance to the links. The longer the distance that is required, the bigger the antenna and more secure fittings that are required. The initial line-of-sight alignment is harder to achieve for longer distances, although better quality units have a field strength indicator built in, which helps to make the alignment easier.

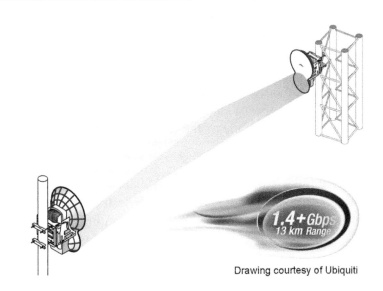

Drawing courtesy of Ubiquiti

Ubiquiti long range high bit-rate digital microwave

Maximum achievable transmitting distances of over 10 km are quoted by most specialized manufacturers. In most cases a typical CCTV application will require only a couple of hundred meters, which is often not a problem as long as there is a line of sight.

The transmitting power and the size of the antenna required for a specific distance need to be confirmed with the manufacturer.

For shorter distances, microwave links may use rod or other types of non-parabolic antennas, which become very practical if dimensions are in question. The obvious security problem in such a case would be the omnidirectional radiation of the signal, but the advantage would be a fairly wide area of coverage. When digital microwave are used high encryption is possible, which increases security.

One very interesting application that was initially developed in Australia was to use an omnidirectional microwave with a transmitting antenna fitted on top of a race car roof, which would send signals to a helicopter above the race track. From there it would be redirected to a TV broadcast van. With such a setup, the so-called "Race-Cam" allowed the television audience to see the driver's view.

In the past, the analog video microwaves would also have PTZ data channel for camera control, but today, with bidirectional digital microwaves and IP PTZ cameras, the control data goes via the Ethernet channel too.

It is important to highlight one more important difference between analog microwaves and digital. With analog, only single analog video link was possible with one pair of microwaves. With the digital bi-directional microwaves, multiple channels can be transmitted between two points, depending on the bandwidth. Such a digital connection appears to the system simply as a part of the network.

RF wireless (open air) video transmission

RF video transmission is similar to the microwave transmission in the way the modulation of signals is done. The major differences, however, are the modulation frequency in the VHF or UHF bands and the transmission of the signal (this is lower than the microwave band), which is usually omnidirectional. When a directional **Yagi** antenna is used (similar to the domestic ones used for reception of a specific channel), longer distances can be achieved and there will be less distraction to the surrounding area. It should be noted, however, that depending on the regulations of your country, the radiated power cannot be above a certain limit, after which you will require approval from your respective frequency regulatory body.

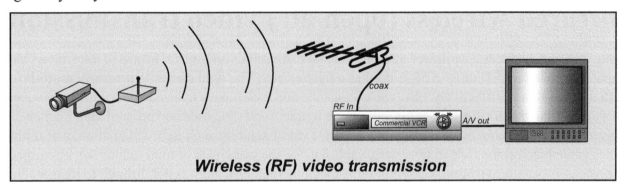

Wireless (RF) video transmission

RF transmitters are usually made with video and audio inputs and the modulation techniques are similar to those of the analog microwaves: video is AM modulated and audio is FM. The spectrum transmitted depends on the make, but generally it is narrower than the microwave. This usually means 5.6 MHz, which is sufficient to have audio and video mixed into one signal.

Consumer products with similar characteristics to those listed above in the past were found in the so-called RF senders, or wireless VCR links. The RF modulator is fed with the audio and video signals of the VCR outputs and re-modulates and then transmits them so they can be picked up by another VCR in the house. Devices like these are not made with CCTV in mind, so the distances achievable are in the vicinity of a household area. Sometimes this might be a cheap and easy way out of a situation where a short-distance wireless transmission is required.

Since VHF and UHF bands are for normal broadcast TV reception, you should check with your local authority and use channels that do not interfere with the existing broadcasting. In most countries, UHF channels 36 to 39 are deliberately not used by the TV stations because they are left for VCR-to-TV conversion, video games, and similar uses.

Wireless RF AV analog transmitters

The downfall of such an RF CCTV transmission is that any TV receiver at a reasonable distance can pick the signal up. Sometimes, however, this might be exactly what is wanted. This includes systems in big building complexes, where the main entrance cameras are injected through the MATV system so that the tenants can call the camera on a particular channel of their TV receivers.

The RF frequency is such that, when compared to the microwave links, it does not require line of sight, as the RF (depending upon whether it is UHF or VHF) can penetrate through brick walls, wood, and other nonmetal objects. How far one can go with this depends on many factors, and the best bet is to test it out in the particular environment (in which the RF transmitter will be used).

Infrared wireless (open air) video transmission

As the heading suggests, an infrared open air video transmission uses optical means to transmit a video signal. An infrared LED or LASER is used as a light carrier. The light carrier is **intensity modulated with a video signal**. Effectively, this type of transmission looks like a hybrid between microwave and fiber optics transmission. Instead of the microwave frequencies being used, the infrared light fraquencies are used (infrared frequencies are higher). And instead of sending such light modulation over a fiber optics cable (such is the case in fiber optic cables, using the principles of total reflection), open space (air, water, or vacuum) is used. Certainly line of sight is required, but it is not difficult to imagine that mirrors may be added to get around physical obstacles. **The obvious advantage of the infrared light radiation is that you do not need a special license.** Another important advantage of such a transmission is immunity to electromagnetic induction (since it is not over copper cables). Understandably, weather conditions like rain, fog, dust, and hot wind will affect infrared links more than microwave transmission.

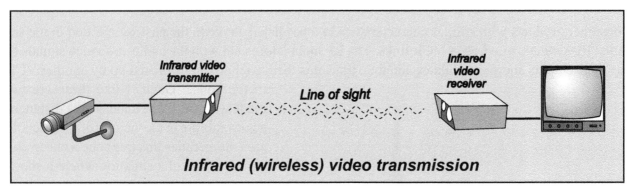

Infrared (wireless) video transmission

In order to have the infrared light concentrated into a narrow beam to minimize dispersion light losses, a lens assembly is required at the transmitter end to concentrate the light into a narrow beam and a lens assembly at the receiver end to concentrate the light onto the photosensitive detector.

Analog video, as well as audio, can be transmitted over distances of more than 1 km. Bigger lens assemblies and more powerful LEDs, as well as a more sensitive receiving end, will provide for even longer distances. In open space applications, precautions have to be taken for high temperatures around the transmitter, as the receiver may detect such infrared frequencies.

There are also some infrared transmitters specifically made for lifts, minimizing typical interference by the lift motors and lights. Video signals coming from a lift cabin is very often affected by the power cables running to the lift itself, but also they are prone to physical wear and tear. Using infrared optical link to transmit the video signal from the lift camera to the outside world eliminates such problems.

And finally, like the digital counterparts with the other analog transmission media, there are digital infra red links too, capable of integrating networks, wirelessly with infrared links. Some are claimed to be capable of delivering even up to 100 Mb/s over 5 km.

IR transmitter   **IR links for lift**

Transmission of images over telephone lines

First there was the "***slow-scan TV***." That was a system that would send video pictures over a telephone line at a very slow speed, usually many tens of seconds for a full-frame B/W picture. Then came up the "***fast scan***" which was a popular alternative to the "slow-scan."

Today, at the time of preparing the third edition of this book, "slow-scan" or "fast-scan" are no longer available. Everything is lightning fast with Asymetric Digital Subscriber Lines (ADSL) when compared to the old PSTN technologies. ADSL connections are also over the telephone copper lines, but they use different technology. We are mentioning the transmission over telephone lines of the old style here, for the sake of being historically correct and complete. But in fact most of today's transmission over telephone lines are in fact a part of the Internet connection over ADSL. Most businesses and households have fast Internet connection, usually using their existing telephone copper lines which offers much faster transmission than the "fast scan."

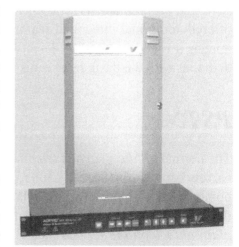

The slow-scan concept originates from the late 1950s, when some amateur radio operators used it. It was later applied to CCTV. The concept is very simple: There are units at both ends of the transmission path, as with any other transmission, a transmitter and a receiver. An analog video signal of a camera is captured and converted into digital format. It is then stored in the random

Fast-scan transmitter and receiver unit

access memory (RAM) of the slow-scan transmitter. This was usually triggered by an external alarm or upon the receiver's request. The stored image, which is at this stage in a digital format, is usually frequency modulated with an audio frequency that can be heard by the receiving phone. This frequency was usually between 1 and 2 kHz, i.e., where the phone line attenuation is lowest. When the receiver receives the signal, it reassembles the picture line by line, starting from the top left-hand corner until the picture at the receiving end is converted into an analog display (a steady picture). This concept was initially very slow, but considering the unlimited distances offered by telephone lines (provided there was a transmitter compatible with the receiver), it was an attractive concept for remote CCTV monitoring.

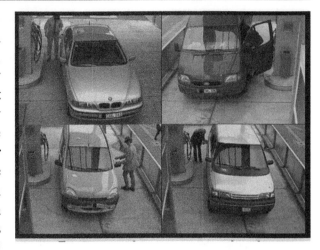

Fast-scan images are often transmitted in quad mode for faster updates

One way of increasing the speed of transmission was to reduce the digitized picture resolution or to use only a quarter of the screen for each camera. So the initial 32 seconds could effectively be reduced to 8 seconds for when one picture update was required, or perhaps, have a 32-second update of a quad screen with blocks of four cameras. Considering that other signals could be added to this, like audio or control signals for remote relay activation, a better picture can be attained for these historic beginnings.

Older generation slow-scan systems would take 32 seconds to send a single low-quality picture from an alarmed site to the monitoring station. Dial-up and connecting time should be added to this, totaling to more than a minute for the completion of the first image transmission. The slow-scan, however, was very popular and ahead of its time. Today we have much more advanced techniques when video signals need to be transmitted over the telephone line.

The more sophisticated fast-scan systems use a method of image updating called ***conditional refresh***. After the initial image is sent, only the portion that changes needs to be resent. A more clever system offered additional integrated feature: video motion detection. The system would automatically send images if motion activity is detected in the video signal. This allowed a much more rapid update rate than that with the basic fast-scan systems.

PSTN

The normal ***public switched telephone network*** (PSTN) line has a narrow bandwidth of usually 300 to 3000 Hz, (measured at 3-dBm points, where dBm is measured relative to 1 mW across 600 ohms, the telephone line impedance). Some people call this type of line ***plain old telephone service*** (POTS). PSTN is analog technology. As such, it is never constant with the bit rate it offers, as it very much depends on the line distance to the switch and the noise.

A plain old telephone

Theoretically, it is impossible to send live video images of 5 MHz over such a narrow channel. It is possible, however, to compress and encode the signal to achieve faster transmission, and this can be done today by most of the fast-scan transmitters. The technological explosion of the PCs, compression algorithms, fast modems, and better telephone lines in the last few years has made it possible to transmit video images over telephone lines at rates unimaginable when the first slow-scan transmitters were developed.

As mentioned earlier, the concept remained the same as that of the slow scan, but the intelligence behind the compression schemes (what and how it transmits) has improved so much that **a color video signal with very good resolution could be sent in less than 1 s per frame**.

In order to understand the PSTN telephone line video transmission rates, let us consider this simplified exercise:

A typical B/W video signal with 256×256 pixels resolution will have $256 \times 256 = 65,536$ bits of information, which is equal to 64 kB of digital information (65,536/1024). (Note the "digital" numbers $64 = 2^6$, $256 = 2^8$.)

To send this amount of uncompressed information over a telephone line using a normal 2400 bits per second modem (as was the case in the early days of slow scan), it would take about 27 s (65,536/2,400).

If the signal is compressed, however (compressions of 10, 20, or even more times are available), by say 10 times, this gets reduced to 3 s. Most fast-scan transmitters will only send the first image at this speed, after which they only send the difference in the pictures, thus dramatically reducing the subsequent images' update time to less than a second.

A color picture with the same resolution will obviously require more. A high-resolution picture of quality better than S-VHS is usually digitized in a 512×512 frame with 24-bit colors (8 bits of each R, G, and B) and will equal $512 \times 512 \times 3 = 786,432$ bytes, or 768 kB. If this is compressed by 10 times, it becomes 76 kB, which is not that hard to transmit with a 14,400 bps modem at approximately $76,000/14,400 = 5$ s. It all depends on the compression algorithm.

In practice, add another few seconds to the dialing time, which is faster with DTMF (dual tone multi-frequency) and slower with pulse dialing lines.

Most of the high security systems have a dedicated telephone line, which means that once the line is established it stays open and there is no further time loss for the modem's handshake and initial picture update delay.

In the end, we should emphasize that the theoretical maximum speed of transmission that can be achieved with a POTS or PSTN line is somewhere around 56 kb/s. In practice, it is rarely over 32 kb/s, and if the telephone line is old, or far from the exchange, it could be as low as 19 kb/s.

Analog modem

ISDN

ISDN lines were proposed and started to be implemented in the mid-1970s, almost at the same time when CCD chips appeared. Today, ISDN is already obsolete, but it is important to highlight the historical importance of ISDN because ISDN was actually the first digital communication network which opened doors of the new digital world we are enjoying today.

There was a basic ISDN channel which offered a bit-rate of 64 kb/s, which dramatically improved the speed of analog PSTN. The ISDN was truly a **digital network and transmitted signals in digital format**; hence the bandwidth was not given in Hz but in b/s. For special purposes, like video conferencing and cable TV (available via telephone lines), ISDN was combined (agregated) into dual-ISDN channels on bundled multiple 64 kb/s channels, called **broadband ISDN** (B-ISDN) links,

ISDN modem

where higher speed rates (multiples of 64 kb/s) could be achieved by intelligent multiplexing of more channels into one.

The unit used to connect a device to an ISDN line was usually called the **Terminal Adaptor** (TA); the function, as well as the appearance of such a device, is very similar to a modem with PSTN lines. Intelligent TA for B-ISDN links were also known as Aggregating Terminal Adaptors.

Cellular network

Transmitting images over mobile phones was an attractive possibility from the introduction of the mobile (cellular) phones technology. With the Smart phones today, this is not only possible, but expected from all major players in our industry. Modern smart phones are in essence pocket computers, powerful enough to perform video decoding and even local recording. With the always evolving wireless technology such as 3G (third generation of mobile telecommunication technology), and now even 4G, wireless data streaming is incredibly faster then in the good old days of analog PSTN modems. The theoretical maximum speed of 3G is several Mb/s, while 4G can go even up to 100 Mb/s.

Many IP CCTV manufacturer have come up with applications for both major categories of smart phones (Apple's iOS and Google's Android) which offer remote monitoring and control of their IP systems. The quality and speed of such connection depends on many factors, but clearly a digital data connectivity (Wi-Fi, 3G, or 4G) is required, and, depending on the reception area, it could vary from satisfactorily fast to slow.

A smartphone CCTV App

Fiber optics

Fiber optics, if correctly installed and terminated, are the best quality and most secure form of transmission. Even though they have been used in long distance telecommunications, even across oceans, for over 30 years, fiber optics have been neglected or avoided in CCTV.

The main excuse installers have used was the fear of unknown technology, often labeled as "touchy and sensitive" and also considered "too expensive."

Fiber optics, however, offers many important advantages over other media and although it used to be very expensive and complicated to terminate, it is now becoming cheaper and simpler to install.

The most important advantages of all are immunity to electromagnetic interference, more secure transmission, wider bandwidth, and much longer distances without amplification. We will, therefore, devote more space to it.

Why fiber?

Fiber optics uses light as a carrier of information, be it analog or digital. This light is usually infrared, and the transmission medium is fiber.

Fiber optic signal transmission offers many advantages over the existing metallic links. These are:

Fiber cables are tiny and fragile but enclosed in tough jackets

• It provides very wide bandwidth.

• It achieves very low attenuation, on the order of 1.5 dB/km compared to over 30 dB/km for RG-59 coax (relative to 10 MHz signal).

• The fiber (which is dielectric) offers electrical (galvanic) isolation between the transmitting and receiving end; therefore, no ground loops are possible.

• Light used as a carrier of the signal travels entirely within the fiber. Therefore, it causes no interference with the adjacent wires or other optical fibers.

• The fiber is immune to nearby signals and electromagnetic interferences (EMI);

therefore, it is irrelevant whether the fiber optics passes next to a 110 VAC, 240 VAC, or 10,000 VAC, or whether it is close to a megawatt transmitter. Even more important, lightning cannot induce any voltage even if it hits a centimeter from the fiber cable.

• A fiber-optic cable is very small and light in weight.

• It is impossible to tap into the fiber-optic cable without physically intercepting the signal, in which case it would be detected at the receiving end. This is especially important for security systems.

• Fiber is becoming less expensive every day. A basic fiber optics cable costs anywhere from $1 to $5 per meter, depending on the specific type and construction used.

Fiber optics also have some not so attractive features, but they are being improved:

• Termination of fiber optics requires special tools and better precision of workmanship than with any other media.

• Switching and routing of fiber-optic signals are difficult.

Fiber optics offer more advantages than other cables.

Fiber optics have been used in telecommunications for many years; they are becoming more popular in CCTV and security in general.

As the technology for terminating and splicing fiber improves and at the same time gets cheaper, there will be more CCTV and security systems with fiber optics.

The concept

The concept of fiber optics lies in the fundamentals of light refraction and reflection.

To some it may seem impossible that a perfectly clear fiber can constrain the light rays to stay within the fiber as they travel many kilometers, and yet not have these rays exit through the walls along the trip. In order to understand this effect, we have to refresh our memory about the physical principle of total reflection.

Physicist Willebrord Snell laid down the principles of refraction and reflection in the early seventeenth century. When light enters a denser medium, not only does the speed reduce, but the direction of travel is also corrected in order for the light to preserve the wave nature of propagation (see Chapter 3). Basically, the manifestation of this is a light ray sharply bent when entering different media. We have all seen the "*broken straw effect*" in a glass of water. That is refraction.

A typical glass has an index of refraction of approximately 1.5. The higher the index, the slower the speed of light will be, thus the bigger the angle of refraction when the ray enters the surface.

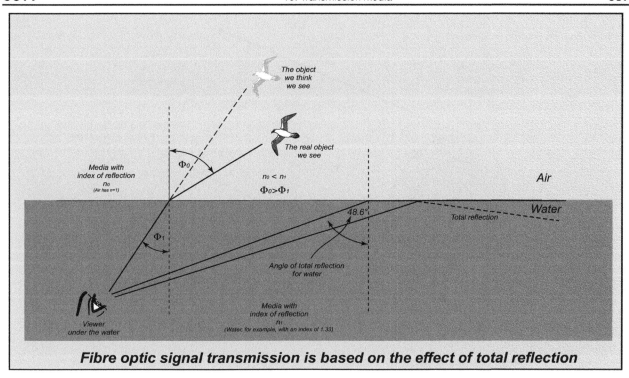

Fibre optic signal transmission is based on the effect of total reflection

The beauty of a diamond comes primarily from the rainbow of colors we see due to its high index of refraction (2.42). This is explained by the fact that a ray of light (natural light) has all the colors (wavelengths) a white light is composed of.

Fiber optics uses a special effect of refraction under a maximum incident angle; hence, it becomes a **total reflection**. This phenomenon occurs at a certain angle when a light ray **exits** from a dense medium to a sparser medium.

The accompanying drawing shows the effect of a diver viewing the sky from under the water. There is an angle below which he can no further see above the water surface. This angle is called the **angle of total reflection**. Beyond that point he will actually see the objects inside the water, and it will seem to him like looking through a mirror (assuming the water surface is perfectly still).

For the index of refraction of water (1.33), using Snell's Law, we can calculate this angle:

$$\sin \Phi_T = 1.00/1.33 = 0.752 \;\rightarrow\; \Phi_T = 48.6° \tag{51}$$

The concept of fiber-optic transmission follows the very same principles.

The core of a fiber-optics cable has an index of refraction higher than the index of the cladding. Thus, when a light ray travels inside the core it cannot escape it because of the total reflection.

So, what we have at the fiber optics transmitting end is an LED (*light-emitting diode*) or LD (*laser diode*) that is modulated with the transmitted signal.

In the case of CCTV the signal will be video, but similar logic applies when the signal is digital, like a PTZ control, network, or other security data. So, when transmitting, the infrared diode is **intensity**

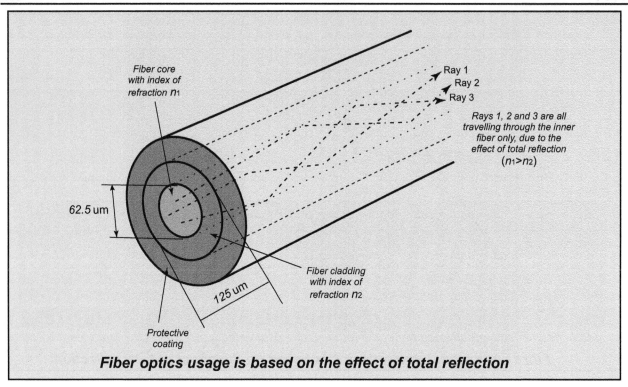

Fiber optics usage is based on the effect of total reflection

modulated and pulsates with the signal variations. At the receiving end, we have basically a photo-detector that receives the optical signal and converts it into electrical.

Fiber optics used to be very expensive and hard to terminate, but that is no longer the case, because the technology has improved substantially. Optical technology has long been known to have many potential capabilities, but major advancements are achieved when mass production of cheap fundamental devices like semiconductor light-emitting diodes, lasers, and optical fibers are made.

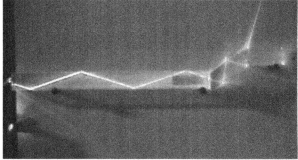

Laser light total reflection inside a fiber channel

Nowadays, we are witnessing a conversion of most terrestrial hard-wired copper links to fiber.

Types of optical fibers

There are a few different types of fiber optics cables. This division is based on the path light waves take through the fiber.

As mentioned in the introduction, the basic idea is to use the total reflection effect that is a result of the different indices of refraction ($n_2 > n_1$) where n_2 is the index of the internal (***core***) fiber and n_1 is the index of the outer (***cladding***) fiber.

A typical representation of what we have just described is the ***step index*** fiber optics cable. The index

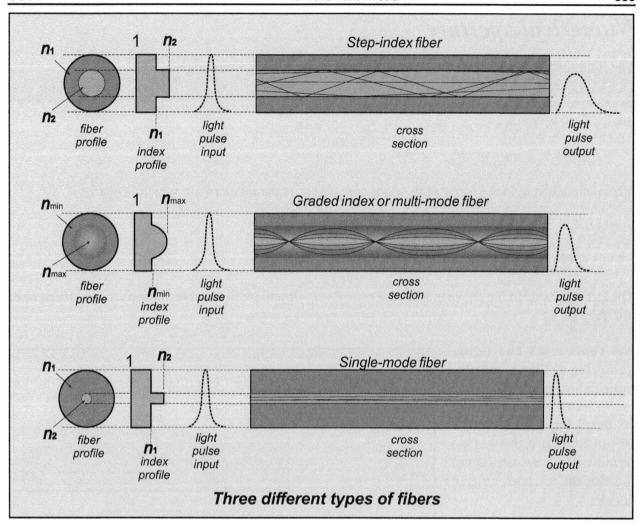

Three different types of fibers

profile is shown here, as well as how light travels through such a cable. Note the input pulse deformation caused by the various path lengths of the light rays bounced from the cylindrical surface that divides the two different index fibers. This is called a ***modal distortion***.

In order to equalize the path lengths of different rays and improve the pulse response, a ***graded index*** (or ***multi-mode***) fiber optics cable was developed. Multi-mode fiber makes the rays travel more or less at an equal speed, causing the effect of optical standing waves.

And finally, a ***single-mode*** fiber cable is available with even better pulse response and almost eliminated modal distortion.

This latter one is the most expensive of all and offers the longest distances achievable **using the same electronics**. For CCTV applications, the multi-mode and step index are adequate.

The index profiles of the three types are shown above.

Numerical aperture

The light that is injected into the fiber cable may come from various angles.

Because of the different indices of the air and the fiber, we can apply the theory of refraction where Snell's Law gives us:

$$\sin\phi_0 \, n_0 = \sin\phi_1 \, n_1 \tag{52}$$

Understandably, n_1 is the index of the fiber core and n_0 is the index of air, which is nearly 1.

Furthermore, this gives us:

$$\sin\phi_0 = \sin\phi_1 \, n_1 \tag{53}$$

The left-hand side of the above is accepted to be a very important fiber cable property, called ***numerical aperture*** (**NA**).

NA represents the light-gathering ability of a fiber-optic cable.

In practice, NA helps us to understand how two terminated fibers can be put together and still make a signal contact.

The realistic value of a typical NA angle for a step index fiber cable is shown on the drawing.

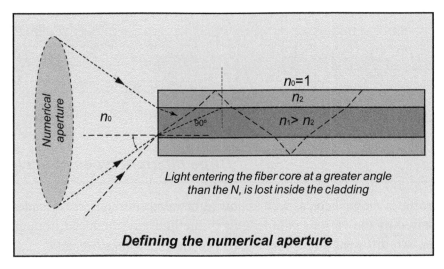

Light entering the fiber core at a greater angle than the N, is lost inside the cladding

Defining the numerical aperture

To calculate NA (basically the angle ϕ_0), it is not necessary to know the angle ϕ_1.

The following are some basic trigonometric transformations that will express NA using only the fiber indices.

Applying Snell's Law and using the drawing, we get:

$$\sin(90° - \phi_1) \, n_1 = sin(90° - \phi_2) \, n_2 \tag{54}$$

For a total reflection we have $\phi_2 = 0°$; therefore, the above becomes:

$$\sin(90° - \phi_1) \, n_1 = n_2 \tag{55}$$

Since $\sin(90° - \phi_1) = cos\phi_1$, we can write:

$$\cos\phi_1 = n_2 / n_1 \tag{56}$$

Knowing the basic rule of trigonometry,

$$\sin^2\phi + \cos^2\phi = 1 \tag{57}$$

and using equation (56), we can convert (53) into a more acceptable relation without sine and cosine:

$$\sin^2\phi_0 / n_1^2 + n_2^2/n_1^2 = 1 \tag{58}$$

$$\sin^2\phi_0 = n_1^2 - n_2^2 \tag{59}$$

$$NA = \sin\phi_0 = SQRT\,(n_1^2 - n_2^2) \tag{60}$$

Formula (60) is the well-known formula for calculating the numerical aperture of a fiber cable, based on the two known indices, the core and the cladding. SQRT stands for square root.

Obviously, **the higher this number is, the wider the angle of light acceptance will be of the cable.**

A realistic example would be with $n_1 = 1.46$ and $n_2 = 1.40$ which will give us NA = 0.41, that is, $\phi_0 = 24°$.

For a graded-index fiber, this aperture is a variable, and it is dependent on the radius of the index which we are measuring, but it is lower than the step index multi-mode fiber. A single-mode 9/125 μm fiber has NA = 0.1.

Light levels in fiber optics

Light output power is measured in watts (like any other power), but since light sources used in fiber optics communications are very low, it is more appropriate to compare an output power relative to the input one, in which case we get the well-known equation for *decibels*:

$$A_R = 10 \log (P_O/P_I) \qquad [\text{dB}] \tag{61}$$

However, if we compare a certain light power relative to an absolute value, like 1 mW, then we are talking about dBm-s, that is:

$$A_A = 10 \log (P/1 \text{ mW}) \qquad [\text{dBm}] \tag{62}$$

Working with decibels makes calculation of transmission levels much easier.

Negative decibels, when A is calculated, mean loss and positive decibels mean gain.

In the case of A_A a negative number of dBm represents power less than 1 mW and a positive number is more than 1 mW.

The definition of dB, when comparing power values, is as shown in equation (61), but as noted earlier, there is a slightly different definition when voltage or current is compared and expressed in decibels:

$$B_R = 20 \log (V_O/V_I) \qquad \text{[dB]} \qquad\qquad\qquad (63)$$

Without going into the theory, it should be remembered that power decibels are calculated with 10 and voltage (and current) decibels are calculated with 20 times in front of the logarithm.

Light, when transmitted through a fiber cable, can be lost due to:

- Source coupling

- Optical splices

- Attenuation of the fiber due to inhomogenity

- High temperature and so on

When designing a CCTV system with fiber optics cables, the total attenuation is very important to know since we work with very small signals. It is therefore better to work with worst case estimates rather than using average values, which will help design a safe and quality system.

For this purpose it should be known that in most cases an 850 nm LED light power output is between 1 dBm and 3 dBm, while a 1300 nm LED has a bit less power, usually from 0 dBm to 2 dBm (note: the power is expressed relative to 1 mW).

The biggest loss of light occurs in the coupling between the LED and the fiber.

It also depends on the NA number and on whether you use step or graded index fiber.

A realistic number for source coupling losses is around –14 dB (relative to the source power output).

Light sources in fiber optics transmission

The two basic electronic components used in producing light for fiber-optic cables are:

- LEDs

- LDs

Both of these produce frequencies in the infrared region, which is above 700 nm.

Photo courtesy of Laser Diode
Laser diode

The light-generating process in both LEDs and LDs results from the recombination of electrons and holes inside a P-N junction when a forward bias current is applied. This light is actually called *electroluminescent*.

The recombined electron/hole pairs have less energy than each constituent had separately before the recombination. **When the holes and electrons recombine, they give up this surplus energy difference, which leaves the point of recombination as photons (basic unit carriers of light).**

The wavelength associated with this photon is determined by the equation:

$$\lambda = hc/E \qquad\qquad (64)$$

where:

> h is the Planck's constant, a fundamental constant in physics: 6.63×10^{-34} Joules

> c is the speed of light (300×10^6 m/s)

> E is the band gap energy of the P-N material

Since h and c are constant, it means that **the wavelength depends solely on the band gap energy, that is, the material in use**. This is a very important conclusion.

For pure gallium arsenide (GaAs) λ is 900 nm. For example, by adding some small amounts of aluminium, the wavelength can be lowered to 780 nm. For even lower wavelengths, other material, such as gallium arsenide phosphate (GaAsP) or gallium phosphate (GaP), is used.

The basic differences between an LED and an LD are in the generated wavelength spectrum and the angle of dispersion of that light.

An LED generates a fair bit of wavelength around the central wavelength as shown below. An LD has a very narrow bandwidth, almost a single wavelength.

An LED P-N junction not only emits light with more frequencies than an LD, but it does so in all directions, that is, with no preferred direction of dispersion. This dispersion will greatly depend on the mechanical construction of the diode, its light absorption and reflection of the area. The radiation, however, is omnidirectional, and in order to narrow it down, LED manufacturers put a kind of focusing lens on top. This is still far too wide an angle to be used with a single-mode fiber cable. So this is the main reason LEDs are not used as transmitting devices with single-mode fiber cables.

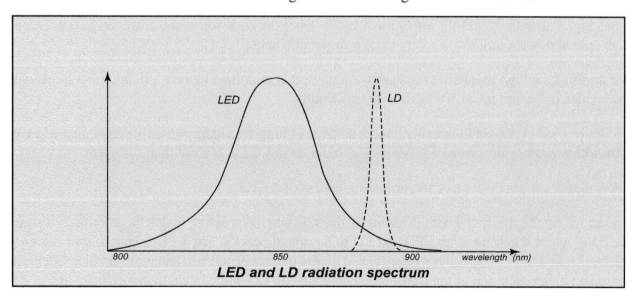

LED and LD radiation spectrum

An LD is built of a similar material as an LED and the light-generating process is similar, but the junction area is much smaller and the concentration of the holes and electrons is much greater. The generated light can only exit from a very small area. At certain current levels, the photon generation process gets into a **resonance** and the number of generated photons increases dramatically, producing more photons with the same wavelength and in phase. Thus, **the optical gain is achieved in an organized way and the generated light is a coherent (in phase), stimulated emission of light**. In fact, the word "laser" is an abbreviation for *light amplification by stimulated emission of radiation.*

In order to start this stimulated emission of light, an LD requires a minimum current of 5 to 100 mA, which is called a *threshold current*. This is

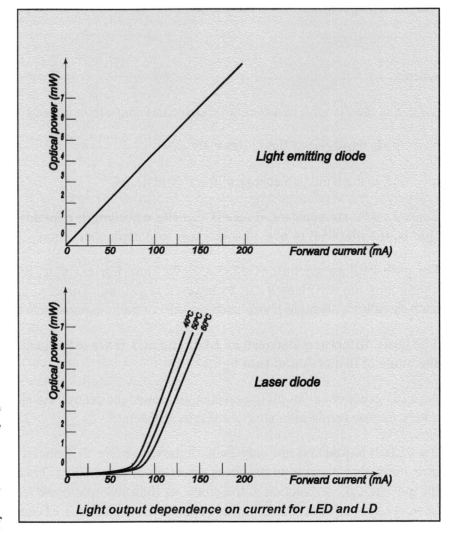

Light output dependence on current for LED and LD

much higher than the threshold with normal LEDs. Once the emission starts, however, LDs produce a high optical power output with a very narrow dispersion angle.

For transmitting high frequencies and analog signals, it is important to have a light output linear with the applied drive current, as well as a wide bandwidth.

LEDs are good in respect to linearity but not so good in high-frequency reproduction compared to the LDs, although, they do exceed 100 MHz, which, for us in CCTV, is more than sufficient.

Laser diodes can easily achieve frequencies in excess of 1 GHz.

The above can be illustrated with the same analogy as when discussing magnetic recording: Imagine the light output spectrum of an LED and LD to be tips of pencils. The LED spectrum will represent the thicker and the LD the thinner pencil tip. With the thinner pencil you can write smaller letters and more text in the same space; the signal modulated with an LD will contain higher frequencies.

LEDs, however, are cheaper and linear and require no special driving electronics. An LED of 850 nm costs around \$10, whereas 1300 nm is around \$100. Their MTBF is extremely high ($10^6 - 10^8$ hrs).

LDs are more expensive, between \$100 and \$15,000. They are very linear once the threshold is exceeded. They often have a temperature control circuit because the operating temperature is very important, so feedback stabilization for the output power is necessary. Despite all of that, they have a higher modulation bandwidth, and a narrower carrier spectral width, and they launch more power into small fibers. Their MTBF is lower than the LEDs', although still quite high ($10^5 - 10^7$ hrs).

Recently, a new LED called a ***super luminescent diode*** (SLD) has been attracting a great deal of attention. The technical characteristics of the SLDs are in between those of the LEDs and LDs.

For CCTV applications, LEDs are sufficient light sources. LDs are more commonly used in multichannel wide bandwidth multiplexers or very long run single-mode fibers.

Light detectors in fiber optics

Devices used for detecting the optical signals on the other side of the fiber cable are known as ***photo diodes***. This is because the majority of them are actually one type of a diode or another.

The basic division of photo diodes used in fiber technology is into:

- P-N photo diodes (PNPD)

- PIN photo diodes (PINPD)

- Avalanche photo diodes (APD)

The PNPD is like a normal P-N junction silicon diode that is sensitive to infrared light. Its main characteristics are low respondence and high rise time.

The PINPD is a modified P-N diode where an intrinsic layer is inserted in between the P and N types of silicon. It possesses high response and low rise times.

The APD is similar to the PINPD, but it has an advantage that almost each incident photon produces more than one electron/hole pair, as a result of an internal chain reaction (avalanche effect). Consequently, the APD is more sensitive than the PIN diode, but it also generates more noise.

All these basic devices are combined with amplification and "trans-impedance" stages that amplify the signal to the required current/voltage levels.

APD diode

Frequencies in fiber optics transmission

The attenuation of the optical fibers can be grouped in attenuation due to material and external influences.

Material influences include:

> • Rayleigh scattering. This is due to the inhomogeneities in the fiber glass, the size of which is small compared to the wavelength. At 850 nm, this attenuation may add up to 1.5 dB/km, reducing to 0.3 dB/km for a wavelength of 1300 nm and 0.15 dB/km for 1550 nm.

> • Material absorption. This occurs if hydroxyl ions and/or metal ions are present in the fiber. Material absorption is much smaller than the Rayleigh scattering and usually adds up to 0.2 dB/km to the signal attenuation.

The external effects that influence the attenuation are:

> • Micro-bending. This is mainly due to an inadequate cable design inconsistency of the fiber cable precision along its length. It can amount up to several dB/km.

> • Fiber geometry. This is similar to the above, but is basically due to the poor control over its drawn diameter.

The accompanying diagram on the next page shows a very important fact: that not all the wavelengths (frequencies) have the same attenuation when sent through a fiber cable.

The wavelengths around the areas indicated with the vertical dotted lines are often called *fiber optics windows*. There are three windows:

> • First window at 850 nm

> • Second window at 1300 nm

> • Third window at 1550 nm

The first window is not really with minimum attenuation compared to the higher frequencies, but this frequency was first used in fiber transmission. The LEDs produced for this use were reasonably efficient and easy to make.

For short-distance applications, such as CCTV, this is still the cheapest and preferred wavelength.

The 1300 nm wavelength is becoming more commonly used in CCTV. This is the preferred wavelength for professional telecommunications as well as CCTV systems with longer cable runs, where higher light source cost is not a major factor. The losses at this frequency are much lower, as can be seen from the diagram. The difference between 850 nm and 1300 nm in attenuation is approximately 2 to 3 dB/km.

The 1550 nm wavelength has even lower losses; therefore, more future systems will be oriented toward this window.

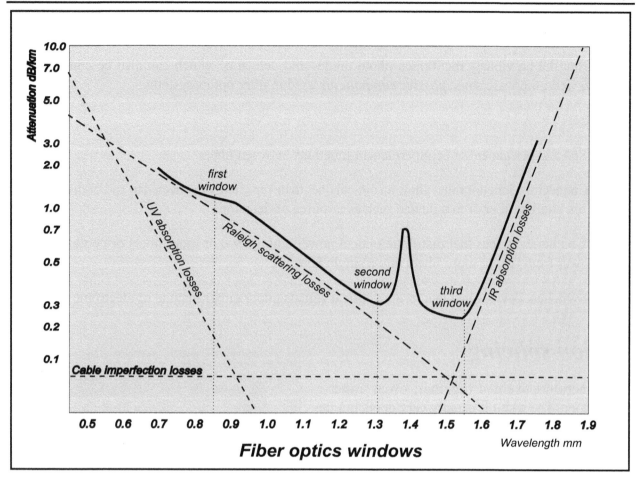

Fiber optics windows

For illustration purposes, a typical attenuation figure of a multi-mode 62.5/125 μm fiber cable, for an 850 nm light source, is less than 3.3 dB per kilometer. If a 1300 nm source is used with the same fiber, attenuation of less than 1 dB can be achieved. Therefore, **longer distances can be achieved with the same fiber cable, by just changing the light source**. This is especially useful with analog signals, such as the video.

When an 850 nm light source is used with 62.5/125 μm cable, we can easily have a run of at least a couple of kilometers, which in most CCTV cases is more than sufficient. However, longer distances can be achieved by using graded multi-mode fiber and even longer when a 1300 nm light source is used instead of 850 nm.

The longest run can be achieved with a single-mode fiber cable and light sources of 1300 nm and 1550 nm.

A typical attenuation figure for a 1300 nm light source is less than 0.5 dB/km, and for 1550 nm it is less than 0.4 dB/km.

Passive components

Apart from the previously mentioned photo diodes and detectors, which can also be considered as *active devices*, there are some *passive components* used in fiber optics systems.

These are:

- Splices: permanent or semipermanent junctions between fibers.

- Connectors: junctions that allow an optical fiber to be repeatedly connected to and/or disconnected from another fiber or to a device such as a source or detector.

- Couplers: devices that distribute optical power among two or more fibers or combine optical power from two or more fibers into a single fiber.

- Switches: devices that can reroute optical signals under either manual or electronic control.

Fusion splicing

Two fibers are welded together, often under a microscope. The result is usually very good, but the equipment might be expensive.

The procedure of fusion splicing usually consists of cleaning the fiber, cleaving, and then positioning the two fibers in some kind of mounting blocks.

The precision of this positioning is improved by using a microscope, which is quite often part of the machine. When the alignment is achieved, an electric arc is produced to weld the two fibers. Such a process can be monitored and repeated if an unsatisfactory joint is produced.

Losses in fusion splicing are very low, usually around 0.1 dB.

Photo courtesy of Orionics

A fusion splicing machine

Mechanical splicing

This is probably the most common way of splicing, owing to the inexpensive tools used with relatively good results.

Fibers are mechanically aligned, in reference to

An ST connector and mechanical splicing joint

their surfaces and (usually) epoxied together. The performance cannot be as good as fusion splicing, but it may come very close to it. More importantly, the equipment used to perform the mechanical splicing is far less expensive.

Losses in a good mechanical splicing range between 0.1 and 0.4 dB.

The mechanical splicing is based on two principles:

- V groove

- Axis alignment

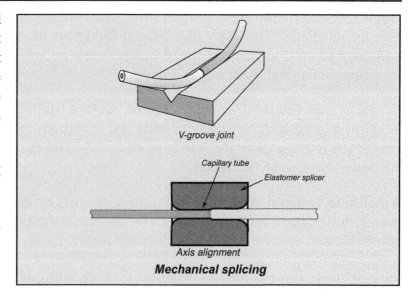

V-groove joint

Capillary tube

Elastomer splicer

Axis alignment

Mechanical splicing

Both of these are shown in the diagrams at right.

For a good connection, the fiber-optic cable needs a good termination, which is still the hardest part of a fiber optics installation. It needs high precision and patience and a little bit of practice. Anyone can learn to terminate a fiber cable; in cases where they have no such skills, installers can hire specialized people who supply the terminals, terminate the cable, and test it. The latter is the most preferred arrangement in the majority of CCTV fiber optics installations.

Fiber-optic multiplexers

These multiplexers are different from the VCR multiplexers described earlier. **Fiber-optic multiplexers combine more signals into one, in order to use only a single fiber cable to simultaneously transmit several live signals**. They are especially practical in systems with an insufficient number of cables (relative to the number of cameras).

There are a few different types of fiber multiplexers. The simplest and most affordable multiplexing for fiber optics transmission is by use of *wavelength division multiplexing* (WDM) couplers. These

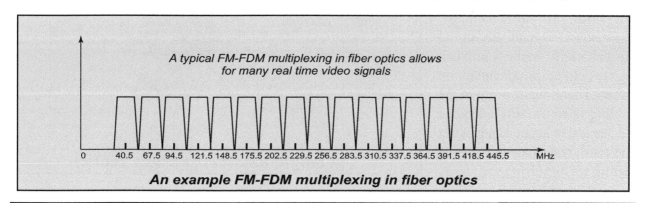

A typical FM-FDM multiplexing in fiber optics allows for many real time video signals

0   40.5 67.5 94.5 121.5 148.5 175.5 202.5 229.5 256.5 283.5 310.5 337.5 364.5 391.5 418.5 445.5   MHz

An example FM-FDM multiplexing in fiber optics

are couplers that transmit optical signals from two or more sources, operating at different wavelengths, over the same fiber. This is possible because **light rays of different wavelengths do not interfere with each other**. Thus, the capacity of the fiber cable can be increased, and if necessary, bidirectional operation over a single fiber can be achieved.

Frequency-modulated frequency division multiplexing (FM-FDM) is a reasonably economical design with acceptable immunity to noise and distortions, good linearity, and moderately complex circuitry. A few brands on the market produce FM-FDM multiplexers for CCTV applications. They are made with 4, 8, or 16 channels.

Amplitude vestigial sideband modulation, frequency division multiplexing (AVSB-FDM) is another design, perhaps too expensive for CCTV, but very attractive for CATV, where with high-quality optoelectronics up to 80 channels per fiber are possible.

Fully digital **pulse code modulation, time division multiplexing** (PCM-TDM) is another expensive multiplexing, but of digitized signals, which may become attractive as digital video gains greater acceptance in CCTV.

Combinations of these methods are also possible.

In CCTV we would most often use the FM-FDM for more signals over a single fiber. The WDM type of multiplexing is particularly useful for PTZ, or keyboard control with matrix switchers. Video signals are sent via separate fibers (one fiber per camera), but only one fiber uses WDM to send control data in the opposite direction.

Even though fiber-optic multiplexing is becoming more affordable it should be noted that in the planning stage of fiber installation it is still recommended that at least one spare fiber is run in addition to the one intended for use.

Fiber-optic cables

The fiber optics themselves is very small in size. The external diameter, as used in CCTV and security in general, is only 125 μm (1 μm = 10⁻⁶ m). Fiberglass, as a material, is relatively strong but can easily be broken when bent to below a certain minimum radius. Therefore, the aim of the cabling is to provide adequate mechanical protection and impact and crush resistance to preserve minimum bending radius as

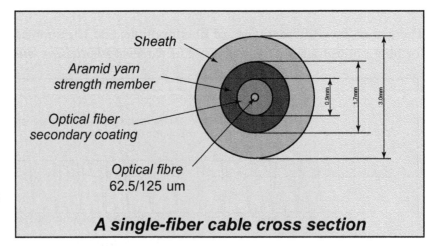

Sheath
Aramid yarn strength member
Optical fiber secondary coating
Optical fibre 62.5/125 um

A single-fiber cable cross section

well as to provide easy handling for installation and service and to ensure that the transmission properties remain stable throughout the life of the system.

The overall design may vary greatly and depends on the application (underwater, underground, in the air, or in conduit), the number of channels required, and similar. Invariably, it features some form of tensile strength member and a tough outer sheath to provide the necessary mechanical and environmental protection.

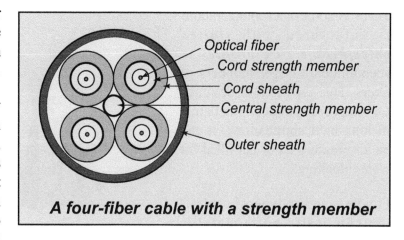

A four-fiber cable with a strength member

Fiber-optic cables have various designs, such as a simple single fiber, loose tube (fiber inserted into a tube), slotted core (or open channel), ribbon, and tight buffer.

We will discuss a few of the most commonly used designs in CCTV.

Single-fiber and dual-fiber cables usually employ a fibrous strength member (aramid yarn) laid around the secondary coated fiber. This is further protected by a plastic outer sheath.

Multi-fiber cables are made in a variety of configurations.

The simplest involves grouping a number of single-fiber cables with a central strength member within the outer jacket. The central strength member can be high tensile steel wire, or a fiberglass reinforced plastic rod. Cables with this design are available with 2 to 12 or more communication fibers. When the plastic rod is applied to the central strength member, it becomes a **metal-free** optical fiber cable. Constructed entirely from polymeric material and glass, these cables are intended for use in installations within buildings. They are suitable for many applications including CCTV, security, computer links, and instrumentation. These heavy-duty cables are made extremely rugged to facilitate pulling through ducts.

Loose tube cables are designed as a good alternative to the single core and slotted cables. The optical fibers are protected by water-blocking gel-filled polyester tubes. This type of multi-fiber cable is designed for direct burial or duct installation in long-haul applications. It can be air pressurized or gel-filled for water-blocking.

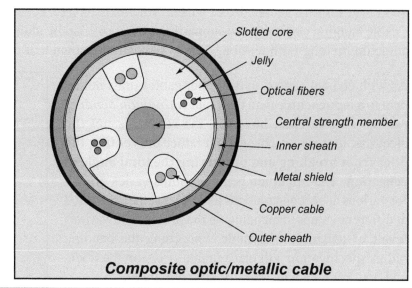

Composite optic/metallic cable

There are some other configurations manufactured with *slotted polyethylene core profiles* to accommodate larger numbers of fibers. This type is also designed for direct burial or duct installation in long-haul applications. It can be air pressurized or gel filled for water blocking.

Finally, another type of cable is the **composite optic/metallic** cable. These cables are made up of a combination of optical fibers and insulated copper wire and

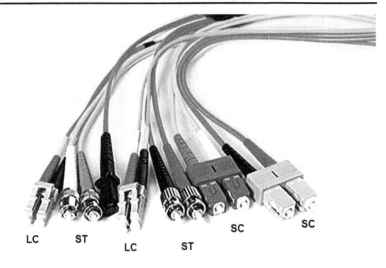

are designed for both indoor and outdoor use. These cables can be fully filled with a water-blocking compound to protect the fibers from moisture in underground installations, for example.

Since the fiber cable is much lighter than other cables the installations are generally easier compared to an electrical cable of the same diameter.

The protection they have will allow fiber cables to be treated in much the same way as electrical cables. However, care should be taken to ensure that the manufacturer's *recommended maximum tensile and crushing force* are not exceeded.

Within a given optical cable, tension is carried by the strength members, usually fiber-reinforced plastic, steel, kevlar, or a combination, that protect the comparatively fragile glass fibers. If the cable tension exceeds the manufacturer's ratings, permanent damage to the fibers can result.

The rating to be observed, as far as installation tension is concerned, is the *maximum installation tension*, expressed in Newtons or kilo-Newtons (N or kN). A typical cable has a tension rating of around 1000 N (1 kN). To get an idea of what a Newton feels like, consider that 9.8 N of tension is created on a cable hanging vertically and supporting a mass of 1 kg. In addition, manufacturers sometimes specify a maximum long-term tension. This is typically less than half of the maximum installation tension.

As with coaxial cables, optical fiber cables must not be bent to a tighter curve than their rated *minimum bending radius*. In this case, however, the reason is not the electrical impedance change, but rather **preventing the fiber from breaking and preserving the total angle of reflection**. The minimum bending radius varies greatly for various cable constructions and may even be specified at different values, depending on the presence of various levels of tension in the cable. Exceeding the bending radius specification will place undue stress on the fibers and may even damage the stiff strength members.

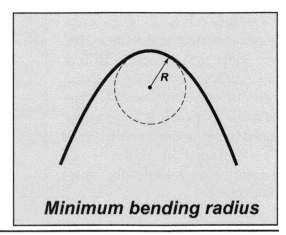

Minimum bending radius

Whenever a cable is being handled or installed, it is most important to keep the curves as smooth as possible.

Often, during an installation, the cable is subject to crush stresses such as being walked on or, even worse, driven over.

Although great care should be taken to avoid such stresses, the cable is able to absorb such forces up to its rated **crush resistance** value. Crush resistance is expressed in N/m or kN/m of cable length. For

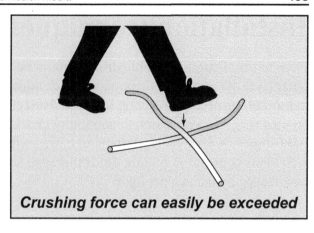

Crushing force can easily be exceeded

example, a cable with a specified crush resistance of 10 kN/m can withstand a load of 1000 kg spread across a full 1 m of cable length (10 Newtons is approximately the force that results from a 1 kg mass). If

we consider a size 9 boot (European 42) to be 100 mm wide, then the cable will support a construction worker who weighs 100 kg standing on one foot squarely on the cable. However, a vehicle driving over this cable may exceed the crush resistance spec and probably damage the cable.

Be careful if a cable has one loop crossing over another, then the forces on the cable due to, say, a footstep right on the crossover will be **greatly magnified because of the smaller contact area**. Likewise in a crowded duct, a cable can be crushed at localized stress points even though the weight upon it may not seem excessive.

An optical cable is usually delivered wound onto wooden drums with some form of heavy plastic protective layer or wooden cleats around the circumference of the drums. When handling a cable drum, due consideration should be given to the mass of the drum. The most vulnerable parts of a cable drum are the outer layers of the cable. This is especially the case when the cable drums are vertically interleaved with each other. Then damage due to local crushing should be of concern. To alleviate such problems, the drums should be stacked either horizontally, or, if vertically, with their rims touching. Do not allow the drums to become interlocked. Also, when lifting drums, with, for example, a forklift do not apply force to the cable surface. Instead, lift at the rims or through the center axis.

Different types of fiber connectors

Installation techniques

Prior to installation, the cable drums should be checked for any sign of damage or mishandling. The outer layer of a cable should be carefully examined to reveal any signs of scratching or denting. Should a drum be suspected of having incurred damage, then it should be marked and put aside. For shorter lengths (i.e., < 2 km) a simple continuity check can be made of the whole fiber using a penlight as a light source. A fiber cable, even though used with infrared wavelengths, transmits normal light just as well. This is useful in finding out if there are serious breaks in the cable. Continuity of the fiber can be checked by using a penlight.

The following precautions and techniques are very similar to what was said earlier for coaxial cable installations, but since it is very important we will go through it again.

Before cable laying commences, the route should be inspected for possible problems such as feed-through, sharp corners, and clogged ducts. Once a viable route has been established, the cable lengths must be arranged so that should any splices occur, they will do so at accessible positions.

At the location of a splice it is important to leave an adequate overlap of the cables so that sufficient material is available for the splicing operation. Generally, the overlap required is about 5 m when the splice is of the in-line type. A length of about 2.5 m is required where the cable leaves the duct and is spliced.

Note that whenever a cable end is exposed it must be fitted with a watertight end cap. Any loose cable should be placed to avoid bending stress or damage from passing traffic. At either end of the cable run, special lengths are often left depending on the configuration planned.

Foremost consideration, when burying a cable, is the prevention of damage due to excessive local points loading. Such loading occurs when backfill material is poured onto or an uneven trench profile digs into the cable, thus either puncturing the outer sheath or locally crushing the cable. The damage may become immediately obvious, or it may take some time to show itself. Whichever, the cost of digging up the cable and repairing it makes the expenditure of extra effort during laying well worthwhile.

When laying cables in trenches, a number of

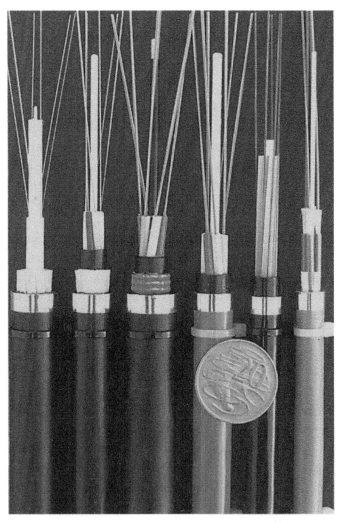

Various types of multi-fiber cables

precautions must be taken to avoid damage to the cable or a reduction in cable life expectancy.

The main protection against cable damage is laying the cable on a bed of sand approximately 50 to 150 mm deep and backfilling with another 50 to 150 mm of sand. Due care needs to be taken in the digging of the trench so that the bottom of the trench is fairly even and free of protrusions. Similarly, when backfilling, do not allow rock soil to fall onto the sand because it may put a rock through the cable.

Trench depth is dependent upon the type of ground being traversed as well as the load that is expected to be applied to the ground above the cable. A cable in solid rock may need a trench of only 300 mm or so, whereas a trench crossing a road in soft soil should be taken down to about 1 m. A general-purpose trench in undemanding situations should be 400 to 600 mm deep with 100 to 300 mm total sand bedding.

The most straightforward technique is to lay the cable directly from the drum into a trench or onto a cable tray. For very long cable runs, the drum is supported on a vehicle and allowed to turn freely on its axis, or it can be held and rested on a metal axis. As the vehicle (or person) advances, the cable is wound off the drum straight to its resting place. Avoid excessive speed and ensure that the cable can be temporarily tied down at regular intervals prior to its final securing.

Placing an optical cable on a cable tray is not particularly different from doing so with conventional cables of a similar diameter. The main points to observe are, again, minimum bending radius and crush resistance.

The minimum bending radius must be observed even when the cable tray does not facilitate this. The tendency to keep it neat and bend the optical cable to match other cables on the tray must be avoided.

Crush loading on cable trays can become a critical factor, where the optical cable is led across a sharp protrusion or crossed over another cable. The optical cable can then be heavily loaded by further cables being placed on top of it or personnel walking on the tray. Keep the cable as flat as possible and avoid local stress points.

The pulling of optical cables through ducts is no different from conventional cabling. At all times use only the amount of force required but stay below the manufacturer's ratings.

The types of hauling eyes and cable clamps normally used are generally satisfactory, but remember that the strength members, not the outer sheath, must take the load.

If a particular duct requires a lubricant, then it is best to obtain a recommendation from the cable manufacturer. Talcum powder and bean bag polystyrene beans can also be quite useful in reducing friction.

In some conditions, the cable may already be

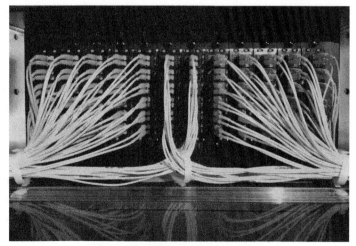

Fibre optic cabling rack

terminated with connectors. These must be heavily protected while drawing the cable. The connectors themselves must not be damaged or contaminated, and the cable must not experience any undue stress around the connectors or their protective sheathing.

Once the cable is installed it is often necessary to tie it down. On a cable tray, the cable can be held down simply with nylon ties. Take particular care to anchor the cable runs in areas of likely creep. On structures that are unsuited to cable ties, some form of saddle clamp is recommended. Care is required in the choice and use of such devices so that the cable crush resistance is not exceeded and so that the outer jacket is not punctured by sharp edges. Clips with molded plastic protective layers are preferred and only one clip should be used for each cable. Between the secured points of a cable it is wise to allow a little slack rather than to leave a tightly stretched length that may respond poorly to temperature variations or vibration.

If the cable is in some way damaged during installation, then leave enough extra cable length around the damaged area so that an additional splice can be inserted.

The conclusion is that **installation of fiber cables is not greatly different from conventional cables, and provided a few basic concepts are observed, the installation should be trouble free.**

Fiber optic link analysis

Now that we have learned the individual components of a fiber-optic system – the sources, cables, detectors, and installation techniques – we may use this in a complete system. But before the installation, we first have to do a link analysis, which shows how much signal loss or gain occurs in each stage of the system. This type of analysis can be done with other transmission media, but it is especially important with fiber optics because the power levels we are handling are very small. They are sufficient to go over many kilometers, but can easily be lost if we do not take care of the microscopic connections and couplings.

The goal of the link analysis is to determine the signal strength at each point in the overall system and to calculate if the power at the receiver (the detector) are sufficient for acceptable performance. If it is not, each stage is examined and some are upgraded (usually to a higher cost), or guaranteed performance specifications (distance, speed, errors) are reduced.

For fiber-optics system, the link analysis should also include unavoidable variations in performance that occur with temperature as a result of component aging and from manufacturing tolerance differences between two nearly identical devices. In this respect, fiber-optic systems need more careful study than all-electronic systems, as there is greater device-to-device variation, together with larger performance changes due to time and temperature.

As a practical example, the diagram on the next page shows a basic point-by-point fiber-optic system, which consists of an electrical data input signal, a source driver, an optical source, a 1 km optical fiber with realistic maximum attenuation of 4 dB/km, an optical detector, and the receiver electronics.

We have assumed that the system is handling digital signals, as is the case with PTZ control, but the logic will be very similar when analog signal budgeting is calculated.

The calculation begins with the optical output power of the source (−12 dBm in this case) and ends with the power that is seen by the detector.

This analysis looks at each stage in the system and shows both the best and worst case power loss (or gain) for each link as a result of various factors, such as coupling losses, path losses, normal parts tolerance (best and worst for a specific model), temperature, and time.

The analysis also allows for an additional 5 dB signal loss that will occur if any repairs or splices are made over the life of the system.

The conclusion of the example is that the received optical power, for a signal to be recognized, can be anywhere between +7 dB (in the best case) and −23 dB (in the worst case) relative to the nominal source value. Technically, +7 dB would mean amplification, which is not really what we have, but rather it refers to the possible tolerance variations of the components. Therefore, the receiving detector must handle a dynamic range of optical signals from −5 dBm (−12 dBm + 7 dB = −5 dBm) to −35 dBm (−12 dBm − 23 dB = −35 dBm), representing a binary 1. Of course, when the source is dark (no light, which means binary 0), the received signal is also virtually zero (except for the system noise).

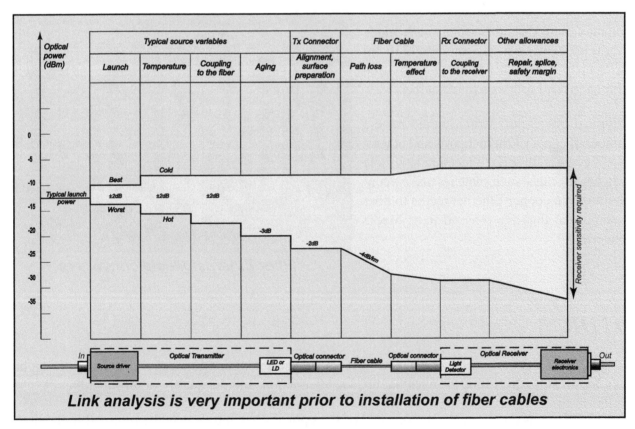

Link analysis is very important prior to installation of fiber cables

It is understandable that a digital signal can go further in distance, using the same electronics and fiber cable, than an analog video signal, simply because of the big error margins digital signals have. Nevertheless, a similar analysis can be performed with analog signals. If however, we are not prepared or do not know how to, we can still get an answer to the basic question, Will it work? Unfortunately, the answer can only be obtained once the fiber is installed. To do so we need an instrument that measures cable continuity as well as attenuation. This is the optical time domain reflectometer.

Fiber Ethernet

As discussed under the Networking chapter, today fiber is used for digital Ethernet too, not just for analog signals. See more for fiber Ethernet standards under this chapter. In fact, the fiber Ethernet is the fastest, widest bandwidth and longest reaching media in today's digital networks. Large IP CCTV systems, like the ones in casinos, with thousands of cameras can only be made possible with fiber optic Ethernet. The reason is simple, and the same as with analog - fiber allows for the widest bandwidth and lowest attenutation per length.

Although the principles of total reflection are the same as for analog signals, the termination of fiber cable is much easier than in the past, and the fiber cable is cheaper than what it used to be. One more point of difference today is with the type of multi-mode fiber becoming more popular, and this is the 50/125 um, as opposed to 62.5/125 um. This improves the maximum length achievable, for 62.5 / 125 um is about 220 m when 1 Gb/s is used, and more than double, 550 m with 50/125 um.

Similar to the analog fiber, the launching light sources can be inra red LEDs or Lasers.

The maximum distances when single mode fiber is used can be from 5 km with 1310 nm Laser, up to 70 km when double Fiber is used with 1550 nm Laser as a light source.

A typical fiber optic switch would have a number of copper Ethernet ports and at least two fiber ports for interconnecting to remote switches. Because such switches make a conversion from copper Ethernet media to fiber media, often they are referred to as Media convertors.

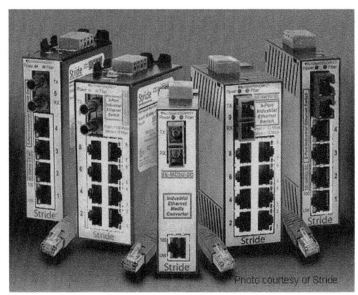

Fiber Ethernet Media converters

OTDR

An **optical time domain reflectometer** (OTDR) is an instrument that can test a fiber cable after it has been installed, to determine the eventual breaks, attenuation, and the quality of termination.

The OTDR sends a light pulse into one end of the optical fiber and detects the returned light energy versus time, which corresponds directly to the distance of light traveled.

It requires connection to only one end of the cable and it actually shows the obvious discontinuity in the optical path, such as splices, breaks, and connectors.

It uses the physical phenomenon known as **Rayleigh back-scattering**, which occurs within the fiber, to show the signal attenuation along the fiber's length. As a light wave travels through the fiber, a very small amount of incident light in the cable is reflected and refracted back to the source by the atomic structure and impurities within the optical fiber. This is then measured and shown visually on a screen and/or printed out on a piece of paper as evidence of the particular installation. Eventual breaks in a fiber cable are found most easily with an OTDR. Being an expensive instrument, an OTDR is usually hired for a fiber optics installation evaluation or used by the specialized people that terminate the cable.

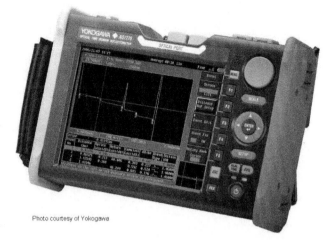

Photo courtesy of Yokogawa

Optical TDR

11. Networking in CCTV

The Information Technology era

Today's world is, without any doubt, the world of *Information Technology*, or, as many would refer to it, the **IT world**.

In CCTV, we are usually looking for visual information about an event, an intruder, or a procedure, such as who entered the building before it caught fire, or what is the procedure during the heart surgery, or what is the license plate of a car involved in a collision.

So, how is information defined, and why is it so important?

Information is any communication or representation of knowledge such as facts, data, or opinions, in any medium or form, including textual, numerical, graphic, cartographic, narrative, or audiovisual forms.

Human knowledge grows exponentially, and what has been achieved only in the last few decades, for example, far exceeds the knowledge accumulated through thousands of years before that. The amount of information in each and particular human activity is so large that without proper understanding and management of such information we would lose track of what we know and where we are heading. Because the information grows exponentially people have seen a need for a complete new subject (IT) that deals with such a large amount of information.

IT is part of the larger scope of things that are especially interesting for us in the CCTV industry, and it is concerned with the hardware and software that processes information, regardless of the technology involved, whether this is digital video recorders, computers, wireless telecommunications, or others.

Because of the large amount of information recorded in our daily lives, **reliable, fast, and efficient access** to such information is of paramount importance.

Filing cabinets and mountains of papers have given way to computers that store and manage information electronically. Colleagues and friends thousands of kilometers apart can share information instantaneously, just as hundreds of workers in a single location can simultaneously review research data maintained on-line. Students, doctors, and scientists can study, research, and exchange information even if they are continents apart. **Computer networks are the glue that binds these elements together.**

The large number of such networks forms the global network called the *World Wide Web*, or as we all know it, *the Internet*. This only started in the 1980s, not much more than 30 years before the writing of this book, and yet most of the research and study I had to do for this book was done using the Internet.

The Internet is probably one of the most important human achievements ever. It is truly a global network, a community of knowledge and information where everybody can join in without a passport, regardless of their skin color, age, agenda, or religion.

Internet and the networking technologies today play very important role in digital CCTV too. In order to understand modern CCTV, we have to dedicate some pages to the networking fundamentals.

Computers and networks

Before defining a network, we have to define the basic intelligent device that is a main part of any network. This is the computer.

Computers are so much in our daily lives that not only can we not live without them, but they are spread around so much that it is difficult to define them accurately. One of the many definitions of a computer is as **an electronic device designed to accept digital information (data input), perform prescribed mathematical and logical operations at high speed (processing), and supply the results of these operations (output).** Now, this could easily refer to a digital calculator, and it would probably be correct, but in CCTV we will use the term *computer* to define **an electronic device that is composed of hardware (main processor–CPU, memory, and a display output) and software (Operating System and applications) and which executes a set of instructions as defined by the software.**

In the early years of the computers, numbers and high-speed calculations were the primary area of computer usage. As the processing speed and computer power grew, the processing of images, video, and audio became more frequent and this is the area of our interest.

Initially, in CCTV, computers were used most often in video matrix switchers to intelligently switch cameras onto monitors based on logical processing of external alarm inputs as well as manual selection. Computers are also used in monitoring stations where thousands and thousands of alarms are processed and logged.

These days computers are used in many new CCTV products where digital video capturing, processing, compression, and archiving are done. The vast majority of these devices are digital video recorders (DVRs), but also network video recorders (NVRs) which are used to record a bunch of IP streams. Such IP streams might come directly from a digital (IP) camera, or from a streamer (sometimes called server) which converts analog camera into a digital stream. Sometimes, NVRs often come with multiple hard

A typical NVR in CCTV use built-in RAID

disks configured in one of the redundant (RAID) modes. Also, IP cameras, even though small, have hardware and software with equivalent functionality of a computer. All such computers can work on their own, but their real power comes to effect when they are put in a network environment. In actual fact, the only way computers, IP cameras and NVRs can communicate among each other is via a network.

A network is simply a group of two or more computers linked together with an Ethernet medium.

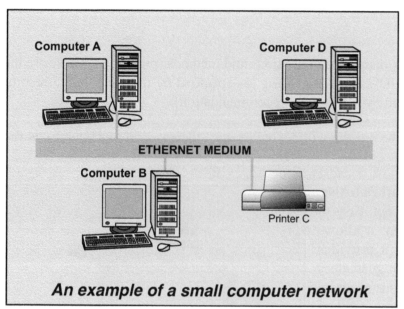

An example of a small computer network

Networking allows one computer to send information to and receive information from another. We may not always be aware of the numerous times we access information on computer networks. The Internet is the most conspicuous example of computer networking, linking millions of computers around the world, but smaller networks play a role in information access on a daily basis. Most libraries and book shops have replaced their paper catalogues with computer terminals that allow patrons to search for books far more quickly and easily. Many companies exchange all their internal information using their own LANs; product leaflets and CCTV system designs are quoted electronically using networks. Internet search engines offer access to almost any information, from anywhere in the world in a second. Billions of people find the information they need, any time in the day, in any language. In each of these cases, networking allows many different computers in multiple locations to access a shared database.

Computers in CCTV are becoming more dominant, regardless of whether they run on a full-blown operating system (OS), such as Windows or Linux, or on an embedded OS residing on a chip. One of the main and indispensable features of computers is their ability to connect to other computers and share information via networks. The fact is networks are already in place in many businesses, organizations, and even homes. Fitting a CCTV system to such networks is just a matter of connecting the LAN cable to a digital video recorder, to a network-ready camera, or perhaps to a computer fitted with a special video capturing card. With some minor network settings the CCTV system can be up and running in a very short period of time.

This ease of network retrofitting and installation is one of the major attractions (though not the only one) of networks for CCTV.

This is not to say that the modern network CCTV systems have to be installed on existing networks. Many designers would actually create a complete new and separate, parallel network, simply because then the system becomes even more secure, dedicated, and, most importantly, does not affect the data traffic of the normal, everyday business usage network.

Once we get to this stage of having networked CCTV, there are many new issues and limitations we face and need to understand in order to further improve or modify our system design.

Later in the book we are going to get deeper into each of these issues, but first, let us start with the basics of the networking and then clarify some of the key concepts and terminology used.

LAN and WAN

There are a few types of network transmission configurations and methods (protocols). These are the *Fiber Distributed Data Interface* (FDDI), the *Token Ring* (as specified by the IEEE 802.5 recommendation), and the *Ethernet* (specified by IEEE 802.3 recommendation).

Of the three, **the most popular**, and the one we will devote most of the space in this book to, **is the Ethernet.**

Ethernet is used in over 85% of the world's LANs, primarily because it is a simple concept, easy to understand, implement, and maintain; it allows low-cost network implementations; it provides extensive topological flexibility; and it guarantees interconnection and operation of various products that comply with the Ethernet standards, regardless of the manufacturer and operating system used on the computer.

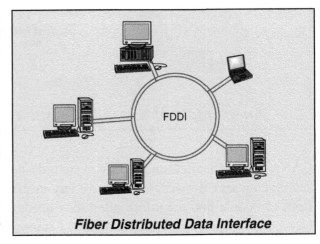

Fiber Distributed Data Interface

Depending upon the scale of such network configurations, we have two major groups: *Local Area Networks* (LANs) and *Wide Area Networks* (WANs).

The Local Area Network (LAN) connects many devices that are relatively close to each other, usually in the same building. A typical example is a business or a company with at least a couple of computers. Sometimes this configuration is called the Intranet.

In a classic LAN configuration, one computer is nominated as the server. It stores all of the software that controls the network, including the software that can be shared by the computers attached to the network. Computers connected to the server are referred to as clients (or workstations). On most LANs, cables are used to connect the network interface cards (NIC) in each computer.

The Wide Area Network (WAN) connects a number of devices that can be many kilometers

IBM's Token Ring

apart. For example, if a company has offices in two major cities, hundreds of kilometers apart, each of their LANs would most likely be configured in WAN formation, using dedicated fiber lines, or creating Virtual Private Network (VPN) over leased ADSL lines.

WANs connect larger geographic areas, such as interstate or country to country. Satellite uplinks or dedicated transoceanic cabling can be used to connect this particular type of network. WANs can be highly complex systems, as they may be connecting local and

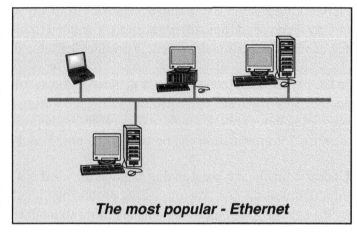

The most popular - Ethernet

metropolitan networks to global communications networks like the Internet. To the user, however, a WAN will appear no more complex than a LAN.

In comparison to WANs, LANs are faster and more reliable, but improvements in technology continue to blur the line of demarcation and have allowed LAN technologies to connect devices tens of kilometers apart, while at the same time greatly improving the speed and reliability of WANs. This in turn blurs the line between the WANs and the Internet.

The Internet can be considered the largest global WAN.

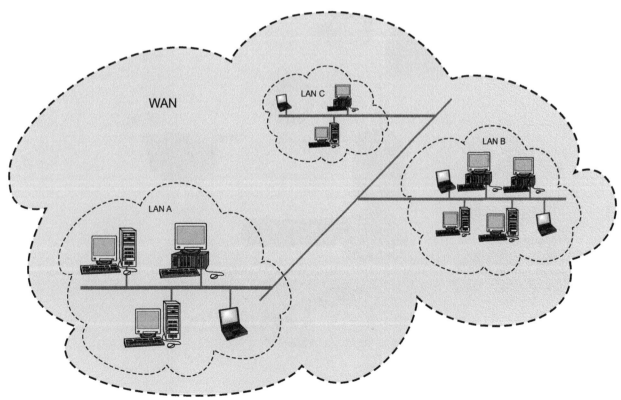

LAN and WAN example

Networking enables the user to access data from any location. This means an employee of a company in City A can send (upload) or receive (download) a file to a colleague in City B in a few seconds. The file can be a quotation document, a product leaflet, a digital photo or video stream.

In CCTV, we are interested mostly in video images (still images or a motion sequence), but other details such as audio, alarms, and various data logged during the system operation, can also be accessed. The principles in a network are the same. With the appropriate security level and a correct password, all such CCTV information can be accessed, copied, and displayed from anywhere on the network.

Understandably, if typical analog cameras are used in a CCTV system, they have to be first converted to digital (unless the cameras are already producing digital output) in order to be recognized and processed by the network computers. We will talk more about the digitization and compression of the video later in the book, but it is important to note that once the information is converted to digital, it is easily shared, copied, printed, transferred, stored, and used, providing one has the correct authorization levels.

This is a very important advantage in security applications since remote sites can be monitored and systems can be controlled from anywhere at any time. Most digital video recorders, or network cameras, are designed to allow you to view information from a remote location as if you were physically there.

Photo courtesy of Axis

Typical network cameras and a PTZ controller

Ethernet

First, let us present a little bit of history.

In 1973, at Xerox Corporation's Palo Alto Research Center (more commonly known as PARC), researcher **Bob Metcalfe** designed and tested the first Ethernet network. While working on a way to link Xerox's "Alto" computer to a printer, Metcalfe developed the physical method of cabling that connected devices on the Ethernet as well as the standards that governed communication on the cable. The data rate of such connection was around 3 megabits per second (3 Mb/s).

The father of modern Ethernet: Bob Metcalfe

Metcalfe's original paper described Ethernet as *"a branching broadcast communication system for carrying digital data packets among locally distributed computing stations. The packet transport mechanism provided by Ethernet has been used to build systems which can be viewed as either local computer networks or loosely coupled multiprocessors. An Ethernet's shared communication facility, its Ether, is a passive broadcast medium with no central control. Coordination of access to the Ether for packet broadcasts is distributed among the contending transmitting stations using controlled statistical arbitration. Switching of packets to their destinations on the Ether is distributed among the receiving stations using packet address recognition."*

After this event, a consortium of three companies – Digital Equipment Corporation (DEC), Intel, and Xerox – produced a joint development around 1980, which defined the 10 Mb/s Ethernet version 1.0 specification. In 1983, the **Institute of Electrical and Electronic Engineers** (IEEE) produced the IEEE 802.3 standard, which was based on, and was very similar to, the Ethernet version 1.0.

In 1985 the official standard **IEEE 802.3** was published, which **marks the beginning of the new era**, leading the way to the birth of the Internet a few years later.

Ethernet has since become the most popular and most widely deployed network technology in the world. Many of the issues involved with Ethernet are common to many network technologies, and understanding how Ethernet addressed these issues can provide a foundation that will improve your understanding of networking in general.

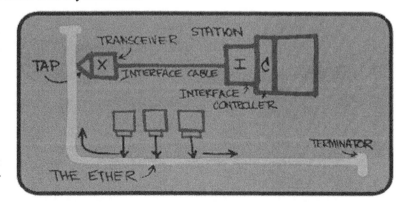

The original Metcalfe Ethernet concept drawing

The main Ethernet IEEE standards

The IEEE stands for Institute of Electrical and Electronic Engineers. This is a standard creating body which has developed the "802" series of standards dealing with networking. The number 802 was simply the next free number IEEE could assign, though "802" is sometimes associated with the date the first meeting was held — February 1980.

The "802" standards are now present in our daily life when working with internet, mobile communications and certainly CCTV, so we will mention here the most important ones.

802.3   *= 10 Mb/s Ethernet (1.25 MB/s)*

802.3u   *= 100 Mb/s Fast Ethernet (12.5 MB/s) w/autonegotiation*

802.3ab   *= 1 Gb/s Gigabit Ethernet (125 MB/s) over copper*

802.3z   *= 1 Gb/s Gigabit Ethernet (125 MB/s) over fibre*

802.3an   *= 10 GBase-T Ethernet 10 GB/s (1250 MB/s) over copper twisted pair (UTP)*

802.3ae   *= 10 Gbit/s Ethernet (1250 MB/s) over single-mode fibre*

802.3aq   *= 10 Gbit/s Ethernet (1250 MB/s) over multi-mode fibre*

802.11a/b/g/n Wireless Ethernet Wi-Fi = 11 Mb/s, 54 Mb/s and 100 Mb/s

802.3af   *= Power over Ethernet (12.95W)*

802.3at   *= Power over Ethernet (25.5W)*

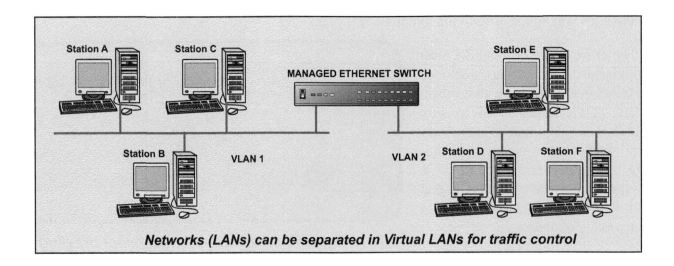

Networks (LANs) can be separated in Virtual LANs for traffic control

10 Mb/s Ethernet (802.3)

This Ethernet category refers to the original shared media Local Area Network (LAN) technology running at 10 Mb/s. Since 8 bits make one Byte, the 10 Mb/s is equivalent to 1.25 MB/s which could be used for a quick and approximate calculation of the time required to send certain number of MB over 10 Mb/s. We say approximate, because certain overheads have to always be considered, as they depend on the quality of the network itself, length, the type of data transmitted, as well as switchers quality. Ethernet can run over various media such as twisted pair and coaxial. *10 Mb/s Ethernet* is distinct from other higher speed Ethernet technologies such as FastEthernet, Gigabit Ethernet, and 10 Gigabit Ethernet.

Depending on the media, 10 Mb/s Ethernet can be subdivided to:

- 10BaseT – Ethernet over Twisted Pair Media

- 10BaseF – Ethernet over Fiber Media

- 10Base2 – Ethernet over Thin Coaxial Media

- 10Base5 – Ethernet over Thick Coaxial Media

Twisted Pair media connector

Fast Ethernet (802.3u)

The ***IEEE 802.3u (Fast Ethernet)*** covers a number of 100 Mb/s Ethernet specifications. It offers a speed increase 10 times that of the 10BaseT Ethernet specification, while preserving such qualities as frame format, MAC mechanisms, and MTU. Such similarities allow the use of existing 10BaseT applications and network management tools on Fast Ethernet networks. A 100 Mb/s is equal to 12.5 MB/s. One important limit of the 802.3u over copper is the distance limit of 100 m (approximately 330 feet) for the guaranteed bandwidth.

Gigabit Ethernet (802.3ab/z)

Gigabit Ethernet (IEEE 802.3ab and 802.3z) builds on top of the Ethernet protocol but increases speed tenfold over Fast Ethernet to 1000 Mb/s, or 1 gigabit per second (Gb/s). This is equivalent to 125 MB/s. Gigabit Ethernet allows Ethernet to scale from 10/100 Mb/s at the desktop computer to 100 Mb/s up the riser to 1000 Mb/s in the data center. The IEEE 802.3 standard of Gigabit Ethernet over copper is denoted with "ab" and with "z" over fiber.

By leveraging the current Ethernet standard as well as the installed base of Ethernet and Fast Ethernet switches and routers, IT engineers do not need to retrain and relearn a new technology in order to provide support for Gigabit Ethernet.

Gigabit Ethernet over copper (802.3ab) specifies operation   **Ethernet fiber connectors**

over the Category 5e/6 cabling systems. As a result, most copper-based environments that run Fast Ethernet can also run Gigabit Ethernet over the existing network infrastructure.

Many large scale digital CCTV networks use switchers with individual ports of 100 Mb/s per channel and then such switchers interconnect between themselves using the Gigabit Ethernet. Very large digital CCTV systems, such as casinos, typically use copper between IP cameras and network switches, and fiber Ethernet between main network switches. This is not only because of the guaranteed bandwidth of gigabit, but also due to need for longer distances than the limit of 100 m with copper.

10 Gigabit Ethernet (802.3an/ae/aq)

10 Gigabit Ethernet (10 GbE) is even faster version of Ethernet. Not very often used in CCTV to date, as it is more demanding and more expensive to implement. Like previous versions of Ethernet, 10 GbE supports both copper and fiber cabling. However, due to its higher bandwidth requirements, higher-grade copper cables are required, such as Category (Cat)-6a and Cat-7 cables for distances of up to 100 m. Unlike previous Ethernet standards, 10 gigabit Ethernet defines only full duplex point-to-point links which are generally connected by network switches. Half duplex operation and hubs to not exist in 10 GbE. Because 10 Gigabit Ethernet is still Ethernet, it supports all intelligent

10 GbE switcher

Ethernet-based network services such as multi-protocol label switching (MPLS), Layer 3 switching, quality of service (QoS), caching, server load balancing, security, and policy-based networking. And it minimizes the user's learning curve by supporting familiar management tools and architectures. With data rates of 10 Gb/s, which is equivalent to 1.25 GB/s, it offers a low-cost solution to the demand for higher bandwidth in the LAN, MAN, and WAN.

Wireless Ethernet (802.11 a/b/g/n/ad)

The **IEEE 802.11**, or Wi-Fi, denotes a set of Wireless LAN standards developed by working group 11 of the IEEE 802 group. Although Wi-Fi is not as widely spread in CCTV as in home use, there are a number of projects where wireless cameras are being installed for practical reasons, where cabling is not possible.

The 802.11 family currently includes five (more to come) separate protocols that focus on data encoding (a, b, g, n, and d). The 802.11b was the first widely accepted wireless networking standard, followed, paradoxically, by 802.11a and 802.11g. The frequencies

The Wi-Fi logo

used by the 802.11 are in the microwave range (2.4GHz and 5GHz) and most are subject to minimal governmental regulation. Licenses to use this portion of the radio spectrum are not required in most locations.

Typical maximum distances in open space are up to 100 m for 802.11a, up to 140 m for 802.11b/g, and up to 250 m for 802.11n. Typical maximum bandwidth depends on many factors, but starts from around few Mb/s for 802.11a, up to 150 Mb/s for 802.11n running on 5 GHz.

Power over Ethernet (PoE) (802.3af/at)

The *IEEE 802.3af/at* standards for Power over Ethernet (PoE) denotes the procedure by which various network components (e.g. an IP camera) can provide Ethernet data communication and be powered with electricity using the same copper Cat cable. The switches capable of powering devices connected to it are typically called PoE switches. They are made to be backwards compatible with non-PoE devices, which means if a non-PoE camera is connected to a PoE switch it should work as normal. On the other hand, if a PoE camera is connected to PoE switch, the first function of the PoE switch is to determine if the terminal device (PD = powered device) requires power, by performing a resistance test. In the next step, the maximum amount of electricity the device is capable of consuming is determined based on the PoE class.

With PoE standard *802.3af*, the maximum output power of a PSE (Power Supply Equipment) is limited to approximately 15 Watt. As a result of power losses and cable length, which have to be taken into account, the end consumer (the network component) is allowed to

A small PoE switch

draw a maximum of 12.95 W. This is more than sufficient for any IP camera. Some manufacturers even allow for infra red LED illuminators, in addition to the IP camera, to be powered of the same PoE switch from this class.

With the newer PoE+ *802.3at* standard, an output power of up to 30 W is possible. The maximum power draw a device can have from this (due to losses) is 25.5 W. In CCTV, this is sufficient for some IP PTZ cameras. Since there are only very few cases in which a PoE switch can provide all ports with maximum PoE power, the configuration of the switch may change and this is defined by the respective connecting devices. Some PoE switches only offer a restricted number of PoE outlets, others limit the performance by a PoE budget, yet others allow for a detailed power configuration via the switch management culminating in a shut down priority. If the PoE budget is all used, further ports are no longer supplied with electricity. For some switches, this already applies as soon as the budget of the maximum theoretical power is reached.

In some digital systems with PoE cameras, where the switch is not PoE capable, it is possible to use PoE injectors which combine external power supply with the Ethernet cat cable.

Data speed and types of network cabling

By definition, Ethernet is a local area technology and works with networks that traditionally operate within a single building, connecting devices in **close proximity**. In the beginning coaxial cable was used for most Ethernet networks, but twisted pair, first Category 3, then Category 5 and Category 6 became the preferred medium for small LANs.

Ethernet uses a bus or star topography (or mix of the two) and supports data transfer rates of 10, 100, 1000, or 10,000 Mb/s. The Ethernet standard has grown to encompass new technologies as computer networking has matured, but the mechanics of operation for every Ethernet network today stem from Metcalfe's original design. The original Ethernet described communication over a single cable shared by all devices on the network. Once a device attached to this cable, it had the ability to communicate with any other attached device. **This allows the network to expand to accommodate new devices without requiring any modification to those devices already on the network.**

One of the most common questions in CCTV today is how quick an image update is over a network or how long a download of certain footage will take.

In order to be able to understand and calculate this the readers should be reminded that there is a difference between bits (marked with the lower case "b") and Bytes (marked with capital "B"). Typically, **there are 8 bits in one Byte**. Therefore, when making a rough calculation as to how long it would take to download a file over a particular data link connection, the Mb/s data transfer rate needs first to be converted to MB/s by dividing it by 8; also, allowance should be made for traffic collision losses and noise, which could vary anywhere from 10 to 50%. So in many worst case scenario calculations, 50% of the data transfer rate should be used.

For example, if we have a dial-up Internet connection with a typical modem of 56 kb/s, the maximum transfer rate will be around 6 to 7 kB/s, as the best, and around 3 kB/s as the worst case scenario. With a dial-up PSTN connection, we still use analog modulation techniques, the quality of which can vary greatly, depending on the line noise, distance and hardware quality, so it is possible that the worst case scenario can be even lower than 3 kB/s. So, when using 56 kb/s modem, there is no guarantee that the established connection will be 56 kb/s, but that represents the maximum achievable data transfer rate

File size	Approximate time taken to download the file size on the left using the speed below						
	52 kb/s (PSTN)	64 kb/s ISDN	128 kb/s (2xISDN)	512 kb/s	1 Mb/s	10 Mb/s	100Mb/s
1 MB	2.5min	2 min	1min	16s	8s	1s	0s
10 MB	24min	21 min	10.5min	2min 36s	1min 18s	8s	1s
100 MB	4hrs 16min	3hrs 30min	1hr 45min	26min	13min	1min 20s	8s
1 GB	1day 18hrs 40min	1day 11hrs	17.5hrs	4hrs 21min	2hrs 10min	13min 20s	1min 20s
10 GB			7days 7hrs	1day 19hrs	22hrs	2hrs 13min	13min 20s
100 GB				18days	9days 8hrs	22hrs 10min	2hrs 13min
1 TB						9days 8hrs	22hrs
10 TB							9days 8hrs

Download time for various files with various data speeds

when conditions are excellent. Going back to our example, if we have for example, a 1 MB file to download, it would take at least 150 seconds, which is around 2.5 minutes, when good-quality PSTN dial-up connection is used. For the same file to be transferred over an ADSL Internet connection of 512 kb/s, it would take much faster, but at least 16 seconds (512 kb/s = 64 kB/s; 1024 kB divided by 64 kB = 16 sec) and can go up to 32 seconds, if equipment and lines are of low quality. This is still much faster compared to over 2.5 minutes for a 56 kb/s PSTN modem connection.

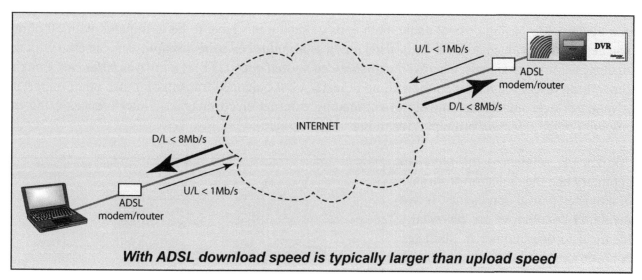

With ADSL download speed is typically larger than upload speed

In calculations as illustrated on the previous page, consideration should be given to the fact that the speed of download of a file is as fast as the slowest speed in the chain. This means that if a computer you download from has a limited upload speed which is much lower than your download speed, then that will define your download time.

More importantly, when using ADSL for connecting to the internet, the download speed is typically larger than the upload speed at the modem itself. This is actually the reason why ADSL is called *"asymmetric digital subscriber line."* Asymmetric denotes inequality of the download and upload. This doesn't affect a normal usage of internet, as most of the time we download data, rather than upload. However, if we want to connect to a remote CCTV, the upload at the remote ADSL will be the bottleneck.

The same principles of data speed calculations apply to a variety of network communication and storage devices. Each component in a computer and a network has its own limitation imposed by the component. This is a very important consideration especially in the modern CCTV digital system designs where there is an ever growing need for faster transmission, more cameras recorded, and more pictures per second required. All components in such a chain of video streaming influence the total performance result. The network is not always the bottleneck. If, for example, we have a Gigabit Ethernet in place (with matching network cards and network switchers or routers), it is quite possible that a computer we have as a video recorder uses SATA hard drives which could be slower than the network itself, and as such becomes a bottleneck in playing back multiple cameras on multiple operators' consoles.

Being aware of the totality of a digital networked system and of each single component making such system is the key to a successful implementation of this new technology we have embraced in CCTV.

Ethernet over coax, UTP and fiber cables

In the past, the first and most common Ethernet media was coaxial cable. The Ethernet coaxial cable uses 50 ohms impedance and cable (RG-58), as opposed to 75 ohms for analog video (RG-59). Very often, due to similar sizes, errors have been made where RG-58 was used for CCTV, and RG-59 for Ethernet. So, care should be taken that these not be mixed. **When using Ethernet coaxial cables, correct terminations are as important, if not more so, as in analog video.** If a network is configured in a bus topology using coaxial cable, both ends of such a bus have to be terminated with 50 ohms. The networking with coaxial cables is also known as ***unbalanced transmission***, as is the case with the analog CCTV video signals, whereas the ***unshielded twisted pair*** (UTP) is known as ***balanced***. Coaxial networking offers longer distances with no repeaters, but balanced transmission has other important advantages over unbalanced mostly in eliminating external electromagnetic interferences using the ***common mode rejection*** principles (as in twisted pair video).

"Balanced" relates to the physical geometry and the dielectric properties of a twisted pair of conductors. If two insulated conductors are physically identical to one another in diameter, concentricity, and dielectric material, and are **uniformly twisted with equal length of conductor**, then the pair is **electrically balanced** with respect to its surroundings. The degree of electrical balance depends on the design and manufacturing process. **For balanced transmission, an equal voltage of opposite polarity is applied on each conductor of a pair.** The electromag-

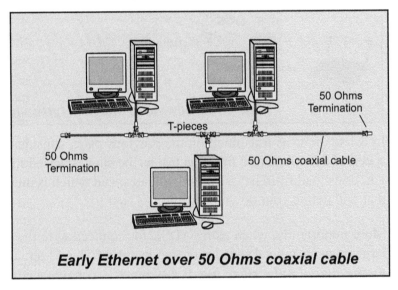

Early Ethernet over 50 Ohms coaxial cable

netic fields created by one conductor **cancel out** the electromagnetic fields created by its "balanced" companion conductor, leading to very little radiation from the balanced twisted pair transmission line. **The same concept applies to external noise that is induced on each conductor of a twisted pair.** A noise signal from an external source, such as radiation from a radio transmitter antenna, generates an equal voltage of the same polarity, or "common mode voltage," on each conductor of a pair. The

difference in voltage between conductors of a pair from this radiated signal (the "differential voltage") is effectively zero.

Since the desired signal on the pair is the differential signal, the interference practically does not affect balanced transmission.

High quality Cat-6a cable offers gigabit Ethernet

The degree of electrical balance is determined by measuring the "differential voltage" and comparing it to the "common mode voltage" expressed in decibels (dB). For most CCTV LANs the Cat-5/5e and Cat-6/6a are most common, as they are easy to prepare. They offer good-quality networking with maximum allowed distance between source and destination of not more than 100 m (330 ft).

For longer distances, fiber Ethernet is recommended.

The term "Cat" refers to *classifications category* of UTP (unshielded twisted pair) cables. The difference in classifications of the cables is based mainly on the bandwidth, copper type, size and electrical performance.

- **Cat-1** – Traditional telephone cable.

- **Cat-3** – Balanced 100 Ohms cable and associated connecting hardware whose transmission characteristics are specified up to 16 MHz. It is used for 10 Mb/s and to an extent 100 Mb/s. Category 3 is the most common type of previously installed cable found in corporate wiring schemes, and it normally contains four pairs.

- **Cat-5** – Balanced 100 Ohms cable and associated connecting hardware whose transmission characteristics are specified up to 100 MHz. This bandwidth allows for data speeds of up to 100 Mb/s over one single run of 100 m (330 ft) distance.

- **Cat-5e** – The Cat-5e specification improves upon the Cat-5 specification by tightening crosstalk specifications and improving it by at least 3 dB. The bandwidth of category 5 and 5e is the same - 100 MHz and it guarantees data rate of 100 Mb/s over 100 m (330 ft). With Cat-5e it is possible to even run 1 Gb/s at shorter distances.

- **Cat-6** – Cat-6 features more stringent specifications for crosstalk and system noise compared to previous categories. The cable standard provides performance of up to 250 MHz and guarantees up to 1 Gb/s over 100 m. When used for 10 Gb/s, the Cat-6 cable's maximum length is 55 meters (180 ft) in a favorable environment, but only 37 meters (121 ft) when many cables are bundled together.

- **Cat-6a** – Cat-6a is defined at frequencies up to 500 MHz, twice that of Cat-6. Cat-6a performs better, and if the environment is favorable, it guarantees data rate of up to 10 Gb/s for up to 100 meters (330 ft).

- **Cat-7** – This is also balanced 100 Ohms twisted pair cable, but with much more strict specifications for crosstalk and system noise than Cat-6. To achieve this, individual pairs of wires are shielded, and the cable as a whole. In addition to the foil shield, the pairs twisting per length is increased, protecting external interference and crosstalk. Cat-7 cable allows 600 MHz of bandwidth, which passes up to 10 Gb/s over 100 m. All four twisted pairs are used.

- **Cat-7a** – This is a similar category cable as Cat-7, just further improved to allow for a 1 GHz bandwidth to pass through over 100 m, which converts to bit rates of over 40 Gb/s propagated to at least 50 m, and 100 Gb/s up to 15 m.

The cable termination (connector) is known as **RJ-45** and resembles a large version of the RJ-11 telephone line connector.

Electrical signals propagate along a cable very quickly (typically 65% of the speed of light), but even for digital signals, as is the case with analog, the same electrical laws apply—they **weaken** as they travel, and **electrical interference** from neighboring electromagnetic devices **affects the signal**. The effects of voltage drop, combined with the effect of inductance and capacitance for high-frequency signals (high bit rate) and external electromagnetic interferences, impose physical limitations on how far a certain cable can carry data before it gets to a repeater (switch or router). A network cable must be short enough that devices at opposite ends can receive each other's signals clearly and with minimal delay. This places a **distance limitation** on the maximum separation between two devices. This is called **network diameter** of the Ethernet network, which for copper is around 100 m (330 feet).

An RJ-45 connector

Limitations apply for other Ethernet media as well, such as wireless or fiber optic, although the minimum distances are different from copper.

The most common network cable, Cat-5, uses AWG24 wires (with approximate diameter of 0.2 mm²) and has 100 ohm impedance. Readers are reminded that AWG (*American Wire Gauge*) is a system that specifies wire size. The gauge varies inversely with the wire diameter size, which defines the electrical resistance (the smaller the AWG number, the larger the conductor diameter, the smaller the resistance).

Twisted pair cable comes in two main varieties, solid and stranded. Solid cable supports longer runs and works best in fixed wiring configurations like office buildings. Stranded cable, on the other hand, is more pliable and better suited for shorter-distance, movable cabling such as "patch" cables.

A variation on Cat-5, called *Cat-5e*, **is even better performing network cable**. It was ratified in 1999, formally called ANSI/TIA/EIA 568A-5, or simply Category 5e (the e stands for **enhanced**). Cat-5e is also 100 ohm impedance cable and is completely

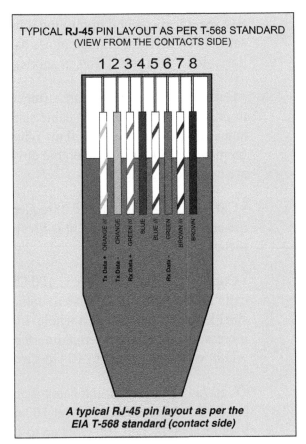

A typical RJ-45 pin layout as per the EIA T-568 standard (contact side)

backward compatible with the Cat-5 equipment. **The enhanced electrical performance of Cat-5e ensures that the cable will support applications that require additional bandwidth, such as gigabit Ethernet or analog video (if used in twisted pair video transmission).**

Cat-5e has an incremental improvement designed to enable cabling to support full-duplex Fast Ethernet operation and Gigabit Ethernet. The main differences between Cat-5 and Cat-5e can be found in the specifications where performance requirements have been raised slightly.

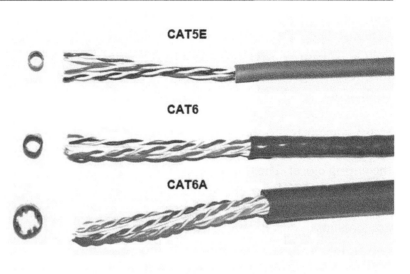

Some of the better Cat cables

While Cat-5 components may function to some degree in a Gigabit Ethernet (at shorter distances), they perform below standard during high data transfer scenarios. Cat-5e cables work better with gigabit speed products. So, when using a 100 Mb/s switch it is better to get Cat-5e cable instead of Cat-5.

The next level in the cabling hierarchy is Category 6 (ANSI/TIA/EIA-568-B.2-1), which was ratified by the EIA/TIA in June 2002. Also built to have 100 ohm impedance, Cat-6 cable requires a greater degree of precision in the manufacturing process compared to Cat-5. Similarly, a Cat-6 connector requires a more balanced circuit design. **Cat-6 provides higher performance than Cat-5e and features more stringent specifications for crosstalk and system noise. All Cat-6 components are backward compatible with Cat-5e, Cat-5, and Category 3.**

Category cables (copper) typical specifications						
	Cat-3	**Cat-5**	**Cat-5e**	**Cat-6**	**Cat-6a**	**Cat-7**	**Cat-7a**
Impedance	100 Ω	100 Ω	100 Ω	100 Ω	100 Ω	100 Ω	100 Ω
NEXT (Near End Cross Talk)	29 dB	32.3 dB	35.3 dB	44.3 dB	47.3 dB	62.1 dB	62.1 dB
Bandwidth	16 MHz	100 MHz	100 MHz	250 MHz	500 MHz	600 MHz	1 GHz
Maximum distances	100 m (< 10 Mb/s)	100 m (< 100 Mb/s)	100 m (< 100 Mb/s)	100 m (< 1 Gb/s) 55 m (< 10 Gb/s)	100 m (< 10 Gb/s)	100 m (< 10 Gb/s)	50 m (< 40 Gb/s) 15 m (< 100 Gb/s)

This table shows typical Category Unshielded Twisted Pair specifications.

The quality of the data transmission depends upon the performance of the components of the channel. So to transmit according to Cat-6 specifications, connectors, patch cables, patch panels, cross-connects, and cabling must all meet Cat-6 standards. The channel basically includes everything from the wall plate to the wiring closet. The Cat-6 components are tested both individually and together for performance. If different category components are used with Cat-6 components, then the channel will achieve the transmission performance of the lower category. For instance, if Cat-6 cable is used with Cat-5e connectors, the channel will perform at a Cat-5e level.

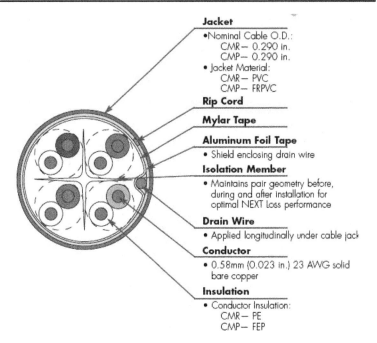

Jacket
- Nominal Cable O.D.:
  CMR— 0.290 in.
  CMP— 0.290 in.
- Jacket Material:
  CMR— PVC
  CMP— FRPVC

Rip Cord

Mylar Tape

Aluminum Foil Tape
- Shield enclosing drain wire

Isolation Member
- Maintains pair geometry before, during and after installation for optimal NEXT loss performance

Drain Wire
- Applied longitudinally under cable jack

Conductor
- 0.58mm (0.023 in.) 23 AWG solid bare copper

Insulation
- Conductor Insulation:
  CMR— PE
  CMP— FEP

Cat-6a cable description

Cat-6 cable contains four pairs of copper wire and, unlike Cat-5, utilizes all four pairs; the communication speed it supports is more than twice the speed of Cat-5e. **As with all other types of twisted pair EIA/TIA cabling, Cat-6 cable runs are limited to a maximum recommended run rate of 100 m (330 ft).**

Because of its improved transmission performance and superior immunity from external noise, systems operating over Category 6 cabling will have fewer errors than Category 5e for current applications. This means fewer re-transmissions of lost or corrupted data packets under certain conditions, which translates into higher reliability.

The "fastest" copper cable currently covered by the EIA/TIA standards is the Category 7 targeting Gigabit networks. Cat-7 is fully compatible with previous standards. The Cat-7 is no longer unshielded. The specification requirements are so high that each pair has to be shielded, and in addition all four pairs have to be shielded again, making the Cat-7 the most expensive Cat cable. Also, Cat-7 no longer suggest RJ-45 connectors.

Cat-7a is a similar category cable as Cat-7, just further improved to allow for a 1 GHz bandwidth to pass through over 100 m, which converts to bit rates of over 40 Gb/s propagated to at least 50 m, and 100 Gb/s up to 15 m. Many users will argue that fiber optics is a better choice once you see a need for such a high-performance network cable, so we will leave this categorization for further reading elsewhere in more up-to-date books and manuals, but readers should be aware that new cable categories have been developed.

The Category Ethernet cables have 8 wires (4 pairs), of which 4 (2 pairs) are used for data on Cat-5/5e and all 8 wires (4 pairs) are used on Cat-6 /6a and Cat-7/7a.

Patch and crossover cables

Two kinds of wiring schemes are available for Ethernet cables: **patch (or straight)** and **crossover** cables.

Patch (or straight) cables are used for connecting computers with switches or routers.

The crossover cables are normally used to connect two PCs without the use of a switcher in between. Many modern switchers and computers have auto-sensing circuitry and if a crossover cable is accidentally used instead of a straight one the communications can still be established.

A crossover cable is a segment of cable that crosses over pins 1&2 and 3&6, which the Tx and Rx pins in order to be able to have two computers exchange information. If a cable does not say crossover, it is a standard patch cable.

If you are not sure what type of cable you have, you can put the two RJ-45 connectors next to each other from the same side (as shown on the photos here) and **if the wiring colors are in identical order from left to right, then it is a patch cable. If pins 1&2 have reversed color wires order, then it is a crossover**. A good practice is to always have the crossover cable color different from the color of the majority of the patch cables used – for example, a yellow crossover cable amongst blue-colored patch cables.

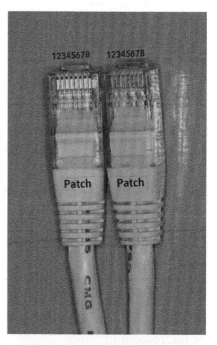

Stranded cable, as opposed to solid core, has several small gauge wires in each separate insulation sleeve. Stranded cable is more flexible, making it more suitable for patch cords. When using **patch cables, the recommended maximum lengths are around 10 m** (30 ft). This construction is great for the flexing and the frequent changes that occur at the wall outlet or patch panel. **The stranded conductors do not transmit data signals as far as solid cable**. The EIA/TIA 568A standard limits the length of patch cables to 10 meters in total length. It does not mean you cannot use stranded cable for longer runs; it is just not recommended. Some installations have stranded cable running over 30 meters with no problems, but care should be taken not to use stranded cable in larger installations.

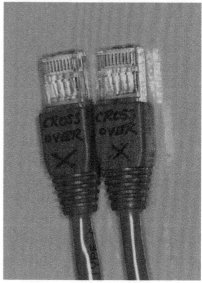

Solid copper cable has one larger gauge wire in each sleeve. Solid cable has **better** electrical performance than stranded cable and is traditionally used for inside walls and through ceilings, or any type of longer run of cable. **All such Category network cables (using solid core) are specified for a maximum length of around 100 m (330 ft) before a repeater is needed**.

Patch and crossover

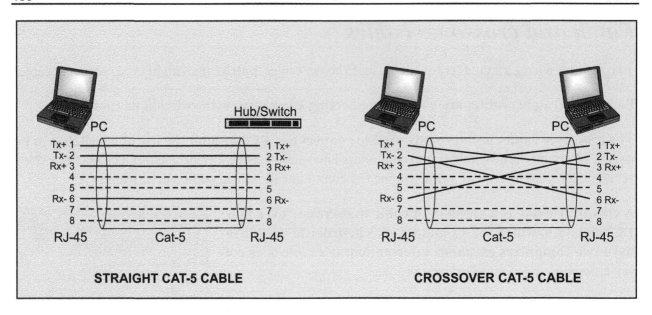

STRAIGHT CAT-5 CABLE   **CROSSOVER CAT-5 CABLE**

This is not to say that longer distances are not possible, but this very much depends on the cable quality and the intended network bandwidth. For example, if Cat-6 cable is used for up to 100 Mb/s, then longer distances than 100 m can be achieved since Cat-6 is very stringent in its design and it targets the Gigabit network speeds. How much longer the cable can be run without a repeater (router/switch) can only be determined by test.

One of the main sources of problems for any copper cabling, including the "Cat" types of cable, are the ***electromagnetic interferences***. Electromagnetic interferences (EMIs) are potentially harmful to your communications system because they can lead to signal loss and degrade the overall performance of high-speed Cat cabling. EMI interference in signal transmission or reception is caused by the radiation of electrical or magnetic fields which are present nearby all power cables, heavy electric machinery, or fluorescent lighting.

This is unfortunately the nature of electrical current flowing through copper cable, and it is the basics of electromagnetic interdependence. We say "unfortunately" in this case, when discussing the unwanted interference to signal cables, but in fact the same concept is used for generating electric power and moving electric motors, in which case the EMI (read it as *electromagnetic inductance* in such case) is a highly desirable effect.

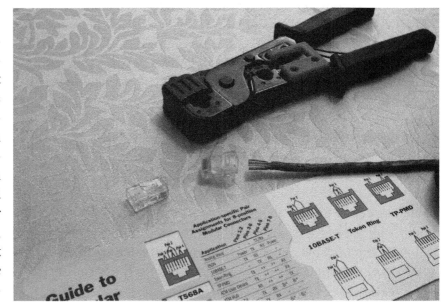

A variety of RJ-45 crimping tools are available.

Avoiding EMI is as simple as not laying the network cable within 30 cm (1 ft) of electrical cable, or, if needed, switching from UTP to more expensive shielded cable. These are basic rules that should be applied at all times.

A variety of expensive and cheaper tools are available to verify the patch or crossover cable quality, and it is recommended that every network cable installer should have at least the basic one.

The only time EMI is not an issue is when using fiber cables. This is simply because fiber does not conduct electricity but uses light as a transmission media. All longer distance and wider bandwidth communications are usually achieved with fiber cables, for they offer not only longer distances (a couple of kilometers) but much wider bandwidth. Most importantly, they are not subject to EMI.

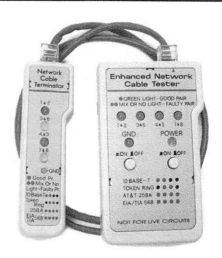

LAN cable tester

Fiber-optic network cabling

As was the case in analog video transmission, fiber optics has some significant technological advantages over copper when used for networking.

Fiber optics can transmit wider bandwidth data and longer distances than copper. Large digital CCTV systems are are almost impossible without fiber.

This means less equipment and infrastructure (such as switches and wiring cabinets) is needed, thereby lowering the overall cost of the LAN. Fiber optics is physically much thinner and more durable than copper, taking up less space in cabling ducts and allowing for a greater number of cables to be pulled through the same duct. New developments in fiber optics cabling also allow it to be tied in a knot and still function normally. As already described under the analog video transmission over fiber section in Chapter 10, fiber optics completely encloses the light pulses within the outer sheath, making it impervious to outside interference or eavesdropping.

Various types of fiber cables

Another very important property **of fiber optic is its immunity to any electromagnetic interference, including lightning induction**. You can submerge it in water, and it is less susceptible to temperature fluctuations than copper. All these qualities make fiber optics cable the ultimate choice.

Fiber provides higher bandwidth (approximately 50 Gb/s, that is 50 gigabits per second, over multi-mode and even higher over single-mode fiber), and it "future proofs" a network's cabling architecture against copper upgrades.

Although users currently do not require speeds faster than Fast Ethernet in small to medium-size CCTV projects, the cost differential between copper and fiber optics will become less and less significant, making fiber optics a compelling option for any size system. Fiber optics infrastructures are still more expensive than copper. Fiber optics switch ports and adapter cards cost, on average, approximately 50% more than comparable copper products. However, when you factor in the cost savings associated with fiber

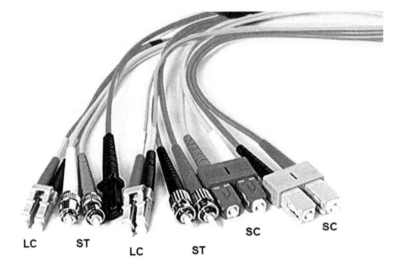

Various fiber connector standards

(such as the need for fewer repeaters and switches, wider bandwidth), the overall cost of a fiber optics system drops comparable to one with a copper-based LAN.

When you eliminate the expense of creating and maintaining extra wiring cabinets, a fiber optics LAN costs about the same, or even less, than a copper LAN. In the past, fiber optics' lofty price had little to do with the medium itself – most of the expense lay in transceivers and connectors. Due to new products in each of these areas, costs have been decreasing, pushing fiber optics use upward.

Maximum distances achievable with a single run of fiber depends on the type of fiber (multi-mode or single-mode) as well as the transmitting and receiving equipment. The accurate distances can only be found after testing an installation with an OTDR (***Optical Time Domain Reflectometer***), which will naturally consider the quality of terminations, cable, and equipment.

Typical maximum distances with fiber depend on the type of fiber used (multi-mode or single-mode) as well as the light launcher (LED or laser). The following are typical distances for gigabit fiber ethernet as per the IEEE 802.3ap recommendations:

- **1000Base-SX** – 62.5 μm/125 μm multi-mode fiber 220 m, and 50 μm/125 μm 550 m

- **1000Base-LX/MMF** – Multi-mode fiber with laser 550 m

- **1000Base-LX/SMF** – Single-mode fiber with laser 5 km

- **1000Base-LX10** – Pair of single-mode fiber with 1310 nm laser, 10 km

- **1000Base-EX** – Pair of high quality single-mode fiber with 1310 nm laser, 40 km

- **1000Base-ZX** – Single-mode fiber with 1550 nm laser, 70 km

Because of the high bandwidth and distances it can handle, fiber optics is most often used as a network backbone, where more network segments are connected in a larger network, typical for the digital CCTV systems of casinos and large shopping centers. In such system design, fiber to copper media converters are used. Many different makes and models are available on the market; they can be stand alone, or multiple converters can be housed in 19" racks.

It is important here to highlight here again the importance of proper tools for testing and fault-finding fiber networks. If you consider fiber networks a serious part of your CCTV business, investing in good-quality instruments and tools is always a wise thing to do. If, however, this is beyond the reach of your budget, you can always hire specialized fiber optics businesses that can perform most of the tasks on your behalf. If the fiber cable is already installed, they would typically charge per fiber connection termination, which would include an OTDR report.

Courtesy of Fluke

We have already explained and described a few fiber cable termination methods in this book, under the analog video transmission media.

With proper tools everything is known – OTDR for networks.

It has, however, become much easier to terminate fiber today, compared to about ten years ago. By learning the basic procedures and a little bit of practice, with the help of right tools and instruments, it is not difficult to master this modern media. Many fiber manufacturers conduct regular training and this is the best starting point.

Photo courtesy of Advantech

A network switch with fiber link

Wireless LAN

An increasing number of CCTV products and projects are starting to use wireless LAN (WLAN). The acceptance and practicality of wireless communications between computers, routers, or digital video devices is becoming so widespread that manufacturers started to bring out even better and cheaper devices. After many years of proprietary products and ineffective standards, the industry has decided to back one set of standards for wireless networking: the 802.11 series from the Institute of Electrical and Electronics Engineers (IEEE). These standards define wireless Ethernet, or wireless LAN (WLAN), also referred to as Wi-Fi (Wireless Fidelity).

With the electronic technology development WLAN standards develop too, so that today there are many "flavors" of the 802.11 standards, and even more versions come out. These are typically denoted with a letter suffix after the 802.11.

Wi-Fi is really only suitable for smaller IP CCTV systems, as the number of channels 802.11 uses and the distance it can go to is limited. It does help however in projects where running cables is limited or impossible. Clearly, when using radio frequencies for communications, the power of the antennas as well as the environment these frequencies are used in, determine the distances one can go to. **Typical indoor distances with the latest 802.11n can be anything up to 70 m (230 ft), and outdoor up to 250 m (820 m), in clean and unpolluted environments.**

If longer distances are needed, solution should be sought in more powerful wireless solutions than Wi-Fi, which fall in the category of generic wireless broadband with radiation power and antenna shapes and sizes designed for open space.

There have been a number of wireless IP CCTV system designs around the world, using the so-called wireless mesh topology. Such solutions might be attractive for urban city areas, where physical cabling might be very difficult, expensive, or impossible.

Photo courtesy of Ubiquiti

Wireless network cameras

What is 802.11?

IEEE 802.11, or Wi-Fi, denotes a set of Wireless LAN standards developed by working group 11 of the IEEE 802 group. The term is also used specifically for the original version; to avoid confusion, that is sometimes called "802.11 legacy."

The 802.11 family currently includes three separate protocols that focus on encoding (a, b, g); security was originally included but is now part of other family standards (e.g., 802.11i). Other standards in the family (c-f, h-j, n) are service enhancement and extensions, or corrections to previous specifications.

802.11b was the first widely accepted wireless networking standard, followed, paradoxically, by 802.11a and 802.11g.

The frequencies used by the 802.11 are in the microwave range and most are subject to minimal governmental regulation. Licenses to use this portion of the radio spectrum are not required in most locations.

802.11 Wi-Fi network standards (source: wikipedia.org)

802.11 protocol	Release	Freq. (GHz)	Bandwidth (MHz)	Data rate per stream (Mb/s)	Allowable MIMO streams	Modulation	Approximate indoor range (m)	Approximate indoor range (ft)	Approximate outdoor range (m)	Approximate outdoor range (ft)
—	01/06/1997	2.4	20	1, 2	1	DSSS, FHSS	20	66	100	330
a	01/09/1999	5	20	6, 9, 12, 18, 24, 36, 48, 54	1	OFDM	35	115	120	390
a	01/09/1999	3.7	20	6, 9, 12, 18, 24, 36, 48, 54	1	OFDM	—	—	5,000	16,000
b	01/09/1999	2.4	20	1, 2, 5.5, 11	1	DSSS	35	115	140	460
g	01/06/2003	2.4	20	6, 9, 12, 18, 24, 36, 48, 54	1	OFDM, DSSS	38	125	140	460
n	01/10/2009	12/25	20	7.2, 14.4, 21.7, 28.9, 43.3, 57.8, 65, 72.2	4	OFDM	70	230	250	820
n	01/10/2009	12/25	40	15, 30, 45, 60, 90, 120, 135, 150	4	OFDM	70	230	250	820
ac (DRAFT)	~Feb 2014	5	20	up to 87.6	8	OFDM				
ac (DRAFT)	~Feb 2014	5	40	up to 200	8	OFDM				
ac (DRAFT)	~Feb 2014	5	80	up to 433.3	8	OFDM				
ac (DRAFT)	~Feb 2014	5	160	up to 866.7	8	OFDM				
ad	01/12/2012	2.4/5/60		up to 7000						

802.11 (legacy)

The original version of the standard IEEE 802.11 released in 1997 and sometimes called "802.1y" specifies two data rates of 1 and 2 megabits per second (Mb/s) to be transmitted via infrared (IR) signals or in the industrial scientific medical frequency band at 2.4 GHz.

IR has been dropped from later revisions of the standard because it could not succeed against the well established IrDA protocol and has had no actual implementations.

Legacy 802.11 was rapidly succeeded by 802.11b.

802.11b

802.11b has a range of about 35 meters indoor, with the low-gain omnidirectional antennas typically used in 802.11b devices. 802.11b has a maximum throughput of up to 11 Mb/s; however, a significant percentage of this bandwidth is used for communications overhead. In practice, the maximum throughput is about 5.5 Mb/s. Metal, water, and thick walls absorb 802.11b signals and decrease the range drastically. 802.11 runs in the 2.4 GHz spectrum and uses Carrier Sense Multiple Access with Collision Avoidance (CSMA/CA) as its media access method.

Extensions have been made to the 802.11b protocol (e.g., channel bonding and burst transmission techniques) in order to increase speed to 22 Mb/s, 33 Mb/s, and 44 Mb/s, but the extensions are proprietary and have not been endorsed by the IEEE. Many companies call enhanced versions "802.11b+".

The first widespread commercial use of the 802.11b standard for networking was made by Apple Computer under the trademark AirPort.

802.11a

In 2001, 802.11a (which came after 802.11b), a faster related protocol started shipping even though the standard was ratified in 1999. The 802.11a standard uses the 5 GHz band, and operates at a raw speed of up to 54 Mb/s and more realistic net achievable speeds in the mid-20 Mb/s. The speed is reduced to 48, 36, 34, 18, 12, 9, and then 6 Mb/s if required. 802.11a has 12 non-overlapping channels, 8 dedicated to indoor and 4 to point to point.

Different countries have different ideas about regulatory support, although a 2003 World Radio-telecommunications Conference made it easier for use worldwide.

802.11a has not seen wide adoption because of the high adoption rate of 802.11b and because of concerns about range: at 5 GHz, 802.11a cannot reach as far as 802.11b, other things (such as same power limitations) being equal. It is also absorbed more readily.

Most manufacturers of 802.11a equipment countered the lack of market success by releasing dual-band/ dual-mode or tri-mode cards that can automatically handle 802.11a and b or a, b, and g as available. Access point equipment that can support all these standards simultaneously is also available.

802.11g

In June 2003, a third standard for encoding was ratified - the 802.11g. This flavor works in the 2.4 GHz band (like 802.11b) but operates at up to 54 Mb/s raw, or about 24.7 Mb/s net, throughput like 802.11a. It is fully backwards compatible with standard 802.11b and uses the same frequencies. Details of making b and g work together well occupied much of the lingering technical process. However, the presence of an 802.11b participant reduces an 802.11g network to 802.11b speeds.

The 802.11g standard swept the consumer world of early adopters starting in January 2003, well before ratification. Most of the dual-band 802.11a/b products became dual-band / tri-mode, supporting a, b, and g in a single card or access point. A new feature called Super G is now integrated in certain access points. These can boost network speeds up to 108 Mb/s by using channel bonding. This feature may interfere with other networks and may not support all b and g client cards. In

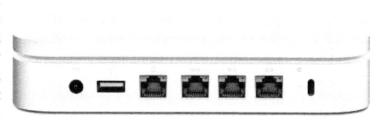

Apple Airport Extreme

addition, packet bursting techniques are also available in some chipsets and products which will also considerably increase speeds. Again, they may not be compatible with some equipment.

802.11b and 802.11g divide the spectrum into 14 overlapping, staggered channels of 22 megahertz (MHz) each. Channels 1, 6, 11, and 14 have minimal overlap, and those channels (or other sets with similar gaps) can be used where multiple networks cause interference problems. Channels 10 and 11 are the only channels which work in all parts of the world, because Spain and France have not licensed channels 1 to 9 for 802.11b operation.

The first major manufacturer to use 802.11g was Apple, under the trademark AirPort Extreme.

802.11n

In January 2004 IEEE announced that it will develop a new standard for wide-area wireless networks. This new standards was introduced in 2009, after the previous edition of this book. Today, at the time of writing this book, this standard has become one of the most popular, offering the best bandwidth and coverage for indoor use.

The 802.11n can go even up to 150 Mb/s when using the 40 MHz bandwidth.

Maximum indoor distances with 802.11n are quoted to be around 70 m (230 ft) and outdoor up to 250 m (820 ft).

An 802.11n modem

Wi-Fi certification and security

Because the IEEE only sets specifications but does not test equipment for compliance with them, a trade group called the Wi-Fi Alliance runs a certification program that members pay to participate in. Virtually all companies selling 802.11 equipment are members. The Wi-Fi label means compliant with any of 802.11a, b, g or n. It also includes the security standard Wi-Fi Protected Access (WPA/WPA2) which the Wi-Fi Alliance defined in response to serious weaknesses researchers had found in the previous system, Wired Equivalent Privacy (WEP). Products that are Wi-Fi are also supposed to indicate the frequency band in which they operate in 2.4 or 5 GHz.

The Wi-Fi alliance logo

WPA (sometimes referred to as the draft IEEE 802.11i standard) became available in 2003. The Wi-Fi Alliance intended it as an intermediate measure in anticipation of the availability of the more secure and complex WPA2. WPA2 became available in 2004 and is a common shorthand for the full IEEE 802.11i (or IEEE 802.11i-2004) standard.

Wired Equivalent Privacy (WEP) was an encryption algorithm designed to provide wireless security for users implementing 802.11 wireless networks. WEP was developed by a group of volunteer IEEE members. The intention was to offer security through an 802.11 wireless network while the wireless data was transmitted from one end point to another over radio waves. WEP was used to protect wireless communication from eavesdropping (confidentiality), prevent unauthorized access to a wireless network (access control), and prevent tampering with transmitted messages (data integrity). Wireless office networks are often unsecured or secured with WEP, which is easily broken. These networks frequently allow "people on the street" to connect to the Internet. WEP used a 40-bit or 104-bit encryption key that must be manually entered on wireless access points and devices and does not change. Volunteer groups have also made efforts to establish wireless community networks to provide free wireless connectivity to the public.

The ***Wi-Fi Protected Access*** (WPA) is a standards-based interoperable security specification. The specification is designed so that only software or firmware upgrades are necessary for the existing or legacy hardware using WEP to meet the requirements. Its purpose was to increase the level of security for existing and future wireless LANs. WPA however was an interim security solution that targets all known WEP vulnerabilities and pending the availability of the full IEEE 802.11i (WPA2) standard. The WPA protocol implemented much of the IEEE 802.11i standard, specifically, the Temporal Key Integrity Protocol (TKIP). TKIP employs a per-packet key, meaning that it dynamically generates a new 128-bit key for each packet and thus prevents the types of attacks that compromised WEP.

WPA also includes a message integrity check. This is designed to prevent an attacker from capturing, altering and/or re-sending data packets. This replaces the cyclic redundancy check (CRC) that was used by the WEP standard. CRC's main flaw was that it did not provide a sufficiently strong data integrity guarantee for the packets it handled. Well tested message authentication codes existed to solve these problems, but they required too much computation to be used on old network cards. WPA uses a message integrity check algorithm called Michael to verify the integrity of the packets. Michael is much stronger than a CRC, but not as strong as the algorithm used in WPA2. Researchers have since discovered a flaw in WPA that relied on older weaknesses in WEP and the limitations of Michael to retrieve the key-stream from short packets to use for re-injection and spoofing.

The *Wi-Fi Protected Access II* (WPA2) has replaced WPA. WPA2, which requires testing and certification by the Wi-Fi Alliance, implements the mandatory elements of IEEE 802.11i. In particular, it introduces CCMP, a new AES-based encryption mode with strong security. Certification began in September, 2004; from March 13, 2006, WPA2 certification is mandatory for all new devices to bear the Wi-Fi trademark. This is currently the most common security for Wi-Fi access points.

What about Bluetooth?

If asked to construct a Wireless Local Area Network (WLAN) most IT managers would think of 802.11b/g/n wireless Ethernet technology. Few would consider using another short-range radio technology, *Bluetooth*, on its own or in combination with 802.11-based equipment.

The reason for its neglect is that Bluetooth has been marketed as a technology for linking devices such as smart phones, headsets, PCs, digital cameras, and other peripherals, rather than as a technology for LANs.

However, Bluetooth could become a serious WLAN option, partly because a lot more Bluetooth devices have been released lately. But IT managers may think twice before supporting this technology because 802.11 and Bluetooth use the same 2.4 GHz spectrum to transmit data, so interference is a real possibility.

Bluetooth is also closing the gap in signal range. Some companies are testing new ceramic antennas that will boost the range of Bluetooth to around 50 meters, up from the 10 meters currently specified and on a par with the maximum range offered by some 802.11 components.

Network concepts and components

Ethernet networking follows a simple set of rules and components that govern its basic operation.

The Ethernet basically uses the CSMA/CD access method to handle simultaneous demands. It is one of the most widely implemented LAN standards. The acronym **CSMA/CD** signifies *carrier-sense multiple access with collision detection* and describes how the Ethernet protocol regulates communication among nodes. Although the term may seem intimidating, if we break it apart into its component concepts we will see that it describes rules very similar to those that several people use in polite conversation. If one talks at the dinner table, for example, the other listens until he or she stops talking. In the moments of silence when somebody decides to say something, the rest of the listeners wait again until the second person finishes talking. If in the moments of pause two or more people start to talk simultaneously, a collision occurs. In networking, this is equivalent to data collision between two computers. The CSMA/CD protocol states that in such cases both computers maintain silence briefly and wait a **random** time until they start talking again. Whichever randomness is shorter becomes the first "speaker" of the two and the others wait until he or she finishes. The random time gives all participants (computer stations) an equal chance in the conversation (data exchange) at the dinner table (Ethernet network) to have their say.

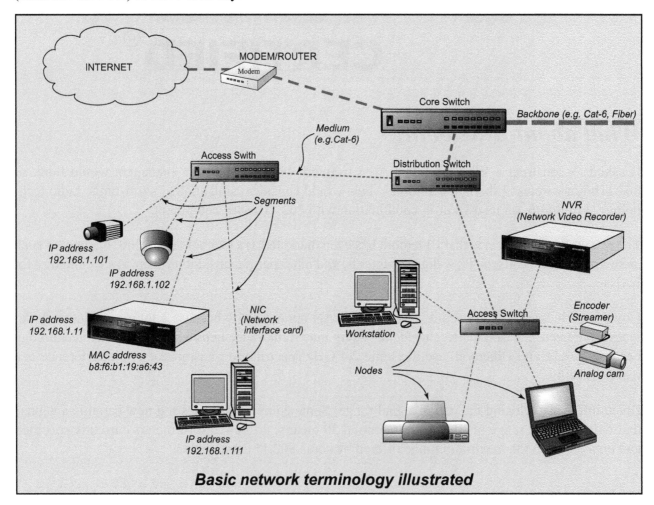

Basic network terminology illustrated

To better understand these rules and components, it is important to understand the basic terminology, so here we are going to introduce the most common ones, with a short description of what they mean.

This book is intended for the CCTV industry, and, as such, the Ethernet basics are somewhat condensed. Readers interested in more details are referred to more extensive books dedicated to networking.

- ***Network*** – A network is a group of computers connected together in a way that allows information to be exchanged between the computers.

- ***Local Area Network*** **(LAN)** – A LAN is a network of computers that are in the same general physical location, usually within a building or a campus. Computers belonging to one LAN, use IP addresses from the same range. If the computers are far apart (such as across town or in different cities), then a Wide Area Network (WAN) is typically used.

- ***Node*** – A node is anything that is connected to the network. Although a node is typically a computer, it can also be something like a printer, an IP camera, or a DVR.

- ***Broadcast domain*** – A broadcast domain is a logical division of a computer network, in which all nodes can reach each other. A broadcast domain can be within the same LAN segment or it can be bridged to other LAN segments.

- ***Virtual LAN (VLAN)*** – A subdivision of a network LAN into little virtual LANs where multiple distinct broadcast domains are created, which are mutually isolated so that packets can only pass between them via one or more routers.

- ***OSI layers*** – An Open System Interconnection reference model was introduced by the International Standards Organization (ISO), which defines ***seven layers*** of networking (will be detailed further in this chapter).

- ***Network Interface Card*** **(NIC)** – Every computer (and most other devices) is connected to a network through a NIC. In most computers, this is an Ethernet card (normally 100 Mb/s or 1 Gb/s) connected to the computer's motherboard.

- ***Media Access Control*** **(MAC)** ***address*** – This is a unique physical address of any device – such as the NIC in a computer – on the network. The MAC address, which is made up of two equal parts, is six bytes long. The first three bytes identify the company that made the NIC, the second three are the serial number of the NIC itself. If an IP address of a particular device is unknown there are software tools for searching devices on the network using company's MAC addresses.

- ***Segment*** – A segment is any portion of a network that is separated from other parts of the network by a switch, bridge, or router.

- ***Backbone*** – The backbone is the main cabling of a network to which all of the segments are connected. Typically, the backbone is capable of carrying more information than the individual segments.

• ***Hub*** – Hub is the simplest device. It only works with the electrical signals (OSI level 1) and knows nothing about the addressing. When the signal comes to the hub, it just duplicates it simultaneously to all its ports without worrying about who the signal is in.

• ***Switch*** – A network switch is an "intelligent" data communication controlling device that is most important in a network. Typically a switch belongs to OSI layer 2, but the more intelligent one to OSI layer 3. A low-quality switch can have major effect on a video network, which then often results in various problems, such as delays, recording gaps, connection losses, codec failures, jitter and/or image artifacts during the transmission. Depending upon where the switch is in the system topology, it can be considered as an ***access*** switch, ***distribution*** or a ***core*** switch. Access switches are the ones to which IP cameras, encoders, digital recorders, or workstations connect. In a small network that is all that might be needed. Based on the number of network sockets, a network switch can have anything from 4, 8, 24, even as many as 48 ports. Also, in addition to the "normal" switches, there are PoE ones, which offer Power over Ethernet to an IP device that might be designed to work with one. This is especially useful in CCTV, where PoE cameras can be powered from such a switch, simplifying the cabling and eliminating the need for separate camera power supplies.

• ***Router*** – Routers are specialized network devices that forward data packets between computer networks. They are connected to two or more different networks with different domains. When a data packet comes in one of the lines, the router reads the address information in the packet to determine its ultimate destination. Then, using information in its routing table or routing policy, it directs the packet to the next network on its journey. Most common and typical representative of routers are ADSL modems/routers connecting home computer network to the Internet. Routers perform the "traffic directing" functions on the Internet. Routing occurs at layer 3 of the OSI model (explained further in the book).

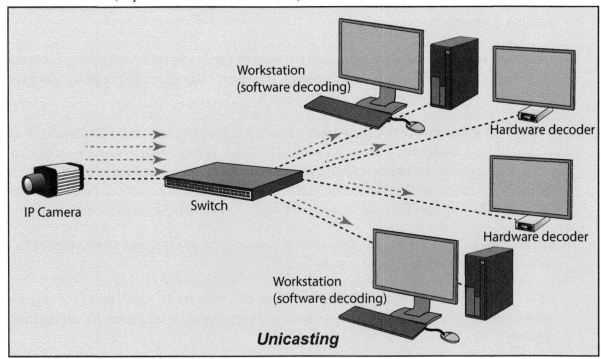

• *Unicasting* – A unicast is a transmission from one node addressed specifically to another node. In a small digital CCTV system, typically operator gets unicast connection when viewing live or playback a footage. Unicast transmission involves acknowledging of good reception by the receiver. If there are more operators wanting to see the same camera live using unicast transmission, the camera encoder needs to be capable of producing additional streams. Unicasting connections lead quickly to a network overload in larger systems.

• *Multicasting* – Multicast is a connection established between a sender (encoder) and several receivers (decoders). One multicast stream can be received by multiple operators without increasing the traffic. Multicasting uses a special range of addresses from the multicast range and the participants log in using the Internet Group Messaging Protocol (IGMP). The involved network components, such as routers or switches, have to ensure that, if possible, only the desired and required multicast streams are transmitted. If multiple operators in a digital CCTV system want to all view the same live camera, the only practical way to do this is by use of multicasting. Multicasting doesn't wait for the acknowledgement by the receiver. Because of this, more intelligent and more expansive switchers are used, capable of snooping and determining which of the multicast streams is no longer needed, in order to be shut down.

• *QoS (Quality of Service)* – Quality of Service is a packet prioritization which ensures that time-critical or essential applications are treated preferentially in terms of receiving their data over the network. There are two mechanisms for QoS: Integrated Services and Differentiated Services. With Integrated Services the required bandwidths are already pre-reserved on the individual network devices. With the Differential Services data packets are marked and processed by the network according to the configuration. Today, the latter seems to be prevailing QoS as it offers better scalability and a higher compatibility.

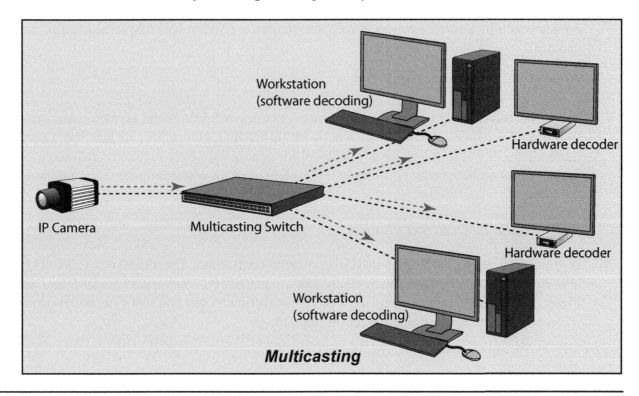

Multicasting

Networking software

The Internet protocols

In order for various computers to talk to each other via any network, there must be a common language of understanding, a common protocol. In networking, **the term *protocol* refers to a set of rules that govern communications.** Protocols are to computers what language is to humans. Since this book is in English, to understand it you must be able to read English. Similarly, for two devices on a network to communicate successfully, they must both understand the same protocols.

Various protocols belonging to various OSI layers (see next chapter), are used in today's world of the Internet, and this book would not be complete without listing most popular ones and describing them briefly.

TCP/IP – Transmission Control Protocol / Internet Protocol

Two of the most popular suites of protocols used in the Internet today. They were introduced in the mid-1970s by Stanford University and Bolt Beranek and Newman (BBN) after funding by DARPA (Defence Advanced Research Projects Agency) and appeared under the Berkeley Software Distribution (BSD) Unix.

TCP is reliable; that is, packets are guaranteed to wind up at their target, in the correct order.

IP is the underlying protocol for all the other protocols in the TCP/IP protocol suite. IP defines the means to identify and reach a target computer on the network. Computers in the IP world are identified by unique numbers, which are known as IP addresses (explained further in this chapter).

PPP – Point-to-Point Protocol

A protocol for creating a TCP/IP connection over both synchronous and asynchronous systems. PPP provides connections for host to network or between two routers. It also has a security mechanism. PPP is well known as a protocol for connections over regular telephone lines using modems on both ends. This protocol is widely used for connecting personal computers to the Internet.

SLIP – Serial Line Internet Protocol

A point-to-point protocol to be used over a serial connection, a predecessor of PPP. There is also an advanced version of this protocol known as CSLIP (*Compressed Serial Line Internet Protocol*) which reduces overhead on a SLIP connection by sending just a header information when possible, thus increasing packet throughput.

FTP – File Transfer Protocol

A protocol that enables the transfer of text and binary files over a TCP connection. FTP allows for files transfer according to a strict mechanism of ownership and access restrictions. It is one of the most commonly used protocols over the Internet today.

Telnet

A terminal emulation protocol, defined in RFC854, for use over a TCP connection. It enables users to log in to remote hosts and use their resources from the local host.

SMTP – Simple Mail Transfer Protocol

A protocol dedicated for sending e-mail messages originating on a local host over a TCP connection to a remote server. SMTP defines a set of rules that allows two programs to send and receive mail over the network. The protocol defines the data structure that would be delivered with information regarding the sender, the recipient (or several recipients), and, of course, the mail's body.

HTTP – Hyper Text Transport Protocol

A protocol used to transfer hypertext pages across the World Wide Web.

SNMP – Simple Network Management Protocol

A simple protocol that defines messages related to network management. Through the use of SNMP, network devices such as routers can be configured by any host on the LAN.

UDP – User Datagram Protocol

A simple protocol that transfers packets of data to a remote computer. UDP does not guarantee that packets will be received in the same order they were sent. In fact, it does not guarantee delivery at all. UDP is one of the most common protocols used in multicasting.

ARP – Address Resolution Protocol

In order to map an IP address into a hardware MAC address the computer uses the ARP protocol which broadcasts a request message that contains an IP address, to which the target computer replies with both the original IP address and the hardware MAC address.

NNTP – Network News Transport Protocol

A protocol used to carry USENET posting between News clients and USENET servers.

The OSI seven-layer model of networking

The basics of networking revolves around understanding the so-called seven-layer OSI model. Proposed by the ISO (International Standards Organization) in 1984, the OSI acronym could be read as ISO backwards, but it actually means Open System Interconnection reference model.

The OSI model describes how information from a software application in one computer moves through a network medium to a software application in another computer. The OSI model is considered the primary architectural model for inter-computer communications.

The idea behind such a layered model is to simplify the task of moving information between networked computers and make it manageable. Within each layer, one or more entities implement its functionality. Each entity interacts directly only with the layer immediately beneath it, and provides facilities for use by the layer above it. A task, or group of tasks, is then assigned to each of the seven OSI layers. Each layer is reasonably self-contained, so that the tasks assigned to each layer can be implemented independently.

OSI has two major components:

> • An abstract model of networking (the Basic Reference Model, or seven-layer model)
> • A set of concrete protocols

Parts of OSI have influenced Internet protocol development, but none more than the abstract model itself, documented in OSI 7498 and its various addenda. In this model, **a networking system is divided into layers**. Within each layer, one or more entities implement its functionality. Each entity interacts directly only with the layer immediately beneath it, and provides facilities for use by the layer above it. Protocols enable an entity in one host to interact with a corresponding entity at the same layer in a remote host.

The OSI seven-layer model of networking					
No.	**Layer**	**Data unit**	**Side of the layer**	**Function**	
7	Application	Data	Host	Network process to application	
6	Presentation	Data	Host	Data representation, encryption and decryption	
5	Session	Data	Host	Inter-communication, managing sessions between applications	
4	Transport	Segments	Host	End-to-end connections, reliability and flow control	
3	Network	Packet/Datagram	Media	Path determination and logical addressing	
2	Data link	Frame	Media	Physical addressing	
1	Physical	Bit	Media	Media, signal and binary transmission	

The seven layers of the OSI Basic Reference Model are (from bottom to top):

Layer 7 – Application

Layer 6 – Presentation

Layer 5 – Session

Layer 4 – Transport

Layer 3 – Network

Layer 2 – Data link

Layer 1 – Physical

Many prefer to list the seven layers starting from layer one down to layer seven, but it does not really matter, as long as they are remembered as the basic building blocks of the whole networking technology. A handy way to remember the layers is the sentence "*All people seem to need data processing*" and each first letter of that sentence corresponds to the first letter of the layers starting from layer seven going to layer one.

The seven layers can be grouped into two main groups: ***upper or host layers*** and ***lower or media layers***. The upper layers of the OSI model deal with application issues and **generally are implemented in software only. The top layer, seven, is the closest to the computer user as it represents the software application passing the information to the user.** Basically, both the user and the application layer processes interact with software application that contains a communication component.

As we go down through the layers, we get closer to the physical medium. So, the lower layers of the OSI are closer to the hardware (although do not exclude software) and handle the data transport issues. **The lowest layer is closest to the physical medium, that is, network cards and network cables, and they are responsible for actually placing information on the network medium.**

Let us now explain the meaning of each layer.

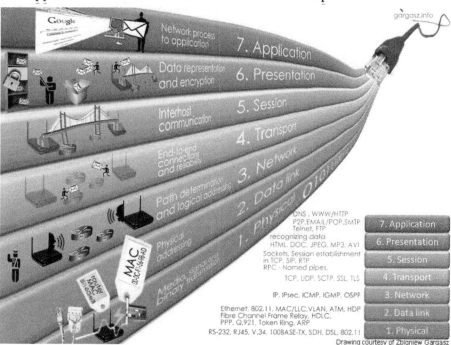

The seven layers model illustrated

1. The Physical layer

The Physical layer describes the physical properties of the various communications media, as well as the electrical properties and interpretation of the exchanged signals. For example, this layer defines the size of Ethernet cable, the type of connectors used, and the termination method.

The Physical layer is concerned with transmitting raw bits over a communication channel. The design issues have to do with making sure that when one side sends a 1 bit, it is received by the other side as a 1 bit, not as a 0 bit. Typical questions here are how many volts should be used to represent a 1 and how many for a 0, how many microseconds a bit lasts, whether transmission may proceed simultaneously in both directions, how the initial connection

Network Interface Card (NIC)

is established, how it is torn down when both sides are finished, how many pins the network connector has, and what each pin is used for. The design issues here deal largely with mechanical, electrical, and procedural interfaces and the physical transmission medium, which lies below the Physical layer. Physical layer design can properly be considered to be within the electrical engineer's domain.

2. The Data link layer

The Data link layer describes the logical organization of data bits transmitted on a particular medium. **This layer defines the framing, addressing, and check-summing of Ethernet packets.** The main task of the Data link layer is to transform a raw transmission facility into a line that appears free of transmission errors in the Network layer. It accomplishes this task by having the sender break the input data up into data frames (typically, a few hundred bytes), transmit the frames sequentially, and process the acknowledgment frames sent back by the receiver. Since the Physical layer merely accepts and transmits a stream of bits without any regard to meaning of structure, it is up to the Data link layer to create and recognize frame boundaries. This can be accomplished by attaching special bit patterns to the beginning and end of the frame. If there is a chance that these bit patterns might occur in the data, special care must be taken to avoid confusion. The Data link layer should provide error control between adjacent nodes.

Another issue that arises in the Data link layer (and most of the higher layers as well) is how to keep a fast transmitter from "drowning" a slow receiver in data. Some traffic regulation mechanism must be employed in order to let the

Frame of the Data link layer

transmitter know how much buffer space the receiver has at the moment. Frequently, flow regulation and error handling are integrated for convenience.

If the line can be used to transmit data in both directions, this introduces a new complication for the Data link layer software. The problem is that the acknowledgment frames for A to B traffic compete for use of the line with data frames for the B to A traffic. A clever solution in the form of piggybacking has been devised.

3. The Network layer

The Network layer describes how a series of exchanges over various data links can deliver data between any two nodes in a network. **This layer defines the addressing and routing structure of the Internet.** The Network layer is concerned with controlling the operation of the subnet. A key design issue is determining how packets are routed from source to destination. Routes could be based on static tables that are "wired into" the network and rarely changed. They could also be determined at the start of each conversation, for example, a terminal session. Finally, they could be highly dynamic, being newly determined for each packet, to reflect the current network load.

If too many packets are present in the subnet at the same time, they will get in each other's way, forming bottlenecks. The control of such congestion also belongs to the Network layer.

Data Packet in the Network layer

Since the operators of the subnet may well expect remuneration for their efforts, often some accounting function is built into the Network layer. At the very least, the software must count how many packets or characters or bits are sent by each customer, to produce billing information. When a packet crosses a national border, with different rates on each side, the accounting can become complicated.

When a packet has to travel from one network to another to get to its destination, many problems can arise. The addressing used by the second network may be different from that of the first one; the second one may not accept the packet at all because it is too large; the protocols may differ; and so on. It is up to the Network layer to overcome all these problems to allow the interconnecting of the heterogeneous networks.

In broadcast networks, the routing problem is simple, so the network layer is often thin or even nonexistent.

A Layer 3 network switch

4. The Transport layer

The Transport layer describes the quality and nature of the data delivery. **This layer ensures that messages are delivered error-free, in sequence, and with no losses or duplications.** This layer defines if and how retransmissions will be used to ensure data delivery. The basic function of the Transport layer is to accept data from the session layer, split it up into smaller units if need be, pass these to the Network layer, and ensure that all the pieces arrive correctly at the other end. Furthermore, all this must be done efficiently and in a way that isolates the Session layer from the inevitable changes in the hardware technology.

Under normal conditions, the Transport layer creates a distinct network connection for each transport connection required by the Session layer. If the transport connection requires a high throughput, however, the Transport layer might create multiple network connections, dividing the data among the network connections to improve throughput. On the other hand, if creating or maintaining a network connection is expensive, the Transport layer might multiplex several transport connections onto the same network connection to reduce the cost. In all cases, the Transport layer is required to make the multiplexing transparent to the Session layer.

The transport layer also determines what type of service to provide to the Session layer, and ultimately, the users of the network. The most popular type of transport connection is an error-free point-to-point channel that delivers messages in the order in which they were sent. However, other possible kinds of transport, service and transport isolated messages exist, with no guarantee about the order of delivery to multiple destinations. The type of service is determined when the connection is established.

The Transport layer is a true source-to-destination or end-to-end layer. In other words, a program on the source machine carries on a conversation with a similar program on the destination machine, using the message headers and control messages.

Many hosts are multi-programmed, which implies that multiple connections will be entering and leaving each host. There needs to be some way to tell which message belongs to which connection. The transport header is one place where this information could be added.

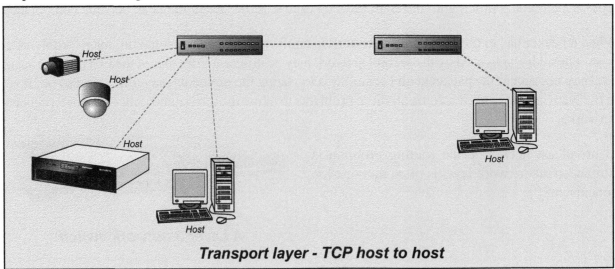

Transport layer - TCP host to host

In addition to multiplexing several message streams onto one channel, the Transport layer must establish and delete connections across the network. This requires some kind of naming mechanism, so that the process on one machine has a way of describing with whom it wishes to converse. There must also be a mechanism to regulate the flow of information, so that a fast host cannot overrun a slow one. Flow control between hosts is distinct from flow control between switches, although similar principles apply to both.

5. The Session layer

The Session layer allows session establishment between processes running on different stations. This layer describes how request and reply packets are paired in a remote procedure call. **The Session layer allows users on different machines to establish sessions between them.** A session allows ordinary data transport, as does the transport layer, but it also provides some enhanced services useful in some applications. A session might be used to allow a user to log into a remote time-sharing system or to transfer a file between two machines.

One service provided by the Session layer is to manage dialogue control. Sessions can allow traffic to go in both directions at the same time or in only one direction at a time. If traffic can only go one way at a time, the Session layer can help keep track of whose turn it is.

A related Session service is token management. For some protocols it is essential that both sides do not attempt the same operation at the same time. To manage these activities, the Session layer provides tokens that can be exchanged. Only the side holding the token may perform the critical operation.

Another Session service is synchronization. Consider the problems that might occur when trying to complete a two-hour file transfer between two machines on a network with a 1 hour mean time between crashes. After each transfer is aborted, the whole transfer will have to start over again, and will probably fail again with the next network crash. To eliminate this problem, the Session layer provides a way to insert checkpoints into the data stream, so that after a crash, only the data after the last checkpoint has to be repeated.

6. The Presentation layer

The Presentation layer describes the syntax of data being transferred. This layer describes how floating point numbers can be exchanged between hosts with different math formats. The Presentation layer performs certain functions that are requested sufficiently often to warrant finding a general solution for them, rather than letting each user solve the problems. In particular, unlike all the lower layers, which are just interested in moving bits reliably from here to there, the Presentation layer is concerned with the syntax and semantics of the information transmitted.

A typical example of a Presentation service is encoding data in a standard, agreed-upon way. Most user programs do not exchange random binary bit strings; they exchange things such as people's names, dates, amounts of money, and invoices. These items are represented as character strings, integers, floating point numbers, and data structures composed of several simpler items. Different computers have different codes for representing character strings, integers, and so on. In order to make it possible for computers with different representation to communicate, the data structures to be exchanged can be defined in an abstract way, along with a standard encoding to be used "on the wire." The Presentation layer handles the job of managing these abstract data structures and converting from the representation used inside the computer to the network standard representation.

The Presentation layer is also concerned with other aspects of information representation. For example, data compression can be used here to reduce the number of bits that have to be transmitted, and cryptography is frequently required for privacy and authentication.

7. The Application layer

The Application layer describes how real work actually gets done. This layer would implement file system operations. The Application layer contains a variety of protocols that are commonly needed. For example, there are hundreds of incompatible terminal types in the world. Consider the plight of a full-screen editor that is supposed to work over a network with many different terminal types, each with different screen layouts, escape sequences for inserting and deleting text, moving the cursor, and so on. One way to solve this problem is to define an abstract network virtual terminal for which editors and other programs can be written to. To handle each terminal type, a piece of software must be written to map the functions of the network virtual terminal onto the real terminal. For example, when the editor moves the virtual terminal's cursor to the upper left-hand corner of the screen, this software must issue the proper command sequence to the real terminal to get its cursor there too. All the virtual terminal software is in the Application layer.

Another Application layer function is file transfer. Different file systems have different file naming conventions, different ways of representing text lines, and so on. Transferring a file between two different systems requires handling these and other incompatibilities. This work, too, belongs to the Application layer, as do e-mail, remote job entry, directory lookup, and various other general-purpose and special-purpose facilities.

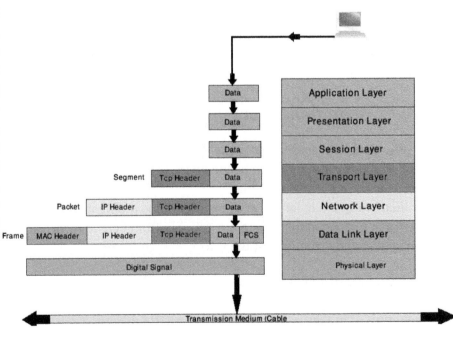

IP addresses

The *Internet Protocol* (IP) was created in the 1970s to support early computer networking with the Unix operating system. Today, IP has become a standard for all modern network operating systems to communicate with each other. Many popular, higher-level protocols such as HTTP and TCP rely on IP.

The Internet Protocol (IP) address uniquely identifies the node or Ethernet device, just as a name identifies a particular person. No two Ethernet devices on the same network should have the same address.

Two versions of IP exist in production use today. The first one is called *IPv4* addressing. It consists of 32 bits, but for human readability it is written in a form consisting of four decimal octets separated by full stops (dots), called *dot-decimal notation*. Most networks use IPv4, but because IPv4 has only 4.3 billion possible addresses (256^4) they start to become insufficient for the amount of IP devices on the planet, so an increasing number of educational and research networks have adopted the **next generation IP version 6 (IPv6)**. The number of possible addresses with IPv6 is well over the maximum number of devices one can imagine with today's technology.

Since a signal on the Ethernet medium reaches every attached node, the destination address is critical to identify the intended recipient of the data frame. For example, when computer B transmits to printer C, computers A and D will still receive and examine the data frame. However, when a station first receives a frame it checks the destination address to see if the frame is intended for itself. If it is not, the station discards the frame without even examining its contents.

Understanding the IP addressing is especially important for the CCTV technical guys who visit various sites that have their own networks. In order to connect to an IP camera, set-up a DVR, or evaluate the network one may need to get an approval from the appropriate IT personnel (if the digital CCTV network is mixed with the corporate network). It should also be clear how a technician needs to set up his/her own PC to become part of the customer's network without intruding or affecting it. Sometimes it

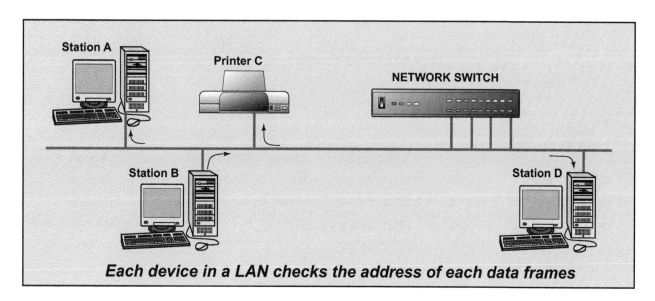

Each device in a LAN checks the address of each data frames

is possible, and perhaps much easier and safer, to connect to a DVR directly by using a crossover Cat-5 cable (that is, if you are physically close to it), it is still important to know how such an IP address can be accessed from one's own PC (not a part of the network one is visiting). The few "classic" network ping-commands mentioned at the end of this chapter may help establish the validity of certain addresses.

IPv4 addressing notation

The most common IP address type is the IPv4, which consists of 4 bytes (32 bits).

These bytes are also known as octets.

For purposes of readability, humans typically work with IP addresses in a decimal notation that uses periods to separate each octet. For example, the IP address

       11000000 10101000 1100110 1011010

shown in the binary system has the first 8 bits (octet) equivalent to the decimal representation of:

$$1\times2^7 + 1\times2^6 + 0\times2^5 + 0\times2^4 + 0\times2^3 + 0\times2^2 + 0\times2^1 + 0\times2^0 = 128 + 64 + 0 + 0 + 0 + 0 + 0 + 0 = 192$$

Similar logic applies to the other three octets, so that the decimal equivalent representation of the previous binary IP address is:

       192.168.102.90

Because each byte is 8 bits in length, each octet in an IP address ranges in value from a minimum of 0 to a maximum of 255 (2^8).

Therefore, the full range of IP addresses in IPv4 annotation is from 0.0.0.0 through 255.255.255.255.

This represents a total of $256\times256\times256\times256 = 256^4 = 4,294,967,296$ possible IP addresses.

IPv4 addressing classes				
Class	Leftmost bits	Start address	End address	Private range	Note
A	0xxx	0.0.0.0	127.255.255.255	10.0.0.0 ~ 10.255.255.255	OK to use
B	10xx	128.0.0.0	191.255.255.255	172.16.0.0 ~ 172.31.255.255	OK to use
C	110x	192.0.0.0	223.255.255.255	192.168.0.0 ~ 192.168.255.255	OK to use
D	1110	224.0.0.0	239.255.255.255		Reserved for Multicasting
E	1111	240.0.0.0	255.255.255.255		Reserved for research

One could say that there are enough IP addresses for almost every single person on our planet, but at the time of writing this book there are already over seven billion people. Certainly not every person has an IP device, but in developed countries a person may have even up to three or four devices using an IP address. The growth of the Internet has been so rapid that larger addressing space is seen as inevitable introduction, hence the IPv6.

IPv6 addressing notation

Although this addressing is not widespread as yet, it is no doubt something that future networks will have use of, if nothing else because of the amount of addresses made available under such notation.

The IPv6 addresses are 16 bytes (128 bits) long, rather than 4 bytes (32 bits).

This represents more than 256^{16} possible addresses, which is quite a large number:

300,000,000,000,000,000,000,000,000,000,000,000,000

The preferred IPv6 addressing form is using hexadecimal values of the eight 16-bit pieces:

BA98:FEDC:800:7654:0:FEDC:BA98:7654:3210

In hexadecimal representation of numbers, rather than decimal, the numbers use A as 11 in decimal, B is 12, C is 13, D is 14, E is 15, and F is 16.

Note that it is not necessary to write the leading zeros in an individual field, but there must be at least one numeral in every field.

In the coming years, as an increasing number of cell phones, PDAs, and other network appliances expand their networking capability, this much larger IPv6 address space will probably be necessary.

IPv6 Address Types

IPv6 does not use classes. IPv6 supports the following three IP address types:

- Unicast

- Multicast

- Anycast

Unicast and multicast messaging in IPv6 are conceptually the same as in IPv4.

IPv6 does not support broadcast, but its multicast mechanism accomplishes essentially the same effect. Multicast addresses in IPv6 start with "FF" (255) just like IPv4 addresses.

Anycast in IPv6 is a variation on multicast. Whereas multicast delivers messages to all nodes in the multicast group, anycast delivers messages to any one node in the multicast group. Anycast is an advanced networking concept designed to support the fail over and load balancing needs of applications.

Reserved addresses in IPv6

IPv6 reserves just two special addresses: 0:0:0:0:0:0:0:0 and 0:0:0:0:0:0:0:1.

IPv6 uses 0:0:0:0:0:0:0:0 internal to the protocol implementation, so nodes cannot use it for their own communication purposes.

IPv6 uses 0:0:0:0:0:0:0:1 as its loopback address, equivalent to 127.0.0.1 in IPv4.

IPv4 address classes

Not all IP addresses are free for use in your local LAN, which you will no doubt find from your IT manager. In addition, not all addresses that you could use can be used, for you have to find out what address is free to use and yet belongs to the same address group addressable by your network equipment.

In order to bring some order to the many possible LANs and WANs, there are some agreed-upon rules and address classes that all Ethernet devices obey. These are the IPv4 classes.

The IPv4 address space can be subdivided into five classes: Class A, B, C, D, and E.

Each class consists of a contiguous subset of the overall IPv4 address range.

With a few special exceptions explained later in this chapter, the values of the left-most 4 bits of an IPv4 address determine its class as shown in this table.

	IPv4 addressing classes			
Class	**Leftmost bits**	**Start address**	**End address**	**Private range**
A	0xxx	0.0.0.0	127.255.255.255	10.0.0.0 ~ 10.255.255.255
B	10xx	128.0.0.0	191.255.255.255	172.16.0.0 ~ 172.31.255.255
C	110x	192.0.0.0	223.255.255.255	192.168.0.0 ~ 192.168.255.255

Class A, B, and C

Class A, B, and C are the three classes of addresses used on the Internet, with private addresses exceptions as explained next.

Private addresses

When a computer or a network device resides on a private network (not on the Internet), it should use one of the many private addresses defined by the IP standards. Such devices, when connected to the Internet, via an ADSL modem, for example, are practically invisible to the other Internet devices which use the other ("visible") Class A, B, or C IP addresses.

The IP standard defines specific address ranges within Class A, B, and C reserved for use by private networks (*intranets*). The following table lists these reserved ranges of the IP address space.

Nodes are effectively free to use addresses in the private ranges if they are not connected to the Internet, or if they reside behind firewalls or other gateways that use Network Address Translation (NAT).

IP address Class C

All Class C addresses, for example, have the leftmost 3 bits set to "110," but each of the remaining 29 bits may be set to either "0" or "1" independently (as represented by an x in these bit positions):

      110xxxxx xxxxxxxx xxxxxxxx xxxxxxxx

By converting the above to dotted decimal notation, it follows that all Class C addresses fall in the range from 192.0.0.0 through 223.255.255.255.

IP loopback address

      127.0.0.1 is the *loopback address* in IP.

Loopback is a test mechanism of network adaptors. Messages sent to 127.0.0.1 do not get delivered to the network. Instead, the adaptor intercepts all loopback messages and returns them to the sending application. IP applications often use this feature to test the behavior of their network interface. On some products this address is used to synchronize the time to a master device.

As with broadcast, IP officially reserves the entire range from 127.0.0.0 through 127.255.255.255 for loopback purposes. Nodes should not use this range on the Internet, and it should not be considered part of the normal Class A range.

Zero addresses

As with the loopback range, the address range from 0.0.0.0 through 0.255.255.255 should not be considered part of the normal Class A range.

0.x.x.x addresses serve no particular function in IP, but nodes attempting to use them will be unable to communicate properly on the Internet.

IP address Class D and Multicast

The IPv4 networking standard defines Class D addresses as reserved for multicast.

Multicast is a mechanism for defining groups of nodes and sending IP messages to that group rather than to every node on the LAN (broadcast) or just one other node (unicast).

For larger digital CCTV systems multicasting is very important and it is the only communication method able to deliver video streams to multiple operators, without overloading the network. The same packets of data (video streams) are picked up by various operators, which as a consequence reduces the data traffic. IP cameras and/or NVRs need to be capable of multicasting for this feature to be used. Multicasting is typically used for live viewing only, and not for playback.

Class	IPv4 addressing classes		
	Leftmost bits	Start address	End address
D	1110	224.0.0.0	239.255.255.255

As with Class E, Class D addresses are not used by ordinary nodes on the Internet.

IP address Class E and limited broadcast

The IPv4 networking standard defines Class E addresses as reserved, which means that they should not be used on IP networks.

Some research organizations use Class E addresses for experimental purposes. However, nodes that try to use these addresses on the Internet will be unable to communicate properly.

A special type of IP address is the limited broadcast address 255.255.255.255. A broadcast

Class	IPv4 addressing classes		
	Leftmost bits	Start address	End address
D	1110	224.0.0.0	239.255.255.255
E	1111	240.0.0.0	255.255.255.255

involves delivering a message from one sender to many recipients. Senders direct an IP broadcast to 255.255.255.255 to indicate that all other nodes on the local network (LAN) should pick up that message. This broadcast is "limited" in that it does not reach every node on the Internet, only nodes on the LAN.

Technically, IP reserves the entire range of addresses from 255.0.0.0 through 255.255.255.255 for broadcast, and this range should not be considered part of the normal Class E range.

Network partitioning

Computers in one network should each have one logical address. Usually this address is unique to each device and can either be configured dynamically from a network server, statically by an administrator, or automatically by stateless address auto-configuration. Computer networks consist of individual segments of network cable. The electrical properties of cabling limit the useful size of any given segment such that even a modestly sized local area network (LAN) will require several of them. Devices such as switches and routers connect these segments together, though not in a perfectly seamless way.

Besides partitioning through the use of cable, subdividing of the network can also be done at a higher level. Subnets support virtual network segments that partition traffic flowing through the cable rather than the cables themselves. The subnet configuration often matches the segment layout one to one, but subnets can also subdivide a given network segment.

Network addressing fundamentally organizes hosts into groups. This can improve security (by isolating critical nodes) and can reduce network traffic (by preventing transmissions between nodes that do not need to communicate with each other). Overall, **network addressing becomes even more powerful when introducing sub-netting and/or super-netting.**

IP Subnet addressing

IP networks can be divided into smaller networks called subnetworks (or subnets). Sub-netting provides the network administrator with several benefits, including extra flexibility, more efficient use of network addresses and the capability to contain broadcast traffic. For the purpose of network management, an IP address is divided into two logical parts, the network prefix and the host identifier. All hosts on a subnetwork have the same network prefix. This routing prefix occupies the most-significant bits of the address. The number of bits allocated within a network to the internal routing prefix may vary between subnets, depending on the network architecture. The host part is a unique local identification and is either a host number on the local network or an interface identifier.

Sub-netting			
	Network	**Network**	**Subnet**	**Host**
Binary	11111111	11111111	11111111	00000000
Dotted-decimal	255	255	255	0

This logical addressing structure permits the selective routing of IP packets across multiple networks via special gateway computers, called routers, to a destination host if the network prefixes of origination and destination hosts differ, or sent directly to a target host on the local network if they are the same. Routers constitute logical or physical borders between the subnets, and manage traffic between them. Each subnet is served by a designated default router, but may consist internally of multiple physical Ethernet segments interconnected by network switches.

The routing prefix of an address is written in a form identical to that of the address itself. **This is called the network mask, or netmask, of the address.** For example, a specification of the most-significant 18 bits of an IPv4 address, 11111111.11111111.11000000.00000000, is written as 255.255.192.0. If this mask designates a subnet within a larger network, it is also called the subnet mask. This form of denoting the network mask, however, is only used for IPv4 networks.

The governing bodies that administer Internet Protocol have reserved certain networks for internal uses. In general, intranets utilizing these networks gain more control over managing their IP configuration and Internet access. A subnet allows the flow of network traffic between hosts to be segregated based on a network configuration. By organizing hosts into logical groups, subnetting can improve network security and performance. Subnetting works by applying the concept of extended network addresses to individual computer (and other network device) addresses.

Class B Subnetting reference chart			
Number of bits	**Subnet mask (dotted-decimal)**	**Number of Subnets**	**Number of Hosts**
2	255.255.192.0	2	16382
3	255.255.224.0	6	8190
4	255.255.240.0	14	4094
5	255.244.248.0	30	2046
6	255.255.252.0	62	1022
7	255.255.254.0	126	510
8	255.255.255.0	254	254
9	255.255.255.128	510	126
10	255.255.255.192	1022	62
11	255.255.255.224	2046	30
12	255.255.255.240	4094	14
13	255.255.255.248	8190	6
14	255.255.255.252	16382	2

An extended network address includes both a network address and additional bits that represent the subnet number. Together, these two data elements support a two-level addressing scheme recognized by standard implementations of IP. The network address and subnet number, when combined with the host address, therefore support a three-level scheme.

A subnet address is created by borrowing bits from the host field and designating them as the Subnet field. The number of borrowed bits varies and is specified by the subnet mask. A given network address can be broken up into many subnetworks. For example, 192.168.1.0, 192.168.2.0, 192.168.3.0, and 192.168.4.0 are all subnets within a network 192.168.0.0. All 0s in the host portion of an address specifies the entire network.

Class C Subnetting reference chart			
Number of bits	**Subnet mask (dotted-decimal)**	**Number of Subnets**	**Number of Hosts**
2	255.255.255.192	2	62
3	255.255.255.224	6	30
4	255.255.255.240	14	14
5	255.255.255.248	30	6
6	255.255.255.252	62	2

Subnet masks use the same format and representation technique as IP addresses. The subnet mask, however, has binary 1s in all bits specifying the Network and Subnetwork fields, and binary 0s in all bits specifying the Host field.

For all 0s in the binary representation above it is possible to have that many hosts, where each 0 position goes to the power of two, increased by one as we go from right to left (left is most insignificant bit).

For example, in CCTV a typical default number of bits is 8, as shown in the chart for Class B Subnetting on the left. This gives 254 possible hosts addresses, a total of 256 combinations, less one for the network address and less one for the broadcast address = 254.

This is sufficient in most of the small IP CCTV LANs and hence rarely changed from the default setting.

Virtual LAN (VLAN)

A single LAN, interconnected with a layer 2 network switch may be partitioned in Virtual LANs (VLANs) so that they create multiple distinct broadcast domains. In such a configuration, data packets can only pass between them via one or more routers.

Grouping CCTV elements (hosts) like IP cameras, NVRs or DVRs, with a common set of requirements regardless of their physical location by VLAN can greatly simplify network design. **The advantages of VLANs important for some CCTV projects are creating more bandwidth by segmentation of broadcast domains, additional security and flexibility.**

A VLAN has the same attributes as a physical local area network (LAN), but it allows for end stations to be grouped together more easily even if they are not on the same network switch. Switch ports configured as a member of one VLAN belong to a different broadcast domain, as compared to switch ports configured as members of a different VLAN. If a switch supports multiple VLANs, broadcast within one VLAN never appear in another VLAN. VLAN membership can be configured through software

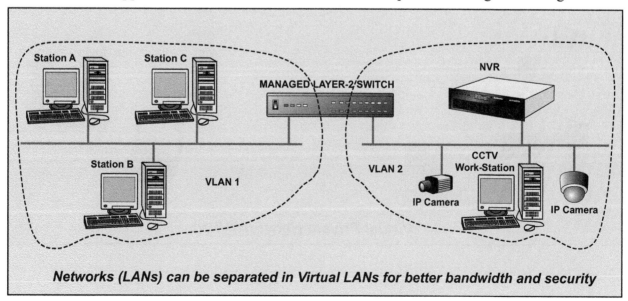

Networks (LANs) can be separated in Virtual LANs for better bandwidth and security

instead of physically relocating devices or connections. Most enterprise-level networks today use the concept of virtual LANs. Without VLANs, a switch considers all interfaces on the switch to be in the same broadcast domain.

Creating VLANs enables administrators to build broadcast domains with fewer users in each broadcast domain. This increases the bandwidth available to users because fewer users will contend for the bandwidth. Traffic can pass from one VLAN to another only through a router. VLANs enable a network administrator to assign users to broadcast domains based upon the user's job need. This provides a high level of deployment flexibility for a network administrator.

Virtual Private Networking (VPN)

Virtual Private Network (VPN) is a technology that allows establishment of an encrypted remote connection between two computers or networks. A VPN utilizes public networks to conduct private data communications. Most VPN implementations use the Internet as the public infrastructure and a variety of specialized protocols to support private communications through the Internet. VPN follows a client and server approach. VPN clients authenticate users, encrypt data, and otherwise manage sessions, with VPN servers utilizing a technique called tunneling. To achieve this, the office IP video surveillance networks connect to the Internet through the VPN gateway, the role of which can be played by both router and computer. Using VPN, the secured connection is established between the office networks, via the Internet using the so-called *VPN tunnel*. Before leaving one office network, the data is encrypted. At the other end of the tunnel, in another office, the data is decrypted.

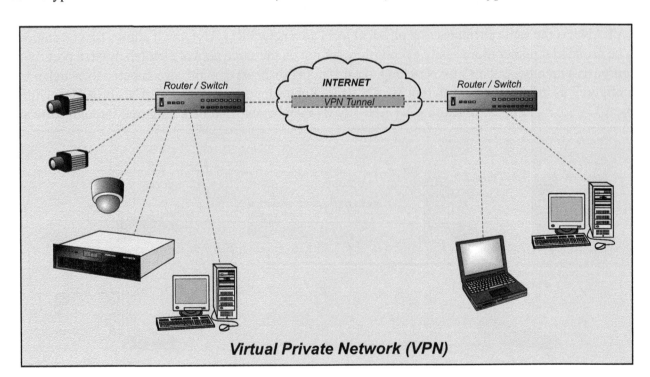

Virtual Private Network (VPN)

DHCP

The Dynamic Host Configuration Protocol (DHCP) is a network protocol used to automatically configure devices that are connected to a network so they can communicate on that network using the Internet Protocol (IP). It involves clients and a server operating in a client-server model. In a typical home local area network, the server is a modem/router, while the clients are the computers, smart phones, or printers. In such a configuration, a user doesn't need to worry about what IP address should his computers have in order to be a part of the same domain. This is typically allocated by the DHCP server. For this to happen in the PC, smart phone or printer, the network settings "Obtain IP address automatically" or "Automatic DHCP" needs to be selected. Then, the PC, the smart phone and the printer, each obtain an IP address from the allocated pool of router's allowed IP addresses.

Automatic IP address via DHCP

The DHCP server maintains a database of available IP addresses and configuration information. When the server receives a request from a client, the DHCP server determines the network to which the DHCP client is connected, and then allocates an IP address or prefix that is appropriate for the client, and sends configuration information appropriate for that client. DHCP servers typically grant IP addresses to clients only for a limited interval. DHCP uses the concept of a "lease" or amount of time that a given IP address will be valid for a computer. Using very short leases, DHCP can dynamically reconfigure networks in which there are more computers than there are available IP addresses. DHCP clients are responsible for renewing their IP address before that interval has expired, and must stop using the address once the interval has expired, if they have not been able to renew it.

DHCP is used for IPv4 and IPv6. While both versions serve the same purpose, the details of the protocol for IPv4 and IPv6 are sufficiently different that they may be considered separate protocols.

Manual IP address setting

Although the same approach with DHCP automatic IP addresses allocation can be used in a CCTV LAN too, in CCTV it is typically preferred that IP addresses are allocated manually by the administrator. The reason for this is more practical, where all IP addresses are known, manual method will also work quicker in regards to video streaming, since no DHCP server needs to be queried about the address associated with a particular MAC address.

When manually configuring a CCTV workstation computer for example, it is important to find out what domain (IP address range) does the CCTV system belong to. Most common method of finding this is by use of so called "IP finder" software which probes the whole network for known IP CCTV devices. Typically all of them should belong to one domain, and if not, they need to be made such. Once this is known, then under the PC "Internet Protocol Version 4 (TCP/IPv4) Properties" and under the "Use the following IP address" a CCTV administrator is supposed to give the PC an IP address that is unique, yet belonging to the same domain range as the other CCTV elements in the network.

Domain Name Systems (DNS)

Although IP addresses allow computers and routers to identify each other efficiently, humans prefer to work with names rather than numbers. The **Domain Name System** (DNS) supports the best of both worlds.

DNS allows nodes on the public Internet to be assigned both an IP address and a corresponding name called a *domain name*. For DNS to work as designed, these names must be unique worldwide. Hence, an entire "cottage industry" has emerged around the purchasing of domain names in the Internet name space. DNS is a hierarchical system and organizes all registered names in a tree structure. At the base or root of the tree are a group of top-level domains including familiar names like *com*, *org*, *net* or *edu,* and numerous country-level domains like *au* (Australia), *ru* (Russia), *mk* (Macedonia), or *uk* (United Kingdom). Below this level are the second-level registered domains such as *vidilabs.com*. These are domains that organizations can purchase from any of numerous accredited registrars. For nodes in the *com*, *org*, and *edu* domains, the Internet Corporation for Assigned Names and Numbers (ICANN) oversees registrations. Below that, local domains like *ip.vidilabs.com* are defined and administered by the overall domain owner. DNS supports additional tree levels as well. The period (the dot '.') always separates each level of the hierarchy in DNS.

DNS is also a distributed system. The DNS database contains a list of registered domain names. It further contains a mapping or conversion between each name and one or more IP addresses. However, DNS requires a coordinated effort among many computers (servers); **no one computer holds the entire DNS database**. Each DNS server maintains just one piece of the overall hierarchy, one level of the tree and then only a subset or zone within that level.

The top level of the DNS hierarchy, also called the root level, is maintained by a set of 13 servers called root name servers. These servers have gained some notoriety for their unique role on the Internet. Maintained by various independent agencies, the servers are uniquely named A, B, C, and so on up to M. Ten of these servers reside in the United States, one in Japan, one in London, and one in Stockholm, Sweden.

DNS works in a client/server fashion. The DNS servers respond to requests from DNS clients called resolvers. ISPs and other organizations set up local DNS resolvers as well as servers. Most DNS servers also act as resolvers, routing requests up the tree to higher-level DNS servers and delegating requests to other servers. DNS servers eventually return the requested mapping (either address-to-name or name-to-address) to the resolver.

Networking hardware

Hubs, switches, bridges, and routers

Hub

Hubs classify as Layer 1 devices in the OSI model. Hubs connect multiple Ethernet devices in a star configuration, so that any device connected to the hub can "see" and talk to any other device in that group (network segment). At the physical layer 1, hubs can support little in the way of sophisticated networking. **Hubs do not read any of the data passing through them and are not aware of their source or destination.** Essentially, a hub simply receives incoming packets, possibly amplifies the electrical signal, and broadcasts these packets out to all devices on the network, including the one that originally sent the packet. Typically, 4, 8, 16, or up to 24 devices can be connected to one hub since there are no hubs with more than 24 ports. If more devices are used, more hubs can be added. The more devices are connected to the hub the **more data packets collision will occur, slowing down the network traffic**. One way to reduce data

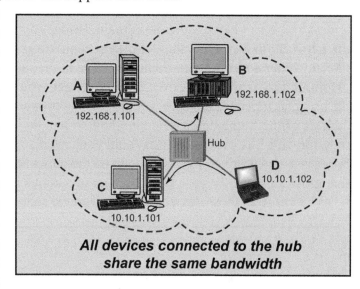

All devices connected to the hub share the same bandwidth

congestion would be to split a single segment into multiple segments, thus creating **multiple collision domains**. This solution creates a different problem, for these now separate segments are not able to share information with each other. This is where network switches and routers are used.

Network switches are devices that link computers, IP cameras, servers and other network devices such as switches and routers. Switches receive messages from any device connected to it and then

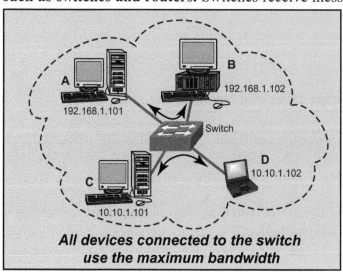

All devices connected to the switch use the maximum bandwidth

transmits the message only to the device for which the message was meant. In the IT world this is described as creating separate collision domains for each port. So, if a computer A intends to communicate with computer B, they will do so, without interfering with the communication between computers C and D on the same switch, hence using the maximum bandwidth. In the case of a hub, they would all share the bandwidth and run in half duplex,

Switch

resulting in collisions, which would then necessitate retransmissions. This makes the switch a more intelligent device than a hub. **Network switches typically work at the data link layer 2 of the OSI model. Switches that additionally process data at the network layer (layer 3) and above are often called layer-3 switches or multi-layer switches.**

Bridge

Network bridges function as agregators of two or more communication networks, or two or more network segments. Bridging is distinct from routing which allows the networks to communicate independently as separate networks. Bridges are devices that forward data packets between separate computer networks that belong or are intended to belong to the same domain. Bridges connect two or more network segments, increasing the network diameter, but they also help **regulate traffic**. They send and receive transmissions just like any other node, but they do not function in the same way as a normal node. **The Bridge does not originate any traffic of its own, it only echoes what it hears from other stations**. So, one goal of the bridge is to reduce unnecessary traffic on both segments. This is done by examining the destination address of the frame before deciding how to handle it. An important characteristic of bridges is that they forward Ethernet broadcasts to all connected segments. This behavior is necessary, as Ethernet broadcasts are destined for every node on the network, but it can pose problems for bridged networks that grow too large. When a large number of stations broadcast on a bridged network, congestion can be as bad as if all those devices were on a single segment.

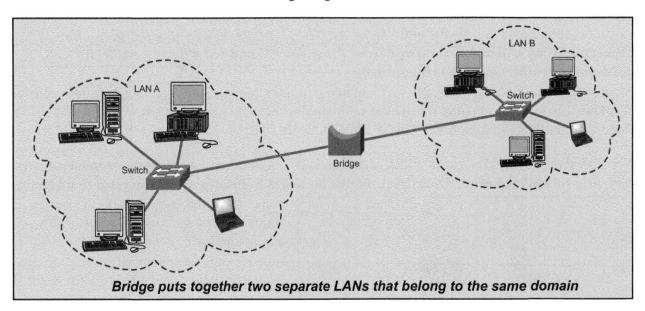

Bridge puts together two separate LANs that belong to the same domain

In the OSI model bridging acts in the first two layers, layer 1 and layer 2, below the network layer. There are four types of network-bridging technologies: simple bridging; multiport bridging; learning, or transparent bridging; and source route bridging.

Router

Routers are "intelligent" networking components that can divide a single network into two logically separate networks. While Ethernet broadcasts across bridges in their search to find every node on the network, they do not cross routers, because **the router forms a logical boundary for the network. Routers operate based on protocols that are independent of the specific networking technology, such as Ethernet or token ring.** This allows routers to easily interconnect various network technologies, both local and wide area, and has led to their widespread deployment in connecting devices around the world as part of the global Internet.

When a data packet comes in one of the lines, the router reads the address information in the packet to determine its ultimate destination. Then, using information in its routing table or routing policy, it directs the packet to the next network on its journey. Routers perform the "traffic directing" functions on the Internet. A data packet is typically forwarded from one router to another through the networks that constitute the internetwork until it reaches its destination node.

The most familiar type of routers are home and small office routers that simply pass data, such as web pages, e-mail, and videos between the home computers and the Internet. An example of a router would be the owner's cable or DSL modem, which connects to the Internet through an ISP. More sophisticated routers, such as enterprise routers, connect large business or ISP networks up to the powerful core routers that forward data at high speed along the optical fiber lines of the Internet backbone. Though routers are typically dedicated hardware devices, use of software-based routers has grown increasingly.

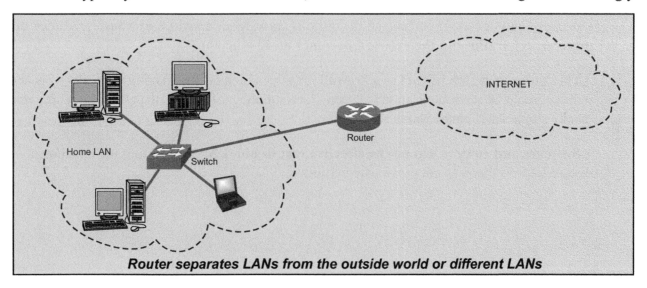

Router separates LANs from the outside world or different LANs

Although switches and routers share most relevant attributes, several distinctions differentiate these technologies. **Routers are generally used to segment a LAN into a couple of smaller segments, whereas switches are generally used to segment a large LAN into many smaller segments. Routers generally have only a few ports for LAN connectivity, and switches generally have many.**

Switches can also be used to connect LANs with different media; for example, a 10 Mb/s Ethernet LAN and a 100 Mb/s Ethernet LAN can be connected using a switch. Some switches support cut-through switching, which reduces latency and delays in the network, whereas routers support only store-and-forward traffic switching. Finally, switches reduce collisions on network segments because they provide dedicated bandwidth to each network segment.

Modern Ethernet implementations often look nothing like their historical counterparts. Where long runs of coaxial cable provided attachments for multiple stations in legacy Ethernet, modern Ethernet networks use twisted pair ("Cat" cable) wiring or fiber optics to connect stations in a *radial pattern* (*star-configuration*). Where legacy Ethernet networks transmitted data at 10 Mb/s, modern networks can operate at 100, 1000 Mb/s, or even 10,000 Mb/s.

Ethernet switching gave rise to another advancement: full-duplex Ethernet. *Full-duplex* is a data communications term that refers to the ability to send and receive data at the same time. Legacy Ethernet is half-duplex, meaning information can move in only one direction at a time. In a totally switched network, nodes only communicate with the switch and never directly with each other. Switched networks also employ either twisted pair or fibre optic cabling, both of which use separate conductors for sending and receiving data. In this type of environment, Ethernet stations can forgo the collision detection process and transmit at will, since they are the only potential devices that can access the medium. This allows end stations to transmit to the switch at the same time that the switch transmits to them, achieving a *collision-free* environment.

By dividing large networks into self-contained units (segments), switches and routers provide several advantages:

> • Because only a certain percentage of traffic is forwarded, **a switch or router reduces the unnecessary traffic and the network becomes more efficient.**

> • **The router or switch will act as a firewall** for some potentially damaging network errors and will accommodate communication between a larger number of devices than would be supported on any single LAN connected to the bridge.

> • **Switches and routers extend the effective length of a LAN**, permitting the attachment of distant stations that was not previously permitted.

Network ports

In addition to the IP address, in computer networking, there also exist *network ports* **as an additional information when communicating with a computer program.** Ports are application-specific or process-specific in the software. They are associated with the IP address of the host, as well as the type of protocol used in the network communication.

The purpose of ports is to uniquely identify different applications or processes running on a single computer and thereby enable them to share a single physical connection to a packet-switched network. **Network ports are usually numbered, and a network implementation (like TCP or UDP) will attach a port number to data it sends. The receiving implementation will use the attached port number to figure out which computer program to send the data to.** The combination of a port and a network address (IP-number) is often called a *socket*.

The protocols that primarily use ports are the Transport Layer protocols, such as the Transmission Control Protocol (TCP) and the User Datagram Protocol (UDP) of the Internet Protocol Suite. **There are a total of 65,536 ports used on a networking device, which comes from the 16 bits allocated to addressing the port numbers (2^{16}).**

The port number, added to a computer's IP address, completes the destination address for a communications session. That is, data packets are routed across the network to a specific destination IP address, and then, upon reaching the destination computer, are further routed to the specific process bound to the destination port number.

Note that it is the combination of IP address and port number together that must be globally unique. Thus, different IP addresses or protocols may use the same port number for communication; e.g., on a given host or interface UDP and TCP may use the same port number, or on a host with two interfaces, both addresses may be associated with a port having the same number.

In some CCTV applications, ports might be unique for the particular program dealing with video streaming for example, and as such might be blocked by a router if accessing from the outside of the CCTV network. This is often the case when remotely connecting to a video server over an ADSL modem/Router for example. In order to make the system working, the router needs to be configured to allow such unique ports to pass through to the computer by so-called *port forwarding*.

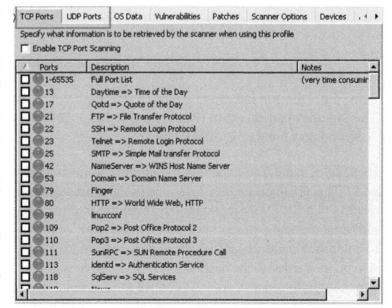

Ports are used in TCP and UDP protocols

Of the 65,536 possible ports, about 250 well-known ports are reserved by convention to identify specific service types on a host. There are others, which are registered and then there are private ports.

So, the general division of ports is into three groups:

- The *Well-Known Ports* are those from 0 ~ 1023.

- The *Registered Ports* are those from 1024 ~ 49151.

- The *Dynamic* and/or *Private Ports* are those from 49152 ~ 65535.

For example, some of the Well-Known Ports are in use daily with the most common programs we use for web browsing, e-mailing or transferring files:

- 20 & 21 – FTP: the file transfer protocol – data & control

- 22 – SSH: secure shell logins, file transfers (scp, sftp), and port forwarding

- 23 – Telnet: non-secure remote login service

- 25 – SMTP: Simple Mail Transfer Protocol (E-mail)

- 53 – DNS: Domain Name Server

- 80 – HTTP: HyperText Transfer Protocol (www)

- 110 – POP3: Post Office Protocol (E-mail)

- 143 – IMAP4: Internet Message Access Protocol (E-mail)

- 443 – HTTPS: used for securely transferring web pages, etc.

- 554 – RTSP: used for Real Time Streaming Protocol

Some times the port indication is written after the colon in the IP address, for example 192.168.1.22:554. This signifies that access to IP address 192.168.1.22 is sought at port 554. This combination of IP address and port number is called a socket.

By default, many ADSL modem/routers are coming with all ports closed except some of the well-known ports. This is important to note, as one of the most common problem in IP CCTV, especially when connecting to a CCTV system remotely, over the Internet, are the blocked ports for live video streaming. Different modems/routers have different methods of forwarding ports, but the typical setting would be to enter the IP address of the video source and tell the router to allow the specific port associated with it to be forwarded to the destination IP address. Sometimes, a few ports need to be forwarded, some of which might be needed for PTZ control for example, or audio communications. Equally, the ports can be closed, depending on the requirement and in order to minimize any risk from external hacker attacks.

A network analogy example

In order to summarize all the aforementioned network concepts and devices, which for many in CCTV might be a bit daunting, let me share with you the following analogy:

Imagine you live in a nice little town. The town could represent your Wide Area Network (WAN), while your own suburb would represent the Local Area Network (LAN) segment. Each house, shop, or object would represent a network device, with its own address, which is basically the IP address in a network. All the houses in your own street have different numbers but carry the name of the street, which is exactly how it is in the Local Area Network domains, where all devices have the first three groups of IP numbers the same and the last is unique to each house. No two houses in the same street have the same number. If one of the houses is better known by its owner's name or the business name residing on that address, that will be equivalent to DNS address allocation instead of the IP number (in our analogy, the house number).

Imagine now that you have a variety of roads in your town, with many vehicles traveling in various directions. In our networking analogy each road would represent the Ethernet media (cable), and each vehicle driving on that road would represent a data packet. Vehicles can be of different sizes, some are small cars, some a larger, which is the case with data packets. Let's assume the roads are narrow (bandwidth) and have traffic flowing in both directions, so that you cannot go any quicker than what the vehicle's speed is in front of you, and you can only use one-half of the road width, which is equivalent to half-duplex data in network communications. If intersections are not regulated with traffic lights, you basically have the equivalent of hub devices in networking, and unless all the drivers are cautious and courteous it easily gets into congestion. They do not regulate the traffic intelligently; they only allow you to get from one street to another, but if you have many cars going in various directions, the waiting time in front of such intersections could be quite long. This is equivalent to the data packets collision in Ethernet terminology. With a traffic light intersection (network switch), traffic is more regulated and flows with higher speed when light is green.

On your way to the chemist shop, you might drive on a brand-new four-lane-wide road (equivalent to 100 Mb/s network), which will get you there quite quickly because there are not that many accidents or stops (data collisions) and because the road is pretty wide and divided (equivalent to full-duplex Ethernet). When you get to the traffic lights before you cross over to the other side where the shopping center might be, it is like getting to a network bridge that separates your traffic from the shopping center traffic. If this were a major traffic intersection with five roads joining, for example, where some roads can take you to other parts of your town, such as the industrial, the traffic intersection and its intelligent traffic light switching would be equivalent to a network switch.

As it happens in real life, each vehicle could have a different size, which is similar to what we have in Ethernet data packets. They could all have a different length and different sizes, and of course different content, which is like having a different number of passengers or items being transported in a vehicle. In order for a vehicle to get from its original location (your home, for example) to the chemist shop which is near the shopping center (your destination, for example), your driver must know the address of the chemist shop, which is the same as having an IP address of the destination.

Let us now assume you want to go into the main shopping center, and let us also assume you have your friend with you, who unfortunately happens to be disabled and uses a wheelchair. In order for you to take him to the shopping center, you would take him via the wheelchair ramp that is designed for such purposes. The wheelchair ramp is another access to the shopping center, which non-disabled people usually do not use. The shopping center is the IP address you know and went to, but the access for your disabled friend is via the wheelchair ramp, for he cannot get up the stairs. This is exactly the same as having a different port on an Ethernet device, designed only for such purposes (i.e., customers). The shopping center delivery docks would be another way into the shopping center, but it is dedicated only to the trucks that deliver goods to the shopping center shops. Again, this is equivalent to another port number of the same IP address (i.e., the shopping center). In CCTV applications, for example, one DVR can have one port for accessing images, and another for accessing the time server function.

Pursuing this analogy, let us now assume that you want to leave your lovely town (your own network) and want to go outside, which happens to be another state, where you have border control, and they will not let you go there unless you have all the right vehicle documents and passport. This would be equivalent to a router device in a network, equipped with a firewall control.

The Wireless LAN is equivalent in our analogy to having helicopters instead of cars, where you can get to any point (in a certain radius, of course, depending upon the helicopter power and fuel) flying in the air, without the need to build roads on the ground (copper or fiber Ethernet). The flight control tower will still have to keep the traffic in order, which is the equivalent of the wireless network bridge or hot-spot.

Hackers or viruses are equivalent to either polite door-knocking salesmen (intruders), or robbers that want to come and steal something from your house and at the same time make a mess, or even burn the house down. Unfortunately, evil people do exist even in the IT world.

A simpler, and similar analogy of how the IT world works is usually made with mailing by post. For example, if you want to send a letter to a friend in another town, you write the letter (data content) take the envelope (data packet), write the address of the recipient (destination IP address) and your return address on it (source IP address). Then, you put the letter in the envelope and take it to the post office (router). The letter goes through a series of intermediate post offices (internet hops) and is sent to the one closest to the recipient and finally delivered to your friend's post office (destination computer).

Now, the letter is still not delivered as it has to be sent to the appropriate box in the post office (port number of a software).Once the letter is inserted in the P.O. box, your friend (program) can accept the letter (data) read it and act upon it. Once he understands what he has received and acknowledges the receipt of your letter, he simply changes places of the source and destination addresses on the envelope and sends the envelope in the opposite direction (this is the acknowledgement of the receipt of your letter).

Post analogy with networking concepts

Putting a network system together

Modern IP CCTV system design will require network knowledge. CCTV integrators, installers and consultants today are expected to know networking and IP principles. Perhaps not in the same depth as an educated and experienced IT person, but sufficient enough to understand the possibilities, limitations of network equipment. It is impossible for a modern digital CCTV system to be designed without this fundamental knowledge. One common advise to anybody involved in IP CCTV design is to attend one of the many Cisco network courses, which give better understanding of networks.

The starting point in putting a CCTV network system is understanding the customer's requirement. This defines the type and number of cameras, the network switches required and the budget allowed for this.

Based on the number of cameras used, and depending on their streaming, for example if it is video compression (H.264 for example) or perhaps image compression (M-JPG), we can an idea on how many ports per switch are required, and how many network switches for the whole system are required. When planning for this, certainly, distances have to be considered. If we use Cat-6 cable for example, a maximum of 100 m from the camera to the first switch should not be exceeded. This distance may have to be lower if the cameras are PTZ and powered by a PoE switch. If longer distances are required, perhaps local camera power supply and additional switches may have to be considered, or alternatively, local power supplies for cameras and fiber connected switches.

Network switches are obviously one of the key component in putting the whole system together. For this reason, it will be very important to choose the correct one, a proven brand and model, capable of continuous high quality video streaming. Like it was the case with lenses, not all switches are made equal. For large and demanding jobs, there are only small number of proven switches on the market, and IT specialist would know them. Sometimes, in a larger corporate environment, network switches may have already been decided on as a preferred equipment on site. Another thing that needs to be considered is that even if the switch make and model are selected, they may still require programming and setting for the CCTV environment. Continuous video streaming, with minimum packet losses

Network switches are the most important hardware devices in IP CCTV

and latency is more demanding then a typical office web browsing and e-mailing network. This may require setting up Virtual LANs, port forwarding on routers, multi-casting on large jobs and Quality of Service implementation.

Multi-casting needs to be considered if there are multiple operators or monitors in the system. As explained previously under the ***Network concepts and components*** heading, multi-casting is very important tool in minimizing the necessary video bandwidth. Instead of streaming individual and separate video to each operator wanting to see the same camera (with unicast connection), multi-casting offers one stream to be accessed by anybody wanting to see the particular camera. It is not only saving the video bandwidth, but it is also reducing the load on the camera encoders. Typically, an IP camera may have only certain encoding capability. One encoder may typically be used for live viewing, another for recording, but very rarely IP cameras have three or more encoders. In a normal unicast connection, each encoder can only be used by one connection, operator or monitor. This is where multicast streaming is the best solution. One stream is available for all that want to view a camera. This is, however, only possible if the IP camera and the network support multi-casting.

Not every camera and every switch supports these functions. The switches that support multicasting are more expensive and more complex to set up and program. The important consideration is to be able to stop the multicasting if nobody watches a particular camera. To do this, network switches with IGMP snooping feature are required. Not every switch is capable of this, and certainly this is the point when getting an IT person's support might be needed.

If the IP CCTV system is small, it may require none of the above, but still consideration should be made on how many operators are connecting to one camera that operates as a normal unicasting streamer.

Another important thing after the network design, is the storage capacity required by the customer. By knowing the video stream produced by each camera, and also whether continuous or motion detection recording is going to be used, it can be calculated how many TB of storage are needed to achieve the required retention. This then, in turn, decides the physical size of the storage equipment, how many racks are needed and where can this equipment be installed. Experienced designer will also consider redundancy of the recording, which will again influence the number of hard disk required and the physical space required for their housing.

Speaking of redundancy in recording, the next redundant thing that needs to be considered is uninter-ruptable power supply (UPS). Some modern DVRs/NVRs these days would have even dual power supply, and the typical usage is one power supply is connected to one phase of the UPS,and the other to the other phase. That way, even if power fails, UPS will support the system for a time defined by the battery capacity, and then if a certain phase fails, the other should take over for the power supply.

Since all of the IP devices will be connected in a LAN, it is also important to pre-allocate IP addresses for all IP devices in the system. This includes IP cameras, recorders, workstations, servers and alike. In most cases, CCTV system is preferred to be separate from any other network. This is for security reasons, but also for network traffic reasons. Video streaming is certainly highly demanding for any network, and if a separate network can not be created, the next best thing is to create VLAN (Virtual LAN) within a larger network. This allows better bandwidth control and improves security. Also, another unwritten rule is that CCTV IP network is typically preferred to have manually allocated IP addresses, rather then automatic, using DHCP server. This helps improve response speed when a camera is called up, but also in programming and planning VMSs. Certainly, in such cases, careful long term planning is needed, foreseeing system growth and making it future-proof.

The IP check commands

Some software commands found on many platforms, Windows and Linux alike, should be known to CCTV users and could help determine what is the IP address of the computer and whether a network device is present on the network and whether it is visible by other computers. These are the "ping" "ipconfig " commands.

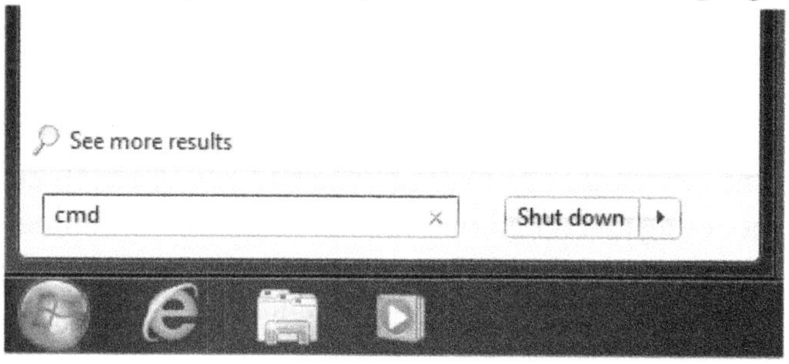

Under Windows => Start => Run type "Command" or "**cmd**."

This will open a DOS window where various commands can be typed in.

To find out the IP address and the MAC address the computer you are on type in "ipconfig."

ipconfig

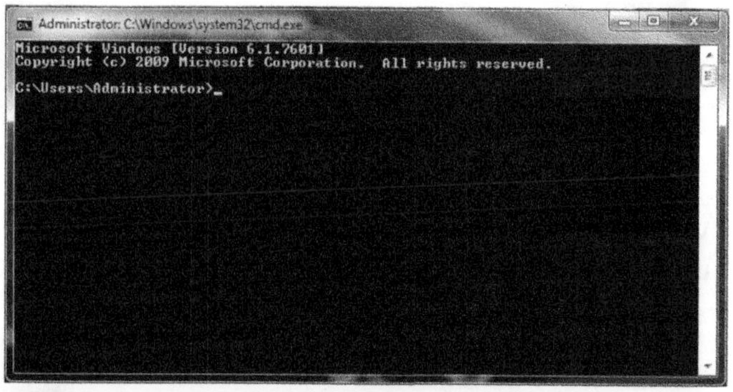

The computer should come with a similar response as shown to the right, with your own numbers. In our example, the computer's IP address is 10.0.0.7. The IP config also shows the MAC address of the device, and if this was an IP camera, the MAC address might be needed if the network was more controlled, or requested by the IT manager.

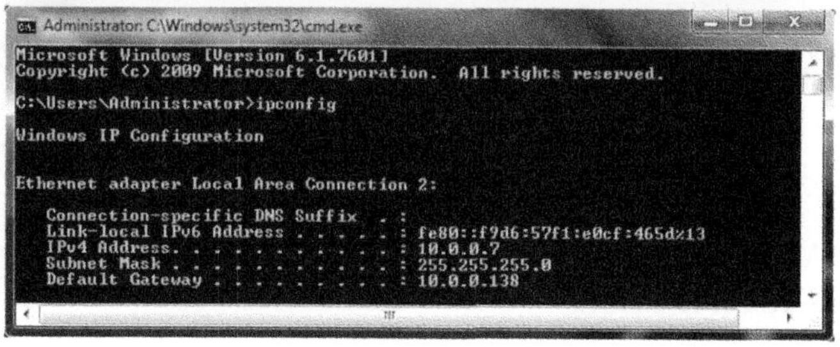

Let's assume we want to make up a little IP CCTV network and we want to give our IP camera an address that belongs to the domain we a part of. One of the easiest ways to check which addresses are used on the LAN is the command "ipconfig /all."

ipconfig /all

This command should list all IP devices on the network, and we can then decide to use an IP address from the same range, but the one that is not in that list. Let's assume we have decided to use 10.0.0.138. Before we do anything, the best suggestion is to test the (non)existence of this address on the network with the "ping" command.

ping <destination address>

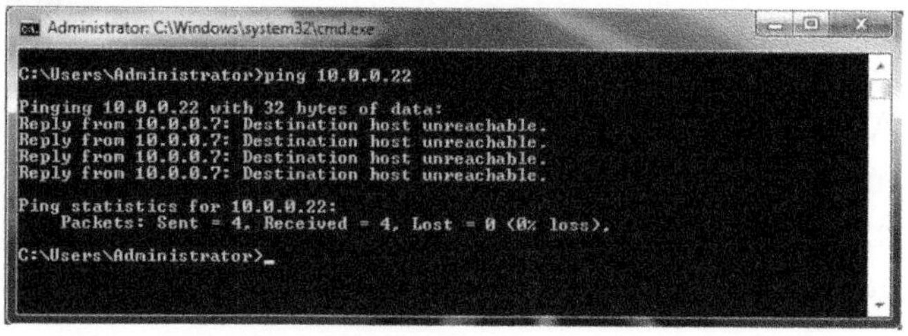

If some device is using this IP address a response should come stating the ping response time. If no device is using this address, the ping should time-out and no response will come back to us, but statement such as "Destination host unreachable."

If the device is connected and has the IP address you have queried (in our example 10.0.0.138), it will respond with something similar to what is shown in the window below. The time taken to respond is shown in milliseconds (ms).

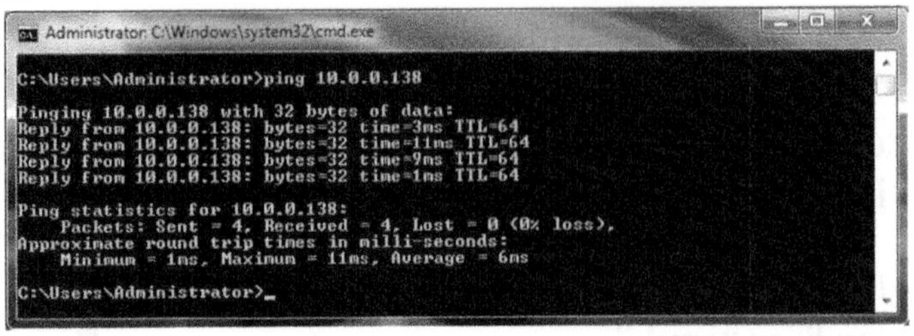

Another useful command is "netstat." The netstat (network statistics) displays network connections (both incoming and outgoing), routing tables, and a number of network interface (network interface controller or software-defined network interface) and network protocol statistics. It is used for finding problems in the network and to determine the amount of traffic on the network as a performance measurement.

netstat

The screen on the right shows the response from my computer when this command is typed in.

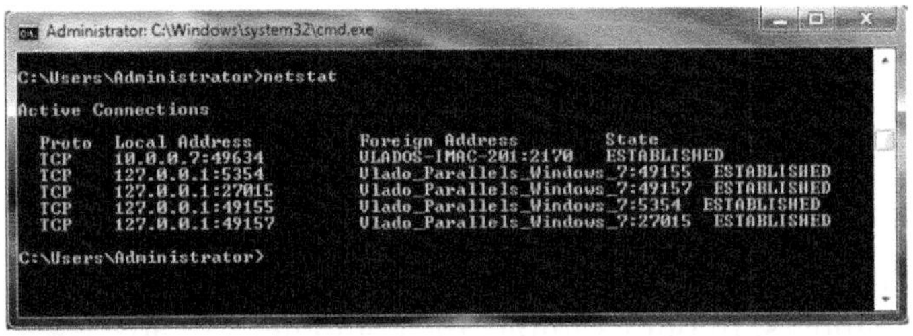

And finally, if there is a problem with the connection, but it is not clear where the network stops going further the "tracert" command may show where the problem is:

tracert <destination address>

For example, if we were to trace how a ping gets to the Google server, the command will be:

tracert www.google.com

And the result may look something like shown here on the right.

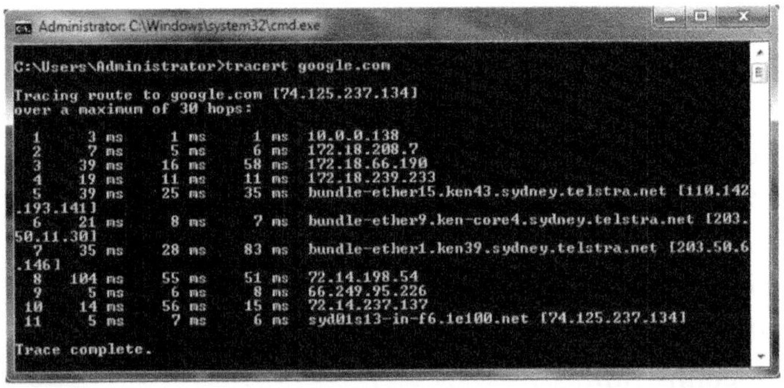

Some times it is necessary to find out the MAC address of all devices connected in the same network. An "arp" (from Address Resolution Protocol) can be used:

arp -a

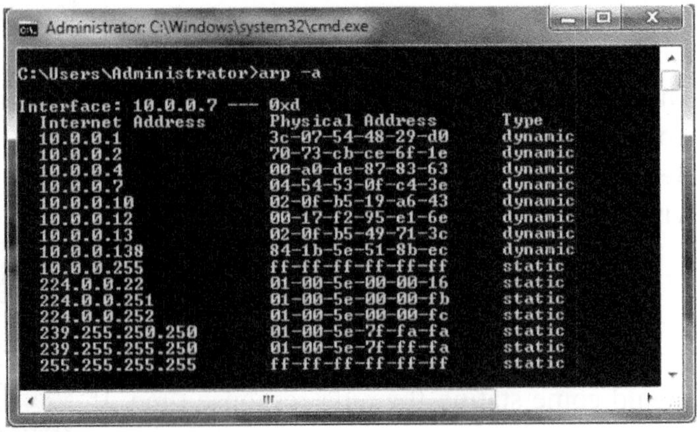

Another variation of the "netstat" command might be useful in IP CCTV is the command that helps find out which ports are opened on a computer and used by which application. This command is written as:

netstat -ano |find /i "listening"

A response from a command like that may look like the one shown on the right.

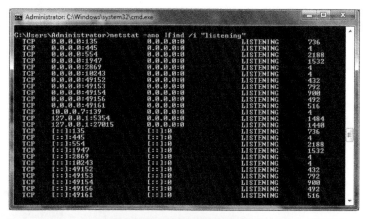

The first column shows if the protocol used is TCP or UDP. The second column refers to the ports used by some of the processes, and the last column on the right shows the ID of the process using the corresponding port. These process IDs in the Windows Operating System are referred to as PID (Process ID) and can be found out via the Windows Task Manager, under the Services tab.

In the example here, the process number 736 that listens on port 135 belongs to the "Remote Procedure Call (RPC)" program.

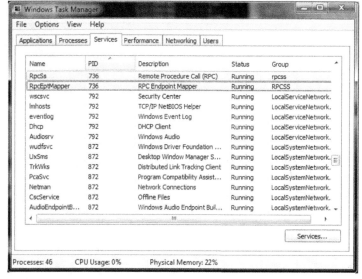

And, finally, a command, or rather graphical representation, showing the CPU load on a Windows based computer can be found under the Windows Task Manager, under the Performance tab. On the screen shot shown to the bottom right it can be seen that this computer has two cores (the two separate windows under the CPU usage history) and that he current CPU usage is only 10% with 47 processes running in the background.

In VMS workstations where multiple streams are being decoded, the best place to see how the computer copes with such a load is to check this window under the Windows Task Manager.

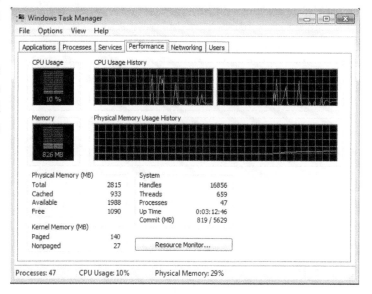

12. Auxiliary equipment in CCTV

Many items in CCTV can be classified as auxiliaries. Some of them are simple to understand and use; others are very sophisticated and complex. We will start with the very popular moving mechanism, usually called *pan and tilt head*.

Pan and tilt heads

When quoting or designing a CCTV system, the first question to ask is how many and what type of cameras: fixed or pan and tilt?

Fixed cameras, as the name suggests, are cameras installed on fixed brackets, using fixed focal length lenses and looking in the one direction without change.

The alternative to fixed are moving (or pan and tilt) cameras. They are placed on some kind of moving platform, usually employing a zoom lens, so the whole set can pan and tilt in virtually all directions and can be zoomed and focused at various distances.

In CCTV terminology, this type of camera is usually referred to as a *PTZ camera.* Perhaps a more appropriate term would be "PTZF camera," referring to Pan/Tilt/Zoom/Focus, or even more precisely in the last few years, a "PTZFI" for an additional iris control. "PTZ camera" is, however, more popularly accepted, and we will use the same abbreviation for a camera that, apart from pan, tilt, and zoom functions, might have focus, or even iris remote control.

A typical outdoor pan/tilt head

A typical P/T head, as shown on the picture, has a side platform for the load (a camera with a zoom lens in a housing). There are pan and tilt heads that have an overhead platform instead. The difference between the two is the load rating that each can have, which depends on the load's center of gravity. This center is lower for side platform pan and tilt heads, which means that of the two types of heads, with the same size motors and torque, the side platform has a better load rating. This should not be taken as a conclusion that the overhead platform P/T heads are of inferior quality, but it is only an observation of the load rating, which in the last few years is not as critical because camera and lens sizes, together with housings, are getting smaller.

On the basis of the application, there are two major subgroups of P/T heads:

- Outdoor

- Indoor

Outdoor P/T heads fall into one of the three categories:

- Heavy duty (for loads of over 35 kg)

- Medium duty (for loads between 10 and 35 kg)

- Light duty (for loads of up to 10 kg)

With the recent camera size and weight reductions, together with the miniaturization of zoom lenses and housings, it is very unlikely that you will need a heavy-duty P/T head these days. A medium-duty load rating will suffice in the majority of applications.

The outdoor P/T heads are weatherproof, heavier, and more robust. The reason for this is that they need to carry heavier housings and quite often additional devices such as wash/wipe assemblies and/or infrared lights.

PTZ on a pole

Indoor P/T heads, as the name suggests, should only be used on premises protected from external elements, especially rain, wind, and snow. Indoor

Classical PTZ and a Dome PTZ camera

P/T heads are usually smaller and lighter and in most cases they fall into the light-duty load category; they can handle loads of no more than a few kilograms. Because of this, indoor pan and tilt heads are often made of plastic molding and have a more aesthetic appearance than the outdoor ones.

In most cases, a typical P/T head is driven by 24 V AC synchronous motors. Mains voltage P/T heads are also available (220/240 V AC or 110 V AC), but the 24 V AC is more popular because of the safety factor (voltages less than 50 V AC are not fatal to the human body) and it is more universal. Most manufacturers have a 24 V AC version of all P/T site drivers.

Some new modern design have appeared, offering not only better esthetical appearance, built-in driver electronics, but also faster and more accurate in movement and repetitive preset positioning.

Photo courtesy of Pelco

A more integrated PTZ camera

It is not unusual for an external PTZ camera to use washer and wiper too, for cleaning dust and improving visibility during or after heavy rain. If dust is to be wiped, usually water sprays are required, which means a water container needs to be considered too. The PTZ sitre driver would also have electronics to control the washer and the wiper. In such a case one has to make sure that the control keyboard and software should include such functions too.

In the early days of pan/tilt technology, perhaps 20 or 30 years ago, the speed of panning and tilting was pretty slow, only a few degrees per second. It took a long time for a PTZ camera to move from one position, to another. The limiting factor were predominantly the technology available at the time, using synchronous AC motors, but also the heavy load of the old cameras, with large housings, possibly infrared illuminators and maybe even washers and wipers. Heavy load requires strong motors, sophisticated electromechanical design, all in strong and weather protected PT housing, which in turns further increases the total weight.

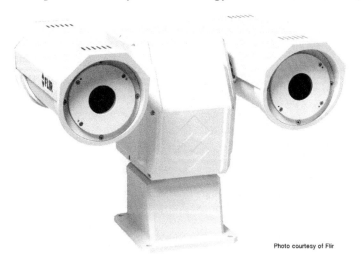

Photo courtesy of Flir

Today, cameras have gotten much smaller and so has the housing. The PTZ controlling electronics became much smarter and faster, using stepper motors or belts. Not only the speed increased dramatically up to hundreds of degrees per second, but the precision increased dramatically.

Some specialized PT heads with dual carriage

Pan and tilt domes

Other divisions of pan and tilt heads, on the basis of physical appearance, are also possible. In the last ten years, **pan and tilt mini-domes** have become more popular. They work in the same way as the heads, only **inside the domes they usually have both the moving mechanism (P/T head) and the control electronics**.

They are usually enclosed in a transparent or semitransparent dome, so they make an acceptable appearance in aesthetically demanding interiors or exteriors, which means they come in ceiling mounted versions, and outdoor in weatherproof, typically IP 66 rated enclosures.

Thanks to the camera and lens size reduction, PTZ domes are getting smaller in diameter. A few years ago, pan and tilt domes up to 1 m in diameter were not rare, while today most of them are even less than 300 mm. Most of the modern HD PTZ cameras are of this type too.

One of the biggest advancements in PTZ domes compared to the earlier bulky models, is their precision and accuracy of repetitive presets movement, but also the optical magnification power they use. This is mostly due to having much smaller cameras, smaller sensors, which means physically smaller, but optically more powerful, lenses. While in the past, 6 or 10 times optical magnification was considered typical or normal, with PTZ dome cameras magnification of 20 times is becoming common, despite the dome being less than 300 mm in diameter.

Most city surveillance utilize PTZ dome cameras

Some of the problems PTZ domes may have is the optical precision of the glass and the light attenuation of tinted domes. So the dome glass can be transparent or slightly tinted. Often, tinted dome is important to have in order to conceal the camera viewing direction. The easiest way to do this is by painting the camera and lens black, and putting a tinted dome on top. In any case, with plastic domes it is very hard to get no distortions at all, especially with heat-blown domes. Much better precision is achieved with injection molded domes, which are more expensive. Thicker glass domes cause more distortion, especially when the lens is zooming. The best optical quality is achieved with thin and injection molded domes.

In order to reduce optical attenuation due to tinting, some manufacturers concentrate the optical precision on a vertical slit with no tinting, the width of the lens, through which the camera can see, and freely tilt up and down and when panning the dome with the slit pans around with the camera. Although this is a clever solution, it can be mechanically troublesome and a limiting factor for faster movements.

Powerful zoom lenses, when fully zoomed in an object, as much as they are useful, may become a nuisance if the PTZ camera is installed on pole or bracket that might oscillate due to length, wind, or as a result of traffic passing by. A simple rule of thumb is, **if the zoom lens magnifies 20 times optically, it will also magnify that many times even the slightest physical movement of the camera itself.** The simple advise one can give is to make sure the mounting poles or brackets are as sturdy as possible. If this cannot be achieved, then the next best thing to do is not save presets with fully zoomed in view, unless it is absolutely necessary. Finally, if this is also not possible, a PTZ camera with *electronic image stabilization* could be considered. These are cameras where shaking image is stabilized by fast image processing. Not all cameras have this feature, and not all features on various cameras perform equally well, but, for some, this might be a solution. An important limitation of image stabilization is that the way it works is - it narrows (crops) the viewing angle by a certain small percentage. This is typically the percentage of movement of the objects in the field of view.

It should be noted that a tinted dome introduces

A typical PTZ dome with built-in camera

The slit hemisphere

light attenuation which is measured in F-stops (like with the lens iris). A typical attenuation is around one F-stop, which means 50% light attenuation. With today's camera technology this is not a threat to the picture quality during the day, but it may reduce the visibility in low light where every single lux is important.

One thing that has not yet been done with PTZ domes, and this is simply because of the spherical shape, is the PTZ domes don't have washers and wipers. Also, PTZ domes can't employ internally infrared illuminators like was the case with the PT heads (because of their size).

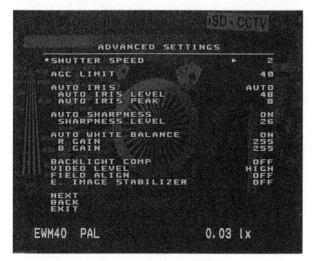

Setup menu of a PTZ dome camera

PTZ site drivers

In the early days of hard-wired PTZ cameras, a common way of controlling a 24 V AC pan and tilt head was by simply applying 24 V to one of the PTZ motors. This means pan and tilt control can be achieved by having voltage applied to a hard-wired connection (for each of the four movement directions) relative to a common wire. So, a total of five wires will give us full control over a typical pan and tilt head. For zoom and focus control as well, another three wires are required (one for zoom, positive and negative voltage, one for focus, and one for common). This gives us **a total of eight wires that would be required for a so-called *hard-wired PTZ controller*.** Maximum distances with such a control were limited, but it was the cheapest way of controlling single PTZ cameras. Today, this is an obsolete method.

Photo courtesy of Pelco

A hard-wired PTZ controller

In the majority of analog CCTV systems, however, we use PTZ data control, which only requires a twisted pair cable through which a matrix switcher or VMS can talk to a number of PTZ devices at the same time. These devices are often called **PTZ site drivers**, **PTZ decoders**, or **PTZ receiver drivers**. They are electronic boxes that receive and decode the instructions of the control keyboard for the camera's movements: pan, tilt, zoom, and focus (and sometimes iris as well).

Typical data communication for analog cameras is RS-422 (4-wires), or RS-485 (two wires), designed to drive up to 32 devices on one data bus, with a maximum total distance over a kilometer. There is unfortunately no standard among manufacturers of the control encoding schemes and protocols, which means the PTZ site driver of one manufacturer cannot be controlled with a PTZ protocol of another. In the last few years however, many PTZ dome cameras come with built-in a couple of different protocols, so that installers can switch between at least two different ones. Pelco protocol seems to be most common among the variety available on the market.

Depending on the site driver design, other functions might also be controllable, such as wash and wipe and turning auxiliary devices on and off. PTZ drivers can also deliver power for the camera, either 12 V DC or 24 V AC.

The P/T heads' movement speed, when driven by 24 V AC synchronous motors (which is most often the case), depends on the mains frequency, the load on the head, and the gearing ratio. Early pan and tilt heads had typical panning speeds are 9°/s and tilting 6°/s. Some designs for lighter loads could reach a faster speed of around 15°/s pan. Most AC-

A typical PTZ site driver (receiver)

driven P/T heads, which are driven by synchronous motors, have fixed speeds because they depend on the mains frequency.

The number of wires required between the site driver and the PTZ head is as follows: five wires are required for the basic pan and tilt functions as described earlier (pan left, pan right, tilt up, tilt down, and common), four wires for pan and tilt preset positioning (positive pot supply, negative pot supply, pan feedback, and tilt feedback), three wires for the zoom and focus functions (sometimes four are required, depending on the zoom lens), and four wires for the zoom and focus preset positioning. This makes a **total of 16 wires**. The thickness of the preset wires and the zoom and focus wires is not critical as we have very low current for these functions, but the pan and tilt wires need to be considered carefully because they depend on the pan and tilt motors' requirements. Typically, PTZ site driver is installed next to the camera. If the situation demands, however, the site driver can be up to a couple of hundred meters away from the camera itself.

Built-in PTZ driver

In the past, when ordering a pan and tilt head, it was expected that preset positioning function needs to be specified, since there were pan and tilt heads without this function. Today most of them have this function built-in. Although there are still PTZ cameras that require a separate driver box, and require elaborate wiring for all functions like pan, tilt, zoom, focus and iris, most of the modern ones have an integrated driver electronics.

Today, faster and more precise DC-driven stepper motors are used, with speeds of over couple of hundred degrees per second for preset positions, and clever zoom proportional speed reduction. The zoom proportional reduction in speed is very handy for when a distant object is followed, when the lens is zoomed in. In order to be able to follow the object more accurately, the P/T movement is slowed down proportionally to the zoom increase.

Photo courtesy of Axis

An IP PTZ dome camera

With the introduction of digital IP PTZ cameras all of the above became much more simplified because the PTZ control data is encoded in the Ethernet communication, over the same Cat-6 cable used for bringing back the digital video stream. So instead of having a coaxial cable for video and a separate twisted pair for PTZ data, for IP cameras all we use one Cat-6 cable. The PTZ site drivers described above, are almost always embedded inside such modern designs. Some IP PTZ domes can also be powered over the same cable using the PoE+ standard. Maximum distances in such a case are limited to the Ethernet recommendation, which is typically around 100 m.

Preset positioning

One of the most common called for function in PTZ cameras is the ***preset positioning***. Basically, when a PTZ camera gets an instruction to go to a preset position, either by an operator or as a response to an alarm trigger, it forces the pan and tilt motors to move (the same applies to zoom and focus) until the position sensors reach a preset value. So, if a certain door is protected, for example, by using a simple reed switch a camera can be triggered to automatically turn in that direction, zooming and focusing on the previously stored view of the door.

The number of preset positions a PTZ camera can have these days is more than sufficient and it can cover over 100 of them. When operator needs to see a particular area, a door or a Help Point for example, all it needs to select is the preset associated with that view. Modern PTZ cameras are fast and it takes less than a second to get to such a pre-saved position, making the surveillance more efficient.

The early PTZ domes had separate camera and lens modules

Most modern PTZ cameras also accept local alarm contacts which can be programmed to trigger a preset. Such local alarms can be as simple as a reed switch from a door, or a button from a Help Point. Many PTZ dome cameras may also have a local relay output which can be used to open or close something remotely.

Preset positions are stored in the PTZ camera non-volatile memory.

A very important question is, "How precisely can the preset positions be repeatedly recalled?" This is defined by the mechanics, electronics, and the design itself. The precision of preset positioning is especially important with the very fast pan and tilt units. An error of only a couple of degrees may not be noticed when the view is wide, but it could be critical when the lens is fully zoomed in. This error is, as we said earlier, magnified with the optical zoom magnification. For this reason mounting PTZ cameras on poles and brackets that are prone to oscillation due to wind or traffic passing by need to be considered carefully.

Photo courtesy of Bosch

Some environments require explosion protection PTZ cameras

Camera housings

In order to protect the cameras from environmental influences and/or conceal their viewing direction, we use camera housings.

Camera housings can be very simple and straightforward to install and use, but they can affect the picture quality and camera lifetime if they are not well protected from rain, snow, dust, and wind or if they are of poor quality.

They are available in all shapes and sizes, depending on the camera application and length. Earlier tube cameras and zoom lenses were much bigger, calling for housings of as much as 1 m in length and over 10 kg in weight. Nowadays, solid state cameras are much smaller, and so are the zoom lenses. As a consequence, housings are becoming smaller too.

A lot of attention in the last few years has been paid to the aesthetics and functionality of the housings, such as easy access for maintenance and concealed cable entries.

With camera size reductions, tinted domes are used often instead of housings, offering much better blending with the interior and exterior.

The glass used for housing is often considered unimportant, but optical distortions and certain spectral attenuation might be present if the glass is unsuitable.

A typical outdoor camera housing

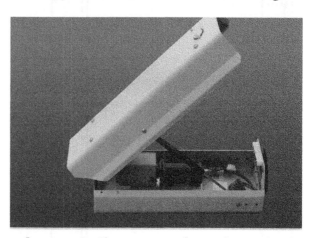

Access to the camera is an important consideration

Corner-mount vandal-resistant housing

Another important factor is the toughness of the glass for camera protection in demanding environments. The optical precision and uniformity are even more critical when domes are used because the optical precision and glass (plastic) distortions are more apparent. Tinted domes are often used to conceal the camera's viewing direction. For tinted domes, light attenuation has to be taken into account. This is usually in the range of one F-stop, which is equal to half the light without the dome.

A lot of housings have provision for heaters and fans. Heaters might be required in areas where a lot of moisture, ice, or snow is expected. Usually, about 10 W of electrical energy is sufficient to produce enough heat for a standard housing interior. The heaters

IP	First number: *Protection against solid objects*	IP	Second number: *Protection against liquids*	IP	Third number: *Protection against mechanical impacts*
0	No protection of the electrical equipment	0	No protection of the electrical equipment	0	No protection of the electrical equipment
1	Protected against solid objects up to 50mm, e.g., accidental touch by hands	1	Protected against vertical falling drops of water	1	Impact 0.225 Joules — 15cm, 150g
2	Protected against solid objects over to 12mm, e.g., fingers	2	Protected against direct sprays of water up to 15° from the vertical	2	Impact 0.375 Joules — 15cm, 250g
3	Protected against solid objects over 2.5mm (tools, wires)	3	Protected against direct sprays of water up to 60° from the vertical	3	Impact 0.50 Joules — 20cm, 250g
4	Protected against solid objects over 1mm (tools, wires, small wires)	4	Protected against water sprayed from all directions – limited ingress permitted	4	N / U
5	Protected against dust-limited ingress (no harmful deposit)	5	Protected against low pressure jets of water from all directions – limited ingress permitted.	5	Impact 2 Joules — 40cm, 500g
6	Totally protected against dust	6	Protected against strong jets of water, e.g., for use on shipdecks – limited ingress permitted	6	N / U
		7	Protected against the effects of immersion between 15cm and 1m — 15cm - 1m	7	Impact 6 Joules — 40cm, 1500g
		8	Protected against long periods of immersion under pressure	8	N / U
				9	Impact 20 Joules — 40cm, 5kg

IP (Ingress Protection) ratings - not to be confused with IP (Internet Protocol)
Ratings for protection from dust, water and ingress of electrical equipment for voltages of up to 1000 V AC and 1200 V DC as per RS data book

can work on 12 V DC, 24 V AC, or even mains voltage. Check with the supplier before connecting. No damage will occur if a 240 V (or even 110 V) heater is connected to 24 V (however, sufficient heat will not be produced), but the opposite is not recommended. Also, avoid improvisations without any calculations, such as connecting two 110 V heaters in series to replace a single 240 V heater (110 + 110 = 220). The little bit of difference (240 V instead of 220 V) is enough to produce excessive heat and cause a quicker burning of the heater and may even cause a fire. Heaters are typically installed to defog the housing glass in cold environments, hence they are mounted close to the glass.

Fans should be used in areas with very high temperatures, and sometimes they can be combined with heaters. The voltage required for the fan to work can also be DC or AC, but be sure to use good-quality fans, as most of the DC fans will sooner or later produce sparks from the brushes when rotating, which will interfere with the video signal. So heaters and fans are an extra obligation, but if you have to have them, make sure you provide them with the correct and sufficient power. They are usually set to work automatically by a thermostat with a rise or fall in temperature (i.e., there is no need for manual control).

Special housings are required if a **wash and wipe** assembly is to be added to the PTZ camera. They are special because of the matching required between the wipe mechanism and the housing window. It should be pointed out that when the wash/wipe assembly is used, the PTZ driver needs to have output controls for these functions as well. Another responsibility when using washers is to make sure that the washer bottle is always filled with a sufficient amount of clean water.

Housings, PTZ cameras domes and boxes that are exposed to environmental influences are rated with the **Ingress Protection** (IP) numbers. The IP rating is shown in the table on the left page. This acronym should not be mixed up with the IP for Internet Protocol. The IP numbers indicate to what degree of shock, dust, and water aggression the box is resistant. For example, a camera housing rated at IP66 is "totally protected against dust" (6) and "protected against strong jets of water with limited ingress" (6).

Most camera housings are well protected from the environment, but in special system designs, even better protection might be required. **Vandal-proof housings** are required in systems where human or vehicle intervention is predicted, so a special, toughened (usually lexan) glass needs to be used, together with special locking screws. Tamper switches may also be added for extra security. In cases like that, the tamper alarm has to come back to the control center, usually through the PTZ driver, providing it has such a facility.

And last, **bullet-proof, explosion-proof,** and **underwater housings** are also available, but they are very rare, specially built, and very expensive. We will therefore not dedicate any space to them in this book, but should you need more details contact your local supplier.

Special liquid-cooled housing for up to 1300° C

Lighting in CCTV

Most of the CCTV systems with outdoor cameras use both day and night light sources for better viewing. Systems for indoor applications use, obviously, indoor (artificial) light sources, although some may mix with daylight, as when sunlight penetrates through a window.

The sun is our daylight source, and as discussed in the ***Light and Television*** chapter, the light intensity can vary from as low as 100 lx at sunset to 100,000 lx at noon. The color temperature of sunlight can also vary, depending on the sun's altitude and the atmospheric conditions, such as clouds, rain, or fog. This might not be critical for B/W cameras, but a color system will reflect these variations.

Artificial light sources fall into three main groups, according to their spectral power content:

   1. Sources that emit **radiation by incandescence**, such as candles, tungsten electric lamps, and halogen lamps.

   2. Sources that emit radiant energy as a result of an **electrical discharge through a gas or vapor**, such as neon lamps and sodium and mercury vapor lamps.

   3. **Fluorescent tubes**, in which a gas discharge emits visible or ultraviolet radiation within the tube, causing phosphors on the inside surface of the tube to glow with their own spectrum.

The light sources of the first group produce a **smooth and continuous light spectrum** as per the Max Planck formula, similar to the black body radiation law. These light sources are very suitable for monochrome cameras because of the similarity in the spectrums, especially on the left side of the CCD/CMOS sensor spectral response.

The second group of light sources produces almost discrete components of particular wavelengths, depending on the gas type.

The third group has a more continuous spectrum than the second one, but it still has components of significant levels (at particular wavelengths only), again, depending on the type of gas and phosphor used.

The last two groups are very tricky for color cameras. Special attention should be paid to the color temperature and white balance capability of the cameras used with such lights.

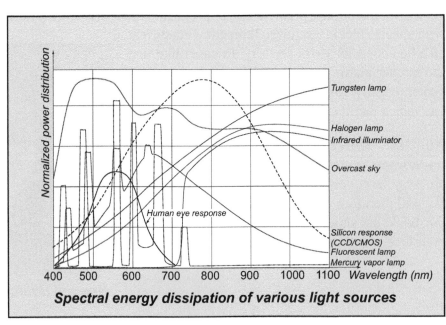

Spectral energy dissipation of various light sources

Infrared lights

When cameras need to see events at night, sensors with removed infra-red cut filter and infra red illuminators should be used. Cameras that have no ***infrared cut-filter*** (IRC) are the B/W (monochrome) models, but color cameras with removable IRC can also be used. The latter ones are typically referred to as Day/Night (D/N) cameras. Not long ago, in CCTV there were B/W and color cameras and one would choose based on whether the main usage would be at night or day-time. Today, all cameras come as color, but some have the D/N capability by automatically removing the IRC filter for improved low light vision and even better with infra red light.

Infrared light is used because the native sensitivity of the Silicon (CCD and CMOS sensors) has very good sensitivity in and near the infrared region. These are the wavelengths longer than 700 nm. As mentioned at the beginning of this book, the human eye can see up to 780 nm, with the sensitivity above 700 nm being very weak, so in general we say that the human eye only sees up to 700 nm.

Sensor without IRC filter see much better in the infrared portion of the spectrum. The reason for this is the nature of the photo-effect itself.

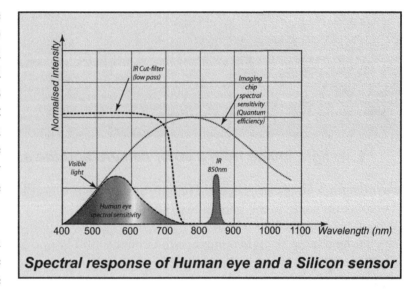

Spectral response of Human eye and a Silicon sensor

Longer wavelength photons (which are usually blocked by the ORC filter in a color camera) penetrate the Silicon structure more deeply. The infrared response is especially high with B/W CCD chips, or color ones without an IRC filter.

A couple of infrared light wavelengths are common to CCTV infrared viewing. Which one is to be used and in what case depends first on the camera's spectral sensitivity (various manufacturers have different spectral sensitivity sensors) and, second, on the purpose of the system.

The two typical infrared wavelengths used with **halogen lamp illuminators** are: one starting from around 715 nm and the other from around 830 nm.

If the idea is to have infrared lights that will be visible to the public, the 715 nm

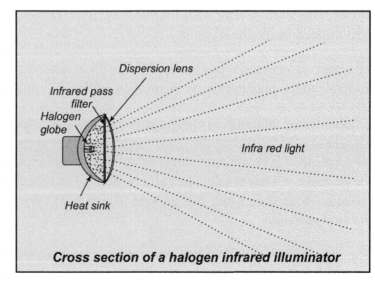

Cross section of a halogen infrared illuminator

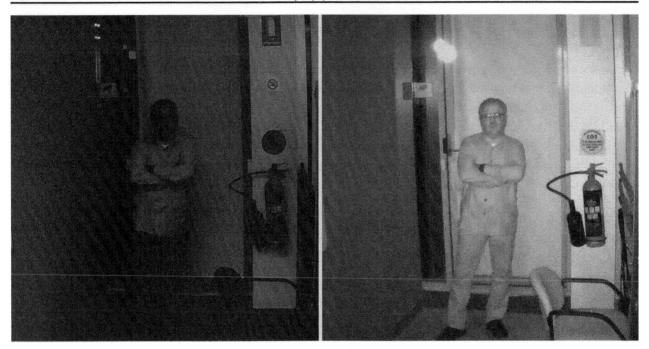

Low light image from a color camera and the same with infrared illuminator

wavelength is the better choice. If night-time hidden surveillance is wanted, the 830 nm wavelength (which is invisible to the human eye) should be used.

The halogen lamp IR light come in two versions: 300 W and 500 W. The principle of operation is very simple: a halogen lamp produces light (with a similar spectrum as the black body radiation), which then goes through an *optical high-pass filter*, blocking the wavelengths shorter than 715 nm (or 830 nm). This is why we say wavelengths **starting** from 715 nm or **starting** from 830 nm. The infrared radiation is **not one frequency only but a continuous spectrum starting from the nominated wavelength.**

The energy contained in the wavelengths that do not pass the filter is reflected back and accumulated inside the infrared illuminator. There are heat sinks on the IR light itself that help cool down the unit, but still, the biggest reason for the short MTBF (1000–2000 hr) of the halogen lamp is the excessive heat trapped inside the IR light.

The same description applies to the 830 nm illuminators; only in this case we have infrared frequencies invisible to the human eye. As mentioned earlier, 715 nm is still visible to many.

These infrared illuminators may pose a certain danger, especially for installers and maintenance people. The reason for this is that the human eye's iris stays open since it does not see any light, so blindness could result. This can happen only when one is very close to the illuminator at night, which is when the human eye's iris is fully opened. The best way to check that the IR works is to feel the temperature radiation with your hand; human skin senses heat very accurately. Remember, heat is nothing but infrared radiation.

The halogen infrared illuminators are mains operated, and photo cells are used to turn them on when light falls below certain lux level.

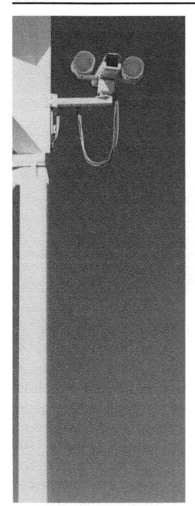

Halogen IR lights

Both types of halogen infrared illuminators mentioned come with various types of dispersion lenses, and it is desirable to know what angle of coverage is best for a situation. If the infrared beam is concentrated to a narrow angle, the camera can see farther, provided a corresponding narrow angle lens is used (or a zoom lens is zoomed in).

Halogen lamp infrared lights offer the best illumination possible for night surveillance, but their short lifespan has initiated new technologies, one of which is the ***solid-state infrared LEDs*** (Light-Emitting Diodes) mounted in the form of a matrix. This type of infrared is made with high-luminosity infrared LEDs, which have a much higher efficiency than standard diodes and radiate a considerable amount of light, yet require much less electrical power. Such infrared lights come with a few different power ratings: 7 W, 15 W, and 50 W. They are not as powerful as the halogen ones, but they are smaller, and their MTTF over 100,000 hr. There are IP HD cameras today which come with built-in high efficiency IR LEDs, able to be powered over a PoE switch and illuminate areas up to 25 m.

How far you can see with such infrareds depends on the camera in use and its spectral characteristics. It is always advisable to conduct a site test at night for the best understanding of distances. The angle of dispersion is limited to the LED's arrangement, and this usually ranges between about 30° and 40°, if no additional optics are placed in front of the LED matrix.

Another type of IR used in applications is an ***infrared LASER diode*** (LASER = light amplification by stimulated emission of radiation). Perhaps not as powerful as the LEDs, but with a laser source, the wavelength is very clean and coherent. A typical LASER diode radiates light in a very narrow angle, so a little lens is used to disperse the beam (usually up to about 30°). Lasers use very little power.

One last technical note is about the focusing point of a projected infra red image on a sensor with IRC filter. Since the infra red wavelength are longer than the visible light, when IRC filter is removed, the focusing point of the infra red wavelengths falls behind the sensor pixel plane. The image may look slightly blurry. In order to fix this, either the lens needs to be re-focused for the night view or ***infra-red corrected lens*** needs to be used.

Photo courtesy of Dallmeier electronic

An IP HD camera with built-in LED IR

Ground loop correctors

Even if all precautions have been taken during installation in an analog system with coaxial cables, problems of a specific nature often occur: *ground loops*.

Ground loops are an unwanted phenomenon caused by the ground potential difference between two distant points. It is usually the difference between the camera and the monitor ground point, but it could also be between a camera and a switcher, or two cameras, especially if they are daisy-chained for synchronization purposes. The picture appears wavy and distorted. Small ground loops may not be noticeable at all, but substantial ones are very disturbing for the viewers. When this is the case, the only solution is to galvanically isolate the two sides. This is usually done with a video **isolation transformer**, sometimes called a *ground loop corrector* or even a *hum bug* unit.

A Ground loop transformer

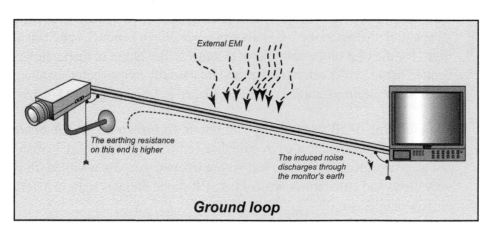

Ground loop

Ground loops can be eliminated, or at least minimized, by using monitors or processing equipment with *DC restoration*. The DC restoration is performed by the input stage of a device that has DC restoration, where the "wavy" video signal is sampled at the sync pedestals so as to regenerate a "straight" DC level video signal. This in effect eliminates low-frequency induction, which is the most common ground loop artifact. A better solution, though a more expensive one, is the use of a fiber optics cable instead of a coaxial, at least between the distant camera(s) and the monitor end.

Ground loops can also be eliminated when using twisted pair video transmission, because such a connection doesn't use a reference to the ground at both ends, but rather a differential signal between two wires. Most common cable used for this is Cat-6, which possesses an excellent twisted pair characteristic.

The appearance of ground loop

Lightning protectors

Lightning is a natural phenomenon about which there is not much we can do to prevent. Lightning induces strong electromagnetic forces in copper cables. The closer it is the stronger the induction is. PTZ sites are particularly vulnerable because they have copper video, power and control cables concentrated in the one area. **A good and proper earthing is strongly recommended** in areas where intensive lightning occurs, and of course surge arresters (also known as spark or lightning arresters) should be put inside all the system channels (control, video, etc.). Most good PTZ site drivers have **spark arresters** built in at the data input terminals and/or galvanic isolation through the communication transformers.

A coax lightning protector

Spark arresters are special devices made of two electrodes, which are connected to the two ends of a broken cable, housed in a special gas tube that allows excessive voltage induced by lightning to discharge through it. They are helpful, but they do not offer 100% protection.

An important characteristic of lightning is that it is dangerous not only when it directly hits the camera or cable but also when it strikes within close range. The probability of having a direct lightning hit is close to zero. The more likely situation is that lightning will strike close by (within a couple of hundred meters of the camera) and induce high voltage in all copper wires in the vicinity. The induction produced by such a discharge is sufficient to cause irreparable damage. Lightning measuring over 10,000,000 V and 1,000,000 A are possible so one can imagine the induction it can create.

Again, as with the ground loops, the best protection from lightning is using a fiber-optic cable; with no metal connection, no induction is possible.

In-line video amplifiers/equalizers

When coaxial cables are used for analog video transmission of distances longer than what is recommended for the particular coax, *in-line amplifiers* (sometimes called *video equalizers* or *cable compensators*) are used.

The role of an in-line amplifier/equalizer is very straightforward: **it amplifies and equalizes the video signal**, so by the time it gets to the monitor end it is restored, more or less, to the levels it should be when a camera is connected to a monitor directly next to it.

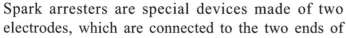

Video equalizer/amplifier

If no amplifier is used on long runs, the total cable resistance and capacitance rise to the values where they affect the video signal considerably, both in level and bandwidth. When using a couple hundred meters of coaxial cable (RG-59), the video signal level can drop from the normal 1 V_{pp} down to 0.2 or 0.3 V_{pp}. Such levels become unrecognizable to the monitor, encoder or DVR. As a result, the contrast is very poor, and the syncs are low, so the picture starts breaking and rolling. In addition, the higher frequencies are attenuated

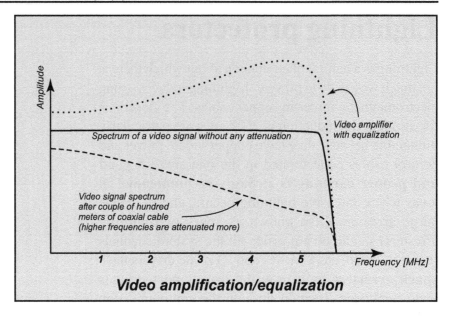

Video amplification/equalization

much more than the lower ones, which is reflected in the loss of fine details in the video signal. From fundamental electronics it is known that higher frequencies are always attenuated more because of various effects such as the skin effect or impedance-frequency relation, to name just two. This is why equalization of the video signal spectrum is necessary and not just the amplification of it.

Obviously, with every amplification stage the noise is also amplified. That is why there are certain guidelines, with each in-line amplifier/equalizer, that need to be followed. Theoretically, it would be best if the amplifier/equalizer is inserted in the middle of the long cable run, where the signal is still considerably high relative to the noise level. However, the middle of the cable is not a very practical place, mainly because it requires power and mounting somewhere in the field or under the ground. This is why most manufacturers suggest one of the two other alternatives.

The first and most common alternative is to install the in-line amplifier at the camera end, often in the camera housing itself. In such a case, we actually do a **pre-amplification and pre-equalization**, where the video signal is boosted up and equalized to unnatural levels, so that by the time it gets to the receiving end (the distance should be roughly known), it drops down to $1V_{pp}$.

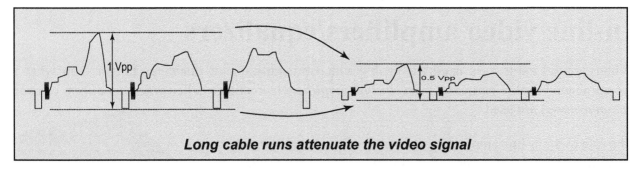

Long cable runs attenuate the video signal

The second installation alternative is to place the amplifier at the monitor end, where more noise will accumulate along the length, but the amplification can be controlled better and needs to be brought up from a couple of hundred millivolts to a standard 1 V_{pp}. This might be more practical in installations where there is no access to the camera itself.

In both of the above alternatives, a potentiometer is usually available at the front of the unit, with calibrated positions for the cable length to be compensated. In any case, it is of great importance to know the length of the cable you are compensating.

A number of in-line amplifiers can be used in series, that is, if 300 m of RG-59/U is the maximum recommended length for a B/W signal, 1 km can be reached by using two amplifiers (some manufacturers may suggest only one for runs longer than a kilometer) or maybe even three. Note: **the noise cannot be avoided, and it always accumulates**. Furthermore, the risk of ground loops, lightning, and other inductions, with more (two or three) in-line amplifiers, will be even greater.

Again, if you know in advance that your installation has to go over half a kilometer, the best suggestion is to use fiber optics. Many would suggest an RG-11/U coaxial cable instead, where a single run, without an amplifier, can go up to 600 – 700 m, but the cost of fiber optics these days is comparable to, if not lower than, the RG-11/U. We have already covered the many advantages of fiber.

Video distribution amplifiers (VDAs)

Very often, a video signal has to be taken to a couple of different users: a switcher, a monitor or a DVR, for example. This may not be possible with all of these units because not all of them have the looping video inputs. Looping BNCs are most common on monitors. Usually, there is a switch near the BNCs, indicated with "75 Ω" and "High" positions. This is a so-called *passive input impedance matching*. If you want to go to another device, a monitor, for example, the procedure is to switch the first monitor

Photo courtesy of Pelco
Video distribution amplifier

to **High Impedance** and loop the coaxial cable to the second monitor, where the impedance setting should be at 75 Ω.

This is important, as we discussed earlier, because a **camera is a 75 Ω source and it has to see 75 Ω at the end of the line in order to have a correct video transmission with 100% energy transfer** (i.e., no reflections).

Now picture a situation, very common in CCTV, where a customer wants to have two switchers at two different locations, showing the same cameras but independently from each other. This can easily be solved by using two video switchers, one looping and the other terminating, where we can use the same logic as with the monitors.

In practice, however, simple and cheap switchers are usually made with just one BNC per video input, which means that they are *terminating inputs* (i.e., with 75 Ω input impedance and looping).

It would be wrong to use BNC T adaptors to loop from one switcher to another, as many installers do. This is incorrect because then we will have two 75 Ω terminations per channel, so the video cameras will see incorrect impedance, causing partial reflection of the signals, in which case the reproduction will be with double imaging and incorrect signal reproduction.

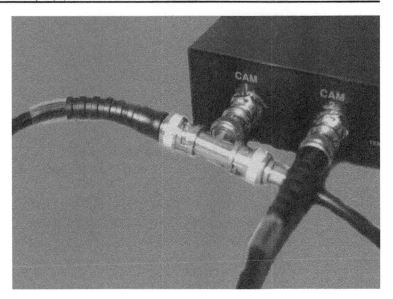

The solution for these sorts of cases is the use of the *video distribution amplifier* **(VDA)**. VDAs do exactly what the name suggests – they distribute one video input to more outputs, preserving the necessary impedance matching. This is achieved with the use of some transistors or op-amp stages. Because active electronics is used (where power needs to be brought to the circuit), this is called *active impedance matching*.

A T-piece looping

A typical VDA usually has one input and four outputs, but models with six, eight, or more outputs are available as well. One VDA is necessary for each video signal, even if not all four outputs are used.

Video matrix switchers use the same concept as the VDAs when distributing a single video signal to many output channels. In such a case, only a limited number of VDA stages can be used in bigger matrix systems. This is because every new stage injects a certain amount of noise, which with analog signals cannot be avoided.

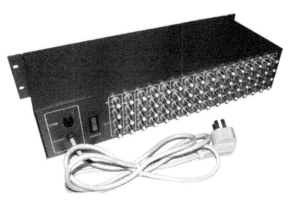

Multi-channel VDA

There is no equivalent concept of VDA in Digital IP CCTV, but the closest function to this are the network switches, which in fact can be considered as repeaters.

The main difference in this analogy between analog and digital is the fact that the distributed signals in analog suffer from noise beeing added at each stage, while in digital the video streams are regenerated exactly as the original stream, practically being immune to external noise. This is certainly a big advantage, although strictly speaking noise can and will influence the digital streams if Ethernet distances are exceeded for the given media.

This page intentionally left blank

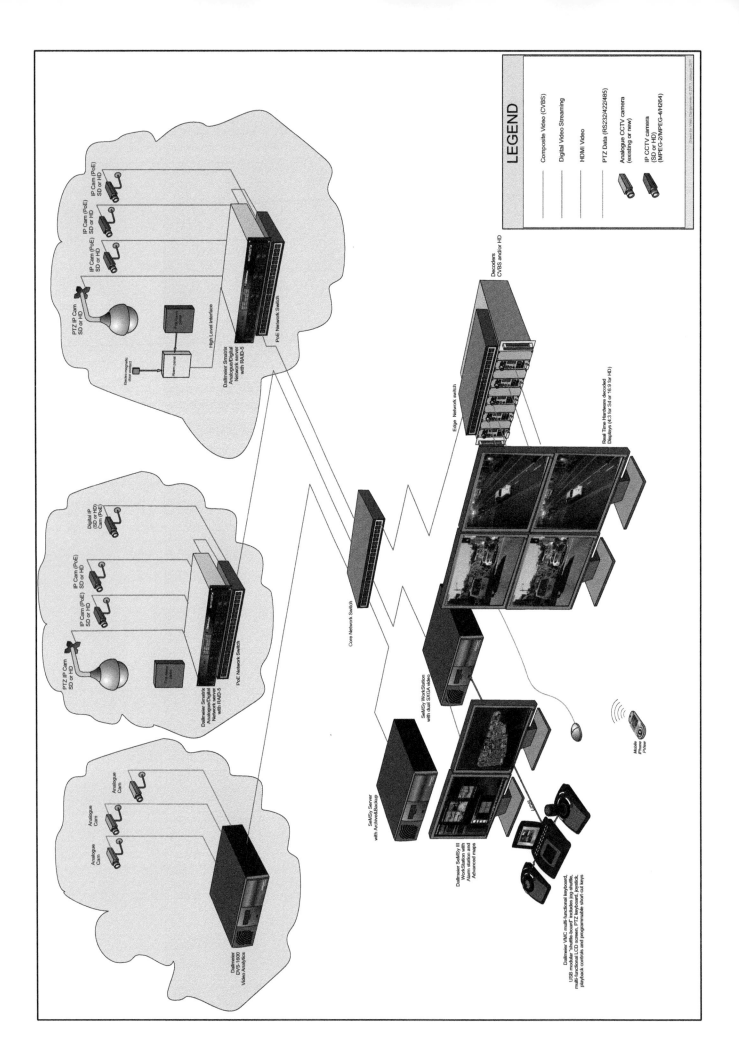

PTZ IP Cam
SD or HD

IP Cam (PoE)
SD or HD

IP Cam (PoE)
SD or HD

IP Cam (PoE)
SD or HD

Electro-magnetic
door contact

Alarm panel

High Level Interface

Dallmeier Smatrix
Analogue/Digital
Network server
with RAID-5

PoE Network Switch

Digital IP
(SD or HD)
Cam (PoE)

IP Cam (PoE) SD or HD

PTZ IP Cam
SD or HD

IP Cam (PoE) SD or HD

Dallmeier Smatrix
Analogue/Digital
Network server
with RAID-5

PoE Network Switch

Analogue
Cam

Analogue
Cam

Analogue
Cam

Dallmeier DVS-1600
Video Analytics

Core Network Switch

Edge Network switch

Decoders
CVBS and/or HD

Real Time Hardware decoded
Displays (4:3 for Std or 16:9 for HD)

SeMSy WorkStation with
dual SXGA video

SeMSy Server
with Archive&Backup

Mobile
Phone
PDA

USB

Dallmeier SeMSy III
WorkStation with
Alarm station and
Advanced maps

USB modular "shuttle-board" includes keyboard,
multi-functional LCD screen, PTZ keyboard, joystick,
playback controls and programmable short cut keys.
Dallmeier VMC multi-functional keyboard,
multi-functional LCD screen, jog-shuttle,

LEGEND

———— Composite Video (CVBS)

———— Digital Video Streaming

———— HDMI Video

———— PTZ Data (RS232/422/485)

Analogue CCTV camera
(existing or new)

IP CCTV camera
(SD or HD)
(MPEG-2/MPEG-4/H264)

13. CCTV system design

Designing a CCTV system is a complex task, requiring at least basic complete, knowledge of all the stages and components in a system. More importantly, prior to designing the system, we need to understand what the customer expects from it, or, if it is a project tender, understand its requirements.

Understanding the system requirements

The first and most important preparation, before commencing a system design and pricing it, is to understand the customer's requirements. Customers can be technically minded people, and many would understand CCTV as well as we do, but most often they may not be aware of the latest technical developments and capabilities of each component.

The starting point is to understand the general concept of the surveillance the customer wants: constant monitoring of cameras and activities undertaken by 24-hour security personnel (*pro-active*), or perhaps just an unattended operation (usually with recording), which is reviewed later (*re-active*), or maybe a mixture of the two. Once this is clarified, it might be a good idea to explain the capabilities and limitations of the equipment you would be proposing. This is reasonably easy done with smaller and simpler systems, but once they grow to a bigger size, include PTZ cameras, multiple monitors, operators, DVRs, and similar, things will get tougher.

Many unknown variables need to be considered and many questions answered: How will the cameras be selected? How to find and playback an incident? What happens when alarm goes off? Which monitor should display the alarms? How many days retention will the recording have? How many hard drives are required to achieve that? What are the light levels of the surveyed areas at night? And so on.

Those are the variables that define the system complexity and as in mathematics, in order to solve a system with more variables, one needs to know more parameters. A customer may be able to answer some of these questions, but a CCTV expert should know better, using the customer's feedback.

Understandably, it is imperative for a CCTV expert to know the components, the hardware, the software and how they interconnect.

You can create a favorable impression in the customer's mind if at the end you give him/her as much as, if not more, than what you have promised. You will prove unsatisfactory if you do not. When the customer is fully satisfied the first time, chances are he or she will come back to do business with you again.

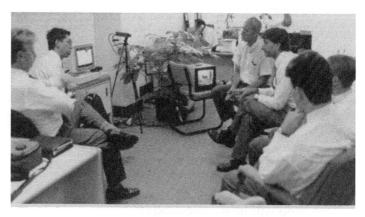

Listening and talking to customers is very important

Designing for a customer should also include the control room

To put it simply: Do not claim the system will do this; if you are not certain, make sure your system delivers what you say it will.

So, to design a good, functional system, one has to know the components used, their benefits and limitations, how they interconnect, and how the customer wants them to be used.

The first few parts are assumed to be fulfilled, since you would not be doing that job unless you knew a few things about CCTV. The last one — what the customer wants — can be determined during the first meeting, over a phone call or e-mail. Usually, the next step is to conduct a site inspection.

Here is a short list of questions you should ask your customer prior to designing the system and before or during the site inspection:

- What is the main purpose of the CCTV system?

If it is a deterrent, you need to plan for cameras and monitors that will be displayed to the public. If it is a concealed surveillance, you will need to pay special attention to the camera type and size, its protection, concealed cabling, and the like, as well as when it is supposed to be installed (after hours perhaps). In many countries a concealed surveillance system may need to be approved by the legal authority.

- Who will be the operator(s)?

If a dedicated 24 hour guard is going to use the system, the alarm response may need to be handled differently from the response expected when unattended, or a partially attended, system operation. Also, how many operators will be on duty at any given time. This defines the number of keyboards and monitors.

• Will it be analog or digital system, and if digital what type: 1080 HD, 720 HD or MegaPixel?

Although most of the new systems today will be digital without any questions asked, it is possible that a customer requires and upgrade to an existing analog system, and may still require analog cameras. The answer to this question will dictate the price, as well as the cabling infrastructure. If analog cameras are proposed, the recorders can still be digital (analogue signal is converted to digital first), but hard disk capacities may be very different if HD cameras are recorded.

• What are the light levels, especially at night?

Often, customers would have no understanding of what light levels are at their premises, hence it is a good idea to plan a site inspection with a lux meter and check the critical low light areas. Different low light performance cameras may be used on one job, but maybe D/N cameras with IR illuminators may need to be considered too. This decision may especially be tricky if using PTZ dome cameras without the possibility to use IR light that will pan and tilt together with the camera.

• How many cameras will be used?

A small system with up to a dozen, or 16 cameras can be easily handled by a single DVR/NVR, and perhaps a simple small VMS. In such a case expected storage capacity can easily be calculated and achieved, but bigger systems may need multiple DVRs/NVRs, and they need to be connected in a well-planned network. If IP cameras are used, allowance for network switcher ports have to be made to cater for all IP cameras, DVRs, NVRs, and workstations.

 • How many of the cameras will be fixed focal length and how many PTZ?

There is a big difference in price between the two because if a PTZ camera is used instead of a fixed one, the extra cost is in the zoom lens, plus the pan and tilt head or dome, the site driver, and the control keyboard to control it (although control can be made from a software keyboard in the VMS). But, the advantages of having a PTZ camera will be manyfold. If on top of this, preset positioning PTZ cameras are used, the system flexibility and efficiency will be too great to be compared with the fixed camera system. If this is the choice, PTZ data transmission needs to be considered, maximum distances and how many cameras on the same data bus. If a digital PTZ camera is proposed, the PTZ data is easy handled over the same Ethernet

Always work as a team

connection used for video streaming. Certainly, if digital system is proposed, network design and switches need to be considered.

• How many monitors and control keyboards are required?

If it is a small system, one monitor and keyboard is the logical proposal, but once you get more operators and/or channels to control and view simultaneously, it becomes harder to plan a practical and efficient system. Then, an inspection of the control room is necessary in order to plan the equipment layout and interconnection. It might be required by the customer for a new control room layout, with the new equipment. If this is the case, some room measurements will be required. Today there are some free 3D CAD programs (like Google SketchUp) that can be learnt very quickly, and are sufficient for a realistic 3D representation of the control room, operator's consoles monitor size and positions, etc.

• Will the system be used for live monitoring, i.e. be "pro-active" (which will require an instant response to alarms), or perhaps recording of the signals for later review and verification, i.e. "re-active"?

The answer to this question will define the number of operators and availability of the control room. If there is 24 hour live monitoring the system alarm response should be designed properly and efficiently. If the system is reactive, intelligent analysis to help find an event quicker, like smart find, line crossing, direction of movement, or similar, should be considered.

3D software like SketchUp can be used for control design

• What transmission media can be used on the premises?

While coaxial cable was the most common media used during the analog CCTV period, today more and more installers are putting Cat-6 even for analog, using twisted pair transmitters and receivers, but also making it ready for when the time comes and the customer wants to switch to fully digital system. So, Cat-6 cables are currently most often recommended and installed as they can be used for both analog and digital CCTV. For larger distances, fiber optics is still the best method, although there is an increasing number of wireless LAN links, especially in urban areas. If the premises are subject to regular lightning activity, fiber optics is again the safest and best media, and although at first may seem most expensive, there will be huge savings in the long run. This needs to be relayed to the customer.

• Lastly, and probably the most important thing to find out, if possible, is what sort of budget is planned for such a CCTV system?

This question will define and clarify some of the previous queries and will force the design to be narrowed down to the type of equipment, the number of cameras, and possibly how the system is going to work. Although this is one of the most important factors, a shortage of it should not force the downgrading of the system to something that may not operate satisfactorily.

If the budget is short of what you may want to propose and cannot allow for the preferred system, it is still beneficial to go back to the customer with your suggested system, despite being over the budget. The first proposal though should come as close under the budget as possible, including all the desired feature that are possible. Perhaps trying to keep the functionality but reduce the number of cameras to bring the proposal under the budget.

Only after presenting the stripped down functionality to meet the budget, try and offer the real system that will cover all of the desired features. The strongest argument you should put forward when suggesting your design is that a CCTV system should be a **secure one and future-proof,** which can only be the case if it is done properly. Thus, the argument could be that by having a well-designed and functional system, that will work for many years to come, bigger savings will be made in the long run.

By presenting a fair and detailed explanation of how **you think the system should work**, the customer will usually accept the proposal.

A well-designed control room

Site inspections

After the initial conversation with the customer and assuming you have a reasonably good idea of what is desired, you have to make a site inspection where you would usually collect the following information:

- Cameras: type (analog (SD) or HD, fixed or PTZ, resolution, etc.).

- Lenses: angles of view, zoom magnification ratio for zoom lenses.

- Camera protection: housing type (standard, weatherproof, dome, discrete, etc.), mounting (wall mount, in ceiling, of a bracket, on a pole,...). An unwritten golden rule for a good picture is to try and keep the camera away from directly facing sunlight.

- Light: levels, light sources in use (fluorescent, mercury vapor, tungsten, etc.), east/west viewing direction. Visualize the sun's position during various days of the year, both summer and winter. This will be very important for overall picture quality that changes during the year.

- Video receiving equipment: location, control room area, physical space, and the console.

- Air-conditioning in the equipment room: clean and cool air is especially important when using DVRs and NVRs with multiple hard disk drives.

- Monitors: resolution, size, position, mounting, and the like. Think about the operators health and comfort during 8 hour shifts.

- Power supply: type, size (always consider more amperes than what are required). Is there a need for an uninterruptable power supply (UPS)? (VA rating in that case).

During site inspections notice the camera views relative to the sun

- Check the available power locations for the cameras and other equipment. Find the power cable trays and plan cabling not too close to the mains cables.

- If installation is in a building locate cable raisers, and see how the cable wiring is done. It will be very useful if a building plan can be obtained.

- If pan/tilt heads are to be used: type, size, IP rating, control type (RS-485, Cat-6, fiber or maybe wireless). Mounting positions? Type of brackets? If pole mounting, avoid or minimise possible oscillation.

• Make a rough sketch of the area, with the approximate initial suggestions for the camera positions. Take into account, as much as possible, the installer's point of view. A small change in the camera's position, which will not affect the camera's performance, can save a lot of time and hassle for the installer and in the end, money for the customer.

• Put down the reference names of areas where the customer wants (or where you have suggested) the cameras to be installed. Also write down the reference names of areas to be monitored because

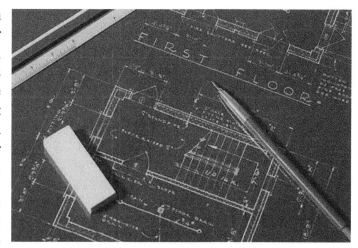

Building plans may be very useful

you will need them in your documentation as reference points. Be alert for obvious "no-nos" (in respect to installation), even if the customer wishes something to be done. Sometimes small changes may result in high installation costs or technical difficulties that would be impossible to solve. It is always easier to deter the customer from making changes by explaining why in the initial stage, rather than having to do so later in the course of installation when additional costs will be unavoidable.

Designing and quoting a CCTV system

With all of the above information, as well as the product knowledge (which **needs constant updating**), you need to sit down and think. Designing a system, like designing anything new, is a form of art. As is true of many artists, your work may not be rewarded immediately, or it may not be accepted for some reason. But think positively and concentrate as if that is to be the best system you can propose. With a little bit of luck you may make it the best, and tomorrow you can use it as a reference or proudly show it to your colleagues and customers.

Different people will use different methods. There is, however, an easy and logical beginning.

Always start with a hand drawing of what you think the system should feature. This could be on paper, or perhaps on a white board when discussing it with your colleagues. Draw the monitors, cameras, housings, interconnecting cables, power supplies, and so on. While drawing you will see the physical interconnection and component requirements. Then you

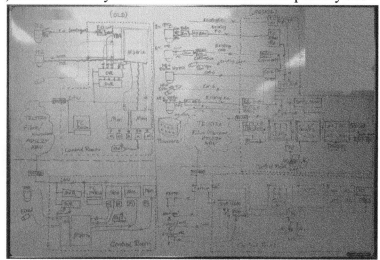

A white board hand-sketch of a system upgrade

will not omit any of the little things that can sometimes be forgotten, such as camera brackets, types of cable used, and cable length. Making even a rough hand sketch will bring you to some corrections, improvements, or perhaps further inquiries to the customer. You may, for example, have forgotten to check what the maximum distance for the PTZ control is or how far the operators are to be from the central video processing equipment, power cable distances, voltage drops, and so on.

Once you have made the final hand drawing, you will know what equipment is required, and it is at this point that you can **make a listing of the proposed equipment**. Then, perhaps, you will come to the stage of matching camera/lens combinations. Make sure that they will fit in the housings or domes you intend to use. This is another chance to glance through the supplier's specifications booklet.

Do not forget to take into account some trivial things that may make installation difficult, like the coaxial cable space behind the camera (remember, it is always good to have at least 50 mm for BNC terminations), the focusing movement of a zoom lens (as mentioned earlier in the chapter on zoom lenses, in a lot of zoom lenses focusing near makes the front optical element protrude for an additional couple of millimeters), and so on.

The next stage is **pricing the equipment** – costs, sales tax and duty, installation costs, profit margins, and, the most important of all (especially for the customer), the **total price**.

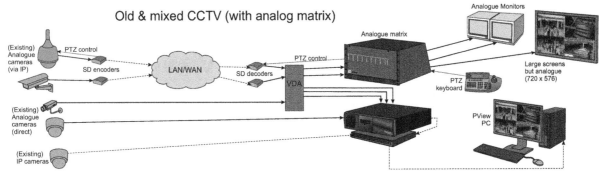

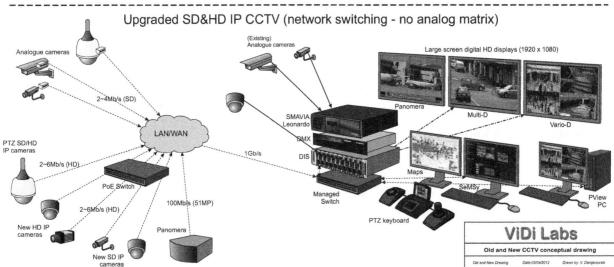

Evolution of a system design to a computer representation

Do not forget to **include commissioning costs** in there, although a lot of people break that up and show the commissioning figure separately. This is more of a practical matter, since the commissioning cost may vary considerably and it could take longer or shorter than planned. General practical experience shows that it will always take at least three times longer than planned. Also, in the commissioning fees, time should be allocated for the CCTV **operator's training**.

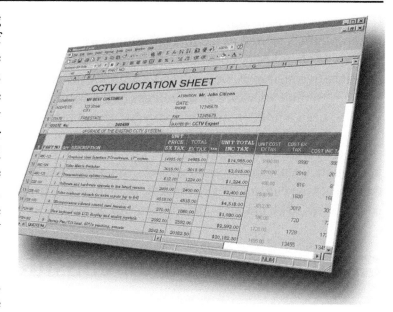

After this step has been completed, you need to make a **final and more accurate drawing** of the system you are proposing. This can be hand drawn, but most CCTV designers these days use computers and CAD programs. It is

A spreadsheet program can help a lot in preparing a precise CCTV quotation

easier and quicker (once you get used to it), and it looks better.

Also, the hand-calculated price needs to be written in **a quotation form, with a basic explanation of how the system will work and what it will achieve**. It is important for this to be written in a concise and simple, yet precise form, because quotations and proposals (besides being read by security managers and technical people) are also read by nontechnical people such as purchasing officers and accountants.

Often, spreadsheet programs are used for the purpose of precise calculation, and this is another chance to double-check the equipment listing with your drawing and make sure nothing has been left out.

If you are an integrator or installer, a project document with a time-line and milestones will also be needed and useful. If the man-hours estimated in the installation costs are close to, or below, the actual cost, it will make the system profitable, but if they are unrealistic or wrong, it will bring the profit margins down. For this reason sitting down and making a good project time-line estimation will help not only sell the project but actually make good profit.

As with any quotation, it is expected and professional to **have a set of brochures enclosed** for the components you are proposing.

In the quotation, you should not forget to include your company's **terms and conditions of sale** which will protect your legal position.

Proper documentation is always important

If the quotation is a response to a **tender** invitation, you will most likely need to submit a **statement of compliance, as well as project milestones and time-line.** This is where you confirm whether your equipment complies or does not comply with the tender requirements. This is where you also have to highlight eventual extra benefits and features your equipment offers. In the tender, you may also be asked to commit yourself to the progress of the work and supply work insurance cover, in which case you will need a little bit of help from your accountant and/or legal advisor.

Many specialized companies only design and supply CCTV equipment, in which case you will need to get a quote from a specialized installer, who, understandably, will need to inspect the site. It is a good practice, at the end, to have all the text, drawings, and brochures bound in a single document, in a few copies, so as to be practical and efficient for reviewing and discussions.

Camera installation considerations

Strapping a PTZ bracket around a concrete column

If you are a CCTV system designer, you do not have to worry about how certain cables will be pulled through a ceiling, raisers, or camera pole mounting; that is the installer's job. But it would be very helpful and will save a lot of money, if you have some knowledge in that area. If nothing else, it is a good practice, before you prepare the final quotation, to take your preferred installer on site, so that you can take into account his or her comments and suggestions of how the practical installation should be carried out.

First, the most important thing to consider is the type of cable to be used for video, power, and data transmission, their distances and protection from mechanical damage, electromagnetic radiation, ultraviolet protection, rain, salty air, and the like. For this purpose it is handy to know the surrounding area, especially if you have powerful electrical machinery next door, which consumes a lot of current and could possibly affect the video and control signals. Powerful electric motors that start and stop often may produce a very strong electromagnetic field and may even affect the phase stability of the mains. This in turn will affect the camera synchronization (if line-locked cameras are used) as well as the monitor's picture display. This is especially the case with analog cameras connected with coaxial cable.

For example, there might be a radio antenna installed in the vicinity, whose radiation harmonics may influence the high-frequency signals your CCTV system uses.

Mounting considerations are also important at both the camera and monitor end. If poles are to be installed, not only the

A hexagonal metal pole is stiffer than a circular pole

height, but also the elasticity of the poles is important. Steel poles, for example, are much more elastic than concrete poles. They may oscillate more when it is windy.

If a PTZ camera is installed, the zoom lens magnification factor will also magnify the pole's movement which could result from wind, or vibrations from the pan/ tilt head movement itself. This magnification factor is the same as the optical magnification (i.e., a zoom lens, when fully zoomed in, may magnify a 1 mm movement of the camera due to wind to a 1 m variation at the object plane).

Hinged PTZ camera pole makes servicing easier

The shape of the pole is also very important – hexagonal poles are less elastic than round ones of the same height and diameter.

Pan/tilt swinging bracket

The same logic applies to camera and pan/tilt head mounting brackets. A very cheap bracket of a bad design can cause an unstable and oscillating picture from even the best camera.

If the system needs to be installed in a prestigious hotel or shopping center, the aesthetics are an additional factor to determine the type of brackets and mounting. It is especially important then not to have any cables hanging.

The monitoring end demands attention to all aspects. It needs to be durable (people will be working with the equipment day and night) or aesthetic (it should look good) and practical (easy to see pictures, without getting tired of too much noise and flashing screens).

Since all of the cables used in a system wind up at the monitoring end and in most cases this is the same room where the equipment is located, special attention needs to be paid to cable arrangement and protection.

A large cable installation can be a challenge

Often, cables lying around on the floor for a few days (during the installation) are subject to people walking on them, which is enough weight to damage the cable characteristics, especially the coaxial cable impedance. Remember, the impedance depends on the physical relation between the center core, the insulation, and the shield. If a bigger system is in question, it is always a better idea to propose a raised floor, where all the cables are installed freely below the raised floor.

Sometimes, if a raised floor is not possible, many cables can be run over a false ceiling, on cable trays. In such cases special care should be taken to secure the cables as they could become very heavy when bundled together.

Many installers fail to get into the habit of labelling or marking the cables properly. Most of them would know all of the cables at the time of installation, but two days later they may not be able to remember. Cable labelling is especially critical with larger and more complex systems. Cables can be marked with permanent markers, although this is not as permanent as some other systems, because it easily fades and erases during the pulling or terminating. Also, if the marking is not consistent and at the same point, with neat and legible writing, it may easily be missed or not found when looking for it. There are proper labelling systems on the market that use various labels, some of them just with numbers, others with

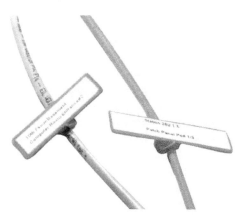

tags that allow for more detailed description of the cable: where is the cable coming from, where it is connected to, and where does it go.

Insist on proper and permanent cable markings as per your drawings. There are plenty of special cable-marking systems on the market. In addition,

One of the recognized labelling

Another labelling type

a listing of all the numbers used on the cables should be prepared and added to the system drawings. Remember, good installers differ from bad ones in the way they terminate, run, arrange, and mark the cables, as well as how they document their work. Larger installations may want a patch panel for the video signals. This is usually housed in a 19" rack cabinet, and its purpose is to break the cables with special coax link connectors so as to be able to reroute them in case of a problem or testing.

Head-end equipment racks

Like with the electronic equipment of other systems in a building (e.g. building management, fire and elevator control), the CCTV and Security equipment at the head-end is commonly installed in cabinets and racks. Cabinets and racks offer physical protection of the equipment, allow for neater and more organized cabling and labelling, and allow for logical separation of the security and CCTV equipment from other systems, allowing for quicker access to the video information when needed by the authority.

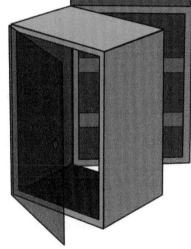

If a system is of a reasonably small size it is possible to have only one or two boxes of digital recording equipment, some camera power supplies and an uninterruptable power supply (UPS), a small lockable cabinet might suffice for all equipment. Such cabinets can be made to sit on the floor on casters (wheels), but some can be made to be installed on the walls, certainly with a proper mounting support. It is not uncommon to have such cabinets hinged so that the whole front part of the cabinet can be opened, thus allowing an easy access to the cabling going at the back of the video equipment.

A small hinged rack

The EIA-310-D standard

The *Electronic Industries Alliance* (EIA) created the EIA-310-D standard in 1992 for cabinets, racks, panels, and associated equipment. This is the most common enclosure when larger systems are used, also popularly known as 19" rack system. The EIA-310-D specification addresses form, fitting, and function for system spacing, mounting, and bezel clearance.

The majority of CCTV head-end equipment is designed to fit neatly and easily in EIA-310-D racks, either by way of mounting special 19" mounting brackets, or by using shelves. The smallest elementary equipment height based on this standard is the *Rack Unit* (RU) which is 1.75" (44.45 mm).

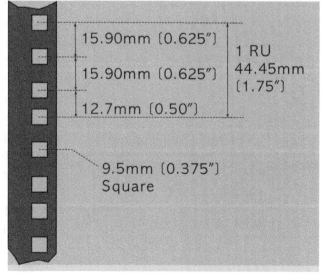

The elementary mounting dimensions

There are different rack heights available, ***up to 45 RU*** height (as can be seen in the table 13.1). The depth of EAI-310-D racks can also be different, but most common are internal depths of 19" (483 mm), 27" (690 mm), 31" (790 mm), and 35" (890 mm).

No matter what size rack is used in a system, the most important aspect are the cabling and the cooling. There are many accessories for proper cable tracing, depending on what type of cable and how many are entering the equipment cabinet. One very common and unforeseen problem is unsecured cables at the entry points into DVRs, NVRs, or matrix switchers — which can physically distort or even break the back panel of the equipment due to its bundled weight. This is the reason that cable support railing is very important. The installer should be aware that such accessories are also available from rack manufacturers.

Cable organizers help make it neat

Cooling is another very important part of the cabinet design. Most computer equipment, DVRs, NVRs, and matrix switchers have suction fans at the front, which suck in the air from the front and push it usually behind the equipment. Considering that most racks are locked up cabinets from all sides, it is important to take such air intake out of the cabinet in order to keep it cool, which as consequence will keep the equipment cool. Obeying the basic low of physics that hot air goes up, cooling fans are typically installed at the top of the racks. It should be ensured that the air taken out of the rack should be taken completely out of the equipment room via the room ventilation.

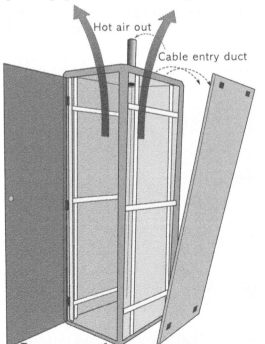

Hot air out

Cable entry duct

Whenever possible, thermometers should be used to monitor the room and rack's inside temperature. There is no clean cut rule for the acceptable temperature levels, but most equipment utilizing processors and hard disks will work better in colder environment. **The colder the better.** This is why many computer server rooms are air conditioned to temperatures preferably below 21°C (70°F). Clean air and cold rooms with low humidity are the best environment for any active electronic equipment.

Cables are always coming from outside the rack(s), and one common method is using a cable entry duct from the ceiling. If equipment room has been planned to have false floor, then another convenient method is by running

Air flow in a cabinet is very important cables from under the cabinet. In each case, suitable

There are various sizes of rack cabinets

cable entry protected by cable duct or raceway needs to be made. Cable hooks and raceways inside the cabinet define the neatness of the installation. Care should always be taken about the minimum recommended bending radius of all cables and not to overtighten with cable ties. Slack cable should be left for testing purposes.

Available access for working with cables and maintenance is another important consideration when positioning racks in a room. Most racks are designed so that equipment access can be made from all sides via removable door-panels, however the most common access required is from the back

of the equipment where most of the video and network cables comes in and terminates.

If more than one rack cabinet needs to be placed next to each other, suitable plans need to be made for access to all cabinets from the back. If video or network patch-panels are to be installed as part of the system design, then appropriate accessories should be sought for such purpose. Some specialized rack manufacturers offer neat protection for various patch panels.

All cables must be marked appropriately, according to the common marking system used throughout the installation. All cables belonging to different purpose should be grouped in their own groups, such as power, network, data, and video. When using variety of network sources and groups, it might be useful to have various colors network cables.

When mounting equipment in the cabinets, whenever possible, allow for some room between items in order to allow for better cooling. Equipment that needs regular visual

Table 13.1: Table of racks standard sizes (EIA 310-D)		
Rack Units (RU)	**Height in mm**	**Height in inches**
1	44.50	1.75
10	445	17.5
15	667	26.25
20	890	35
25	1,112.50	43.75
30	1,335	52.5
31	1,379.50	54.25
32	1,424	56
33	1,468.50	57.75
34	1,513	59.5
35	1,557.50	61.25
36	1,602	63
37	1,646.50	64.75
38	1,691	66.5
39	1,735.50	68.25
40	1,780	70
41	1,824.50	71.75
42	1,869	73.5
43	1,913.50	75.25
44	1,958	77

inspections, such as LEDs showing the working status, should be placed at, or close to, eye level. The same logic should be applied when installing patch-panels, switches and equipment that may require changing cables or connectors—they should be installed near eye-level or near it for easier and quicker identification.

At the end of the rack/cabinet installation **it should be a common practice to prepare and leave documentation (typically on the inside of the cabinet doors) with all the layout of the equipment and its position in the rack, cables that go to it and come out of it. Contact details of the installer or system integrator in charge of the installation should also be made available.**

Patch panels and protectors

CCTV and security installation concentrate all of the signals into these cabinets and racks, thus this is the most representative part of someone's work. It makes sense to pay the biggest attention of workmanship at such places.

Drawings

There is no standard for drawing CCTV system block diagrams, as there is in electronics or architecture. Any clear drawing should be acceptable as long as you have clearly shown the equipment used (i.e., cameras, monitors, or VCRs) and their interconnection.

Many people use technical drawing programs, such as AutoCAD, Visio, CorelDRAW, Illustrator, or other PC or Mac-based drawing packages. Depending on the system size, it might be necessary to have two different types of drawings: one of a CCTV block diagram showing the CCTV components' interconnection and cabling requirements, while the other could be a site layout with the camera positions and coverage area. In smaller installations, just a block diagram may be sufficient.

The CCTV block diagram needs to show the system in its completeness, how the components are interconnected, which part goes where, what type of cable is used, and where it is used.

While the block diagram explains on a higher level how are the components interconnected, and it is typically drafted before the installation commences, or even before the job is awarded, there is a need for a detailed drawing after the installation is completed. Such a drawing should show all the hardware components, their interconnections, labels they were given and possibly IP addresses of each DVR or NVR, as well as cameras. This is typically referred to *as installed drawings*. If the site layout drawing is well prepared, it can later be used as a reference by the installer, as well as by your customer and yourself when reviewing camera locations, reference names, and discussing eventual changes.

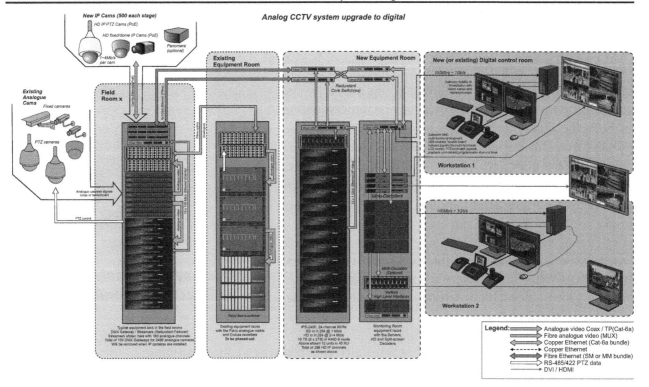

An example system drawing made by the author, using CorelDraw

When the CCTV system is installed and the job is finished, drawings may need small alterations, depending on the changes made during the installation. After the installation, the "as installed" drawings are usually enclosed with the final documentation, which should also include manuals, brochures, and other relevant documentation.

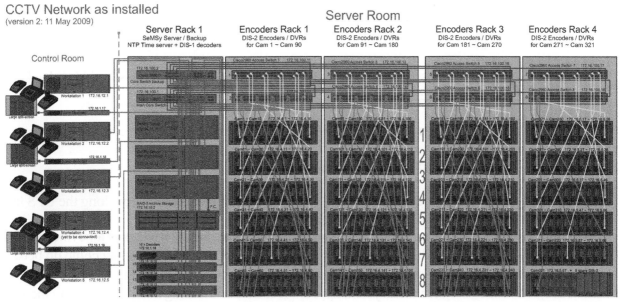

A section of an "as installed" drawing with the network switches and IP addresses

Commissioning

Commissioning is the last and most important procedure in a CCTV system design before handing it over to the customer. It involves great knowledge and understanding of both the customer's requirements and the system's possibilities. Quite often, CCTV equipment programming and setup are also part of this. It includes IP address allocation to all IP devices, VMS programming, DVRs, and NVRs recording configuration, camera views setup, response to alarms and so on.

Detailed system analysis during commissioning

Commissioning is usually conducted in close cooperation with the consultant and customer's system manager and/ or operator(s), since a lot of settings and details are made to suit their work environment.

The following is a typical list of what is usually checked when commissioning:

- All wiring is correctly terminated and labelled.

- Supply voltage is correct to all appropriate parts of the system (local power or PoE).

- Camera type and lens fitted are correct for each viewing position.

- If PTZ cameras are used, their presets and views are as requested.

- Low light level and very bright light levels are handled satisfactory.

- If IP cameras are used, their IP addresses, compression and streaming setting (uni-casting, multi-casting).

- If DVRs/NVRs are installed, the compression used and image quality should be checked in order to achieve the desired days of recording.

- All system controls are properly functioning (pan/tilt, zoom, focus, relays, etc.)

- The VMS pre-set screens and functions.

- Check the system operation under UPS (disconnect mains and check how long the UPS last).

Commissioning larger systems may take a bit longer than the smaller ones. This is an evolution from the system on paper to the real thing, where a lot of small and unplanned things may come up because of new variations in the system concept. Customers, or users, can suggest the way they want things to be done, only when they see the initial system appearance. Commissioning in such cases may therefore take up to a few days.

Training and manuals

After the initial setup, programming, and commissioning are finished, the operators, or system users, will need some form of training.

For smaller systems this is fairly straightforward and simple. Just a verbal explanation may be sufficient, although every customer deserves a written user's manual. This can be as simple as a laminated sheet of paper with clearly written instructions.

Every piece of equipment should come with its own User's Manual, but the key components such as the cameras, DVRs, VMS, keyboards, switches and monitors should be included as a minimum. They should be put together in a system with all their interconnections and this is what has to be shown to the customer. Every detail should be covered, especially alarm response and the system's handling in such cases.

User manuals are often compiled from manufacturers instructions

Often, a large system may have more than one operator, and operators may change with time. So, for a more complex system it is expected that a simplified manual, or synopsis, is produced for quick reference. This is perhaps the most important piece of information to the operators.

For larger systems, it is a good idea to bind all the component manuals, together with the system drawings, wiring details, and operator's instructions, in a separate folder or a binder. Naturally, for systems of a larger size, training can be a more complex task. It may even require some special presentation with slides and drawings so as to cover all the major aspects.

Good systems are recognized not only by their functionality but also by their documentation.

Handing over

When all is finished and the customer is comfortable with what he or she is getting, it is time to hand over the system. This is an official acceptance of the system as demonstrated and is usually backed by the signing of appropriate documents.

It is at this point in time that the job can be considered finished and the warranty begins to be effective. From now on, the customer takes over responsibility for the system's integrity and operation.

If customers are happy with the job they usually write an official note of thanks. This may be used later, together with your other similar letters, as a reference for future customers.

Preventative maintenance

Regardless of the system's use, the equipment gets old and dirty, and faults may develop due to various factors. It benefits the customer if you suggest preventative maintenance of the system after the warranty expires.

This should be conducted by appropriately qualified persons and most often it is the installer that can perform this task successfully. However, a third party can also do this provided the documentation gives sufficient details about the system's construction and interconnection.

The system should be inspected at least twice a year, or in accordance with the manufacturer's recommendations and depending on the environmental aggressiveness. Where applicable, the inspection should be carried out in conjunction with a checklist, or equipment schedule, and should include inspections for loosened or corroded brackets, fixing and cleaning of the cameras housings or domes, monitor screens, checking of DVR/NVR hard drives health status and replacing if necessary, length of recording, air filters replacement (if any), compression versus length of recording correction, improving back-focus on some cameras, and so on.

Preventative maintenance

Larger systems with VMS may require reprogramming of some functions, depending on the customer's suggestions. This is especially required at the first preventative maintenance because the customer/operators would already be familiar with the new system and would have a good idea what is missing or what needs to be added to the functionality of the system. This is a great point in time to learn from the experience of real operators, or perhaps clarify some trivial functionality of the system that have not been picked up or understood during the initial training.

Preventative maintenance should be reinforced as much as possible, especially with the new and digital CCTV systems. It is better to replace a suspicious hard drive before it fails, than after it fails.

Unfortunately, in our industry, the preventative maintenance is seen as unnecessary until things fail.

Experience shows that failure is usually discovered in the worst moment, when a serious incident happens. Somebody gets stabbed, a house was set on fire, or even worst, a bomb was planted in a public place and suddenly the CCTV footage is needed, only to be discovered that it is missing because operators did not know that a hard drive failed, and they did not notice the HD failure alarm. Events like this may happen, or have happened, and it puts nothing but bad name to the CCTV efficiency.

Preventative maintenance should be seen the same as regular scheduled service a new car needs to go through. The car is not necessarily broken, but to keep it running most efficiently and faultlessly a regular service is recommended. If a CCTV system costs equal or more than an expensive car, preventative maintenance should be seen as extension of its life expectancy and efficiency, not as unnecessary cost.

This page intentionally left blank

Please handle your Test Chart with care.

The ViDi Labs SD/HD Test Chart was designed primarily for indoor use.

If used outdoor, avoid direct exposure to rain, snow, dust, or long periods of exposure to direct sunlight.

Although the ViDi Labs Test Chart has been designed specifically for the CCTV Industry,

it can be used to verify the quality of other visual, transmission, encoding and recording systems.

ViDi Labs Pty. Ltd. has designed this chart with the best intentions to offer objective and independent measurement of various video signal characteristics, and although all the details are as accurate as we can make them, we do not take any responsibility for any damage or loss resulting from the use of the chart.

This chart is copyrighted and cannot be copied or reproduced without a specific written permission.

The chart design is subject to change without notice due to ongoing product improvements

14. Video testing

The very first edition of the CCTV book, which was published in 1995 in Australia, introduced a test chart dedicated to the CCTV industry, the first of its kind. This was an attempt to help the readers and CCTV users evaluate their cameras, monitors and systems. This test chart was initially thought of as an integral part of the book, appearing on the back covers. With every new edition of the book there was an improved version of it, with some new and added features. It was found that a separate chart on a larger format can be printed with better quality and can be handled easier if attached on a light but sturdy backing. Over 1,000 CCTV companies, manufacturers, consultants, integrators, and installers have purchased and used some of these test charts throughout the years.

With this last book edition we introduce the latest version of it, which is now an evolved test chart designed for both SD and HD video cameras and systems. Unfortunately, due to the limited and un-controlled print quality of the book covers, we decided to make this version of the test chart available only as a large format, by ordering it through our web site *www.vidilabs.com*.

This last chapter will attempt to explain how the ViDi Labs test chart can be used to make certain system evaluation and measurement. There other test charts available, but the ViDi Labs SD/HD test chart was specifically developed for the CCTV industry, and they are readily available.

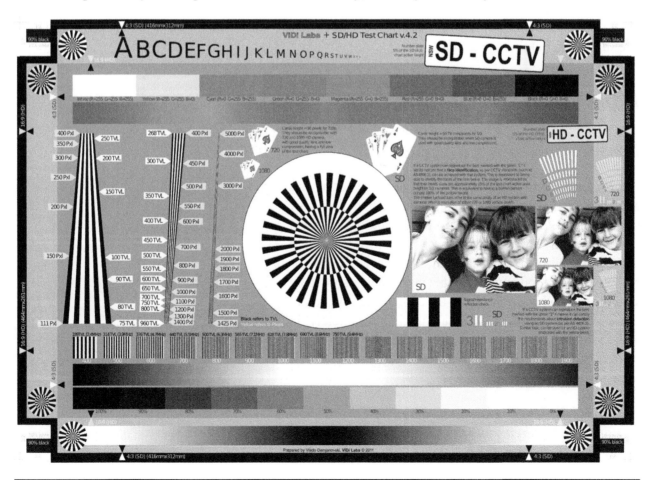

The ViDi Labs SD/HD test chart

In order to help you determine your camera resolution, as well as check other video details, ViDi Labs Pty. Ltd. has designed this special test chart in A3+ format, which combines three charts in one: for testing standard definition (SD) with 4:3 aspect ratio, high definition (HD) with 16:9 aspect ratio and mega pixel (MP) cameras, and systems with 3:2 aspect ratio.

We have tried to make it as accurate and informative as possible and although it can be used in the broadcast applications it should not be taken as a substitute for the various broadcast test charts. Its primary intention is to be used in the CCTV industry, as an objective guide in comparing different cameras, encoders, transmission, recording, and decoding systems.

Using our experience and knowledge from the previously designed CCTV Labs test chart, as well as the feedback we have had from the numerous users around the world, we have designed this SD/HD/MP test chart from ground up adding many new and useful features, but still tried to preserve the useful ones from the previous design. We kept the details used to verify face identification as per VBG (Verwaltungs-Berufsgenossenschaft) recommendation Installationshinweise für Optische Raumüberwachungs-anlagen (ORÜA) SP 9.7/5, and compliant with the Australia Standard AS 4806.2.

With this chart you can check a lot of other details of an analogue or digital video signal, primarily the resolution, but also bandwidth, monitor linearity, gamma, color reproduction, impedance matching, reflection, encoders and decoders quality, compression levels, details quality in identifying faces, playing cards, numbers, and characters.

Before you start testing

Lenses

For the best picture quality you must first select a very good lens (that has equal or better resolution than the CCD/CMOS chip itself). In order to minimise opto-mechanical errors, typically found in vari-focal lenses, we suggest to use good quality fixed focal length manual iris lens or perhaps a very good manual zoom lens. The lens should be suitable to the chip size used on the camera, i.e. its projection circle should cover the imaging chip completely, in addition to offering superior resolution for the appropriate camera. Avoid vari-focal lenses, especially if resolution is to be tested.

Start with a finest lens

NOTE: If your lens has lower resolution then the camera you are testing, then you will have a false conclusion about the camera and the system you are testing.

Shorter focal lengths, showing angles of view wider than 30°, should usually be avoided because of the spherical image distortion they may introduce. A good choice for 1/2" CCD cameras would be an 8 mm, 12 mm, 16 mm, or 25 mm lens. For 1/3" CCD cameras a good choice would be when 6 mm, 8 mm, 12 mm, or 16 mm lens is used. Since the introduction of mega-pixel and HD cameras there are "mega-pixel" lenses that can be used. Although the name "mega-pixel" on the lens may not necessarily be a guarantee for superior optical quality, one would assume this would be better quality than just an average "vari-focal" CCTV lens.

Monitors

Displaying the best and most accurate picture is of paramount importance during testing.

If you are using the chart for testing only analogue SD cameras and systems, it is recommended that you use a high resolution interlaced scanning CRT monitor, with CVBS (Composite Video Signal) input and underscanning feature. High resolution in the analogue world means - a monitor that could offer at least 500TVL. Colour monitors are acceptable only if they are of broadcast, or near-broadcast, quality. Such monitors are harder to find these days, as many monitor manufacturers have ceased their production, but good quality CVBS monitors can still be found. A good try could be the supplier of broadcast equipment close to you.

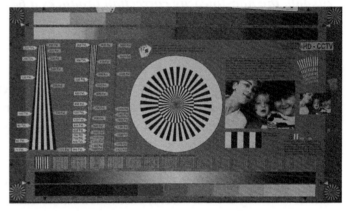

Screen shot of an actual HD camera output

Understandably, cameras having over 500 TV lines of horizontal resolution cannot have their resolution tested with such monitors, but even higher quality are needed, which are only found in the broadcast industry. In the past, when testing camera resolution the best choice were high quality monochrome (B/W) monitor since their resolution, not having RGB mask, reaches 1000 TV lines. Unfortunately, they are almost impossible to find these days.

The next and more common option for good quality display nowdays are LCD and plasma screens. Some CCTV suppliers offer LCD monitors with BNC or RCA inputs for composite video signals too. If an LCD monitor is used for your analogue cameras testing, it is important to understand that the only way to have a composite video displayed on an LCD monitor is by way of converting the analogue signal to digital (A/D) by the monitor itself. The quality of such a conversion, combined with the quality of the LCD display, define how fine details you can see. The LCD monitor may have a high resolution (pixel count) which might be good for your computer image, but may fail when showing composite video. The reason for this could be the A/D conversion itself and the quality of this circuit (A/D precision and the upsampling quality). So, caution has to be excercised if using LCD monitors with CVBS input for measuring analogue camera resolution.

When using LCD monitors for testing your digital SD, HD, and MegaPixel cameras, the first rule is to use the video driver card (of the computer that decodes the video) to run in the native resolution mode of the monitor. The native resolution of the monitor is usually shown in the pixel count specification of the monitor.

Furhermore, if an SD video signal is for example decoded and displayed on an LCD monitor with higher pixel count than the SD signal itself (e.g. PAL digitised signal of 768x576 pixel count is displayed on an XGA monitor of 1024x768) then the best image quality of the SD signal would be if it is shown in the native resolution of that image, e.g. 768x576.

NOTE: The PAL digitisation as per ITU-601 produces a digital picture frame of 720x576 (also known as D1), but when this is decoded, there is a correction of the "non-square pixels", so that 720x576 gets converted to 768x576 pixels with 4:3 aspect ratio. Similar applies to NTSC signals, where from 720x480, the non-linearity correction produces a 640x480 signal.

Tripod

If you are using longer focal length lens on your camera, this will force you to position the camera further away from the test chart. For this purpose it is recommended that you use a photographic tripod.

Some users prefer to use "vertical" setup rather than "horizontal," whereby the test chart is positioned on the floor and the camera up above looking down. This might be easier for the perpendicularity setup. In this case, a larger tripod is recommended with adjustable mounting head so that a camera can be positioned vertically, looking at the test chart. There are photographic tripods on the market which are very suitable for such mounting.

A good tripod may offer more degrees of freedom

Light

When the test chart is positioned for optimal viewing and testing, controlled and uniform light is needed for illuminating the test chart surface.

One of the most difficult things to control is the color temperature of the source of light used in the testing. This is even more critical if color white balance is tested. Caution has to be exercised if correct color and white balance is tested as there are many parameters influencing this values.

Traditionally, tungsten light are used to illuminate the chart from either sides, at a steep angle enough so as to not cause reflection from the test chart surface, but at the same time illuminate the chart uniformly. Tungsten light has different color temperatures depending on the wattage and the type of light. Typically, a 100W tungsten light globe would produce an equivalent color temperature of 2870°K, whilst a professional photographic lights are designed to have around 3200°K.

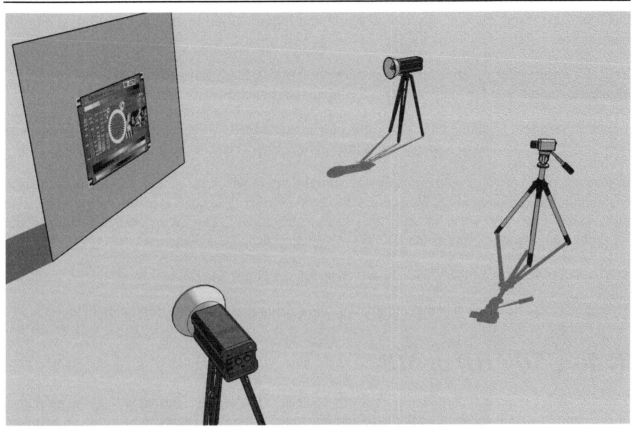

Photographic controlled lighting should be used

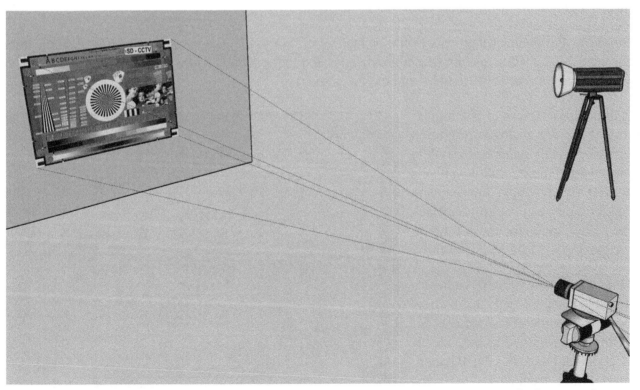

Perpendicular positoning of camera is the most important for resolution testing

Per broadcast standards, a resolution measurement requires illumination of 2000 lux at 3100°K light source.

In CCTV we allow for variation on these numbers because we rarely have controlled illumination as in broadcast, but it is good to know what are the broadcast standards.

With the progress of lighting technology it is now possible to get solid state LEDs light globes with very uniform distribution of light (which is the important bit when checking camera response).

In practice, many of you would probably use natural light, in which case the main consideration is to have uniform distribution of the light across the chart's surface. The chart is made of a matte finish in order to minimise reflections, but still care should be taken not to expose the chart to direct sunlight for prolonged periods of time as the UV rays may change the colour pigments.

The overall reflectivity of the ViDi Labs test chart v.4.1 is 60%.

This number can be used when making illumination calculations, especially at low light levels.

Testing SD / HD or MP

This test chart has actually three aspect ratios on one chart. The aspect ratios, as well as the resolution of each part, has been accurately calculated and fine tuned so that the chart can be used as a Standard Definition test chart, with aspect ratio of 4:3, as a High Definition test chart with aspect ratio of 16:9 and as a MegaPixel with aspect ratio of 3:2.

Since most of the measurements would be made with SD and HD cameras, we have made indicators for the SD to be white in color (the edge triangles and focus stars), and the indicators for the HD to be yellow in color (the edge triangles and the focus stars).

Similar logic refers to the indicators of the SD analogue resolution in TVL (usually black text on white or gray) and the resolution in pixels for HD shown with yellow numbers (under the sweep pattern), or black on yellow area (for the resolution wedges).

Only when analogue (SD) camera is adjusted to view exactly to the white/black edges the measurements for resolution, bandwidth, face identification, and the other detail parameters will be accurate.

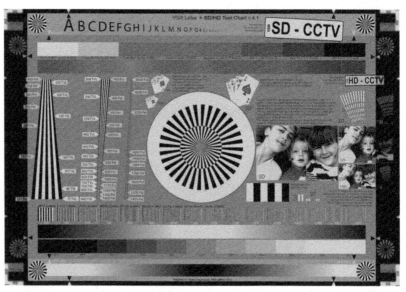

The SD section of the chart (black/white pointers)

The measurements for resolution, pixels count, face identification, and the other detailed parameters will only be accurate when the HD camera is adjusted to view exactly to the yellow/black edges.

Finally, a MegaPixel camera with 3:2 aspect ratio can also be tested, and in this case the complete chart has to be in view, up to the white/black arrows top and bottom, and up to the yellow/black arrows left and right. In such a case, an approximate pixel count can be measured using the yellow/black numbers.

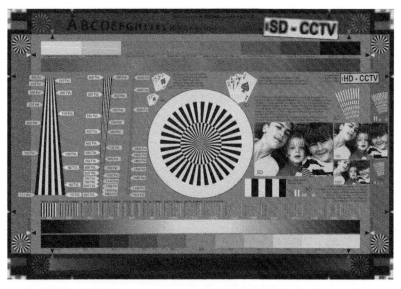

The HD section of the chart (black/yellow pointers)

NOTE: Since there are two HD standards, the SMPTE296 (usually referred to as HD720p) and the SMPTE 274 (usually referred to as HD1080 or True HD) in this test chart we have inserted indicators for details for both of these standards. They are depicted as HD720 and HD1080.

Setup procedure

Position the chart horizontally and perpendicular to the optical axis of the lens.

When testing SD systems - the camera has to see a full image of the SD chart exactly to the white triangles around the black frame. To see this you must switch the CVBS monitor to underscan position so you can see 100% of the image. Without having under-scanning monitor it is not possible to test resolution accurately.

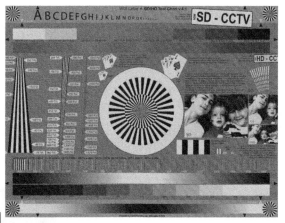

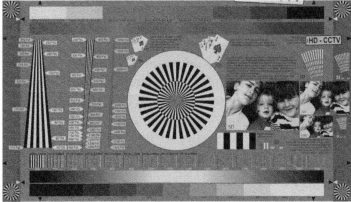

When testing HD systems the camera has to see a full image of the HD chart exactly to the yellow triangles/arrows around the black frame.

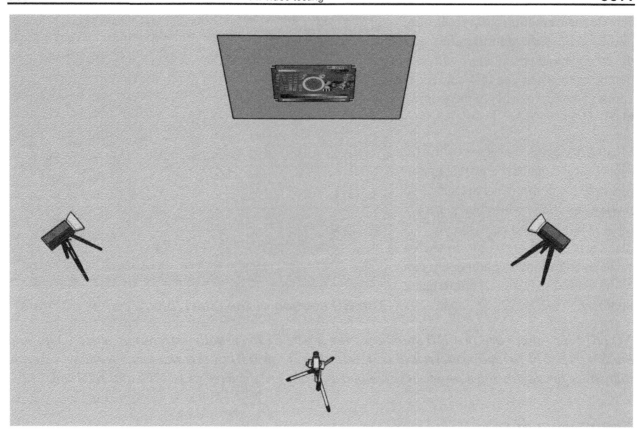

NOTE: The accurate positioning of the camera for SD and HD systems respectively, refers to testing the resolution, pixel count, bandwidth, face identification, playing cards and number-plate detection. Other visual parameters, such as colour, linearity, A/D conversion and compression artefacts can be determined/measured without having to worry about the exact positioning.

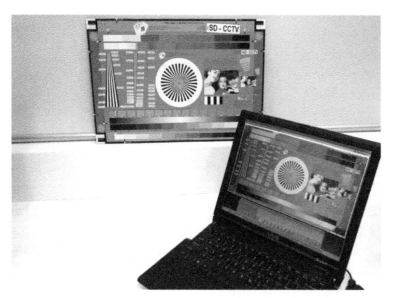

HD reproduction must be seen on equal or better monitor resolution (>1920x1080 pixels)

Illuminate the chart with two diffused lights on both sides, while trying to avoid light reflection off the chart. For more accurate resolution test, the illumination of the test chart, according to broadcast standards, should be around 2000 lux, but anything above 1000 lux may still provide satisfactory results as long as this illumination is constant.

It would be an advantage to have the illuminating lights controlled by a light dimmer, because then you can also test the camera's minimum illumination. Naturally, if this needs to be tested, this whole operation would need to be conducted in a room without any additional light. Also, if you want to check the low-light level performance of your camera you would need to obtain a precise lux-meter.

When using color cameras, please note that most cameras have better colors when switched on after the lights have been turned on, so that the color white balance circuit detects its white point.

Position the camera on a tripod, or a fixed bracket, at a distance which will allow you to see a sharp image of the full test chart. The best focus sharpness can be achieved by seeing the centre of the "focus target" section.

SD reproduction must be seen on equal or better monitor(>600 TVL)

Set the lens' iris to the middle position (F/5.6 or F/8) as this is the best optical resolution in most lenses and then adjust the light dimmer to get a full dynamic range video signal. In order to see this an oscilloscope will be necessary for analog cameras.

For digital, or IP, cameras, good quality computer with viewing/decoding software will be needed. Care should be taken about the network connection quality, such as the network cable, termination, and the network switch.

For analog cameras, make sure that all the impedances are matched, i.e., the camera "sees" 75 Ohms at the end of the coaxial line.

When measuring minimum illumination of a camera, it is expected that all video processing circuitry inside the camera electronics are turned off, e.g. AGC, Dynamic Range, IR Night Mode, CCD-iris, BLC, and similar.

What you can test

To check the camera resolution you have to determine the point at which the five wedge lines converge into four or three. That is the point where the resolution limits can be read off the chart, but only when the camera view is positioned exactly to the previously discussed white/black or yellow/black arrows. The example on the right shows a horizontal resolution of approximately 1,300 pixels when HD camera is tested. If, hypothetically, this image was from a 4:3 aspect ratio camera, it would have had an equivalent analogue resolution of approximately 900TVL.

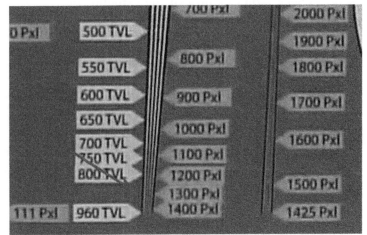

Visual resolution at the merging lines point

If you want to check the video band-width of the signal read the megahertz number next to the finest group of lines where black and white lines are distinguishable. On the right example, one can notice that the analogue band-width indication ends up at 9.4MHz (or 750TVL) which is sufficient to cover what an analogue SD camera can produce. The real example to the right shows blurring of the 1,400 pixels pattern. This is the same consistent camera result as in the example explaining the wedges above.

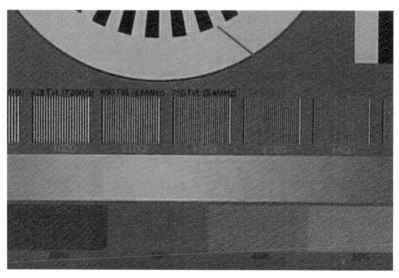

The resolution has an equivalent pixel count

NOTE: A camera that is designed to produce a True HD signal (1920x1080) doesn't necessarily mean it will show 1,920 pixels on the test chart. Details can be lost due to compression, misalignment, low light or bad lens/focus.

The concentric star circle in the middle, as well as around the chart corners, can be used for easy focusing and/or back-focus adjustments. Prior to doing this, you should check the exact distance between the camera and the test chart. In most cases, the distance should be measured to the plane where the CCD chip resides. Some lenses though, may have the indicator of the distance referring to the front part of the lens.

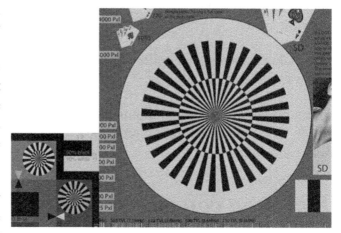

Circle patterns for easy focusing

The main circle may indicate the non-linearity of (usually) a CRT monitor, but it can also be used to check A/D circuitry of the cameras, or monitor stretching, like in cases when there is no pixel for pixel mapping. The imaging CCD/CMOS chips, by design, have no geometrical distortions, but it is possible that A/D or compression circuitry introduces some non-linearity and this can be checked with the main circle in the middle. The big circle in the centre can also be used to see if a signal is interlaced or progressive (progressive would show smoother lines).

The smaller starred circles around the corners of the chart can be used not only for focus and back-focus adjustments, but also for checking the lens distortions, which typically appears near the corners. On some cameras it is possible to have lens optical axis misaligned, i.e. the lens not being exactly per-pendicular to the imaging chip, in which case the four small starred circles around the test chart will not appear equally sharp.

The wide black and white bars on the right-hand side have two-fold function. Firstly, they will show you if your impedances are matched properly or if you have signal reflection, i.e. if you have a spillage of the white into the black area (and the other way around), which is a sign of reflections from the end of the line of an analogue camera. The same clean black/white bars can show you the quality of a long cable run (when analogue cameras are tested), or, in the case of a DVR/encoder/decoder, its decoding/playback quality.

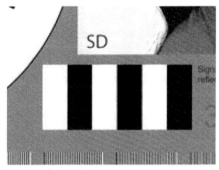

Black/white bars for reflection

The children's head shots, as well as the white and yellow patterns on the right-hand side, can be used to indicate face identification as per Australian Standard AS4806.2 where a person's head needs to occupy around 15% of the SD test chart height. This is equivalent to having 100% person's height in the filed of view, as per AS4806.2. The equivalent head dimensions have been calculated and represented with another two smaller shots of the same, one referring to HD720 and the other to HD1080 when using the 16:9 portion of the chart.

The same can be measured by using the white and yellow patterns, as per VBG (Verwaltungs-Berufsgenossenschaft) recommendations. The white pattern refers to SD cameras with 4:3 aspect ratio and the yellow ones refer to HD720 and HD1080 respectively.

Face identification

If you can distinguish the pattern near the green letter C, then you can identify a face with such system. If your system can further distinguish B, or, even better, A pattern, then the performance of such a system exceeds when compared to a system where only C can be distinguished. However, distinguishing the C pattern only is sufficient to comply with the standards.

NOTE: It is the total system performance that define the measured details. This includes the lens optical quality and sharpness, the angle of coverage (lens focal length), the camera in general (imaging chip size, number of pixels, dynamic range, noise), the illumination of the chart, the compression quality, the decoder quality, and, finally, the monitor itself. This is why this whole testing refers to system measurement rather than camera only. We the observer has 20/20 vision.

Furthermore, the skin color of the three kids' faces will give you a good indication of the Caucasian flesh colors. If you are testing cameras for their color balance you must consider the light source color temperature and the automatic white balance of the camera, if any. In such a case you should take into account the color temperature of your light source, which, in the case of tungsten globes, is around 2800° K. Simplest and easiest to use is the daylight for such testing. Avoid testing color performance of a camera under fluoro-lights or mixed sources of light (e.g. tungsten and fluoro).

Television colors in order of appearance

The color scale on the top of the chart is made to represent typical broadcast electronic color bars consisting of white, yellow, cyan, green, magenta, red, blue, and black colors. These colors are usually reproduced electronically with pure saturated combinations of red, green, and blue, which is typically expressed with intensity of each of these primary colors from 0 to 255 (8-bit colors). Such coordinates are shown under each of the colors. If you have a vectorscope you can check the color output on one of the lines scanning the color bar. Like with any color reproduction system, the color temperature of the source is very important and, in most cases, it should be a daylight source.

NOTE: The test chart is a hard copy of the computer created artwork. Since the test chart is on a printed medium, it uses different color space then the computer color space (subtractive versus additive color mixing). Because of these differences it is almost impossible to replicate 100% accurately these colors on paper. We have certainly used all available technology to make such colors as close as possible, by using Spyder3 color matching system, but with time, and different exposure of the chart to UV and humidity, the color pigment ages and changes. For these reasons we do not recommend using the color bars as an absolute color reference.

The continuous color change can be used to check compression artefacts

The RGB continuous colour strip below the colour bars shown above, is made to have gradual change of colours, from red, through green and blue at the end. This can be used to check how good a digital circuitry, or how high compression, an encoder uses. If the end result shows obvious discountinuity in this gradual change of colors it would indicate that either the encoder or the level of compression is not at its best.

Similarly, using the gray-scales at the bottom of the chart, a few things can be checked and/or adjusted. Using the 11 gray-scale steps Gamma response curve of camera/monitor can be checked. All 11 steps should be clearly distinguished. Monitors can also be adjusted using these steps by tweaking the contrast/brightnest so as to see all steps clearly. When doing so, analogue cameras have to be set so that the video signal is 1 Vpp video signal, while viewing the full image of the test chart. Observe and note the light conditions in the room while setting this up, as this dictates the contrast/brightness setting combination.

Discrete and continuous grey level change for quantization and Gamma

The continuous changing strips, from black to white and from white to black, are made so that their peak of white, and their peak of black, respectively, comes in the middle of the strips. These can also be used to verify and check A/D conversion of a streamer, encoder or compression circuitry. The smoother these strips appear when displayd on a screen the better the system.

Always use minimum amount of light in the monitor room so that you can set the monitor brightness pot at the lowest position. When this is the case the sharpness of the electron beam of the monitor's CRT is maximum since it uses less electrons. The monitor picture is then, not only sharper, but the lifetime expectancy of the phosphor would be prolonged.

Modern day displays (LCD, plasma and similar) do not use electronic beam which could be affected by the brightness/contrast settings, but they will also display better picture, and better dynamic range if the brightness/contrast are set correctly, using the 11 steps mentioned previously.

Lately, there are an increasing number of LCD monitors with composite video inputs, designed to combine a CCTV analogue display, as well as HD. As noted in the very beginning of this manual, under the Monitor heading, please be aware of the re-sampling such monitors perform in order to fill-up a composite analogue video into a (typically) XGA screen (1024X768 pixels). Because of this, LCD monitors are not recommended for resolution testing.

If image testing needs to be done using a frame grabber board on a PC use the highest possible resolution you can find, but not less than the full ITU-601 recommendation (720X576 for PAL, and 720X480 for NTSC). Again, in such a case, native camera resolution testing can not be performed accurately as signal is digitised by the frame grabber. If, however, various digital video recorders are to be compared then the "artificial (digitised) resolution" can be checked and compared.

Sometimes using various fonts sizes may be more useful

The "ABC" fonts are designed to go from 60 points font size for "A" down to 4 points for "Z."

This could also be used for some testing and comparisons amongst systems.

For the casino players, we have inserted playing cards as visual references. The cards may be used to see how well your system can see card details. If you can recognize all four of the cards (the Ace, the King, the Queen, and the Jack) then your system is pretty good.

Casino playing cards in various resolutions

Similar logic was used with the playing card setup for SD, HD720 and HD1080 standards, hence there are three different cards sizes. It goes without saying that when the SD cards are viewed the CCTV camera should be set to view the 4:3 portion of the chart, exactly to the white/black arrows. Similarly, when the HD cards are viewed, the camera has to be set to view the 16:9 portion of the test chart, exactly to the yellow/black arrows.

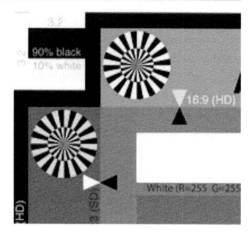

Typically, playing cards height should not be smaller than approximately 50 pixels on the screen, irrespective of whether this is an SD, HD720 or HD1080 system. The playing cards different sizes in this test chart are calculated so that they are displayed on your screen at approximately 50 pixels height.

The corners with 10% and 90%

NOTE: In casino systems the illumination levels are very low, typically around 10 lux or lower. Such a low illumination may influence cards recognition too, so, if realistic testing is to be done, the chart illumination should be around the same low levels.

Finally, the four corners of the test chart have 90% black and 10% white areas. Although these corners fall outside the 4:3 and the 16:9 chart areas, they may still be used to check on system reproduction quality. If you can distinguish these areas from the 100% black border or 100% white (0% black) frame with the "3:2" numbers it suggests your system overall performance is keeping such details and can be classified as very good. If this is not the case, adjustments need to be done either in the camera A/D section (typically where Gamma or brightness/contrast settings are) or where encoder/compression section is.

IMPORTANT NOTE: The lifetime expectancy of the color pigments on the test chart is approximately two years, and it can be shorter if exposed longer periods to sunlight or humidity. It is therefore advised that a new copy is ordered after two years. A special discount is available for existing customers.

Some real-life examples

Here are some more snap-shots from various real camera testing. The quality of this reproduction is somewhat reduced by the PDF/JPG compression of images, as well as the quality of this print in the book, but the main idea can still be seen.

For example, the four snippets below are from a same HD camera, using H.264 video compression, set to 4 Mb/s, using four different lenses, all of which are claimed to be "mega-pixel" lenses.

The snippets shown below are color and HD resolution (1920 pixels wide), but this book print is in B/W so the readers may not see the real difference, but anybody interested in it can e-mail me and I wil make the actual files available for inspection.

It is quite obvious that the first lens has the worst optical quality, and the last has the best. If you would to use this camera/lens combination for face identification, the first one would hardly qualify. The smallest pattern in the yellow group of patterns is not distinguishable, which means face identification will hardly happen. Yet, as shown in the bottom snippet, the same camera with the same compression settings and another (better) lens will pass.

Four different "mega-pixel" lenses produce different results on the same sensor
using the same compression settings

Another real example snapshots are shown below, where two different lenses are tested and compared. Namely, of the two lenses, the top one obviously looks sharp throughout the test chart, while the bottom one appears reasonably sharp in the middle, but becomes blurry as you go towards the periphery of the test chart.

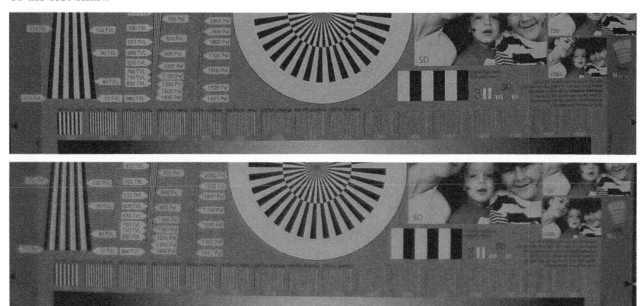

Lens with non-uniform sharpness (bottom)

A test chart at 1.1 lux illumination is below. Resolution in low light cannot be measured due to high noise content.

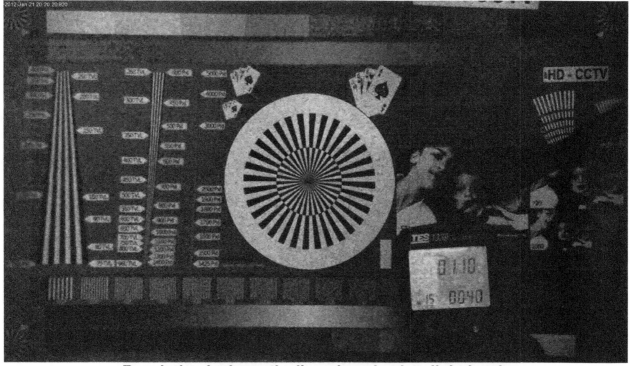

Resolution is dramatically reduced at low light levels

As indicated below, lenses viewing the test chart from a very close distance usually produce geometric "barrel" distortion. It is not recommended for accurate resolution measurement, but, if there is no choice, it may still allow for reasonably good quality measurement

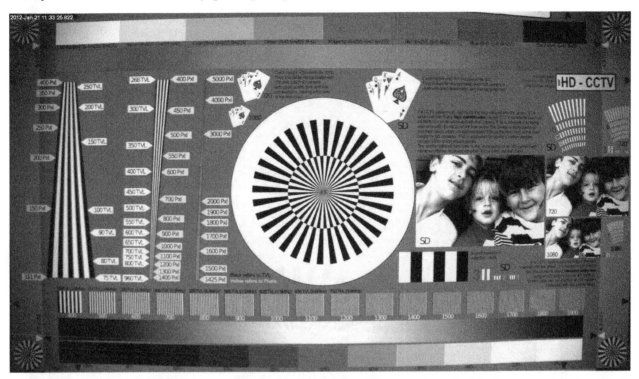

Short focal length lenses produce barrel distortion, not suitable for resolution test

The ViDi Labs test chart can be used for other various testing too, precise focusing at certain distance for example, or just simply to compare details and video reproduction in various circumstances.

The example on the right shows just one more way of using the ViDi Labs test chart for checking dynamic range of various cameras, as conducted by IPVM (*www.ipvm.com*).

And last, but not least, on the next page there is a graphical summary of the key measuring points in the test chart and their meaning.

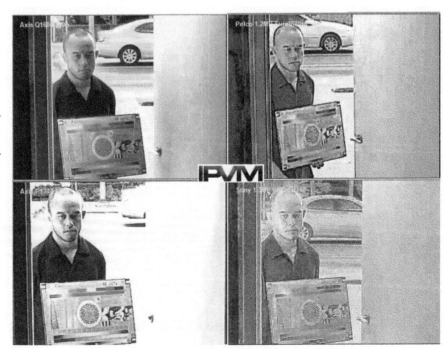

Dynamic range testing by IPVM

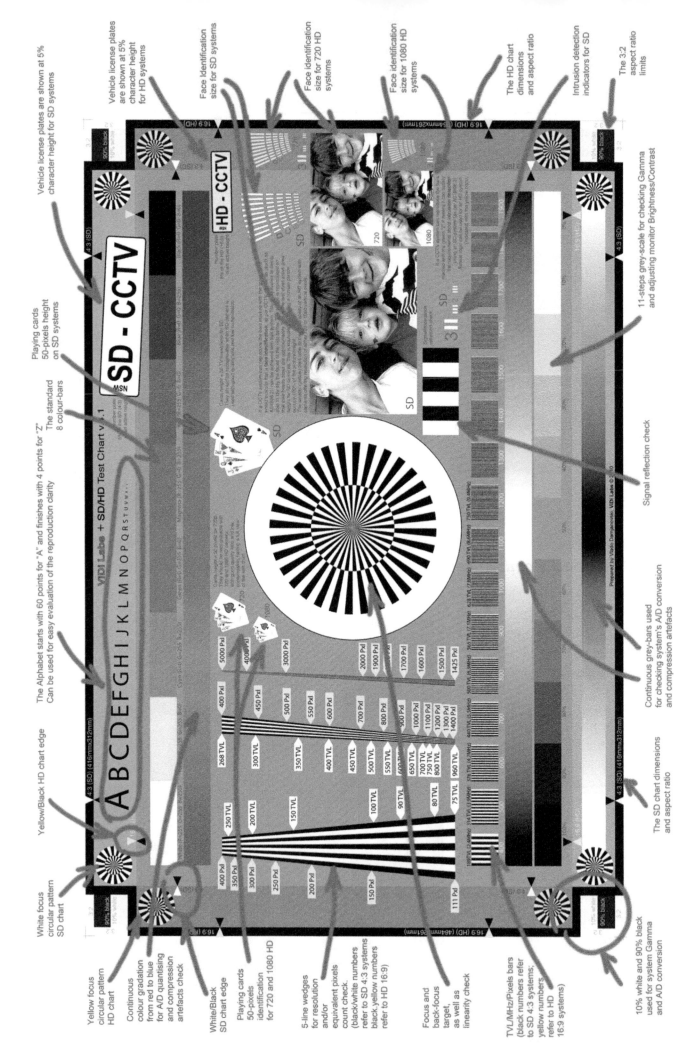

This page intentionally left blank

Appendix A

Common terms used in CCTV

1080i. One of the resolution specs used in the HDTV. 1080i stands for resolution of 1920 × 1080 pixels, and the little "i" means that the video is being interlaced. Other common HDTV resolutions are 720i and 720p.

1080p. Same as above but with progressive scanning.

16:9. Aspect ratio of the High Definition (HD) television standard.

4:3. Aspect ratio of the Standard Definition (SD) television standard.

4k. A new and higher definition television standard with four times the resolution of HD (3,840 x 2,160 pixels).

8k. Experimental higher resolution of four times the 4k, or 16 times the HD (7,680 x 4,320 pxels), Also known as Super High Vision (SHV).

720p. One of the resolution specs used in the HDTV. 720p stands for resolution of 1280 × 720 pixels, and the magic little "p" means that the video is in progressive format. The 720HD format comes only in "p" mode, not "i."

802.11. A range of IEEE standards covering the communication in networking.

Aberration. A term from optics that refers to anything affecting the fidelity of the image in regards to the original scene.

AC. Alternating current.

Activity detection. Refers to a method built into some multiplexers for detecting movement within the camera's field of view (connected to the multiplexer), which is then used to improve camera recording update rate.

AC/DC. Alternating current/direct current.

A/D (AD). Usually refers to analog-to-digital conversion.

ADC. Analog-to-digital conversion. This is usually the very first stage of an electronic device that processes signals into digital format. The signal can be video, audio, control output, and the like.

AGC. Automatic gain control. A section in an electronic circuit that has feedback and regulates a certain voltage level to fall within predetermined margins.

ALC. Automatic light control. A part of the electronics of an automatic iris lens that has a function similar to backlight compensation in photography.

Aliasing. An occurrence of sampled data interference. This can occur in CCD image projection of high spatial frequencies and is also known as Moiré patterning. It can be minimized by a technique known as optical low-pass filtering.

Alphanumeric video generator (also text inserter). A device for providing additional information, normally superimposed on the picture being displayed; this can range from one or two characters to full-screen alphanumeric text. Such generators use the incoming video signal sync pulses as a reference point for the text insertion position, which means if the video signal is of poor quality, the text stability will also be of poor quality.

Amplitude. The maximum value of a varying waveform.

Analog signal. Representation of data by continuously varying quantities. An analog electrical signal has a different value of volts or amperes for electrical representation of the original excitement (sound, light) within the dynamic range of the system.

ANSI. American National Standards Institute.

Anti-aliasing. A procedure employed to eliminate or reduce (by smoothing and filtering) the aliasing effects.

Aperture. The opening of a lens that controls the amount of light reaching the surface of the pickup device. The size of the aperture is controlled by the iris adjustment. By increasing the F-stop number (F/1.4, F/1.8, F/2.8, etc.), less light is permitted to pass to the pickup device.

Apostilb. A photometric unit for measuring luminance where, instead of candelas, lumens are used to measure the luminous flux of a source.

Archive. Long-term off-line storage. In digital systems, pictures are generally archived onto some form of hard disk, magnetic tape, floppy disk, or DAT cartridge.

ARP. Address Resolution Protocol.

Artifacts. Undesirable elements or defects in a video picture. These may occur naturally in the video process and must be eliminated in order to achieve a high-quality picture. The most common are cross-color and cross-luminance.

ASCII. American Standard Code for Information Interchange. A 128-character set that includes the upper-case and lower-case English alphabet, numerals, special symbols, and 32 control codes. A 7-bit binary number represents each character. Therefore, one ASCII-encoded character can be stored in one byte of computer memory.

Aspect ratio. This is the ratio between the width and height of a television or cinema picture display. The present aspect ratio of the television screen is 4:3, which means four units wide by three units high. Such an aspect ratio was elected in the early days of television, when the majority of movies were of the same format. The new, high-definition television format proposes a 16:9 aspect ratio.

Aspherical lens. A lens that has an aspherical surface. It is harder and more expensive to manufacture, but it offers certain advantages over a normal spherical lens.

Astigmatism. The uneven foreground and background blur that is in an image.

Asynchronous. Lacking synchronization. In video, a signal is asynchronous when its timing differs from that of the system reference signal. A foreign video signal is asynchronous before a local frame synchronizer treats it.

ATM. Asynchronous transfer mode. A transporting and switching method in which information does not occur periodically with respect to some reference such as a frame pattern.

ATSC. Advanced Television System Committee (think of it as a modern NTSC). An American committee involved in creating the high definition television standards.

Attenuation. The decrease in magnitude of a wave, or a signal, as it travels through a medium or an electric system. It is measured in decibels (dB).

Attenuator. A circuit that provides reduction of the amplitude of an electrical signal without introducing appreciable phase or frequency distortion.

Auto iris (AI). An automatic method of varying the size of a lens aperture in response to changes in scene illumination.

AWG. American wire gauge. A wire diameter specification based on the American standard. The smaller the AWG number, the larger the wire diameter (see the reference table in Chapter 5).

Back-focus. A procedure of adjusting the physical position of the CCD-chip/lens to achieve the correct focus for all focal length settings (especially critical with zoom lenses).

Back porch. 1. The portion of a video signal that occurs during blanking from the end of horizontal sync to the beginning of active video. 2. The blanking signal portion that lies between the trailing edge of a horizontal sync pulse and the trailing edge of the corresponding blanking pulse. Color burst is located on the back porch.

Balanced signal. In CCTV this refers to a type of video signal transmission through a twisted pair cable. It is called balanced because the signal travels through both wires, thus being equally exposed to the external interference; thus, by the time the signal gets to the receiving end, the noise will be canceled out at the input of a differential buffer stage.

Balun. This is a device used to match or transform an unbalanced coaxial cable to a balanced twisted pair system.

Bandwidth. The complete range of frequencies over which a circuit or electronic system can function with minimal signal loss, usually measured to the point of less than 3 dB. In PAL systems the bandwidth limits the maximum visible frequency to 5.5 MHz, in NTSC to 4.2 MHz. The ITU 601 luminance channel sampling frequency of 13.5 MHz was chosen to permit faithful digital representation of the PAL and NTSC luminance bandwidths without aliasing.

Baseband. The frequency band occupied by the aggregate of the signals used to modulate a carrier before they combine with the carrier in the modulation process. In CCTV the majority of signals are in the baseband.

Baud. Data rate, named after Maurice Emile Baud, which generally is equal to 1 bit/s. Baud is equivalent to bits per second in cases where each signal event represents exactly 1 bit. Typically, the baud settings of two devices must match if the devices are to communicate with one another.

BER. Bit error rate. The ratio of received bits that are in error relative to the total number of bits received, used as a measure of noise-induced distortion in a digital bit stream. BER is expressed as a power of 10. For example, a 1 bit error in 1 million bits is a BER of 10^{-6}.

Betamax. Sony's domestic video recording format, a competitor of VHS.

B-frame. Bidirectionally predictive coded frame (or picture). This terminology is used in MPEG video compression. The B pictures are predicted from the closest two I (intra) or P (predicted) pictures, one in the past and one in the future. They are called bi-directional because they refer to using the past and future images.

Bias. Current or voltage applied to a circuit to set a reference operating level for proper circuit performance, such as the high-frequency bias current applied to an audio recording head to improve linear performance and reduce distortion.

Binary. A base 2 numbering system using the two digits 0 and 1 (as opposed to 10 digits [0–9] in the decimal system). In computer systems, the binary digits are represented by two different voltages or currents, one corresponding to zero and another corresponding to one. All computer programs are executed in binary form.

Bipolar. A signal containing both positive-going and negative-going amplitude. May also contain a zero amplitude state.

B-ISDN. Broadband Integrated Services Digital Network. An improved ISDN, composed of an intelligent combination of more ISDN channels into one that can transmit more data per second.

Bit. A contraction of binary digit. Elementary digital information that can only be 0 or 1. The smallest part of information in a binary notation system. A bit is a single 1 or 0. A group of bits, such as 8 bits or 16 bits, compose a byte. The number of bits in a byte depends on the processing system being used. Typical byte sizes are 8, 16, and 32.

Bitmap (BMP). A pixel-by-pixel description of an image. Each pixel is a separate element. Also a computer un-compressed image file format.

Bit rate. B/s = Bytes per second, b/s = bits per second. The digital equivalent of bandwidth, bit rate is measured in bits per second. If expressed in bytes per second, multiplied with 8 gives bits per second. It is used to express the data rate at which the compressed bitstream is transmitted. The higher the bit rate, the more information that can be carried.

Blackburst (color-black). A composite color video signal. The signal has composite sync, reference burst, and a black video signal, which is usually at a level of 7.5 IRE (50 mV) above the blanking level.

Black level. A part of the video signal, close to the sync level, but slightly above it (usually 20 mV–50 mV) in order to be distinguished from the blanking level. It electronically represents the black part of an image, whereas the white part is equivalent to 0.7 V from the sync level.

Blanking level. The beginning of the video signal information in the signal's waveform. It resides at a reference point taken as 0 V, which is 300 mV above the lowest part of the sync pulses. Also known as pedestal, the level of a video signal that separates the range that contains the picture information from the range that contains the synchronizing information.

Blooming. The defocusing of regions of a picture where brightness is excessive.

Bluetooth. A wireless data standard, used in a variety of electronic devices for close proximity interconnection (see Chapter 11).

BNC. Bayonet-Neil-Concelman connector. It is the most popular connector in CCTV and broadcast TV for transmitting a basic bandwidth video signal over an RG-59 type coaxial cable.

B-picture. Bidirectionally predictive coded picture. This terminology is used in MPEG video compression. The B pictures are predicted from the closest two I (intra) or P (predicted) pictures, one in the past and one in the future. They are called bi-directional because they refer to using the past and future images.

Braid. A group of textile or metallic filaments interwoven to form a tubular structure that may be applied over one or more wires or flattened to form a strap.

Bridge (network). A more "intelligent" data communications device that connects and enables data packet forwarding between homogeneous networks.

Brightness. In NTSC and PAL video signals, the brightness information at any particular instant in a picture is conveyed by the corresponding instantaneous DC level of active video. Brightness control is an adjustment of setup (black level, black reference).

Burst (color burst). Seven to nine cycles (NTSC) or ten cycles (PAL) of subcarrier placed near the end of horizontal blanking to serve as the phase (color) reference for the modulated color subcarrier. Burst serves as the reference for establishing the picture color.

Bus. In computer architecture, a path over which information travels internally among various components of a system and is available to each of the components.

Byte. A digital word made of 8 bits (zeros and ones).

Cable equalization. The process of altering the frequency response of a video amplifier to compensate for high-frequency losses in coaxial cable.

CAD. Computer-aided design. This usually refers to a design of system that uses computer specialized software.

Candela [cd]. A unit for measuring luminous intensity. One candela is approximately equal to the amount of **light energy** generated by an ordinary candle. Since 1948 a more precise definition of a candela has become: "the luminous intensity of a black body heated up to a temperature at which platinum converges from a liquid state to a solid."

CATV. Community antenna television.

C-band. A range of microwave frequencies, 3.7–4.2 GHz, commonly used for satellite communications.

CCD. Charge-coupled device. The new age imaging device, replacing the old tubes. When first invented in the 1970s, it was initially intended to be used as a memory device. Most often used in cameras, but also in telecine, fax machines, scanners, and so on.

CCD aperture. The proportion of the total area of a CCD chip that is photosensitive.

CCIR. Committée Consultatif International des Radiocommuniqué or, in English, Consultative Committee for International Radio, which is the European standardization body that has set the standards for television in Europe. It was initially monochrome; therefore, today the term *CCIR* usually refers to monochrome cameras that are used in PAL countries.

CCIR 601. An international standard (now renamed to *ITU 601*) for component digital television that was derived from the SMPTE RP1 25 and EBU 3246E standards. ITU 601 defines the sampling systems, matrix values, and filter characteristics for Y, Cr, Cb, and RGB component digital television. It establishes a 4:2:2 sampling scheme at 13.5 MHz for the luminance channel and 6.75 MHz for the chrominance channels with 8-bit digitizing for each channel. These sample frequencies were chosen because they work for both 525-line 60 Hz and 625-line 50 Hz component video systems. The term 4:2:2 refers to the ratio of the number of luminance channel samples to the number of chrominance channel samples; for every four luminance samples, each chrominance channels is sampled twice.

CCIR 656. The international standard (now renamed to ITU 656) defining the electrical and mechanical interfaces for digital television equipment operating according to the ITU 601 standard. ITU 656 defines both the parallel and serial connector pinouts, as well as the blanking, sync, and multiplexing schemes used in both parallel and serial interfaces.

CCTV. Closed circuit television. Television system intended for only a limited number of viewers, as opposed to broadcast TV.

CCTV camera. A unit containing an imaging device that produces a video signal in the basic bandwidth.

CCTV installation. A CCTV system, or an associated group of systems, together with all necessary hardware, auxiliary lighting, etc., located at the protected site.

CCTV system. An arrangement comprised of a camera and lens with all ancillary equipment required for the surveillance of a specific protected area.

CCVE. Closed circuit video equipment. An alternative acronym for CCTV.

CD. Compact disc. A media standard as proposed by Philips and Sony, where music and data are stored in digital format.

CD-ROM. Compact disk read only memory. The total capacity of a CD-ROM when storing data can be 640 MB or 700 MB.

CDS. Correlated double sampling. A technique used in the design of some CCD cameras that reduces the video signal noise generated by the chip.

CFA. Color filter array. A set of optical pixel filters used in single-chip color CCD cameras to produce the color components of a video signal.

Chip. An integrated circuit in which all the components are micro-fabricated on a tiny piece of silicon or similar material.

Chroma crawl. An artifact of encoded video, also known as dot crawl or cross-luminance, Occurs in the video picture around the edges of highly saturated colors as a continuous series of crawling dots and is a result of color information being confused as luminance information by the decoder circuits.

Chroma gain (chroma, color, saturation). In video, the gain of an amplifier as it pertains to the intensity of colors in the active picture.

Chroma key (color key). A video key effect in which one video signal is inserted in place of areas of a particular color in another video signal.

Chrominance. The color information of a color video signal.

Chrominance-to-luminance intermodulatlon (crosstalk, cross-modulation). An undesirable change in luminance amplitude caused by superimposition of some chrominance information on the luminance signal. Appears in a TV picture as unwarranted brightness variations caused by changes in color saturation levels.

CIE. Commission Internationale de l'Eclairagé. This is the International Committee for Light, established in 1965. It defines and recommends light units.

CIF. Common Interchange Format, refers to digitized image with pixel count of 352×288 (or 240) pixels.

Cladding. The outer part of a fiber optics cable, which is also a fiber but with a smaller material density than the center core. It enables a total reflection effect so that the light transmitted through the internal core stays inside.

Clamping (DC). The circuit or process that restores the DC component of a signal. A video clamp circuit, usually triggered by horizontal synchronizing pulses, reestablishes a fixed DC reference level for the video signal. A major benefit of a clamp is the removal of low-frequency interference, especially power line hum.

Clipping Level. An electronic limit to avoid over-driving the video portion of the television signal.

C-mount. The first standard for CCTV lens screw mounting. It is defined with the thread of 1" (2.54 mm) in diameter and 32 threads/inch, and the back flange-to-CCD distance of 17.526 mm (0.69"). The C-mount description applies to both lenses and cameras. C-mount lenses can be put on both, C-mount and CS-mount cameras; only in the latter case an adaptor is required.

CMOS. Complimentary Metal Oxide Semiconductor. Micro-electronic technology for producing semiconductor components, including imaging sensors.

CMYK. Cyan, magenta, yellow, and black. A color encoding system used by printers in which colors are expressed by the "subtractive primaries" (cyan, magenta, and yellow) plus black (called K). The black layer is added to give increased contrast and range on printing presses.

Coaxial cable. The most common type of cable used for copper transmission of video signals. It has a coaxial cross section, where the center core is the signal conductor, while the outer shield protects it from external electromagnetic interference.

Codec. Code/Decode. An encoder plus a decoder is an electronic device that compresses and decompresses digital signals. Codecs usually perform A/D and D/A conversion.

Color bars. A pattern generated by a video test generator, consisting of eight equal-width color bars. Colors are white (75%), black (7.5% setup level), 75% saturated pure colors red, green, and blue, and 75% saturated hues of yellow, cyan, and magenta (mixtures of two colors in 1:1 ratio without third color).

Color carrier. The subfrequency in a color video signal (4.43 MHz for PAL) that is modulated with the color information. The color carrier frequency is chosen so that its spectrum interleaves with the luminance spectrum with minimum interference.

Color difference signal. A video color signal created by subtracting luminance and/or color information from one of the primary color signals (red, green, or blue). In the Betacam color difference format, for example, the luminance (Y) and color difference components (R–Y and B–Y) are derived as follows:

Y = 0.3 Red + 0.59 Green + 0.11 Blue

R–Y = 0.7 Red – 0.59 Green – 0.11 Blue

B–Y = 0.89 Blue – 0.59 Green – 0.3 Red

The G-V color difference signal is not created because it can be reconstructed from the other three signals. Other color difference conventions include SMPTE, EBU-N1 0, and MII. Color difference signals should not be referred to as component video signals. That term is reserved for the RGB color components. In informal usage, the term *component video* is often used to mean color difference signals.

Color field. In the NTSC system, the color sub-carrier is phase-locked to the line sync so that on each consecutive line, the sub-carrier phase is changed 180° with respect to the sync pulses. In the PAL system, color sub-carrier phase moves 90° every frame. In NTSC this creates four different field types, while in PAL there are eight. In order to make clean edits, alignment of color field sequences from different sources is crucial.

Color frame. In color television, four (NTSC) or eight (PAL) properly sequenced color fields compose one color frame.

Color phase. The timing relationship in a video signal that is measured in degrees and keeps the hue of a color signal correct.

Color sub-carrier. The 3.58 MHz for NTSC, and 4.43 MHz for PAL signal that carries color information. This signal is superimposed on the luminance level. Amplitude of the color sub-carrier represents saturation, and phase angle represents hue.

Color temperature. Indicates the hue of the color. It is derived from photography where the spectrum of colors is based on a comparison of the hues produced when a black body (as in Physics) is heated from red through yellow to blue, which is the hottest. Color temperature measurements are expressed in Kelvin degrees.

Comb filter. An electrical filter circuit that passes a series of frequencies and rejects the frequencies in between, producing a frequency response similar to the teeth of a comb. Used on encoded video to select the chrominance signal and reject the luminance signal, thereby reducing cross-chrominance artifacts or, conversely, to select the luminance signal and reject the chrominance signal, thereby reducing cross-luminance artifacts. Introduced in the S-VHS concept for a better luminance resolution.

Composite sync. A signal consisting of horizontal sync pulses, vertical sync pulses, and equalizing pulses only, with a no-signal reference level.

Composite video signal. A signal in which the luminance and chrominance information has been combined using one of the coding standards NTSC, PAL, SECAM, and so on.

Concave lens. A lens that has negative focal length – the focus is virtual, and it reduces the objects.

Contrast. A common term used in reference to the video picture dynamic range – the difference between the darkest and the brightest parts of an image.

Convex lens. A convex lens has a positive focal length – the focus is real. It is usually called magnifying glass, since it magnifies the objects.

CPU. Central processing unit. A common term used in computers.

CRO. Cathode ray oscilloscope. See Oscilloscope.

Crosstalk. A type of interference or undesired transmission of signals from one circuit into another circuit in the same system. Usually caused by unintentional capacitance (AC coupling).

CS-Mount. A newer standard for lens mounting. It uses the same physical thread as the C-mount, but the back flange-to-CCD distance is reduced to 12.5 mm in order to have the lenses made smaller, more compact, and less expensive. CS-mount lenses can only be used on CS-mount cameras.

CS-to-C-mount adaptor. An adaptor used to convert a CS-mount camera to C-mount to accommodate a C-mount lens. It looks like a ring 5 mm thick, with a male thread on one side and a female on the other, with 1" diameter and 32 threads/inch. It usually comes packaged with the newer type (CS-mount) of cameras.

CVBS. Composite video bar signal. In broadcast television this refers to the video signal, including the color information and syncs.

D/A (also DA). Opposite to A/D, that is, digital to analog conversion.

Dark current. Leakage signal from a CCD sensor in the absence of incident light.

Dark noise. Noise caused by the random (quantum) nature of the dark current.

DAS (Direst Attached Storage). A digital storage system directly attached to a server or workstation, without a network in between.

DAT (digital audio tape). A system initially developed for recording and playback of digitized audio signals, maintaining signal quality equal to that of a CD. Recent developments in hardware and software might lead to a similar inexpensive system for video archiving, recording, and playback.

Datagram. A basic transfer unit associated with a packet-switched network in which the delivery, arrival time, and order of arrival are not guaranteed by the network service

dB. Decibel. A logarithmic ratio of two signals or values, usually refers to power, but also voltage and current. When power is calculated, the logarithm is multiplied by 10, while for current and voltage by 20.

DBS. Direct broadcast satellite. Broadcasting from a satellite directly to a consumer user, usually using a small aperture antenna.

DC. Direct current. Current that flows in only one direction, as opposed to AC.

DCT. Discrete cosine transform. Mathematical algorithm used to generate frequency representations of a block of video pixels. The DCT is an invertible, discrete orthogonal transformation between the time and frequency domain. It can be either forward discrete cosine transform (FDCT) or inverse discrete cosine transform (IDCT).

Decoder. A device used to recover the component signals from a composite (encoded) source.

Degauss. To demagnetize. Most often found on CRT monitors.

Delay line. An artificial or real transmission line or equivalent device designed to delay a wave or signal for a specific length of time.

Demodulator. A device that strips the video and audio signals from the carrier frequency.

Depth of field. The area in front of and behind the object in focus that appears sharp on the screen. The depth of field increases with the decrease of the focal length – the shorter the focal length the wider the depth of field. The depth of field is always wider behind the objects in focus.

DHCP. Dynamic Host Configuration Protocol. Protocol by which a network component obtains an IP address from a server on the local network.

Dielectric. An insulating (nonconductive) material.

Differential gain. A change in the subcarrier amplitude of a video signal caused by a change in the luminance level of the signal. The resulting TV picture will show a change in color saturation caused by a simultaneous change in picture brightness.

Differential phase. A change in the subcarrier phase of a video signal caused by a change in the luminance level of the signal. The hue of colors in a scene changes with the brightness of the scene.

Digital disk recorder. A system that allows the recording of video images on a digital disk.

Digital signal. An electronic signal whereby every different value from the real-life excitation (sound, light) has a different value of binary combinations (words) that represent the analog signal.

DIN. Deutsche Industrie-Normen. Germany's standard.

Disk. A flat circular plate, coated with a magnetic material, on which data may be stored by selective magnetization of portions of the surface. May be a flexible, floppy disk or a rigid hard disk. It could also be a plastic compact disk (CD) or digital video disk (DVD).

Distortion. Nonproportional representation of an original.

DMD. Digital micro-mirror device. A new video projection technology that uses chips with a large number of miniature mirrors, whose projection angle can be controlled with digital precision.

DNS. (Domain Name System) system that translates Internet domain names into IP addresses.

DOS. Disk operating system. A software package that makes a computer work with its hardware devices such as hard drive, floppy drive, screen, and keyboard.

Dot pitch. The distance in millimeters between individual dots on a monitor screen. The smaller the dot pitch the better, since it allows for more dots to be displayed and better resolution. The dot pitch defines the resolution of a monitor. A high-resolution CCTV or computer monitor would have a dot pitch of less than 0.3 mm.

Drop-frame time code. SMPTE time code format that continuously counts 30 frames per second but drops two frames from the count every minute except for every tenth minute (drops 108 frames every hour) to maintain the synchronization of time code with clock time. This is necessary because the actual frame rate of NTSC video is 29.94 frames per second rather than an even 30 frames.

DSP. Digital signal processing. It usually refers to the electronic circuit section of a device capable of processing digital signals.

Dubbing. Transcribing from one recording medium to another.

Duplex. A communication system that carries information in both directions is called a duplex system. In CCTV, duplex is often used to describe the type of multiplexer that can perform two functions simultaneously, recording in multiplex mode and playback in multiplex mode. It can also refer to duplex communication between a matrix switcher and a PTZ site driver, for example.

D-VHS. A new standard proposed by JVC for recording digital signals on a VHS video recorder.

DV-Mini. Mini digital video. A new format for audio and video recording on small camcorders, adopted by the majority of camcorder manufacturers. Video and sound are recorded in a digital format on a small cassette (66 × 48 × 12 mm), superseding S-VHS and Hi 8 quality.

Dynamic range. The difference between the smallest amount and the largest amount that a system can represent.

EBU. European Broadcasting Union.

EDTV. Enhanced (Extended) definition television. Usually refers to the progressive scan transmission of NTSC (also referred to as 480p) and PAL (also referred to as 576p).

EIA. Electronics Industry Association, which has recommended the television standard used in the United States, Canada, and Japan, based on 525 lines interlaced scanning. Formerly known as RMA or RETMA.

Encoder. A device that superimposes electronic signal information on other electronic signals.

Encryption. The rearrangement of the bit stream of a previously digitally encoded signal in a systematic fashion to make the information unrecognizable until restored upon receipt of the necessary authorization key. This technique is used for securing information transmitted over a communication channel with the intent of excluding all other than authorized receivers from interpreting the message. Can be used for voice, video, and other communications signals.

ENG camera. Electronic News Gathering camera. Refers to CCD cameras in the broadcast industry.

EPROM. Erasable and programmable read only memory. An electronic chip used in many different security products that stores software instructions for performing various operations.

Equalizer. Equipment designed to compensate for loss and delay frequency effects within a system. A component or circuit that allows for the adjustment of a signal across a given band.

Ethernet. A local area network used for connecting computers, printers, workstations, terminals, and so on, within the same building. Ethernet operates over twisted wire and coaxial cable at speeds up to 10 Mbps. Ethernet specifies a CSMA/CD (carrier sense multiple access with collision detection). CSMA/CD is a technique of sharing a common medium (wire, coaxial cable) among several devices.

External synchronization. A means of ensuring that all equipment is synchronized to the one source.

FCC. Federal Communications Commission (US).

FFT. Fast Fourier Transformation.

Fiber optics. A technology designed to transmit signals in the form of pulses of light. Fiber optic cable is noted for its properties of electrical isolation and resistance to electrostatic and electromagnetic interference.

Field. Refers to one-half of the TV frame that is composed of either all odd or even lines. In CCIR systems each field is composed of $625/2 = 312.5$ lines, in EIA systems $525/2 = 262.5$ lines. There are 50 fields/second in CCIR/PAL and 60 in the EIA/NTSC TV system.

Film recorder. A device for converting digital data into film output. Continuous tone recorders produce color photographs as transparencies, prints, or negatives.

Fixed focal length lens. A lens with a predetermined fixed focal length, a focusing control, and a choice of iris functions.

Flash memory. Nonvolatile, digital storage. Flash memory has slower access than SRAM or DRAM.

Flicker. An annoying picture distortion, mainly related to vertical syncs and video fields display. Some flicker normally exists due to interlacing; more apparent in 50 Hz systems (PAL). Flicker also shows when static images are displayed on the screen such as computer-generated text transferred to video. Poor digital image treatment, found in low-quality system converters (going from PAL to NTSC and vice versa), creates an annoying flicker on the screen. There are several electronic methods to minimize flicker.

F-number. In lenses with adjustable irises, the maximum iris opening is expressed as a ratio (focal length of the lens)/(maximum diameter of aperture). This maximum iris will be engraved on the front ring of the lens.

Focal length. The distance between the optical center of a lens and the principal convergent focus point.

Focusing control. A means of adjusting the lens to allow objects at various distances from the camera to be sharply defined.

Foot-candela. An illumination light unit used mostly in American CCTV terminology. It equals 10 times (more precisely, 9.29) the illumination value in luxes.

Fourier Transformation. Mathematical transformation of time domain functions into frequency domain.

Frame (in television). Refers to a composition of lines that make one TV frame. In CCIR/PAL TV system one frame is composed of 625 lines, while in EIA/NTSC TV system of 525 lines. There are 25 frames/second in the CCIR/PAL and 30 in the EIA/NTSC TV system. (See also Field.)

Frame (in data networking). Data structure that collectively represents a transmission stream including headers, data, and the payload and provides information necessary for the correct delivery of the data.

Frame-interline transfer (FIT). Refers to one of the few principles of charge transfer in CCD chips. The other two are interline and frame transfer.

Frame store. An electronic device that digitizes a TV frame (or TV field) of a video signal and stores it in memory. Multiplexers, fast scan transmitters, Quad compressors, DVRs, and even some of the latest color cameras have built-in frame stores.

Frame switcher. Another name for a simple multiplexer, which can record multiple cameras on a single VCR (and play back any camera in full screen) but does not have a mosaic (split-screen) image display.

Frame synchronizer. A digital buffer that, by storage and comparison of sync information to a reference and timed release of video signals, can continuously adjust the signal for any timing errors.

Frame transfer (FT). Refers to one of the three principles of charge transfer in CCD chips. The other two are interline and frame-interline transfer.

Frequency. The number of complete cycles of a periodic waveform that occur in a given length of time. Usually specified in cycles per second (Hertz).

Frequency modulation (FM). Modulation of a sine wave or carrier by varying its frequency in accordance with amplitude variations of the modulating signal.

Front porch. The blanking signal portion that lies between the end of the active picture information and the leading edge of horizontal sync.

FTP. File Transfer Protocol.

Gain. Any increase or decrease in strength of an electrical signal. Gain is measured in terms of decibels or number of times of magnification.

Gamma. A correction of the linear response of a camera in order to compensate for the monitor phosphor screen nonlinear response. It is measured with the exponential value of the curve describing the nonlinearity. A typical monochrome monitor's gamma is 2.2, and a camera needs to be set to the inverse value of 2.2 (which is 0.45) for the overall system to respond linearly (i.e., unity).

Gamut. The range of voltages allowed for a video signal, or a component of a video signal. Signal voltages outside of the range (i.e., exceeding the gamut) may lead to clipping, crosstalk, or other distortions.

Gen-lock. A way of locking the video signal of a camera to an external generator of synchronization pulses.

GB. Gigabyte. Unit of computer memory consisting of about one thousand million bytes (a thousand megabytes). Actual value is 1,073,741,824 bytes.

GHz. GigaHertz. One billion cycles per second.

GND. Ground (electrical).

GOP (Group of Pictures). Used in video (temporal) compressions. Refers to a group of pictures composed of Intra, Bi-directional, and Predicted pictures, which make up a logical group that an encoder uses to compress video data.

Gray scale. A series of tones that range from black to white, usually expressed in 10 steps.

Ground loop. An unwanted interference in the copper electrical signal transmissions with shielded cable, which is a result of ground currents when the system has more than one ground.

GUI. Graphical user interface.

H.261. A video conferencing compression standard, typically used in ISDN communications. Works with CIF size images.

H.263. This is a further development of H.261 and has been especially optimized for low data transfer rates below 64 kb/s within the H.320 standard.

H.264. A video compression which is a combined effort of the ITU and MPEG, offering a superior video performance especially for HDTV. Also known as AVC, or MPEG4-part 10.

H.265. A new video compression offering further improvement on H.264. Also known as High Efficiency Video Coding (HEVC). Officially published as ISO/IEC 23008-2.

HAD. Hole accumulated diode. A type of CCD sensor with a layer designed to accumulate holes (in the electronic sense), thus reducing noise level.

HDD. Hard disk drive. A magnetic medium for storing digital information on most computers and electronic equipment that process digital data.

HDTV. High-definition digital television. The new standard of high resolution broadcast television with 1920 × 1080 or, 1280 × 720 pixels, and aspect ratio of 16:9. Either one of the two can be an interlaced or progressive scanning, which is indicated by the letter "i" or "p" next to the vertical pixels count, that is, 1080i or 1080p, and, 720i or 720p.

Headend. The electronic equipment located at the start of a cable television system, usually including antennas, earth stations, preamplifiers, frequency converters, demodulators, modulators, and related equipment.

Helical scan. A method of recording video information on a tape, most commonly used in home and professional VCRs.

Herringbone. Patterning caused by driving a color-modulated composite video signal (PAL or NTSC) into a monochrome monitor.

Hertz. A unit that measures the number of certain oscillations per second.

HEVC. High Efficiency Video Coding. See H.265.

Horizontal drive (also horizontal sync). This signal is derived by dividing sub-carrier by 227.5 and then doing some pulse shaping. The signal is used by monitors and cameras to determine the start of each horizontal line.

Horizontal resolution. Chrominance and luminance resolution (detail) expressed horizontally across a picture tube. This is usually expressed as a number of black to white transitions or lines that can be differentiated. Limited by the bandwidth of the video signal or equipment.

Horizontal retrace. At the end of each horizontal line of video, a brief period when the scanning beam returns to the other side of the screen to start a new line.

Horizontal sync pulse. The synchronizing pulse at the end of each video line that determines the start of horizontal retrace.

Host. Computer on a network that is a repository for services available to other components on the network.

Housings, environmental. Usually refers to cameras' and lens's containers and associated accessories, such as heaters, washers, and wipers, to meet specific environmental conditions.

HS. Horizontal sync.

HTML (Hyper Text Mark-up Language). Coding language used to create Hypertext documents for use on the World Wide Web.

HTTP (Hyper Text Transport Protocol). Connection oriented protocol for transmitting data over a network or protocol for moving hyper text files across the Internet

Hub. Hub connects multiple computers and devices into a LAN.

Hue (tint, phase, chroma phase). One of the characteristics that distinguishes one color from another. Hue defines color on the basis of its position in the spectrum, that is, whether red, blue, green, or yellow.

Hum. A term used to describe an unwanted induction of mains frequency.

Hum bug. Another name for a ground loop corrector.

Hyper-HAD. An improved version of the CCD HAD technology, utilizing on-chip micro-lens technology to provide increased sensitivity without increasing the pixel size.

ICMP (Internet Control Message Protocol). Error protocol indicating, for instance, that a requested service is not available or that a host or router could not be reached.

IDE. Interface device electronics. Software and hardware communication standard for interconnecting peripheral devices to a computer.

I-frames (or pictures). Intra-frames. Used in MPEG video compression and refers to the reference pictures, usually compressed with JPEG image compression.

IGMP (Internet Group Management Protocol). A communications protocol used by hosts and adjacent routers on IP networks to establish multicast group memberships. IGMP is an integral part of IP multicast.

I/O. Input/Output.

I/P. Input. A signal applied to a piece of electric apparatus or the terminals on the apparatus to which a signal or power is applied.

I^2R. Formula for power in watts (W), where I is current in amperes (A), and R is resistance in ohms (Ω).

IEC. International Electrotechnical Commission (also CEI).

Imaging device. A vacuum tube or solid-state device in which the vacuum tube light-sensitive face plate or solid-state light-sensitive array provides an electronic signal from which an image can be created.

Impedance. A property of all metallic and electrical conductors that describes the total opposition to current flow in an electrical circuit. Resistance, inductance, capacitance, and conductance have various influences on the impedance, depending on frequency, dielectric material around conductors, physical relationship between conductors and external factors. Impedance is often referred to with the letter Z. It is measured in ohms, whose symbol is the Greek letter omega – Ω.

Input. Same as I/P.

Inserter (also alphanumeric video generator). A device for providing additional information, normally superimposed on the picture being displayed; this can range from one or two characters to full-screen alphanumeric text. Usually, such generators use the incoming video signal sync pulses as a reference point for the text insertion position, which means that if the video signal is of poor quality, the text stability will also be of poor quality.

Interference. Disturbances of an electrical or electromagnetic nature that introduce undesirable responses in other electronic equipment.

Interlaced scanning. A technique of combining two television fields in order to produce a full frame. The two fields are composed of only odd and only even lines, which are displayed one after the other but with the physical position of all the lines interleaving each other, hence interlace. This type of television picture creation was proposed in the early days of television to have a minimum amount of information and yet achieve flickerless motion.

Interline transfer. One of the three principles of charge transferring in CCD chips. The other two are frame transfer and frame-interline transfer.

IP (in networking - Internet Protocol). The principal communications protocol in the Internet protocol suite for relaying datagrams across network boundaries. Its routing function enables internetworking, and essentially establishes the Internet.

IP (mechanical - Ingress Protection or, Index of Protection). A numbering system that describes the quality of protection of an enclosure from outside influences, such as moisture, dust, and impact.

IP Address (Internet Protocol Address). Address of a host computer used in the Internet Protocol

IPv4. The most common IP address type is the IPv4 which consists of 4 bytes (32 bits).

IPv6. The new IP address type which offers a much larger number of addresses, and consists of 16 bytes (128 bits) long.

IRE. Institute of Radio Engineers. Units of measurement dividing the area from the bottom of sync to peak white level into 140 equal units. 140 IRE equals $1V_{PP}$. The range of active video is 100 IRE.

IR light. Infrared light, invisible to the human eye. It usually refers to wavelengths longer than 700 nm. Monochrome (B/W) cameras have extremely high sensitivity in the infrared region of the light spectrum.

Iris. A means of controlling the size of a lens aperture and therefore the amount of light passing through the lens.

ISDN. Integrated Services Digital Network. The newer generation telephone network, which uses 64 kb/s digital transmission. Being a digital network, the signal bandwidth is not expressed in kHz, but rather with a transmission speed.

ISO. International Standardization Organization.

ITU. International Telecommunications Union.

JPEG. Joint Photographic Experts Group. A group that has recommended a compression algorithm for still digital images that can compress with ratios of over 10:1. Also the name of the format itself.

kb/s (kilobits per second). 1024 bits per second. Also written as kbps.

kB/s (kilobytes per second). 1024 Bytes per second.

Kelvin. One of the basic physical units of measurement for temperature. The scale is the same as the Celsius, but the 0° K starts from -273° C, called *absolute zero*. Also the unit of measurement of the temperature of light is expressed in Kelvins or K. In color recording, light temperature affects the color values of the lights and the scene that they illuminate.

K factor. A specification rating method that gives a higher factor to video disturbances that cause the most observable picture degradation.

kHz. Kilohertz. Thousand Hertz.

Kilobaud. A unit of measurement of data transmission speed equaling 1000 baud.

Lambertian source or surface. A surface is called a Lambert radiator or reflector (depending on whether the surface is a primary or a secondary source of light) if it is a perfectly diffusing surface.

LAN. Local Area Network. A short-distance data communications network (typically within a building or campus) used to link together computers and peripheral devices (such as printers, CD ROMs, and modems) under some form of standard control.

Laser. Light amplification by stimulated emission of radiation. A laser produces a very strong and coherent light of a single frequency.

LCD. Liquid crystal display. A screen for displaying text/graphics based on a technology called liquid crystal, where minute currents change the reflectiveness or transparency of the screen.

LED. Light-emitting diode. A semiconductor that produces light when a certain low voltage is applied to it in one direction.

Lens. An optical device for focusing a desired scene onto the imaging device in a CCTV camera.

Level. When relating to a video signal, it refers to the video level in volts. In CCTV optics, it refers to the auto iris level setting of the electronics that processes the video signal in order to open or close the iris.

Line-locked. In CCTV, this usually refers to multiple cameras being powered by a common alternative current (AC) source (either 24 V AC, 110 V AC, or 240 V AC) and consequently having their field frequencies locked to the same AC source frequency (50 Hz in CCIR systems and 60 Hz in EIA systems).

Lumen [lm]. A light intensity produced by the luminosity of 1 candela in one steradian of a solid angle (approximately 57° squared).

Luminance. 1. Refers to the video signal information about the scene brightness. The color video picture information contains two components, luminance (brightness and contrast) and chrominance (hue and saturation). 2. It also refers to the photometric quantity of light radiation.

LUT. Look-up table. A cross-reference table in the computer memory that transforms raw information from the scanner or computer and corrects values to compensate for weakness in equipment or for differences in emulsion types.

Lux [lx]. Light unit for measuring illumination. It is defined as the illumination of a surface when the luminous flux of 1 lumen falls on an area of 1 m². It is also known as lumen per square meter, or meter-candelas.

MAC Address (Media Access Control Address). Unique identifier attached to network adapters i.e a name for a particular adapter. In computer networking, this is the physical address of each network card.

Manual iris. A manual method of varying the size of a lens's aperture.

Matrix. A logical network configured in a rectangular array of intersections of input/outputs.

Matrix switcher. A device for switching more than one camera, VCR, DVR, video printer, and the like, to more than one monitor, VCR, DVR, video printer and so on.

MATV. Master antenna television.

MB. Megabyte. Unit of measurement for computer memory consisting of approximately one million bytes. Actual value is 1,048,576 bytes. Kilobyte × Kilobyte = Megabyte.

Mb/s. Megabits per second. Million bits per second. Also written as Mbps.

MB/s. Megabytes per second. Million bytes per second or 8 million hits per second. Also written as MBps.

MHz. Megahertz. One million hertz.

Microwave. One definition refers to the portion of the electromagnetic spectrum that ranges between 300 MHz and 3000 GHz. The other definition is when referring to the transmission media where microwave links are used. Frequencies in microwave transmission are usually between 1 GHz and 12 GHz.

MOD. Minimum object distance. Feature of a fixed or a zoom lens that indicates the closest distance an object can be from the lens's image plane, expressed in meters. Zoom lenses have MOD of around 1 m, while fixed lenses usually have much less, depending on the focal length.

Modem. This popular term is made up of two words: modulate and demodulate. The function of a modem is to connect a device (usually computer) via a telephone line to another device with a modem.

Modulation. The process by which some characteristic (i.e., amplitude, phase) of one RF wave is varied in accordance with another wave (message signal).

Moiré pattern. An unwanted effect that appears in the video picture when a high-frequency pattern is looked at with a CCD camera that has a pixel pattern close (but lower) to the object pattern.

Monochrome. Black-and-white video. A video signal that represents the brightness values (luminance) in the picture but not the color values (chrominance).

MPEG. Motion Picture Experts Group. An ISO group of experts that has recommended manipulation of digital motion images. Today there are a couple of MPEG recommendations, of which the most well known are MPEG-1, MPEG-2, and MPEG-4. The MPEG-2 is widely accepted for digital broadcast television, as well as DVD. MPEG-4 is popular for Internet video streaming and CCTV remote surveillance.

MPEG-1. Video compression standard, progressive scanned images with audio. Bit rate is from 1.5 Mbps up to 3.5 Mbps.

MPEG-2. The standard for compression of progressive scanned and interlaced video signals with high-quality audio over a large range of compression rates with a range of bit rates from 1.5 to 30 Mb/s. Accepted as a HDTV and DVD standard of video/audio encoding.

MPEG-4. Modern video compression, uses wider bit rates than MPEG-2, and starts as low as 9.6 kb/s. It works with objects as a new category in video processing.

MPEG-7. Not really a video compression as such, but rather defines an interoperable framework for content descriptions way beyond the traditional "metadata."

MPEG-21. The MPEG-21 goal is to describe a "big picture" of how different elements to build an infrastructure for the delivery and consumption of multimedia content relate to each other.

Multicast. Network streaming technology that reduces bandwidth usage by delivering a single stream of video information to multiple network recipients.

Multicasting (in networking). Multicasting is an operational mode, whereby a node sends a packet addressed to a special group address.

NAS (Network Attached Storage). A computer data storage connected to a network, providing data access to various group of clients.

NIC (Network Interface Card). Every computer (and most other devices) connects to a network through a NIC.

NNTP. Network News Transport Protocol.

Node (in networking). A communication device attached to a network or end point of a network connection such as a workstation, IP camera, printer, etc.

Noise. An unwanted signal produced by all electrical circuits working above the absolute zero. Noise cannot be eliminated but only minimized.

Non-drop frame time code. SMPTE time code format that continuously counts a full 30 frames per second. Because NTSC video does not operate at exactly 30 frames per second, non-drop-frame time code will count 108 more frames in one hour than actually occur in the NTSC video in one hour. The result is incorrect synchronization of time code with clock time. The drop-frame time code solves this problem by skipping or dropping 2 frame numbers per minute, except at the tens of the minute count.

Noninterlaced. The process of scanning whereby every line in the picture is scanned during the vertical sweep.

NTP (Network Time Protocol). Standard for synchronizing computer system clocks in packet-based communication networks.

NTSC. National Television System Committee. American committee that set the standards for analog color television as used in the United States, Canada, Japan, and parts of South America. NTSC television uses a 3.57945 MHz subcarrier whose phase varies with the instantaneous hue of the televised color and whose amplitude varies with the instantaneous saturation of the color. NTSC employs 525 lines per frame and 59.94 fields per second. NTSC is now renamed to ATSC, for Advanced Television System Committee.

Numerical aperture. A number that defines the light-gathering ability of a specific fiber. The numerical aperture is equal to the sine of the maximum acceptance angle.

O/P. Output.

Objective. The very first optical element at the front of a lens.

Ocular. The very last optical element at the back of a lens (the one closer to the CCD chip).

Ohm. The unit of resistance. The electrical resistance between two points of a conductor where a constant difference of potential of 1 V applied between these points produces in the conductor a current of 1 A, the conductor not being the source of any electromotive force.

Oscilloscope (also CRO, from cathode ray oscilloscope). An electronic device that can measure the signal changes versus time. A must for any CCTV technician.

OSI layers (in networking). An Open System Interconnection reference model introduced by ISO, which defines seven layers of networking.

Overscan. A video monitor condition in which the raster extends slightly beyond the physical edges of the CRT screen, cutting off the outer edges of the picture.

Output impedance. The impedance a device presents to its load. The impedance measured at the output terminals of a transducer with the load disconnected and all impressed driving forces taken as zero.

PAL. Phase alternating line. Describes the color phase change in a PAL color signal. PAL is a European color TV system featuring 625 lines per frame, 50 fields per second, and a 4.43361875-MHz subcarrier. Used mainly in Europe, China, Malaysia, Australia, New Zealand, the Middle East, and parts of Africa. PAL-M is a Brazilian color TV system with phase alternation by line, but using 525 lines per frame, 60 fields per second, and a 3.57561149 MHz subcarrier.

Pan and tilt head (P/T head). A motorized unit permitting vertical and horizontal positioning of a camera and lens combination. Usually, 24 V AC motors are used in such P/T heads, but also 110 V AC (i.e., 240 V AC units can be ordered).

Pan unit. A motorized unit permitting horizontal positioning of a camera.

Peak-to-peak (pp). The amplitude (voltage) difference between the most positive and the most negative excursions (peaks) of an electrical signal.

Pedestal. In the video waveform, the signal level corresponding to black. Also called setup.

P-frames (or pictures). Prediction-coded pictures used in MPEG video compression. It describes pictures that are coded using motion-compensated prediction from the past I reference picture. See also B-frames; I-frames.

Phot. A photometric light unit for very strong illumination levels. One phot is equal to 10,000 luxes.

Photodiode. A type of semiconductor device in which a PN junction diode acts as a photosensor.

Photo-effect. Also known as photoelectric-effect. This refers to a phenomenon of ejection of electrons from a metal whose surface is exposed to light.

Photon. A representative of the quantum nature of light. It is considered as the smallest unit of light.

Photopic vision. The range of light intensities, from 10^5 lux down to nearly 10^{-2} lux, detectable by the human eye.

Pinhole lens. A fixed focal-length lens, for viewing through a very small aperture, used in discrete surveillance situations. The lens normally has no focusing control but offers a choice of iris functions.

Pixel. Derived from picture element. Usually refers to the CCD chip unit picture cell. It consists of a photosensor plus its associated control gates.

Phase locked loop (PLL). A circuit containing an oscillator whose output phase or frequency locks onto and tracks the phase or frequency of a reference input signal. To produce the locked condition, the circuit detects any phase difference between the two signals and generates a correction voltage that is applied to the oscillator to adjust its phase or frequency.

Photo multiplier. A highly light-sensitive device. Advantages are its fast response, good signal-to-noise ratio, and wide dynamic range. Disadvantages are fragility (vacuum tube), high voltage, and sensitivity to interference.

Pixel or picture element. The smallest visual unit that is handled in a raster file, generally a single cell in a grid of numbers describing an image.

Plumbicon. Thermionic vacuum tube developed by Philips, using a lead oxide photoconductive layer. It represented the ultimate imaging device until the introduction of CCD chips.

Polarizing filter. An optical filter that transmits light in only one direction (perpendicular to the light path), out of 360° possible. The effect is such that it can eliminate some unwanted bright areas or reflections, such as when looking through a glass window. In photography, polarizing filters are used very often to darken a blue sky.

Port (in networking). Number or identifier for a particular service on a server, mostly standardized for certain services e.g. RTSP, UPnP, HTTP, etc.

POTS. Plain old telephone service. The telephone service in common use throughout the world today. Also known as PSTN.

PPP. Point-to-Point Protocol.

Preset positioning. A function of a pan and tilt unit, including the zoom lens, where a number of certain viewing positions can be stored in the system's memory (usually this is in the PTZ site driver) and recalled when required, either upon an alarm trigger, programmed or manual recall.

Primary colors. A small group of colors that, when combined, can produce a broad spectrum of other colors. In television, red, green, and blue are the primary colors from which all other colors in the picture are derived.

Principal point. One of the two points that each lens has along the optical axis. The principal point closer to the imaging device (CCD chip in our case) is used as a reference point when measuring the focal length of a lens.

PROM. Programmable read only memory. A ROM that can be programmed by the equipment manufacturer (rather than the PROM manufacturer).

Protocol. A specific set of rules, procedures, or conventions relating to format and timing of data transmission between two devices. A standard procedure that two data devices must accept and use to be able to understand each other. The protocols for data communications cover such things as framing, error handling, transparency, and line control.

PSTN. Public switched telephone network. Usually refers to the plain old telephone service, also known as POTS.

PTZ camera. Pan, tilt, and zoom camera.

PTZ site driver (or receiver or decoder). An electronic device, usually a part of a video matrix switcher, which receives digital, encoded control signals in order to operate pan, tilt, zoom, and focus functions.

Pulse. A current or voltage that changes abruptly from one value to another and back to the original value in a finite length of time. Used to describe one particular variation in a series of wave motions.

QAM. Quadrature amplitude modulation. Method for modulating two carriers. The carriers can be analog or digital.

Quad compressor (also split-screen unit). Equipment that simultaneously displays parts or more than one image on a single monitor. It usually refers to four quadrants display.

Radio frequency (RF). A term used to describe incoming radio signals to a receiver or outgoing signals from a radio transmitter (above 150 Hz). Even though they are not properly radio signals, TV signals are included in this category.

RAM. Random access memory. Electronic chips, usually known as memory, holding digital information while there is power applied to it. Its capacity is measured in kilobytes. This is the computer's work area.

RAID. Redundant arrays of independent disks. This a technology of connecting a number of hard drives into one mass storage device, which can be used, among other things, for digital recording of video images. There are RAID-0 through to RAID-6. See more in Chapter 9.

Random interlace. In a camera that has a free-running horizontal sync as opposed to a 2:1 interlace type that has the sync locked and therefore has both fields in a frame interlocked together accurately.

Registration. An adjustment associated with color sets and projection TVs to ensure that the electron beams of the three primary colors of the phosphor screen are hitting the proper color dots/stripes.

Resolution. A measure of the ability of a camera or television system to reproduce detail. The number of picture elements that can be reproduced with good definition.

Retrace. The return of the electron beam in a CRT to the starting point after scanning. During retrace, the beam is typically turned off. All of the sync information is placed in this invisible portion of the video signal. May refer to retrace after each horizontal line or after each vertical scan (field).

Remote control. A transmitting and receiving of signals for controlling remote devices such as pan and tilt units, lens functions, wash and wipe control, and the like.

RETMA. Former name of the EIA association. Some older video test charts carry the name RETMA Chart.

RF signal. Radio frequency signal that belongs to the region up to 300 GHz.

RG-11. A video coaxial cable with 75 Ω impedance and much thicker diameter than the popular RG-59 (of approximately 12 mm). With RG-11 much longer distances can be achieved (at least twice the RG-59), but it is more expensive and harder to work with.

RG-58. A coaxial cable designed with 50 Ω impedance; therefore, not suitable for CCTV. Very similar to RG-59, only slightly thinner.

RG-59. A type of coaxial cable that is most common in use in small to medium-size CCTV systems. It is designed with an impedance of 75 Ω. It has an outer diameter of around 6 mm and it is a good compromise between maximum distances achievable (up to 300 m for monochrome signal and 250 m for color) and good transmission.

Rise time. The time taken for a signal to make a transition from one state to another; usually measured between the 10% and 90% completion points of the transition. Shorter or faster rise times require more bandwidth in a transmission channel.

RMS. Root mean square. A measure of effective (as opposed to peak) voltage of an AC waveform. For a sine wave it is 0.707 times the peak voltage. For any periodic waveform, it is the square root of the average of the squares of the values through one cycle.

ROM. Read only memory. An electronic chip, containing digital information that does not disappear when power is turned off.

Router. A device that routes information between interconnected networks, able to select the best path to route a message by determining the next network point to where a packet should be forwarded on its way to its final destination.

Routing switcher. An electronic device that routes a user-supplied signal (audio, video, etc.) from any input to any user-selected output. This is a broadcast term for matrix switchers, as we know them in CCTV.

RS-125. A SMPTE parallel component digital video standard.

RS-170. A document prepared by the Electronics Industries Association describing recommended practices for NTSC color television signals in the United States.

RS-232. A serial format of digital communication where only two wires are required. It is also known as a serial data communication. The RS-232 standard defines a scheme for asynchronous communications, but it does not define how the data should be represented by the bits, that is, it does not define the overall message format and protocol. It is often used in CCTV communications between keyboards and matrix switchers or between matrix switchers and PTZ site drivers. The advantage of RS-232 over others is its simplicity and use of only two wires, but it is limited with distance. Typically, maximum 15 m is recommended.

RS-422. A serial data communication protocol which specifies 4-wire, full-duplex, differential line, multi-drop communications. It provides for balanced data transmission with unidirectional, nonreversible, terminated or nonterminated transmission lines. This is an advanced format of digital communication when compared to RS-232. The signal transmitted is read at the receiving end as the difference between the two wires without common earth. So if there is noise induced along the line, it will be canceled out. The RS-422 can drive lines of 1200 m in length and distribute data to up to 10 receivers, with data rate of up to 100 kb/s.

RS-485. This is a more advanced format compared to RS-422. It is an electrical specification of a two-wire, half-duplex, multipoint serial connection. The major improvement is in the number of receivers up to 32 that can be driven with this format. In contrast to RS-422, which has a single driver circuit which cannot be switched off, RS-485 drives need to be put in transmit mode explicitly by asserting a signal to the driver. This allows RS-485 to implement star network topologies using only two lines. RS-485, like RS-422, can be made full-duplex by using four wires, however, since RS-485 is a multipoint specification, this is not necessary in many cases.

RTCP (Real-Time Control Protocol). Supporting protocol for real-time transmission of groups within a network. Quality-of-service feedback from receivers to the multicast group and support for synchronization of different media streams e.g. video, audio, metadata.

RTP (Real-Time Transport Protocol). Internet protocol for transmitting real-time data such as video.

RTSP (Real Time Streaming Protocol). Control protocol standard (RFC 2326) for delivering, receiving and controlling real-time data streams such as video, audio and metadata and starting entry point for negotiating transports such as RTP, multicast and unicast, including the negotiating of Codec's.

SAN (Storage Area Network). High-speed network or sub network whose primary purpose is to transfer data between network devices and storage systems consisting of a communication infrastructure, providing physical connections, a management layer and storage elements.

Saturation (in color). The intensity of the colors in the active picture. The degree by which the eye perceives a color as departing from a gray or white scale of the same brightness. A 100% saturated color does not contain any white; adding white reduces saturation. In NTSC and PAL video signals, the color saturation at any particular instant in the picture is conveyed by the corresponding instantaneous amplitude of the active video subcarrier.

Scanner. 1. When referring to a CCTV device, it is the pan only head. 2. When referring to an imaging device, it is the device with CCD chip that scans documents and images.

Scanning. The rapid movement of the electron beam in the CRT of a monitor or television receiver. It is formatted line-for-line across the photosensitive surface to produce or reproduce the video picture. When referred to a PTZ camera, it is the panning or the horizontal camera motion.

Scene illumination. The average light level incident upon a protected area. Normally measured for the visible spectrum with a light meter having a spectral response corresponding closely to that of the human eye and is quoted in lux.

Scotopic vision. Illumination levels below 10^{-2} lux, thus invisible to the human eye.

SCSI. Small computer systems interface. A computer standard that defines the software and hardware methods of connecting more external devices to a computer bus.

SHV (Super High Vision). Experimental higher resolution of four times the 4k, or 16 times the HD (7,680 x 4,320 pxels), also known as **8k**.

SECAM. Sequential Couleur Avec Memoire, sequential color with memory. A color television system with 625 lines per frame (used to be 819) and 50 fields per second developed by France and the former U.S.S.R. Color difference information is transmitted sequentially on alternate lines as an FM signal.

Serial data. Time-sequential transmission of data along a single wire. In CCTV, the most common method of communicating between keyboards and the matrix switcher and also controlling PTZ cameras.

Serial interface. A digital communications interface in which data are transmitted and received sequentially along a single wire or pair of wires. Common serial interface standards are RS-232 and RS-422.

Serial port. A computer I/O (input/output) port through which the computer communicates with the external world. The standard serial port is RS-232 based and allows bidirectional communication on a relatively simple wire connection as data flow serially.

Sidebands. The frequency bands on both sides of a carrier within which the energy produced by the process of modulation is carried.

Signal-to-noise ratio (S/N). An S/N ratio can be given for the luminance signal, chrominance signal, and audio signal. The S/N ratio is the ratio of noise to actual total signal, and it shows how much higher the signal level is than the level of noise. It is expressed in decibels (dB), and the bigger the value is, the crisper and clearer the picture and sound will be during playback. An S/N ratio is calculated with the logarithm of the normal signal and the noise RMS value.

Silicon. The material of which modern semiconductor devices are made.

Simplex. In general, simplex refers to a communications system that can transmit information in one direction only. In CCTV, simplex is used to describe a method of multiplexer operation whereby only one function can be performed at a time (e.g., either recording or playback individually).

Single-mode fiber. An optical glass fiber that consists of a core of very small diameter. A typical single-mode fiber used in CCTV has a 9 μm core and a 125 μm outer diameter. Single-mode fiber has less attenuation and therefore transmits signals at longer distances (up to 70 km). Such fibers are normally used only with laser sources because of their very small acceptance cone.

Skin effect. The tendency of alternating current to travel only on the surface of a conductor as its frequency increases.

SLIP. Serial Line Internet Protocol.

Slow scan. The transmission of a series of frozen images by means of analog or digital signals over limited bandwidth media, usually telephone.

Smear. An unwanted side effect of vertical charge transfer in a CCD chip. It shows vertical bright stripes in places of the image where there are very bright areas. In better cameras smear is minimized to almost undetectable levels.

SMPTE. Society of Motion Picture and Television Engineers.

SMPTE time code. In video editing, time code that conforms to SMPTE standards. It consists of an 8-digit number specifying hours: minutes: seconds: frames. Each number identifies one frame on a videotape. SMPTE time code may be of either the drop-frame or non-drop-frame type.

SMTP. Simple Mail Transfer Protocol.

S/N ratio. See Signal-to-noise ratio.

SNMP (Simple Network Management Protocol). A set of standards for communication with devices connected to a TCP/IP network for the management of network nodes (servers, workstations, routers, switches and hubs, video transmission devices, etc), enabling network administrators to manage network performance, find, solve network problems and plan network extensions.

Snow. Random noise on the display screen, often resulting from dirty heads or weak broadcast video reception.

SNTP (Simple Network Time Protocol). Adaptation of the Network Time Protocol (NTP) synchronizing computer clocks on a network, when the accuracy of the full NTP implementation is not needed according to IETF RFC 2030

Spectrum. In electromagnetics, spectrum refers to the description of a signal's amplitude versus its frequency components. In optics, spectrum refers to the light frequencies composing the white light which can be seen as rainbow colors.

Spectrum analyzer. An electronic device that can show the spectrum of an electric signal.

SPG. Sync pulse generator. A source of synchronization pulses.

Split-screen unit (quad compressor). Equipment that simultaneously displays parts or more than one image on a single monitor. It usually refers to four quadrants' display.

SSL (Secure Socket Layer). Application layer security protocol to enable encrypted, authenticated communications across networks.

Staircase (in television). Same as color bars. A pattern generated by the TV generator, consisting of equal-width luminance steps of 0, +20, +40, +60, +80, and +100 IRE units and a constant amplitude chroma signal at color burst phase. Chroma amplitude is selectable at 20 IRE units (low stairs) or 40 IRE units (high stairs). The staircase pattern is useful for checking linearity of luminance and chroma gain, differential gain, and differential phase.

Start bit. A bit preceding the group of bits representing a character used to signal the arrival of the character in asynchronous transmission.

Subcarrier (SC). Also known as SC: 3.58 MHz for NTSC, 4.43 MHz for PAL. These are the basic signals in all NTSC and PAL sync signals. It is a continuous sine wave, usually generated and distributed at 2V in amplitude, and having a frequency of 3.579545 MHz (NTSC) and 4.43361875 MHz (PAL). Subcarrier is usually divided down from a primary crystal running at 14.318180 MHz, for example, in NTSC, and that divided by 4 is 3.579545. Similar to PAL. All other synchronizing signals are directly divided down from subcarrier.

Subnet Mask. Subnetting is a method that allows one large network to be broken down into several smaller ones.

SVGA. Super Video Graphic Array offering 800 × 600 pixels of resolution.

S-VHS. Super VHS format in video recording. A newer standard proposed by JVC, preserving the downwards compatibility with the VHS format. It offers much better horizontal resolution up to 400 TV lines. This is mainly due to the color separation techniques, high-quality video heads, and better tapes. S-VHS is usually associated with Y/C separated signals.

Switch (in network). A device that connects network devices to hosts, allowing a large number of devices to share a limited number of ports.

SXGA. Computer screen resolution offering 1400 × 1050 pixels.

Sync. Short for synchronization pulse.

Sync generator (sync pulse generator, SPG). Device that generates synchronizing pulses needed by video source equipment to provide proper equipment video signal timing. Pulses typically produced by a sync generator could be subcarrier, burst flag, sync, blanking, H and V drives, and color black. Most commonly used in CCTV are H and V drives.

T1. A digital transmission link with a capacity of 1.544 Mbps. T1 uses two pairs of normal twisted wires. T1 lines are used for connecting networks across remote distances. Bridges and routers are used to connect LANs over T1 networks.

T1 channels. In North America, a digital transmission channel carrying data at a rate of 1.544 million bits per second. In Europe, a digital transmission channel carrying data at a rate of 2.048 million bits per second. AT&T term for a digital carrier facility used to transmit a DS-1 formatted digital signal at 1.544 Mbps.

T3 channels. In North America, a digital channel that communicates at 45.304 Mbps commonly referred to by its service designation of DS-3.

TBC. Time base correction. Synchronization of various signals inside a device such as a multiplexer or a time base corrector.

TCP/IP (Transmission Control Protocol/Internet Protocol). A suite of protocols that define networks and the Internet in general.

TDG. Time and date generator.

TDM. Time division multiplex. A time-sharing of a transmission channel by assigning each user a dedicated segment of each transmission cycle.

Tearing. A lateral displacement of the video lines due to sync instability. It appears as though parts of the images have been torn away.

Teleconferencing. Electronically linked meeting conducted among groups in separate geographic locations.

Telemetry. Remote controlling system of, usually, digital encoded data, intended to control pan, tilt, zoom, focus, preset positions, wash, wipe, and the like. Being digital, it is usually sent via twisted pair cable or coaxial cable together with the video signal.

Termination. This usually refers to the physical act of terminating a cable with a special connector, which for coaxial cable is usually BNC. For fiber optic cable this is the ST connector. It can also refer to the impedance matching when electrical transmission is in use. This is especially important for high-frequency signals, such as the video signal, where the characteristic impedance is accepted to be 75 Ω.

TFT. Thin-film-transistor. This technology is used mainly for manufacturing flat computer and video screens that are superior to the classic LCD screens. Color quality, fast response time, and resolution are excellent for video.

Time-lapse VCR (TL VCR). A video recorder, most often in VHS format, that can prolong the video recording on a single tape up to 960 hours (this refers to a 180 min tape). This type of VCR is often used in CCTV systems. The principle of operation is very simple – instead of having the videotape travel at a constant speed of 2.275 cm/s (which is the case with the domestic models of VHS VCRs), it moves with discrete steps that can be controlled. Time-lapse VCRs have a number of other special functions that are very useful in CCTV, such as external alarm trigger, time and date superimposed on the video signal, and alarm search.

Time-lapse video recording. The intermittent recording of video signals at intervals to extend the recording time of the recording medium. It is usually measured in reference to a 3-hr (180 min) tape.

Time multiplexing. The technique of recording several cameras onto one time-lapse VCR by sequentially sending camera pictures with a timed interval delay to match the time lapse mode selected on the recorder.

T-pulse to bar. A term relating to the frequency response of video equipment. A video signal containing equal-amplitude T-pulse and bar portions is passed through the equipment, and the relative amplitudes of the T-pulse and bar are measured at the output. A loss of response is indicated when one portion of the signal is lower in amplitude than the other.

Tracking. The angle and speed at which the tape passes the video heads.

Transcoder. A device that converts one form of encoded video to another, for example, to convert NTSC video to PAL. Sometimes mistakenly used to mean translator.

Transducer. A device that converts one form of energy into another. For example, in fiber optics, a device that converts light signals into electrical signals.

Translator. A device used to convert one component set to another (e.g., to convert Y, R-Y, B-Y signals to RGB signals).

Transponder. The electronics of a satellite that receives an uplinked signal from the Earth, amplifies it, converts it to a different frequency, and returns it to the Earth.

TS (Transport stream). Content binary stream usually in reference to an MPEG-2 AV stream format

TTL. 1. Transistor-transistor logic. A term used in digital electronics mainly to describe the ability of a device or circuit to be connected directly to the input or output of digital equipment. Such compatibility eliminates the need for interfacing circuitry. TTL signals are usually limited to two states, low and high, and are thus much more limited than analog signals. 2. Thru-the-lens viewing or color measuring.

Twisted-pair. A cable composed of two small insulated conductors twisted together. Since both wires have nearly equal exposure to any interference, the differential noise is slight.

UDP (User Datagram Protocol). Stateless protocol for the transfer of data without provision for acknowledgement of packets received.

UHF signal. Ultrahigh-frequency signal. In television it is defined to belong in the radio spectrum between 470 MHz and 850 MHz.

Unbalanced signal. In CCTV, this refers to a type of video signal transmission through a coaxial cable. It is called unbalanced because the signal travels through the center core only, while the cable shield is used for equating the two voltage potentials between the coaxial cable ends.

Underscan. Decreases raster size H and V so that all four edges of the picture are visible on the monitor.

UPS. Uninterruptible power supply. These are power supplies used in the majority of high security systems, whose purpose is to back up the system for at least 10 minutes without mains power. The duration of this depends on the size of the UPS, usually expressed in VA (volt-amperes), and the current consumption of the system itself.

URL (Uniform Resource Locator). Previously Universal Resource Locator. The unique address for a file or web site that is accessible on the Internet.

UTC. Universal Time Coordinated

UTP. Unshielded twisted pair. A cable medium with one or more pairs of twisted insulated copper conductors bound in a single sheath. Now the most common method of bringing telephone and data to the desktop.

UXGA. Computer screen resolution offering 1600 × 1200 pixels.

Variable bit rate. Operation where the bit rate varies with time during the decoding of a compressed bit stream.

VDA. See video distribution amplifier.

Vectorscope. An instrument similar to an oscilloscope that is used to check and/or align amplitude and phase of the three color signals (RGB).

Velocity of propagation. Speed of signal transmission. In free space, electromagnetic waves travel at the speed of light. In coaxial cables, this speed is reduced by the dielectric material. Commonly expressed as percentage of the speed in free space.

Vertical interval. The portion of the video signal that occurs between the end of one field and the beginning of the next. During this time, the electron beams in the monitors are turned off (invisible) so that they can return from the bottom of the screen to the top to begin another scan.

Vertical interval switcher. A sequential or matrix switcher that switches from one camera to another exactly in the vertical interval, thus producing roll-free switching. This is possible only if the various camera sources are synchronized.

Vertical resolution. Chrominance and luminance detail expressed vertically in the picture tube. Limited by the number of scan lines.

Vertical retrace. The return of the electron beam to the top of a television picture tube screen or a camera pickup device target at the completion of the field scan.

Vertical shift register. The mechanism in CCD technology whereby charge is read out from the photosensors of an interline transfer or frame interline transfer sensor.

Vertical sync pulse. A portion of the vertical blanking interval that is made up of blanking level. Synchronizes vertical scan of television receiver to composite video signal. Starts each frame at same vertical position.

Vestigial sideband transmission. A system of transmission wherein the sideband on one side of the carrier is transmitted only in part.

VGA. Video graphics array with resolution of 640 × 480 pixels.

Video bandwidth. The highest signal frequency that a specific video signal can reach. The higher the video bandwidth, the better the quality of the picture. A video recorder that can produce a very broad video bandwidth generates a very detailed, high-quality picture on the screen. Video bandwidths used in studio work vary between 3 and 12 MHz.

Video distribution amplifier (VDA). A special amplifier for strengthening the video signal so that it can be supplied to a number of video monitors at the same time.

Video equalization corrector (video equalizer). A device that corrects for unequal frequency losses and/or phase errors in the transmission of a video signal.

Video framestore. A device that enables digital storage of one or more images for steady display on a video monitor.

Video gain. The range of light-to-dark values of the image that are proportional to the voltage difference between the black and white voltage levels of the video signal. Expressed on the waveform monitor by the voltage level of the whitest whites in the active picture signal. Video gain is related to the contrast of the video image.

Video in-line amplifier. A device providing amplification of a video signal.

Video matrix switcher (VMS). A device for switching more than one camera, VCR, DVR, video printer, and similar to more than one monitor, VCR, DVR, video printer, and the like.

Video monitor. A device for converting a video signal into an image.

Video printer. A device for converting a video signal to a hard-copy printout. It could be a monochrome (B/W) or color. They come in different format sizes. Special paper is needed.

Video signal. An electrical signal containing all of the elements of the image produced by a camera or any other source of video information.

Video switcher. A device for switching more than one camera to one or more monitors manually, automatically or upon receipt of an alarm condition.

Video wall. A video wall is a large screen made up of several monitors placed close to one another, so when viewed from a distance, they form a large video screen or wall.

VITS. Video insertion test signals. Specially shaped electronic signals inserted in the invisible lines (in the case of PAL, lines 17, 18, 330, and 331) that determine the quality of reception.

VOD. Video on Demand. A service that allows users to view whatever program they want whenever they want it with VCR-like control capability such as pause, fast forward, and rewind.

VHF. Very high frequency. A signal encompassing frequencies between 30 and 300 MHz. In television, VHF band I uses frequencies between 45 MHz and 67 MHz, and between 180 MHz and 215 MHz for Band III. Band II is reserved for FM radio from 88 MHz to 108 MHz.

VHS. Video home system. As proposed by JVC, a video recording format used most often in homes but also in CCTV. Its limitations include the speed of recording, the magnetic tapes used and the color separation technique. Most of the CCTV equipment today supersedes VHS resolution.

VLF. Very low frequency. Refers to the frequencies in the band between 10 and 30 kHz.

VMD. Video motion detector. A detection device generating an alarm condition in response to a change in the video signal, usually motion, but it can also be a change in light. Very practical in CCTV as the VMD analyzes exactly what the camera sees (i.e., there are no blind spots).

VR. Virtual Reality. Computer-generated images and audio that are experienced through high-tech display and sensor systems and whose imagery is under the control of a viewer.

VS. Vertical sync.

WAN (Wide Area Network). A network connecting computers within large areas, e.g. beyond the limits of a single protected site.

Waveform monitor. Oscilloscope used to display the video waveform.

Wavelet. A particular type of video compression that is especially suitable for CCTV. Offers higher compression ratio with equal or better quality to JPEG.

White balance. An electronic process used in video cameras to retain true colors. It is performed electronically on the basis of a white object in the picture.

White level. This part of the video signal electronically represents the white part of an image. It resides at 0.7 V from the blanking level, whereas the black part is taken as 0 V.

Workstation. A computer connected to a network at which operators interact with the video display.

Wow and flutter. Wow refers to low-frequency variations in pitch while flutter refers to high-frequency variations in pitch caused by variations in the tape-to-head speed of a tape machine.

WSGA. Computer screen format offering 1640 × 1024 pixels resolution.

WUXGA. Computer screen resolution offering 1920 × 1280 pixels. This covers the HDTV.

W-VHS. A new wide-VHS standard proposed by JVC, featuring a high-resolution format and an aspect ratio of 16:9.

XGA. Computer screen format offering 1024 × 768 pixels resolution.

XML (eXtensible Markup Language). Widely used protocol for defining data formats, providing a very rich system to define complex data structures.

Y/C. A video format found in Super-VHS video recorders. Luminance is marked with Y and is produced separate to the C, which stands for chrominance. Thus, an S-VHS output Y/C requires two coaxial cables for a perfect output.

Y, R-Y, B-Y. The general set of component video signals used in the PAL system as well as for some encoder and most decoder applications in NTSC systems; Y is the luminance signal, R-Y is the first color difference signal, and B-Y is the second color difference signal.

Y, U, V. Luminance and color difference components for PAL systems; Y, B-Y, R-Y with new names; the derivation from RGB is identical.

Z. In electronics and television this is usually a code for impedance.

Zoom lens. A camera lens that can vary the focal length while keeping the object in focus, giving an impression of coming closer to or going away from an object. It is usually controlled by a keyboard with buttons that are marked zoom-in and zoom-out.

Zoom ratio. A mathematical expression of the two extremes of focal length available on a particular zoom lens.

Appendix B

Bibliography and acknowledgments

- **1/2.5-Inch 5MP CMOS Digital Image Sensor,** Aptina, 2005, *(www.aptina.com)*

- **802.3af,** IEEE Standards, 2003 (

- **A Brief History of Pixel,** Richard F.Lyon, Foveon Inc., 2006

- **A Broadcasting Engineer's Vade Mecum,** an IBA Technical Review

- **A Guide to Picture Quality Measurements for Modern Television Systems,** Tektronix, 1997 *(www.tek.com)*

- **A Guide to Standard and High-Definition Digital Video Measurements,** Tektronix, 2007 *(www.tek.com)*

- **Advanced Imaging magazines,** 1995–1999, Cygnus Publishing Inc.

- **Angle of View & Image resolution,** white paper by Dallmeier electronic, 2012 *(www.dallmeier.com)*

- **Australian CCTV Standard, AS-4806,** 2004

- **Avigilon,** various products brochures, 2012, *(www.avigilon.com)*

- **Axis, Digital Video Compression,** various white papers, 2012, *(www.axis.com)*

- **Axxon,** various products brochures, 2012, *(www.axxonsoft.com)*

- **Black Magic Design,** various product brochures, *(www.blackmagicdesign.com)*

- **CCD Cameras,** Thesis 1982, by Vlado Damjanovski

- **CCD Data Book, Loral Fairchild,** 1994/1995, Loral Fairchild Imaging Sensors

- **CCD Data Book, Thomson composants militaires at spatiaux,** 1988, Thomson-CSF Silicon Division

- **CCD Image Sensors,** DALSA Inc., 1992

- **CCD vs. CMOS: Facts and Fiction,** Photonic Spectra, Laurin Publishing Co.Inc. 2001

- **CCTV focus magazines,** www.cctv-focus.com, 1999–2004

- **CCVE Catalogues,** Panasonic, 1995–1999

- **Changing the game with JPEG2000,** T-VIPS, *(www.t-vips.com)*

- **Charge-Coupled Devices,** by D.F. Barbe

- **Charge-Coupled Imaging Devices,** IEEE Trans., by G.F. Amelio, W.J. Bertram, and M.F. Tompsett

- **CMOS Chip technology,** by Dr.Hans Stohr, Photonfocus, 2004

- **CMOS vs. CCD,** *Maturing Technologies, Maturing Markets,* Dave Litwiller, Dalsa, 2005

- **Dallmeier electronic,** Panomera brochure, 2013

- **Digital Pixel System Technology,** Pixim, white paper, Platform Overview, 2005

- **Dirac Pro "Light Compression,"** Dr.Tim Borer, BBC R&D, 2009

- **Discrete Cosine Transform,** Nuno Vasconcelos, UCSD

- **Elbex literature on EXYZ matrix switcher,** 1998

- **Electronic Communication Systems,** by William Schweber, 1991, Prentice Hall

- **Envivio,** whitepaper: IP Streaming of MPEG-4: Native RTP vs MPEG-2 Transport Stream by Alex MacAulay, Boris Felts, Yuval Fisher, 2005

- **European CCTV Standards,** EN 50132-7:1996

- **EWW magazines,** 1992–1996

- **Failure Trends in a Large Disk Drive Population,** Eduardo Pinheiro, Wolf-Dietrich Weber and Luiz Andre Barroso, Goodle Inc., 2007

- **Fiber Optics in Local Area Networks,** by John Wise, course organized by OSD

- **Fiber-Optics in Security,** by Vlado Damjanovski, one-day Seminars

- **From MPEG-1 to MPEG-21,** by Rob Koenen, MPEG, 2001

- **Gamma correction,** various authors, Wikipedia *(en.wikipedia.org)*

- **HDD & RAID,** white paper by Dallmeier electronic, 2012 *(www.dallmeier.com)*

- **HEVC Demystified, A Primer on the H.265 Video Codec,** Elemental, 2013 *(www.*

elementaltechnologies.com)

- **H.261 recommendation,** by ITU-T, 1998, *(www.itu.int)*

- **H.263 recommendation,** by ITU-T, 1998, *(www.itu.int)*

- **H.264 recommendation,** by ITU-T, 2003, *(www.itu.int)*

- **H.265 recommendation,** by ITU-T, 2013, *(www.itu.int)*

- **High Definition (HD) Image Formats for Television Production,** EBU-Tech 3299, Specification, 2009

- **High Dynamic Range Image Sensors,** by Abbas El Gamal, Stanford University, 2002

- **IEC,** Project No.62676

- **Image formats for HDTV,** John Ive, Sony Europe, 2004

- **Image Processing Europe magazines,** 1995–1999, PennWell

- **In-depth Technical Review and Fault Finding,** by Vlado Damjanovski, two-day Seminars

- **Information technology - Coding of audio-visual objects, Part 2: Visual,** ISO/IEC 14496-2, 2001

- **Intel DataSheet Volume 1,** October 2011

- **Internetworking Technologies Handbook,** fourth edition, by Cisco Systems

- **IPVM,** various articles, *(www.ipvm.com)*

- **IP Streaming of MPEG-4: Native RTP vs MPEG-2 Transport Stream,** Alex MacAulay, Boris Felts, Yuval Fisher, Envivio, 2005

- **Kodak,** Solid State Image Sensors Terminology, Application note, 2010

- **Kodak,** Image sensors, 2008

- **Kodak,** Fundamental Radiometry and photometry, 2008

- **Light and Color Principles,** an IBA Technical Review, 1999

- **Light measurement guidance notes,** Skye Instruments Ltd., *(www.skyeinstruments.com)*

- **MPEG-1 recommendation,** by MPEG

- **MPEG-2 recommendation,** by MPEG

- **MPEG-4 recommendation,** by MPEG

- **MPEG-4,** by Rob Koenen, IEEE Spectrum, 1999

- **MPEG Solutions, Transition to H.264 Video,** Tektronix, 2008

- **MTBF,** white paper by Dallmeier electronic, 2012 *(www.dallmeier.com)*

- **New opportunities for video communication, H.264, The Advanced Video Coding Standard,** ITU-T, 2003 *(www.itu.int/ITU-T)*

- **NICET Certification - Video Systems Installation Technician,** Vlado Damjanovski and Howard Konstham, SIA publication 2008.

- **NTSC Systems Television Measurements,** Tektronix, 1999

- **Objective perceptual assessment of video quality: Full reference television,** ITU-T, 2004 *(www. itu.int/itu-t)*

- **OmniTek Test Sequence A (TSA),** white paper by Mark Davies, OmniTek, 2003 *(www.omnitek. tv)*

- **ONVIF Profile Policy,** 2011 *(www.onvif.org)*

- **ONVIF Core Specification v.2.2.1,** 2012 *(www.onvif.org)*

- **ONVIF Test Specification v.12.12,** 2012 *(www.onvif.org)*

- **Optical Calculations for Close-Up Applications,** Pentax

- **Osnovi Televizijske Tehnike,** by Branislav Nastic, 1977, Naucna knjiga, Beograd

- **Overview of the High Efficiency Video Coding,** IEEE Trans.VOL. 22, 2012

- **Overview of the H.264/AVC Video Coding Standard,** Thomas Wiegand,Gary J. Sullivan, Gisle Bjontegaard and Ajay Luthra, IEEE Trans., 2003

- **PAL Systems Television Measurements,** Tektronix, 1999

- **Pelco Product Specification Book,** 1995–1998

- **Pentax - CCTV Closed Circuit Television Lenses,** Pentax - Ricoh, 2012

- **Performance analysis and comparison of Dirac Video Codec with H.264/MPEG-4 Part 10 AVC,** by Aruna Ravi, University of Texas, 2009

- **Physics for Scientists and Engineers,** by Raymond Serway, 1992, Saunders College Publishing

- **Pipeline HD,** Seagate, 2011 *(www.seagate.com)*

- **PowerQuest,** Hard disks, 2002

- **Predavanja po Televizija,** by Prof. Dr. Ljupco Panovski, 1979–1981, ETF Skopje

- **PSIA, Physical Security Interoperability Alliance,** Specification Package, 2009

- **PSIA Common Metadat/Event Management Specification, v.1.2.1,** 2012

- **Real Time Streaming Protocol (RTSP),** H.Schulzrinne, A.Rao, R.Lanphier, 1998

- **RTP Payload Format for H.264 Video,** S.Wenger, M.M.Hannuksela, T.Stockhammer, M.Westerlund, D.Singer, The Internet Society - Network Working Group, 2005

- **Signal-to-Noise impact of CCD Operating Temperature,** white paper by John C. Smith, 2008

- **SMPTE: Motion Imaging Journals,** 2010 ~ 2013

- **Societal security - Video surveillance - Export interoperability,** ISO/DIS 22311 - Draft, 2011

- **Solid State Imaging,** by D.F. Barbe & S.B. Campana

- **Sony,** various security brochures, 2011 *(www.sony.com/security)*

- **Switch basics,** white paper by Dallmeier electronic, 2012 *(www.dallmeier.com)*

- **Sybex CCNA 640-802,** Instructor Aamir Mehmood, 2011

- **Spectral Instruments Inc,** various brochures, *(www.specinst.com)*

- **Super Hi-Vision at the London 2012 Olympics,** SMPTE Mot. Imag., 2013 *(journal.smpte.org)*

- **Teledyne Dalsa,** various brochures and white papers, *(www.teledynedalsa.com)*

- **Television measurements in PAL systems,** by Margaret Craig, 1991, Tektronix

- **Television measurements in NTSC systems,** by Margaret Craig, 1994, Tektronix

- **The Art of Electronics,** by Paul Horowitz and Winfield Hill, 1989, Cambridge University Press

- **The Engineer's Guide to Compression,** by John Watkinson, Snell & Wilcox, 1996

- **Two revolutionary optical technologies** *(Scientific background on the Nobel Prize in Physics 2009)*, compilation by the Class for Physics of the Royal Swedish Academy of Sciences, 2009

- **The Secrets of higher sensitivity CCTV cameras,** Nikolai Uvarov, CCTV focus magazine, 2003

- **TV Optics II,** by Shigeru Oshima, 1991, Broadcast Equipment Group, Canon Inc.

- **UB Video Inc. white papers on various compression,** (*www.ubvideo.com*)

- **Ubiquity Networks,** various brochures, *(www.ubnt.com)*

- **Understanding Jitter Measurement for Serial Digital Video Signals,** Tektronix, 2005 *(www.tektronix.com/video)*

- **Video Equipment Catalogue,** 2008, Sony

- **Video Quality Analysis Software,** Tektronix, 2010

- **VBG (Verwaltungs-Berufsgenossenschaft): Installationshinweise für Optische Raumüberwachungs-anlagen (ORÜA) SP 9.7/5**

- **Wikipedia**, various authors, *(en.wikipedia.org)*

This page intentionally left blank

Appendix C

CCTV training based on this book

www.cctvseminars.com & www.vidilabs.com

If you find this book informative and useful, it might be beneficial to know that I traditionally conduct seminars based on the content of this book. In fact, I have a dedicated web site, *www.cctvseminars.com,* which illustrates some of the many seminars I have conducted in the past.

Duration of seminars may vary, depending on the depth of coverage required, but for a full book coverage typically at least two full days are required, although three days is a better choice, and the information presented can be absorbed easier. It is also possible to cover only a part of (or parts) of the book, based on each particular requirement. The advantages of direct seminars like these are manyfold, one of which is the possibility to ask the author questions directly. For the shy ones that could not do that, they can always refer to the book.

These seminars can be in-house, or, we can arrange for a classroom, depending on your requirements.

Anybody interested in such a specialized training, please write to me at *vlado@vidilabs.com,* and I will do my best to fit you in my schedule and offer you the most appropriate training package.

Thank you for buying the book and I hope you found it useful.

The author,

May, 2013

Sydney

This page intentionally left blank

Appendix D

Book co-sponsors

This book has been made possible not only by the publisher, Elsevier (Butterworth-Heinemann), but also by the CCTV manufacturers who have believed in me and co-sponsored this edition.

The following pages are dedicated to their high-quality CCTV products, and they are listed in alphabetical order.

I would like to sincerely express my deep appreciation; they have made the challenge of writing this book easier to handle.

This page intentionally left blank

This page intentionally left blank

This page intentionally left blank

This page intentionally left blank

Multi-sensor system

PANOMERA®

Video surveillance without limits – unprecedented resolution in minute detail

Dallmeier is one of the world leading providers of products for network-based video security solutions. The multi-sensor system Panomera® was specially developed for the all-encompassing video surveillance of expansive areas. With this completely new camera technology a huge area can be surveyed from a single location.

- Zoom right down to the smallest details even at large distances
- Permanent recording of the entire scene
- Lower costs for infrastructure and maintenance

Made in Germany

About the author...

*Photo by **Alison Giles - Damjanovska***

Vlado Damjanovski is an internationally renown CCTV expert. He was born in Skopje, Republic of Macedonia, and since 1987 he lives in Sydney, Australia.

Vlado has a degree in Electronics Engineering from the University "Kiril i Metodij" in Skopje (Macedonia), and he has specialised in Television and CCTV.

His Thesis (Diploma Work) in 1982 was on CCD Cameras, 10 years before they started being widely used commercially.

Vlado is an author, inventor, a lecturer and a CCTV expert, known to the Australian and international CCTV industry.

Through his company ViDi Labs (***www.vidilabs. com***) he performs the following tasks: consulting, system design, commissioning, testing, project management, desk-top publishing, training and publishing.

A summary of the ever-evolving CCTV knowledge that Vlado posses first appeared in 1995 in his first published book - simply called "CCTV." This was, and still is, one of the first and truly complete reference manuals on the subject of CCTV. This book, covering all that a CCTV expert, installer or consultant should know, has already been accepted and approved by many international authorities as an exceptional reference book. Many are referencing it as "The CCTV Bible" so that even the official Russian translation of it carries such a title.

The international publisher Elsevier - Butterworth-Heinemann published another two editions of it, the first in 1999 and the second in 2005, so that Amazon.com always rates this book with 5 stars. The popularity of the previous editions are the reason for this latest update.

Based on the complete content of his best-seller books, Vlado conducts easy-to-understand CCTV seminars, de-mystifying and explaining all tricks and technologies of the CCTV trade. These seminars are conducted world-wide and more information can be found on ***www.cctvseminars.com***.

In addition to all the activity above, Vlado has designed and commissioned a number of CCTV Systems around Australia and overseas. Some of the more interesting CCTV projects include Darling Harbour, Downing Centre Control, Sydney City, Darwin Sky City casino, etc. One of the world first digital CCTV system designs was done by Vlado for the Star City Casino in 1997, where gaming disputes are sorted out immediately at the gaming tables, costing in excess of $1.5M. Then, in 2004, another new design was proposed for the Sands Casino in Macao in 2004, costing in excess of $5M. Vlado was instrumental in offering and winning the system design for the Sands casino - another first in the world utilizing digital matrix switching of over 2,500 digital recorders.

Another publishing project was the international magazine for CCTV called "CCTV focus". This magazine was launched at the ISC in New York, in August 1999. Prepared and published by Vlado's previous company

- CCTV Labs, the "CCTV focus" became the world leading technical magazine for Closed Circuit Television. Sadly, the printed version stopped being produced by the end of 2006 due to insufficient help from the industry.

Vlado is also the chairman of the CCTV Standards sub-committee of Australia and New Zealand, and he has contributed immensely in creating the latest Australian and New Zealand CCTV standards known as AS4806.1, AS4806.2 and AS4806.3.

The CCTV test chart and the programmable test pattern generator are some of the original and unique products developed by Vlado and his company, and they are used in various measurement and quantifying CCTV system performance.

Based on Vlado's immense experience in CCTV and Digital imaging, there have been a number of industry specific products and developments, all aimed at improving the knowledge and understanding.

Vlado is also known for his photographic works, albeit as an amateur, but having prepared a number of sole exhibitions in his native Macedonia, as well as Australia,published his work in many magazines and books and won numerous awards (visit his photographic web site Poems with Pixels at *www.damjanovski.com*).

Index

Q

This page intentionally left blank

This page intentionally left blank

Printed and bound by CPI Group (UK) Ltd, Croydon, CR0 4YY

03/10/2024

01040326-0020